Risk Taking and Decisionmaking

YAACOV Y. I. VERTZBERGER

Risk Taking and Decisionmaking

FOREIGN MILITARY INTERVENTION DECISIONS

STANFORD UNIVERSITY PRESS

STANFORD, CALIFORNIA

1998

Stanford University Press
Stanford, California
©1998 by the Board of Trustees of the
Leland Stanford Junior University
Printed in the United States of America

CIP data appear at the end of the book

ISBN 978-0-8047-2747-1
ISBN 978-0-8047-3168-3

For those who care and dare

Contents

Preface

A few years ago an advertisement for Bankers Trust caught my attention. It warned: "Risk. You have to look at it even when you don't want to." I thought then, as I do now, that as obvious and banal as this advice was, decisionmakers, even astute ones, too often neglect to follow it. Decisionmakers, confronted with complex high-stakes problems, are faced with the tyranny of resultant risks, which are often difficult to identify and define with precision or anticipate with certainty. They find that the combined impacts of these objective difficulties in accurately perceiving and assessing potential risks, as well as the prospects of grave consequences from risky choices, make coping with risk a daunting task that they would rather avoid. Yet there is no escape. Coming to terms with and coping with risks are unavoidable features of effective decisionmaking and policy formulation.

Risk is a generic problem in human affairs, and viewed strategically, the manner of coping with it is for the most part not domain specific but generalized across domains and issue-areas. Responses to risk are driven primarily not by the content of the problem (e.g., whether it is a problem concerning international security, a health hazard, financial investments) but by its structural attributes (e.g., the level of complexity, the degree of uncertainty and ambiguity, the time horizon for gains and losses) and the personal and social attributes of the decisionmakers (e.g., personality, group structure). Risk-taking behavior thus varies mainly as a function of changes in the defining attributes of the problem and the decisionmakers and relatively marginally with problem-domain changes. Because the essence of the risk *problematique* is shared across different decision domains, the study of risk and risk-taking behavior is, by implication, a field in which scholars from different disciplines should find common ground, share findings, and aim for cross-disciplinary fertilization through interdisciplinary research. At the same time, the high salience and policy relevance of the study of risk should provide an incentive for research that aims to be of practical value for decisionmakers. This requires social scientists to come down from the ivory tower of abstract, even if elegant, conceptualization and methodological puritanism into the trenches of pragmatic and often messy praxis. Thus, the quality and relevance of research on risk-taking behavior should be judged by the multiple criteria of analytic comprehensiveness, practical applicability, and positivist scientific validity, rather than by a single criterion given exclusive emphasis.

This book embodies a number of premises and objectives derived with these concerns in mind. First, it is a response to the relative neglect of risk in foreign policy decisionmaking literature. Second, it is a recognition of the critical influence of risk assessments and risk preferences on the nature and quality of the decisionmaking process and outcomes. Third, it provides an emphasis on the essentially multivariate and interdisciplinary nature of research on risk-taking behavior. Fourth, it is a reflection of my dissatisfaction with the conceptual and practical limitations of the standard rational-choice approach to risk that has traditionally dominated the treatment of risk and risk taking in international relations research. For a long time I have been struck by the discrepancy between the tight, parsimonious, and logical elegance of rational choice theory and its inadequacy in capturing how real-life decisionmakers, both individuals and groups, actually respond to risky situations, how they assess and develop dispositions and preferences with regard to risk taking or its avoidance, and how sensitive or insensitive their assessments and preferences are in iterative decision situations to change through learning.

In this book, therefore, I suggest an alternative to the rational choice approach. In my sociocognitive approach—initially developed in an earlier book (Vertzberger, 1990)—decisionmakers are portrayed as human beings, rather than as human computing systems conditioned to relentlessly pursue the strict requirements of rational behavior. Decisionmaking in general and risk taking in particular are conceptualized as the products of individuals who are not only affected by cognitive and motivational drives, biases, and constraints but also embedded in, and affected by, the multilayered social contexts—cultural, organizational, small group—of which they are a part. The result is a comprehensive context-dependent theory that describes and explains the formation of risk perceptions and risk-taking preferences by decisionmaking individuals, groups, and organizations. To test the theory I have applied it to an acute and important issue-area in international politics—foreign military intervention decisions—where it is examined by way of five detailed case histories. The approach combines extensive interdisciplinary theorizing with detailed empirical applications and theory testing in a naturalistic environment.

This book is intended not only for observers and students of decisionmaking, organizational behavior, strategy, and current history but also for practitioners who regularly face the very real tasks of formulating and executing policies and coping with the consequent risks. Although the case studies are drawn from the national security and foreign policy domains, I believe that the book will be of interest and relevance to decisionmakers in all fields of social risk taking where problems are ill defined, information is characterized by uncertainty, policy outcomes involve high stakes, and the risks are complex and difficult to quantify. It is my hope that decisionmakers of all persuasions will, while reading this book, recognize themselves in it and with its help identify their own concerns and vulnerabilities and the constraints they face. In the process they might realize that even though to err is human, so is learning to avoid or at least to minimize errors. By using this book for self-analysis and for critical examination of their decisionmaking systems and processes, they may considerably improve the quality of their decisionmaking performance.

The completion of this book required substantially more time, and turned out to be much more of a challenge, than was anticipated initially. Fortunately, I benefited from the advice, support, and encouragement of colleagues and students, as well as the generous funding and hospitality of institutions, that helped in turning raw, abstract ideas into a book. The least I can do to recognize these debts of gratitude to individuals and institutions is to acknowledge their contributions.

Michael Brecher read the entire manuscript and made detailed and constructive suggestions. Two reviewers for Stanford University Press read the complete final draft, and their careful comments were instrumental in sharpening arguments and clarifying vague points. Very useful advice on portions of the manuscript was provided by Muthiah Alagappa, Richard Baker, George Breslauer, Alexander George, Bernard Gordon, Paul Kreisberg, Charles Morrison, Janice Stein, Blema Steinberg, and the late Aaron Wildavsky. Some of the chapters were presented as conference papers on various occasions. The responses to these papers by Berndt Brehmer, Margaret Hermann, Arie Kacowicz, Max Metselaar, Eric Stern, Bengt Sundelius, Paul 't Hart, and Steve Walker contributed to the improvement of the manuscript. I also had the good fortune of being a member of the Mershon Center Research Training Group during most of the period that the study was in progress, and I consequently had the opportunity of tapping the collective and individual wisdom of the members of this talented multidisciplinary group of scholars.

A project of this magnitude required extensive assistance of various kinds. For research assistance I am indebted to Shimon Arad, Yoav Gortzak, Galia Press-Barnathan, Miriam Fendius-Elman, Sherée Groves, Sabine Grund, Ashley Harshak, Valerie Koenig, Yael Shalgy, and Orly Vered. I greatly appreciate the cheerful efficiency with which Marilu Khudari, Dorin McConnell, and Dorothy Villasenor handled all administrative tasks for the project. Most of the burden of typing the manuscript fell on Cynthia Nakachi, whose unmatched word-processing skills, combined with a commitment to perfection, were responsible for the production of a draft manuscript on time and in superb shape. The penultimate version of the manuscript was prepared by

Kari Druck with her usual strict attention to, and sharp eye for, details and aesthetics. Two skillful professionals, David Hornik and John Thomas, edited the manuscript. Two additional experienced editors, Deborah Forbis and Janis Togashi, helped in improving specific chapters. Paul Oshiro produced the figures.

The research for the project depended to a substantial extent on the enthusiastic cooperation of the librarians at the Social Sciences Library of the Hebrew University (Jerusalem) and dedicated, efficient assistance from the librarians of the Research Information Services at the East-West Center (Honolulu). I am also obliged to John Wilson, of the Lyndon B. Johnson Library (Austin), for providing documents for the Vietnam intervention case and to Malcolm Byrne, of the National Security Archives (Washington, D.C.), for providing documents for the Czechoslovakia intervention case.

This project would not have been possible without the financial assistance of the Leonard Davis Institute of International Relations, the Harry S. Truman Research Institute for Advancement of Peace, the S. A. Schonbrunn Research and Development Fund, and the Authority for Research and Development of the Hebrew University. I also received funding from the Institute for the Study of World Politics (Washington, D.C.) and the Mershon Center Research Training Group (RTG) on the Role of Cognition in Collective Political Decisionmaking at Ohio State University (National Science Foundation Grant DIR-9113599). A short-term residency in 1989 at the Rockefeller Foundation's Bellagio Study and Conference Center, the Villa Serbelloni, allowed me to review and revise a preliminary draft of the theoretical arguments. My largest institutional debt is to the East-West Center and its Program on International Economics and Politics (IEP), which not only provided a three-year fellowship and superb logistic support, allowing me to focus on my research goals with few distractions, but also presented me with a congenial intellectual environment and an opportunity for exchanging views and learning from my colleagues at the center. For these privileges I am most grateful to Charles Morrison, former director of the program, as well as Michel Oksenberg, former East-West Center president, and Vice Presidents Kenji Sumida (now president) and Bruce Koppel.

At Stanford University Press I had the pleasure of working with one of the best teams an author could wish for. Executive editor Muriel Bell was supportive of the project long before it was completed, and oversaw the publishing of the book with utmost professionalism. In-house editor John Feneron handled the nuts and bolts of the process with efficiency and attentiveness to my requests. Copy editor Mary Pasti proved, and not for the first time, her unmatched skills in working miracles with words.

Earlier versions of some of the chapters were published in article form in various scholarly journals and an edited volume: "Foreign Military Intervention and National Capabilities: A Policy-Relevant Theoretical Analysis," *International Interactions*, 17(4), 1992: 349–73; "The International Milieu and Foreign Military Intervention: When and How Much Does the Milieu

Matter?" *Journal of Strategic Studies,* 17(3), September 1994: 139–79; "Rethinking and Reconceptualizing Risk in Foreign Policy Decisionmaking: A Sociocognitive Approach," *Political Psychology,* 16(2), June 1995: 347–80; and "Collective Risk Taking: The Decision-Making Group," in Paul 't Hart, Eric K. Stern, and Bengt Sundelius, eds., *Beyond Groupthink: Political Group Dynamics and Foreign Policy-Making* (Ann Arbor: University of Michigan Press, 1997), pp. 275–307.

Writing may be a highly individualistic experience, but if these acknowledgments prove anything at all, it is that writing also has a socially rewarding side. All of those mentioned above should share in the merits of this project, but I alone remain responsible for the views expressed and the errors or biases left uncorrected.

Yaacov Y. I. Vertzberger
Jerusalem 1997

Abbreviations

ANGLICO	Air Naval Gunfire Laison Company
ARVN	Army of the Republic of Vietnam
C³I	Command, Control, Communications and Intelligence
CARICOM	Caribbean Community
CIA	Central Intelligence Agency
CINCPAC	Commander in Chief, Pacific
CMEA	Council for Mutual Economic Assistance
CPCZ	Communist Party of Czechoslovakia
CPPG	Crisis Pre-Planning Group
CPSU	Communist Party of the Soviet Union
DIA	Defense Intelligence Agency
DRV	Democratic Republic of Vietnam
FRG	Federal Republic of Germany
GDDM	Group Dominated Decisionmaking
GDR	German Democratic Republic
GRU	Soviet military intelligence
GULP	Grenada United Labor Party
IDDM	Individual Dominated Decisionmaking
IDF	Israel Defense Forces
JCS	Joint Chiefs of Staff
JEWEL	Joint Endeavor for Welfare, Education and Liberation
KGB	Committee on State Security
MAAG	Military Assistance and Advisory Group
MAC	Military Airlift Command
MACV	Military Assistance Command, Vietnam
MAP	Movement for Assemblies of the People

NATO	North Atlantic Treaty Organization
NJM	New Jewel Movement
NLF	National Front for the Liberation of South Vietnam
NPT	Nonproliferation Treaty
NSAM	National Security Action Memorandum
NSC	National Security Council
NSDD	National Security Decision Directive
NSPG	National Security Planning Group
NVA	North Vietnam Army
OAS	Organization of American States
OECS	Organization of Eastern Caribbean States
PAVN	People's Army of Vietnam
PDF	Defense Forces of the Republic of Panama
PLO	Palestine Liberation Organization
PPBS	Planning Programming Budget Systems
PRA	People's Revolutionary Army
PRC	People's Republic of China
PRG	People's Revolutionary Government
RDJTF	Rapid Deployment Joint Task Force
RIG	Restricted Interagency Group
RMC	Revolutionary Military Council
SALT	Strategic Arms Limitation Talks
SEAL (team)	Sea, Air and Land (team)
SEATO	Southeast Asia Treaty Organization
SNIE	Special National Intelligence Estimate
SOP	Standard Operating Procedures
SSG	Special Situation Group
StB	Czechoslovak State Security Organization
U.N.	United Nations
UNIFIL	United Nations Interim Force Lebanon
USARSO	U.S. Army South
VC	Vietcong
WTO	Warsaw Treaty Organization

Risk Taking and Decisionmaking

Introduction

The Study of Risk and Foreign Military Intervention Decisions

Although the concept of risk and its behavioral implications have been singled out for extensive social scientific research in a broad range of fields—including medicine, economics, industry, technology, and environmental studies—it has largely gone unnoticed in the field of international politics and specifically in the study of international security issues, where risk is perennial and its consequences are visible and critical.[1] At the same time, some aspects of risk that are of particular interest to students and practitioners of foreign policy fall outside the purview of knowledge accumulated in other disciplines. The very few studies of risk in the field of international relations tend to implement the technical, classic definition of risk that draws its inspiration from the gambling metaphor and treats risk as a probability-centered phenomenon (e.g., Adomeit, 1973, 1982). This approach, though parsimonious and easy to employ in empirical research and measurement, is also inadequate, as will become evident in Chapter 2.[2] Related studies are based on explicit and implicit assumptions whose validity is debatable—for instance, that decisionmaking systems are unitary actors (Bueno de Mesquita, 1981; Huth et al., 1992), that actors are driven by a single motivation, or that preferences are given and fixed, which means ignoring such social processes as group dynamics, which can produce changes in preferences.

A sophisticated attempt to integrate risk taking into the analysis of foreign policy choices by Lamborn (1985, 1991) is of particular interest. In his studies he explores the quality of such choices by examining the link between statecraft and domestic politics. Risk-taking preferences are used to describe how policy factions within governments react to policy and political risks, or, in other words, how they react to the domestic and diplomatic

price of power. Lamborn's work builds on earlier writings by Bueno de
Mesquita, but its focus represents a substantial advance over existing studies
because it disaggregates the state and thus discards unitary-actor assumptions.

Lamborn's treatment of risk and risk taking nonetheless raises questions
and puzzles. To start with, he defines risk in terms of probability and adverse
outcomes, as is commonly the practice. In the next chapter, I argue that this
kind of definition only partially captures the essence of risk found in the turbulent and often opaque environment of international politics. Furthermore,
Lamborn assumes the primacy of domestic politics over other factors that
drive foreign policy behavior, an assumption that is contested, as the author
himself acknowledges (Lamborn, 1991). Also following from the emphasis
on domestic politics is a set of implicit, yet debatable, premises. One is that
dispositions toward risk and perceptions of risk are shaped almost exclusively by decisionmakers' institutional positions. Another is that risk dispositions of foreign policy decisionmakers emerge in what is assumed to be a rational process that aggregates power and position-based interests. It follows
that all members of the same faction, group, or organization are likely to
share similar risk perceptions and risk-taking preferences. Lamborn ignores
the important and well-documented fact that risk perceptions and dispositions are to a large extent individual-based cognitive phenomena, and he
therefore also misses the important cognitive and motivational consequences
of the interaction between individual and collective preferences.

On the whole, a rational choice approach to risk in foreign policy decisionmaking fails to capture its complexity. A more sophisticated, contingent,
and empirically grounded theory of risk judgment and risk-preference formation in foreign policy decisionmaking is required.[3] Such a theory should
address the existence of a variety of paths to risk aversion and risk acceptance.[4]

Although risk in international politics has been given limited scholarly
attention, military intervention in one state by another has generated a fairly
extensive literature, for interventions have been frequent since 1945 (e.g.,
Dunér, 1983b, 1985). As the number of independent state actors has increased, so has the objective probability of interventions. At the same time,
growing interdependence has made the internal affairs of each state a subject
of acute and often active interest to other states. Many of the new actors have,
furthermore, been too weak to offer serious resistance to military power projected by more powerful states (regional or extraregional), although even the
weaker states can make—and have made—intervention very costly to an intervening power. These developments have been augmented by a diffusion,
albeit asymmetrical, of economic and military power throughout the international system, particularly by massive arms transfers to many new actors.
This has increased the number of countries capable of using intervention as
state policy and has reduced the incentive and capability of global and re-

gional powers and organizations to prevent violent domestic and international conflicts or to control their outcomes at an acceptable cost.[5]

I use the term *intervention* to mean coercive military intrusion into the internal or foreign affairs of another state. Intervention can be on behalf of an incumbent government (e.g., Cuba's intervention in Angola in 1975–90), against an incumbent government (e.g., Vietnam's intervention in Cambodia in 1978), or in a situation of political chaos, where there is no effective government (e.g., Israel's and Syria's interventions in Lebanon in the 1970s and 1980s and U.S. intervention in Somalia in 1992). The forms of intervention vary depending on scope, motives, goals, duration, and ramifications for the intervening state, target state, and other states in the international system. Intervention outcomes can be anticipated or unanticipated. Sometimes they can be contained; at other times they become unmanageable. Intervention can be proactive, to preempt events in a target country, or reactive, to cope with events that have already taken place. It is a complex and multifaceted phenomenon.

The literature on foreign military intervention can be divided into four categories. The first comprises the numerous idiographic empirical, detailed analyses of particular cases. Such studies are either narrow in narrative focus (e.g., Kahin, 1987; Adkin, 1989) or, even when shaped by an explicit analytic framework, generalizable only to the investigated country (Barrett, 1993; Burke et al., 1989; Dawisha, 1984). The second category comprises comparative-episodic literature: several cases are briefly discussed and compared, which leads to generalizations regarding the structural attributes of the act of intervention and of the participating actors (e.g., Haass, 1994; MacFarlane, 1985). The third category includes aggregate analyses that draw on a large number of cases. The quantitative studies do not address the questions why and how intervention decisions are made (e.g., Anyanwu, 1976; Dunér, 1985; Pearson, 1974a, 1974b; Pearson and Baumann, 1989; Tillema, 1989, 1994). The reason is that structural attributes of participating actors and situations are the variables that are readily quantifiable, not characteristics of the decisionmaking process. Aggregate studies have therefore primarily reflected the static dimension of intervention, rather than portraying the ebb and flow of intervention as a process that develops through the iterative decisions of politicians and through implementation by military and civilian officials in the field and the outcomes of which feed back into the decisionmaking system. In the fourth category of intervention studies, where the emphasis is on the normative legal aspects of intervention, normative guidelines and prescriptions are provided (e.g., Moore, 1969; O'Brien, 1979; Reed and Kaysen, 1993; Thomas, 1985; Vincent, 1974); these studies are of little direct relevance to explaining the decisionmaking process, which is the focus of this study.[6]

A brief review of three important edited collections with articles falling into one or more of these four categories is illustrative. The volumes edited by Bull (1986) and by Schraeder (1989a) contain studies that cover systemic variables, normative questions associated with intervention, the policies of particular countries, or specific cases. They provide instructive insights into various aspects of foreign military intervention, but they do not construct a theory or general framework that would provide a cohesive and organized structure of knowledge for systematically analyzing cases of intervention. In fact, the case studies that appear in these volumes are unrelated to the more theoretical articles.

The third and more recent volume, edited by Levite, Jentleson, and Berman (1992), is the most systematic comparative study of intervention in terms of both methodology and organization. The editors sensibly treat foreign military intervention within a three-phase framework: getting in, staying in, and getting out. The methodology is essentially inductive: six cases of protracted intervention are used to draw out the common characteristics of each phase. This approach results in a cohesive study. However, the resulting rich and important theoretical generalizations form only a loose theoretical entity and therefore might be difficult to apply systematically to other case studies. The study deals only with protracted interventions and not with quick, decisive ones (e.g., Panama, Grenada). Finally, as in most major studies of foreign military intervention, the decisionmaking process is marginalized. In fact, compared with the overall number of studies on intervention, the paucity of studies focusing on what may be the most policy-relevant aspect—decisionmaking—is puzzling. Only three studies—by Brands (1987–88), Little (1975), and Tillema (1973)—represent serious attempts to conceptualize a generalized decisionmaking process for foreign military intervention.

In light of the limited treatment of risk in general, it is not surprising that the few studies dealing with the process by which intervention decisions are considered and formulated are indifferent to one of the most pertinent aspects of those decisions—risk taking. Tillema (1973), in a detailed study of overt foreign military intervention, addresses systematically the conditions and constraints under which American intervention decisions were typically made. In essence, his argument is that the United States has intervened only when all the following conditions were present: perception of a great threat to U.S. interests, such as the threat of a communist takeover, particularly in a country where the United States had high stakes; the absence of alternatives (other instruments of statecraft or international actors willing to act as proxies); absence of domestic political constraints; a clear sense of moral justification; and a low risk of nuclear confrontation or direct confrontation with Soviet conventional forces. Although the explanation is sensible, its components, particularly the decisionmaking constraints, are not rigorously conceptualized. The process of commitment to and termination of intervention is not explained. And the study lacks external validity; that is, the conclusions cannot be projected to intervention by other, non-American interveners.[7]

Little (1975) explains the commitment process but fails to provide a comprehensive or systematic explanation of the decisionmaking process even as he touches on some aspects of it, such as consensus building in the decisionmaking group and the effects of cognitive dissonance. In fact, both Little and Tillema made very limited use of findings from other social science disciplines and—probably because their publications appeared before the cognitive revolution had its main effects on political science—missed relevant and prominent contributions from cognitive psychology to the literature on decisionmaking in international relations (e.g., Tetlock and McGuire, 1984).

A more recent analytic attempt to deal with U.S. armed intervention was made in Brands's (1987–88) short comparative study of Lebanon in 1958, the Dominican Republic in 1965, and Grenada in 1983. He suggests a list of conditions that have made armed intervention by the United States likely. The list contains a mixture of motivations for intervention and styles of intervention decisions but does not constitute an analysis of the policy formulation process. Although the study concludes with predictive conditions for intervention, the reader remains perplexed about why intervention did not occur in several cases in which these enabling conditions existed.

Except for these few studies, most analytic studies treat the intervening states, tacitly or explicitly, as unitary actors (e.g., MacFarlane, 1985; Scott, 1970). Little effort has been invested in motivational and cognitive-oriented explorations of either the structural attributes of intervention or the way these attributes are perceived and interpreted and the way the social context affects decisionmakers.[8] How were decisions to intervene, to escalate the intervention, and to terminate it formulated by individual decisionmakers and decisionmaking groups? What aspects of uncertainty and risk were perceived, and how were they dealt with? What are the possible and actual consequences of the way decisionmakers cope with risk and uncertainty? Which dimensions of uncertainty and risk are systematically ignored? Nearly three decades ago James Rosenau wrote that "scholarly writings on the problem of intervention are singularly devoid of efforts to develop systematic knowledge on the conditions under which interventionary behavior is initiated, sustained and abandoned," and his words are still largely true (1969: 149; see also Tillema, 1990: 25).

These questions are important for understanding decisions about foreign military intervention because intervention is supposed to have limited costs; that is, no intervention decision is intended to be carried out at any and all costs. At the argumentation stage, the main theme of intervention proponents is always that the cost-benefit ratio is favorable, and the risk involved is acceptable. If this cannot be argued forcefully and with a convincing emphasis on the controllability of risk, a decision to intervene is unlikely to be made; even supporters will find it difficult to accept and defend an open-ended commitment. Thus the question of control is crucial. Control, by definition, entails reduction of risk, uncertainty, and complexity to an acceptable minimum—not an easy task in most intervention decisions, which by their nature require a degree of policy-relevant knowledge that is not readily, if at all, available.

Decisionmakers face additional dilemmas. Is intervention the most appropriate instrument of statecraft? What is the best timing for intervention? What should its scope be? The answers are related. For example, intervention at an early stage of a developing domestic crisis in the target state could mean a short, low-intensity intervention, whereas a less risky option of nonintervention today may entail larger-scale intervention in the future. What the decisionmakers face is choice among different uncertainties and risks.

The calculus of intervention, then, is risk-rich, highly complex, and multivariate. To make an optimal decision, decisionmakers contemplating intervention must take into account, as shall be explicated in Chapters 4 and 5, a broad range of both external and internal variables. The former include the structural attributes of the global system and their implications for the reaction of major powers or other third parties; the target state's government and other local actors' attitudes and ability to support or hinder goal achievement by the intervening state; reactions of other regional powers and of intraregional and extraregional adversaries and allies; and the impact of geographic proximity, that is, whether available technologies will offset the effects of distance. Among the internal variables, decisionmakers have to consider their country's military ability to project power and its economic ability to sustain intervention; domestic political consequences of the intervention policy, including the mobilization of support for it within the political system, the bureaucracy, the general public, and the media; decisionmaking institutions, which will determine whether and how an intervention decision is made and what final form it will take (e.g., because of necessary politicking); and the possibility of adjusting or reversing the decision once it has been made. These considerations must be judged in the context of the national interest in its broadest sense, prevailing ideologies and norms, and accumulated national historical experience with similar decisions.

Even when this complex computation is carried out and a policy preference for intervention emerges, several uncertainty dimensions remain and must be considered by decisionmakers before they operationalize the policy preference. Can the economic, military, and political costs be accurately anticipated, calculated, and controlled? What are the socially acceptable costs? Can the time dimension of the operation be controlled? Can the implementation of the decision be controlled—or, once the operation begins, is it likely to escalate and get out of hand, whether unintentionally or purposefully (by individuals and organizations implementing it)? And how reversible is the decision at each stage? In fact, the two sets of information are interdependent. Accurate information about the external and domestic variables is necessary for dealing effectively with the second set of uncertainties and for allowing decisionmakers to attach valid probability estimates to them.

To make determinations as rationally and optimally as possible, the decisionmakers must (1) know of and understand all variables impinging on the outcomes of intervention, (2) correctly and coherently assimilate these variables to produce an integrated evaluation, and (3) know of potential biases or errors that may affect their assessments of the variables. This is a tall order under any circumstances and particularly when decisions have to be made under the pressure of time and stress. Complexity and uncertainty can then be reduced by, for example, allocating the problem among several organizations, each of which thus sees only part of the picture. Decisionmakers may also use standard operating procedures (SOPs) that apply prepared plans and scenarios, or they may use historical analogies, heuristics, schemata, and other cognitive constructs. Each of these solutions has costs that require trade-off decisions. Because of the high level of risk, indeterminacy, and uncertainty and the difficulty of accurately anticipating all relevant contingencies, prepared plans and SOPs must be continually and rapidly adjusted; otherwise, risk assessments become outdated and irrelevant. Furthermore, because of the prevailing ambiguity and uncertainty, the influence of personality variables on the formulation of intervention decisions and their implementation is likely to be highly pertinent. It seems, then, that decisionmaking combines bureau-organizational politics with cognitive and cybernetic processes. How risk perceptions and risk-taking preferences evolve in the decisionmaking process and affect choice is my main theme.

Methodology

In this book I address three sets of questions:

1. How are risk judgments and preferences formulated, and how do risk calculations influence the treatment of complex, ill-defined problems, such as foreign military intervention decisions? What are the relations among actual risk, perceived risk, and acceptable risk?[9]

2. Taking into account the risks involved, what determines an astute decision on foreign military intervention? An astute decision is one likely to achieve the goals of intervention at a cost that decisionmakers consider acceptable.

3. Are risky decisions responsive to feedback information from their ongoing implementation? What determines the pattern of learning and policy adaptation?

Answering these questions may shed light on how risky, ill-defined, complex decisions are made under conditions of high uncertainty and, more specifically, why an intervention policy is chosen, persists, goes through changes, and is terminated. I also propose a systematic framework for analyzing the chances of success and failure of intervention decisions and suggest measures for increasing the quality of decisions. These suggestions may help decisionmakers avoid adventuristic policies.[10] The emphasis is on risk as *choice* rather than as *circumstance*, although the latter is not ignored.

This study has two dimensions: theoretical and comparative-historical. It thus combines a deductive-theoretical approach with an inductive-empirical one.[11] Three premises inform the methodology. First, to be useful and generalizable, the analysis of risk has to take place in a real-life context. Here it is anchored in a context of foreign military intervention, where the two related subjects of risk and intervention cannot be imagined to be independent. Their interdependence requires a dual theoretical construct, one to provide an organizing framework for discussing the risks of intervention and the other to explain the causes of risk judgments and preferences. Second, to provide a comprehensive explanation of the multiple causal influences on risk judgment and preferences, the theoretical analysis has to be multivariate and interdisciplinary (Vertzberger, 1990).[12] The sociocognitive approach taken here integrates individual-level variables (e.g., belief system, operational code, personality attributes), social-level variables (e.g., group dynamics and organizational structure), and cultural-level variables (e.g., cultural-societal attributes and norms). Third, because explanations of decisions to take risk or to avoid it are subject to equifinality—that is, similar choices can result from different causal paths—a credible theoretical analysis of risk taking should map the spectrum of alternative patterns rather than unrealistically invoking the principle of parsimony and attempting to identify a single causal path.[13] The high complexity of risk and intervention and their interaction defies a parsimonious explanation. I believe that Yale Ferguson and Richard Mansbach are correct in stating that "until theories achieve greater explanatory power, there is no choice but to broaden the search for understanding by painting our pictures with greater detail" (1991: 369). Similarly, Alexander George argues that "scholars should recognize that too strict a pursuit of the scientific criterion of parsimony in their efforts to theorize is inappropriate for developing useful policy-relevant theory and knowledge" (1993: 140).

The organization of this book reflects these methodological and conceptual concerns. The introductory chapter is followed by four chapters that set forth in detail a deductive theoretical analysis of risk and intervention. In Chapter 2, I define the concept of risk as a multifaceted phenomenon. Chapter 3 provides a wide-ranging sociocognitive explanation of determinants of risk assessment and the formation of risk-taking preferences. In Chapter 4, I shift the theorizing to foreign military intervention and the systematic mapping of the risks, opportunities, and incentives associated with the intervener's national capabilities—military, economic, and political. In Chapter 5, I do the same for the risks, opportunities, and incentives posed by the international environment in which the intervener operates. Although the main burden of empirical illustration, evidence, and testing is left to the second part of the book, occasional brief empirical examples are provided in the theoretical analysis to clarify particular points.

In the second part of the book (Chapters 6 and 7), which contains the historical analyses, I apply a structured, focused-comparison method and a process-tracing approach (George, 1979a; George and McKeown, 1985) to five carefully selected case histories in order to test, expand, and modify the deductive theoretical analyses of risk taking and intervention presented in Part I. By infusing findings from experimental behavioral sciences with inferences from historical case studies, it becomes possible to test the validity of highly controlled laboratory results against the real world, to take into account the political context in which decisionmakers operate (Farnham, 1990, 1995), and to enrich our understanding of decisionmaking situations with different levels of risk.[14]

The structure of the comparison requires cases that illustrate escalating levels of risk, thereby allowing us to observe the effect of changing levels of risk on the process and quality of intervention decisions. In this way we construct a quasi-controlled experiment with history that allows for careful manipulation and observation of the main dependent and independent variables—intervention and risk. A detailed discussion of the way the cases are structured follows below. But briefly, the cases included are U.S. interventions in Grenada in 1983 and Panama in 1989 and Soviet intervention in Czechoslovakia in 1968 (low to moderate risks), and U.S. intervention in Vietnam in 1965–68 and Israeli intervention in Lebanon in 1982–83 (high risks). The restriction of the cases to a single issue-area, foreign military intervention, makes the comparison across cases sharper and better focused. The elimination of redundant variance also increases the validity and clarity of the inferences from the comparative analysis.

The strength of this methodology lies in its balance between external and internal validity. With three U.S. case histories as the core of the empirical analysis, the cases in the set contain enough basic necessary similarities to allow for valid comparison. At the same time, the Soviet and Israeli cases add variance in terms of the intervening states' levels of power, decisionmaking institutions, national styles and political cultures, geographic distance to the target states, and types of relations between the intervening states and the regimes of the target states. Each case study provides, in my view, a highly plausible interpretation of the investigated events. Admittedly, each interpretation may not be the only and definitive interpretation because some of the cases involve highly controversial events, and some of the occurrences happened only in the minds of decisionmakers, leaving no documentary or other visible and unambiguous traces to substantiate the superiority of one interpretation over another.[15]

The combination of variety and similarity allows us to discern those patterns of behavior that cut across society and regime types, which can thus be defined as structural attributes of any intervention situation, and those patterns that are context specific. The variety of decisionmaking institutions and styles (presidential, cabinet, and collective leadership) represented by the cases brings to light those attributes of information processing and decisionmaking related to risk taking and intervention that can be explained by the unique characteristics of a decisionmaking system. The cases enable us to observe those patterns of behavior that can be associated with the type of society (open versus closed) and to assess the impact of geographic distance (intervention in neighboring states versus distal intervention), as well as the impact of cooperation or noncooperation by the target state's government with the intervener, on the success or failure of intervention policies.

Several qualifying comments must be made with regard to the claim for external validity. First, the number of cases is small. Whether such a set is large enough to generalize from is a subject of ongoing controversy (Bueno de Mesquita, 1985; Eckstein, 1975; Jervis, 1985b; King et al., 1994; Krasner, 1985; Liberson, 1992). Second, although the cases are not exclusively American, the majority (three of five) are; hence it would seem appropriate to generalize with greater confidence from these findings to other cases of American intervention. The United States is, I might note, one of the two most prominent powers in terms of both the number and the intensity of foreign interventions that it has initiated since 1945. Finally, the availability of the detailed data required by the theoretical framework strictly limit the countries and cases that can be investigated to principally open Western societies, and in particular to the United States. In my view, these qualifying comments do not provide a strong-enough reason to contest a claim for generalizability (see also Kennedy, 1979). But the conclusions should be viewed as plausible rather than definitive. I hope that others will apply the theory proposed here to additional case studies and in this way enlarge the base for generalization.

The study concludes (Chapter 8) with observations, based on case comparisons, concerning risk taking and intervention decisions and with a set of policy-relevant prescriptions for improving the process of judgment, assessment, and choice in diverse risk situations. The policy prescriptions focus on reducing biases in risk assessment and making the choice process more effective by identifying controllable factors that have significant nonrational diversionary effects on the formation of risk-taking preferences.

The Case Studies: Organization and Analysis

Part II, as indicated, consists of five in-depth case studies of foreign military intervention, presented in two chapters. Chapter 6 deals with three cases that represent low- to moderate-risk decisions: U.S. intervention in Grenada in 1983 and in Panama in 1989 and Soviet intervention in Czechoslovakia in 1968. Chapter 7 deals with two cases that represent high-risk decisions: U.S. intervention in Vietnam in 1964–68 and Israeli intervention in Lebanon in 1982–83. The low-, moderate-, and high-risk labels reflect my view of the most appropriate way to classify the cases for the purpose of this study. The labels are not arbitrary, however, but based on decisionmakers' perceptions as inferred from an analysis of historical facts and counterfactuals.

In situations of foreign military intervention, decisionmakers encounter two sets of risks. One includes risks posed by the problem, the other the risks posed by the policy remedy, that is, foreign military intervention. In the case studies, I consider both risk sets but focus on the risks associated with a chosen policy of intervention, tracing the paths and causal factors that take poli-cymakers from the initial problem recognition and definition to risk aversion or risk acceptance in contexts representing the whole spectrum of risk levels from low to high. An observer can define the level of risk of a policy by as-sessing the potential costs if anything goes wrong with the key factors affect-ing the intervention operation. Low-risk intervention policies are those for which the chances of goal achievement at an acceptable cost would not be substantially affected even if many elements went wrong in their implemen-tation. Risk increases as the odds of success are more tightly correlated with the validation of the underlying premises and expectations for the plan or policy; in other words, the more sensitive the predicted policy success is to unforeseen and unexpected misfortunes, the greater the risk. Risk level and the affordable margin of error work in opposite directions. Low-risk to mod-erate-risk decisions allow for a relatively large margin of error in assessment (e.g., Grenada, Panama, Czechoslovakia). As risk increases, the affordable margin of error decreases, because the cost of errors increases in proportion to the riskiness of the decision (e.g., Vietnam, Lebanon).

Consider, for example, the two main options debated by the Johnson ad-ministration in the spring and summer of 1965: taking over the ground war in South Vietnam or getting out of Vietnam altogether. We can compare the perceived implications of a worst-case analysis for each. If the United States took over and lost the ground war, dominoes would fall, the United States would lose its hegemonic position in Asia, it would be humiliated and lose much of its credibility, and its loss of reputation would cause the emergence of communist revolutionary movements that would turn to armed struggle as a means of seizing power. In addition, the defeat would result in heavy fi-nancial costs and numerous casualties. On the other hand, if the United States pulled out of Vietnam altogether, all of the above would happen, but without the material costs and the loss of lives. Thus, according to the sug-gested criterion, the first option is riskier.

The analysis of the case studies in Chapters 6 and 7 follows the struc-tured focused-comparison method. In these case studies I seek to illuminate risk and risk-taking behavior and how they relate to decisionmaking in for-eign military intervention. The chronological sequence of decisions, there-fore, is not a main concern; it is provided only in the narrative section for each case. The cases are interpretive histories. They are not intended to cover all aspects and dimensions of the events or to analyze all the decisions, only those pertaining to the research questions under consideration. The analyses are based on extensive secondary sources, which are complemented by newspaper reports and, when available and necessary, relevant government documents. These provide the raw material for applying and extending the theoretical conceptualization.

To facilitate comparison the discussion of each case is organized in the same way. I begin with a short historical overview and narrative of events leading to the intervention, then I identify the stakes and interests, as perceived by the decisionmakers, which became the incentive for intervention. Next, after discussing the risks and constraints posed by the national military, economic, and political capabilities, I go on to the risks and constraints posed by the international environment at both the systemic and state levels. Finally, each case is identified as belonging to either the individual-dominated decisionmaking (IDDM) mode or the group-dominated decisionmaking (GDDM) mode. The factors motivating decisionmakers to take risks are then traced, elaborated, and related to policy outputs.

The analysis and interpretation of risks and constraints related to national capabilities and the international environment follows the theoretical discussion in Chapters 4 and 5. Similarly, the analysis and interpretation of risk-taking behavior is closely informed by Chapters 2 and 3. I make a conscious effort to avoid repetition of the theoretical arguments in the empirical chapters and to structure these chapters in a manner that stimulates association with the earlier theoretical discussion, so that the cases are also self-contained analyses of important international events. Each chapter in Part II concludes with a comparison of the analyzed cases and a statement of the resulting implications for theory and policy. The conclusions from Chapters 6 and 7 are integrated in the concluding chapter of the book.

To sum up, much of our knowledge of social risk taking is based on findings from experimental laboratory research that deals almost exclusively with discrete, well-defined events. Historical case studies allow for the analysis of risk judgments and preferences that take place in a real-life environment, where policy emerges through decisions iterated over time. The availability of detailed and reliable empirical data on important historical instances of such decisions and their outcomes makes them, metaphorically speaking, an ideal laboratory for applying a process-tracing study of judgment and decision formation under conditions of variable risk. By drawing together previously unrelated multidisciplinary strands of research on risk and thereby providing a complete, contingency-sensitive theoretical framework for explaining and understanding risk-taking behavior, the book also aims to contribute not only to the elucidation of foreign policy decisionmaking but also to an understanding of how and why risky decisions are made in the social domain in general. Finally, applying a sociocognitive analysis of risk to an acute foreign policy problem—specifically, foreign military intervention decisions—makes the analysis comprehensive on multiple aspects of decisionmaking (descriptive, explanatory, predictive, and prescriptive) and therefore more relevant and attuned to the requirements of policymakers.

The reader should be aware that this book is first and foremost about how *judgment* of risk is formed and how *choice* among risk-taking preferences is made. It charts different paths to judgment and choice on the assumption that the paths could vary depending on context.[16] The architecture of the book allows the reader three distinct options. The book can be approached as a map for the labyrinth of contingent situations and contexts in which risk is an important dimension of the opportunity for decision. In the book are identified the diverse variables that affect the experience of risk and the shaping of preferences for risk acceptance or avoidance. The relevance and importance of particular variables differ across contexts and situations. The theoretical analysis of risk (Chapters 2 and 3) integrates and goes well beyond the existing perspectives on risk-taking behavior in a systematic manner that will, I hope, point the way to disciplined inquiry. The book can also be read as an analysis of an important and acute political-strategic problem: foreign military intervention. And finally, it can serve as an interpretive history of five well-known cases of post-1945 foreign military intervention, which can be understood individually or in a comparative perspective. As a whole, the book links the theoretical and the empirical. Theory is used to make sense of evidence; evidence is used to validate, sharpen, and refine theory.

PART ONE

Theory

CHAPTER 2

The Anatomy of Risk

WHAT IS RISK AND HOW IS
IT FRAMED?

Decisions involving risk in politics are different from similar decisions in business, an area that has generated extensive research on risk taking. Political decisionmakers, especially foreign policy decisionmakers, are
not accustomed to defining their level of acceptable risk precisely and systematically prior to making a decision. This is not the case, for example, in
stock market gambles, where investors are expected to specify acceptable
risk levels. The main reason for this difference is that in business there exist,
to a greater extent than in politics, consensual norms of what reasonable
business practices are. In politics, on the other hand, a consensual normative
framework that distinguishes the gambler from the astute, responsible government leader has yet to emerge, let alone become an integral part of decisionmaking culture. The absence is manifested in the ongoing debate over
such questions as: What constitutes good judgment in political issues? Are
there criteria to distinguish good decisions from poor ones? What differenti-
ates reasonable risk taking from adventurous risk taking?

Our current knowledge of risk in international politics is incomplete and
lacking in systematic formalization, so what exactly do we mean by "risky
decisions"? How can we define risk scientifically as an analytic construct,
rather than as a gut feeling of danger or discomfort? Is risk a cognitive, affective, or behavioral construct? Is it a uniform construct? And are risks in all
areas of human life similar in nature? In this chapter I address these questions, define risk, and specify the nature of risk in international politics in a
manner that will be applicable to case-study analyses. Unlike the rational
choice approach to risk, characterized by the simple "black boxing" of risk,
the sociocognitive approach presented here opens the black box, that is, disaggregates risk and details its attributes. This is essential for a more realistic,
accurate, and policy-relevant analysis. In the latter half of the chapter I iden-

tify and specify the nexus between the nature and framing of risk, the communication of risk, and decisionmakers' responses.

A Typology of Risks

To be analytically useful, risk has to be disaggregated into three types: real, perceived, and acceptable. *Real risk* is the actual risk resulting from a situation or behavior, whether decisionmakers are aware or unaware of it. Risk, however, can be subjective in the sense that "it is perceived by the subject; it flows from individual experience and cognitive processes and organizational characteristics, as well as from environmental contingencies" (Kobrin, 1982: 45). Risk perception is socially constructed and differently understood in connection with different activities and is therefore dissimilarly experienced and conceived by different individuals or groups (Blaylock, 1985; Dake, 1992; Otway, 1992; Vlek and Stallen, 1981; Wildavsky and Dake, 1990). *Perceived risk*, then, is the level of risk attributed to a situation or behavior by the decisionmakers. I assume that a standard of absolute risk is impossible and that perceived risk need not be, and often is not, congruent with real risk. Incongruence may be caused by the unavailability of information, by misperception, or by misinterpretation. Thus, the response of different individuals and groups facing the same type and level of real risk may vary because of dissimilar risk perceptions. The more that information is based on comprehensive and independent knowledge, such as valid statistical data or experts' consensus, the more the assessment of risk approximates objectivity.[1]

The third type of risk is *acceptable risk*, which is the level of risk representing the net costs that decisionmakers perceive as sustainable, and are willing to bear, in pursuit of their goals. Acceptable risk does not have to be congruent with either perceived or real risk. The interfaces between the various expressions of risk and their relation to the subject of foreign military intervention decisions are projected in Figure 1 and the table (below).

Figure 1 sets out the causal sequence of challenge, risk, intervention, and feedback. Faced with a challenge to national interests—a challenge is either a threat to those interests or an unexpected opportunity to advance them—decisionmakers search for a way to cope with it. The situation may offer a reasonable opportunity to deal with the challenge with foreign military intervention. But this option almost always carries with it a set of potential political, economic, and military risks. In deciding whether to seize the opportunity and intervene, decisionmakers have to compare the intervention option with other options and assess the balance of risks and gains.

The nature and level of risks associated with the intervention option are recognized and interpreted through the mediating role of various information-processing mechanisms. Once decisionmakers have formed a risk assessment, accurate or not, the decision on how to proceed is shaped by measuring perceived risks against what are presumed to be acceptable risks.

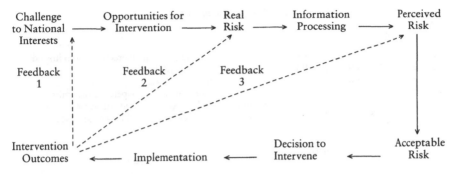

Fig. 1. The Risk-Intervention Nexus

Based on this comparison, the decisionmakers decide whether to accept or reject the risks posed by a policy of intervention, and they decide, too, the type and intensity of intervention congruent with their definition of acceptable risk. For a variety of reasons, actual intervention may or may not be commensurable with the intervention foreseen in the initial decision to intervene. This can be the case when implementation is overzealous, when resources to carry out the decision are lacking, or when enthusiasm by individuals and organizations entrusted to execute the decision is likewise lacking.

The outcomes of the actual intervention have three feedback effects. The first (Feedback 1) may change the nature and intensity of the challenge to the national interest. This change will then affect both opportunities for and risks of intervention. Intervention outcomes can also directly affect the nature and intensity of risks posed by intervention (Feedback 2) by reshaping the environment and affecting the actual level of risk that will face the actors. A third feedback effect (Feedback 3) influences risk perceptions through a process of learning. The revised risk perceptions may lead to reevaluation and possibly revision of the intervention policy to either escalate or de-escalate the current level of intervention. The revised intervention policy will represent a different level of real risk that, when recognized and interpreted, may result in a revised risk perception, and so on.

The Definition of Risk

What is risk? As a real-life construct of human behavior, risk represents a complex interface among a particular set of behaviors and outcome expectations in a particular environmental context.[2] Risk in social life must be approached in a nontechnical manner;[3] hence the common distinction between risk and uncertainty is neither realistic nor practical when applied to the analysis of nonquantifiable and ill-defined problems, such as those posed by important political-military issues. The classic distinction found in economics between risk and uncertainty postulates that risk exists when decisionmakers have perfect knowledge of all possible outcomes associated with an event

Risk Contingencies and Outcomes

	Contingency	Possible Outcome
1.	RR > PR & PR ≤ AR	Risk-seeking policy. Damage from misperception maximized.
2.	RR = PR & PR ≤ AR	Potential for taking optimal risk levels.
3.	RR < PR & PR ≤ AR	Disposition to take moderate risks.
4.	RR > PR & PR > AR	Preferences not clear. Damage from misperception limited.
5.	RR = PR & PR > AR	Risk-averse policy. Likely error on the side of overcaution.
6.	RR < PR & PR > AR	Risk-averse policy. Misperceptions could lead to missed opportunities.

NOTE: RR = real risk; PR = perceived risk; AR = acceptable risk.

and the probability distribution of their occurrence, whereas uncertainty exists when a decisionmaker has neither the knowledge of nor the probability distribution of the outcomes (Kobrin, 1979: 70; Kobrin, 1982: 41–43).

These definitions tend to overlook outcome values. Yet *risk* in everyday language and as commonly understood by decisionmakers has a utility-oriented connotation. "The word *risk* now means danger, *high risk* means a lot of danger. . . . The language of risk is reserved as a specialized lexical register for political talk about undesirable outcomes" (Douglas, 1990: 3; see also March and Shapira, 1987; Shapira, 1995: 43–44). Uncertainty, on the other hand, connotes a state of incomplete information.[4] In the most extreme case, it entails a lack of information even regarding which dimensions are relevant to the description of the risk-set. This is defined as *descriptive uncertainty*. Even when the relevant problem dimensions are known, their values may not be; this is defined as *measurement uncertainty* (Rowe, 1977: 17–18; Yates and Stone, 1992b).

Clearly the deduction from the classic definition of risk—that all outcomes for which probability distributions are similar represent the same level of risk—is not realistic. People tend first and foremost to associate risk with the content and nature of the outcome, that is, with outcome ambiguity (whether the full range of outcomes is known or unknown) and outcome value (whether the outcome is positive or negative, desirable or undesirable). Associating risk with outcome probability (high or low) comes second.[5] In fact, it is sensible to view the classic probability-oriented distinction between risk and uncertainty as merely describing two levels of uncertainty. I define the case in which outcome probabilities are known as *structured uncertainty* and the case in which outcome probabilities are unknown as *unstructured uncertainty*. The term *risk* will be reserved for situations where not only the probabilities of outcomes are uncertain but the situation itself is ambiguous, that is, poses a plausible possibility that at least some outcomes are unknown and will have *adverse* consequences for the decisionmakers. According to this formulation, all risky situations subsume uncertainty, but not all uncertain situations involve risky (ambiguous and adverse) outcomes.[6] Let me note that the notion of adverse outcomes is not used here narrowly to mean only losses. It is conceptualized in comparative terms, the reference point for comparison being a decisionmaker's aspirational outcome level. An outcome that falls substantially short of the decisionmaker's aspiration level, even if it is by itself a gain, will be considered adverse.

A prominent attribute of foreign military intervention decisions is the high level of structured and often unstructured uncertainty that they entail. This uncertainty may be a feature of the event itself, or it may result from difficulties that individuals and organizations have in observing, interpreting, anticipating, and evaluating important aspects of the event and its outcomes. Decisionmakers are often not sure whether their cognitive map of reality is complete, whether those parts of reality that they have observed are the most pertinent ones, or whether their assumptions and beliefs about causal interrelations among the variables discerned are accurate. Decisionmakers are often uncertain about the hierarchy of the relevant values, especially mutually exclusive ones. Consequently, they may also feel uncertain about the goals they want to attain and the priorities among them, and may find it difficult to make policy choices that require value trade-offs (Brunsson, 1985: 40–41; Lipshitz and Strauss, 1997). What further complicates the decisionmakers' tasks is that the risks of foreign military intervention, for example, are complex, not only in that they have multiple dimensions (e.g., political, economic, military) but also in that information about each dimension contains a different degree of uncertainty. It makes combining that information into a unified assessment a formidable problem.

A broad definition of social risk should therefore include uncertainty as an integral element, thus accounting for the predecisional state of knowledge and its impact on the incentive to take or avoid risk. For example, after a policy failure, uncertainty can be used as an excuse to argue that the policy risks could not have been foreseen and thus reduce decisionmakers' accountability. Knowing that the excuse can be used in case of failure, they are likely to be less inhibited in taking risks than they should be (Brunsson, 1985: 51–52). The definition of social risk also has to reflect the subjective validity that this knowledge has been given by its holders. That is, the validity attributed by decisionmakers to their assessments of both the outcomes and the probability distribution of outcomes may range from total confidence to very low confidence. They are keenly aware that the less valid their estimates, the higher the probability that inappropriate choices will be made, resulting in costs and losses.[7] Yet reduction of uncertainty does not in itself reduce risk. Optimal information is not necessarily equivalent to control over adverse consequences. Even with perfect information, control over outcomes may be nonexistent or limited; in such a case, risk is deterministic or at best quasi-deterministic.

The level of risk, from the decisionmaker's vantage point, will thus be defined by the answers to the following questions: (1) What are the gains and losses associated with each known outcome? (2) What is the probability of each outcome? (3) How valid are the outcome probabilities and gain-loss estimates? *Risk is the likelihood that validly predictable direct and indirect consequences with potentially adverse values will materialize, arising from particular events, self-behavior, environmental constraints, or the reaction of an opponent or third party.* Accordingly, risk estimates have three dimensions: outcome values (desired or undesired), the probability of outcomes, and the validity attributed to the estimates of outcome values and probabilities.[8] Although the three components of real risk are independent of each other, that is not true for subjective risks. Payoffs affect probabilities (Einhorn and Hogarth, 1986), and probabilities influence the weight of payoffs in the process of choice. Experimental evidence shows that in a choice between two gambles, when the probability of winning is greater than the probability of losing, choice is more closely associated with the amounts involved in each gamble; but when the probability of losing is greater than the probability of winning, choice is more closely associated with the probabilities. This indicates that choice is not independent of the structure of the alternatives available to the decisionmaker (Payne, 1975).[9]

The validity dimension affects the weight that decisionmakers attribute to probability and payoff estimates in their risk evaluation. Because estimates of probabilities usually raise many more difficulties than do payoff estimates, risk judgment tends to be dominated by payoff estimates. Validity considerations generally bias choice in favor of decisions with known risks over alternatives with unknown risks (Baird and Thomas, 1985; Curley et al., 1986; Ellsberg, 1961; Fischhoff, 1983; Gärdenfors and Sahlin, 1982; Heath and Tversky, 1991; Shapira, 1995).[10] The need for veridicality is most prominent in high-stakes decisions, where the costs of errors, including challenges to the legitimacy of decisions, could prove critical. Validity is thus a particularly important consideration in the formation of risk assessments when decisionmakers have reason to doubt the information available to them.[11]

In foreign policy, information is very often vague or ambiguous (Vertzberger, 1990: 26–34; Wallsten, 1990: 33). In making value and probability estimates, decisionmakers require a certain threshold of confidence before they will consider a risk worth worrying about. As confidence in high-probability high-cost outcome estimates increases, so does the perception of risk; and when confidence declines, so does risk perception. This correlation implies that similar value and probability estimates may result in dissimilar overall risk perceptions and related anxiety with different individuals, because probability and value estimates may be held with dissimilar levels of confidence in their validity. In addition, individuals vary in their minimal confidence threshold and thus in the degree to which anxiety will increase in response to the same incremental rise in the validity of probability and value estimates. For some people in risky situations, a small increase in confidence in threatening information produces a large leap in anxiety; for others, only a large increase in confidence will cause a significant increase in anxiety.

Confidence in knowledge can be epistemic based, person based, belief based, or situation based. For analytic purposes, let us treat each source of validation as an ideal type, then analyze the interaction between two or more sources. In epistemic-based validation, confidence in knowledge is rooted in the manner in which knowledge is generated by following predetermined norms, principles, and rules that are considered guarantees of the validity of knowledge. The user's confidence in the validity of the information depends on evidence that confirms the epistemological correctness of the *process* that generated the knowledge in question. For example, scientists judge the validity of knowledge by its epistemic correctness. Decisionmakers often use a similar criterion for such issues as environmental hazards when they perceive it to be relevant.

In person-based validation, confidence is rooted in the individual who is the source of the knowledge. The user does not care how the knowledge was generated but cares instead about who is disseminating it. Confidence in a particular person may derive from innate qualities (e.g., charisma), affective qualities (e.g., liking), past experience (e.g., the person has formerly proved to be credible), or an established relationship (e.g., a longtime friend), among other sources.

The third source of confidence is belief-based validation. Knowledge that conforms to or is congruent with important beliefs of the user will be considered valid, even if the knowledge is methodologically flawed. The identity of the person delivering the knowledge is of little consequence to the user. Ideologues, such as former President Ronald Reagan, are inclined to use this validation criterion.

The fourth and least important source of validation is the situation. Here context determines the reliability of knowledge; context is considered when the observer distrusts his or her information sources. Situation-based validation is based on the premise that in certain situations information cannot be manipulated and therefore can be trusted or that the information source has no incentive to manipulate the information because the costs are too high or the gains are marginal.

Being cognitive misers, people tend to devise a hierarchy of their most preferred to least preferred validation criteria. Judgment of the reliability of value and probability assessments relates to this hierarchy. Decisionmakers start by applying their most preferred criterion. If the most preferred criterion cannot be applied in their judgment of reliability, they proceed to the next level in the hierarchy, and so on, moving down the list of preferences that are identified at each step with less and less confidence. In the absence of all four validation criteria, confidence in value and probability assessments is minimal.

Choice of one source as a key to validation does not preclude a secondary role for one or more of the other sources of validation. Preference for one source over another has important implications for the increase or decrease of confidence levels over time. Epistemic-based validation has built-in rules for discrediting or falsifying currently held assessments. Person-based validation will change when trust in a particular person is diminished or when a highly regarded person provides invalidating information. Belief-based validation is the most difficult to discredit because beliefs, especially core beliefs, change very slowly. Practically, the only quick way to convince a decisionmaker relying on belief-based validation to change is by reframing the information so that it does not any longer seem to be congruent with his or her beliefs or so that the original preference for reliance on belief-based validation seems to have been an error. A change in situation-based validation will follow from contextual changes.[12]

Confidence in the validity of estimates has a major effect on decisionmakers' adaptation and learning from feedback information. When the stakes are high, low confidence induces alertness to feedback information and could trigger retreat from or a change in the original decision as soon as negative feedback confirms the decisionmaker's low confidence. High confidence, on the other hand, reduces alertness to warning cues and may generate premature cognitive closure and conservatism regarding risk estimates. But for decisionmakers to reliably estimate the validity of their own risk estimates, they need an adequate data base on their past performance in similar tasks.

Such data are rarely available regarding political and military decisions, for two reasons. First, decisions of this type are not routine; each is unique and needs to be dealt with on its own terms. This reduces opportunities for learning and for statistical aggregation of past performance information. Decision outcomes are not quickly known, in any case, and do not provide unequivocal feedback. Second, these decisions involve controversial choices. Decisionmakers are not keen to create an easily accessible data base with reliably complete records of their past performances—they want only their successes to be on record. As a result, they are likely to be overconfident in their judgments and to present overly narrow bounds for their uncertainty (Freudenburg, 1988, 1992; Lichtenstein et al., 1982). All else being equal, group estimates are likely to be held with greater subjective confidence than individual estimates are, for both cognitive and motivational reasons. Group members have more information collectively than each does individually, especially when information is shared through discussion. From the motivational perspective, group decisionmaking involves the investment of more resources (time and effort) than individual decisionmaking does. Group members feel compelled, therefore, to correlate the quantity of resources with the quality of the product and are thus likely to be more confident about their decisions than individuals are (Mayseless and Kruglanski, 1987; Sniezek, 1992).

There are, then, several strategies by which perceived risk can be manipulated. The first is to underestimate the *probability* of undesirable consequences or to overestimate the probability of desirable consequences. The second is to underestimate the *disutilities* of undesired consequences or to overestimate the utilities of desirable consequences. The third is to manipulate the *validity* of the information about both the probabilities and the value of undesired consequences by increasing or decreasing the validity of the probabilities attached to various outcomes or the validity of judgments about the values attached to outcomes.

Risk perception is, in the final account, a scientific question that is affected by nonscientific considerations (e.g., cognitive biases); likewise, risk acceptability is a political question that is affected by nonpolitical considerations (e.g., the nature of risk). Knowledge of acceptable risk levels may bias risk estimation; decisionmakers may avoid or ignore relevant information and scientific methods of estimation in order to match estimated risk to acceptable risk and avoid the dilemma of choice that arises from incongruence. To prevent biased assessments from happening, the two—risk estimation and the setting of risk acceptability—must be delinked, and estimation must happen first. If the level of acceptability is set in advance, it may become the anchoring value for the estimation process and will thereby bias it.

A common error is to identify risk taking exclusively and incorrectly with active policy choices. Passive, "stick in the mud" (Corbin, 1980; Mandel, 1987) policies may also entail risk taking by attempting to preserve the status quo and ignoring environmental signals that indicate a need for initiative and change. In other words, there are no risk-free choices, including the decision not to decide. Even though decisionmakers often equate passivity with risk avoidance, avoiding active decisions may in some cases entail more risk than making an active choice would. It follows that risk taking in intervention is not necessarily congruent with decisions that increase the chance of violence by armed intervention. A broader view of risk should take into account the chance that risk avoidance in the short run (e.g., refusing to deploy troops) may turn out to have been a very risky decision. Consider, for example, the options that the United States faced with respect to military intervention in the Persian Gulf. Intervention involved the risk of military hostilities with Iraq. But nonintervention meant the risk of Iraqi control over the bulk of the world's oil supply or the risk of a future confrontation with a nuclear Iraq. In such cases, the choice is not between risk taking and risk avoidance but between different types of risk.

Risk assessment and acceptability are thus relative and comparative. In judging a policy's riskiness, decisionmakers are concerned not only with the actual content of the policy but about its relation to competing policies offered by others. This comparative judgment of riskiness is embedded in an important defining characteristic of the political context within which foreign policy decisions are made—that is, acceptability by a decisionmaker's constituency.[13] To convince constituencies of the merits of a particular policy, its supporters have to address the way others perceive and compare it to alternative policy options, which enables supporters to place their option within the acceptability domain. Because future outcomes usually seem more abstract and less tangible than the immediate, vivid costs of a policy. the perception of a policy's riskiness is based primarily on its immediately observable costs, although future costs may also be taken into account.

The Texture of Risk and the Taste for Risk

Risk, social risk in particular, is a complex phenomenon with multiple attributes. This has distinct effects on the way decisionmakers compare risks, on the biases that affect risk perceptions, and on decisionmakers' risk preferences. To explain these effects requires two concepts: the texture of risk and the taste for risk.

Comparing risks is an essential task in choosing among policy options. The most obvious approach is to compare the magnitude of risk represented by the competing options. But along with the problem of producing a single measurement of risk in real-life political-strategic contexts—which are mostly ill defined and contain hard-to-quantify variables—this approach poses another difficulty. Although people may make a choice based on magnitude alone, especially when the options are clearly differentiated into high-risk and low-risk ones, they usually do not when the perceived risk-level variance among options is moderate or low. Decisionmakers rely therefore on a more complex and differentiated comparison based on the texture of risk, which is a set of risk-defining attributes.

Here are the attributes that define the texture of risk.

1. *Risk transparency.* How ambiguous or well understood are the risky consequences of a decision? Debates among experts and policy advisers are likely to increase doubts among decisionmakers regarding whether the risks of a particular policy are really understood. Doubt is apt to diminish the perceived validity of value and probability assessments and increase risk aversion.

2. *Risk severity.* How serious and damaging are the perceived consequences of a decision or situation?

3. *Risk certainty.* How certain is any particular adverse outcome to materialize? If risks cannot even be guessed at, the level of perceived risk will be much higher because of the possibility of surprise and the possible lack of resources to cope with whatever risk emerges. Similar situations may by implication represent dissimilar levels of risk, depending on prior expectations and preparation.

4. *Risk horizon.* How close in time are the adverse consequences? The closer in time they are, the more vivid and salient they will seem, and the more weight they will be given. Distant negative consequences are underweighed and perceived as less likely to occur; they therefore have only a minor impact on decisions. Short-term high-risk assessments are given much more weight than assessments of the accumulation of long-term risky outcomes (Milburn and Billings, 1976; Svenson, 1991: 273).

5. *Risk complexity.* The more complex the risk calculus, the more likely the decisionmakers are to ignore most or many of the dimensions and concentrate on one or a few of the most immediate and salient. The complexity of the choice also reduces the decisionmakers' comprehension of the task

and becomes a source of anxiety because of the increased probability of making the wrong decision. Complexity can be assessed using four criteria:

    a. Measurability of risk, with risk dimensions being more elusive and difficult to assess the less quantifiable they are.

    b. Variability of issue dimensions, that is, the range of issue-areas affected by risk dimensions (e.g., economic, military, political).

    c. Multiplicity of time dimensions, that is, whether or not all risky effects are expected to occur within the same time frame (i.e., whether they are all short-term or long-term consequences, or both).

    d. Interactivity of risk dimensions, that is, whether change in one risk dimension affects the level of risk in other dimensions, with the risk calculus becoming more complex with greater interaction among the risk dimensions.

    6. *Risk reversibility.* Are risky decisions reversible once they are made, and at what cost?

    7. *Risk controllability and containability.* Are the risks generated by the decisions controllable and containable? The answer matters even when risk decisions are irreversible. For example, controllable risk is a more acceptable result of an irreversible decision than uncontainable risk would be.

    8. *Risk accountability.* Will decisionmakers be held responsible by the public for adverse consequences? If so, what is the magnitude of the personal political cost that they will have to bear? "As a general rule, the more directly accountable a decisionmaker is to the public, the more likely it is that public perceptions will receive consideration in priority setting" (Kasper, 1980: 79). Choice preferences are thus shaped by the criterion of which policy options are easier to explain and justify (Ranyard, 1976; Tversky, 1972).

Although for analytic purposes these risk attributes are presented as mutually exclusive, clear-cut distinctions are rare in real life, and the relations among attributes are interactive. Complexity could affect transparency, and severity could affect accountability, and so forth. Any combination of attributes affects risk preference either directly or indirectly through its impact on any one or all three components of risk: value, probability, and validity. As a rule, the more transparent, severe, proximate, certain, irreversible, and uncontrollable the risk, the less complex the related risk calculus. Also the more accountable decisionmakers are, the more likely they are to recognize the risks in time, consider them relevant, and give them higher importance in the shaping of decisions. And the more transparent, severe, certain, close in time, complex, irreversible, and uncontrollable the risks are, and the more they require that someone be accountable for them, the more risk dense is the policy associated with these risks considered.

    These multiple attributes, which characterize social and political risk, make it difficult to compare the magnitude of risks, because they are not easily translated into common comparable measures. Attempts to quantify political and social risks are suspect. "This is not to say that quantitative risk assessments are not useful or illuminating, only that it pays to be somewhat skeptical of the quantitative results of risk assessments and to recognize that the appearance of great accuracy that precise numbers in such analyses carry with them is spurious" (Kasper, 1980: 74). It is for this reason that risk assessment in political-military decisions can rarely make effective use of statistical decision theory in its formal mathematical guise. The problems are hard to model and do not have a sufficiently well-defined and quantifiable structure to be. adequately represented by abstract mathematical models. Still, the conceptual principles that underlie statistical decision theory are valid and, when interpreted qualitatively, may act as guidelines for risk assessment even when quantification is not possible (Dror, 1986: 170; Hogarth and Kunreuther, 1995; Lowrance, 1980: 12; Strauch, 1971, 1980).

The set of risk attributes can be found in every risky situation. It provides the coordinates that describe a particular situation and allows comparisons with other risk situations. It also explains differences and similarities in decisionmakers' responses across risky situations; because perceptions of the attributes and the taste for risk may vary among decisionmakers, their mapping of the same situation may result in different definitions of the situation and different risk assessments. The accuracy of decisionmakers' perceptions of these risk attributes determines a number of important issues: (1) their sensitivity to the completeness or incompleteness of their information about the problem, and their ability to understand the risks involved; (2) the likelihood that they will be open-minded and imaginative about those aspects of the problem for which hard data are unavailable; and (3) their motivation to analyze and deal with awkward or unpleasant aspects of the situation. The key, then, to the accuracy and comprehensiveness of risk assessment, especially for ill-defined and hard-to-model problems, can be found in three determinants: information, imagination, and motivation. Information without imagination and motivation is not likely to yield an astute risk assessment. Imagination unconstrained by a broad and valid information base may yield paranoia, self-serving assessments, and hallucination. Motivation without information will produce wishful thinking—or its counterpart, a worst-case analysis of the situation.

Closely related to the concept of risk texture is the concept of the taste for risk. I argue that in most cases decisionmakers—like consumers in economics—have an individual taste for risk. Taste reflects a preference for particular risks over others, even if their magnitude is similar. Taste is an expression of a preference for a particular texture of risk, one with a specific combination of or emphasis on specific risk attributes. Risks of the same magnitude (as measured, for example, in dollars) but representing different combinations of risk attributes are not perceived by different decisionmakers, or even the same ones, as similar, and could engender variable risk preferences.

Emphasizing that each risky problem has a unique texture explains the cross-decisional diversity of risk preference. In decisionmaking by individuals, taste is why the same decisionmaker may seem to be risk acceptant on one occasion and risk averse in another situation with a similar level of risk. Likewise, two people with different tastes will make dissimilar choices given the same risk texture, even if they have the same risk disposition (i.e., risk averse or risk acceptant). The difficulty in aggregating group preferences when a group faces a risky decision is not necessarily intragroup differences in assessing risks but the distribution of differences in the taste for risk within the group. The greater the intragroup variance in taste, the more difficult it is to reach a consensus on a policy option. For example, even if two group members have the same perception of the attributes of risky policy options, their policy choices will be different if one prefers long-term, less transparent chances over short-term chances ("let the future look after itself") and the other prefers short-term, well-specified chances.

Generic Response Patterns

Decisionmakers' response to their risk assessments may be active or passive. An active response is either risk averse (overcautious) or risk seeking. The risk-averse person strives to avoid risky situations and choices and will, accordingly, attend to and process information selectively. The risk seeker tries to get involved in situations that contain a strong element of risk and will likewise attend to and process information selectively, but in this case, to justify risk-taking behavior, even if such behavior is counterproductive.

The passive approach to risk is common in risk-tolerant decisionmakers. They neither search for risk nor avoid it, but accept it and attempt to cope with it on its own terms—for example, through preemptive risk management. In the case of anticipated intervention, the risk-tolerant decisionmaker can reduce in advance the need to actively cope with associated risks by preparing for them. To illustrate, contractual agreements, such as defense treaties with a third party, can increase the target country's certainty if it is attacked that a third party will counterintervene.

A serious deficiency of the rational choice approach to risk taking is that it does not distinguish among different motives for risk taking. Identifying such motives is essential for a realistic determination of likely circumstances for risk-seeking or risk-averse behavior. I suggest a simple typology of motives (discussed in detail in Chapter 3).

1. *Deliberative (reasoned) risk taking.* Here, risk-taking behavior results either from astute judgment, based on available information, that a risk is worth taking (in rational cost-benefit terms) because other alternatives are even worse, or from an error in judgment that leads to underestimation of the actual risk. In international politics, given the lack of a valid data base, risk assessments often have little to do with systematic inferences from the frequency of particular outcomes in similar risky situations in the past. Hence, "assessment of risk must be interpreted as a matter of political judgment: as a

degree of belief based on partial information about objective conditions and potential costs and benefits" (Adomeit, 1973: 3). The lack of base-rate information and the consequent necessary use of political judgment suggest, then, two types of cognitive illusions: judgmental illusions, which involve a misestimation of probabilities and payoffs, and cognitive mirages, which involve a misinterpretation of the situation that leads to a belief in the existence of nonexistent options (Thaler, 1983). The result in either case is lack of both procedural and substantive rationality.

2. *Dispositional risk taking.* Here, risk-taking behavior is driven by personality attributes (e.g., Bromiley and Curley, 1992; Plax and Rosenfeld, 1976) or by core values and beliefs that prevail over other situational considerations. These cause decisionmakers to consistently prefer risk-acceptant behavior to risk-averse behavior (or vice versa), irrespective of situational context. Their risk taking is not issue dependent, and their responses to risk across issue-areas are similar and undifferentiated. Such decisionmakers are likely to be impervious to learning.

3. *Socially driven risk taking.* Here, the preference for risk has its sources in the social context—in peer pressure (e.g., Dion et al., 1970) or prior commitment to a particular line of action—which results in either premature cognitive closure concerning other options or the assumption that the social costs of not carrying out the commitment (e.g., losing credibility) are much higher than the costs stemming from the risk of carrying out the commitment, even if there is an awareness of other, better response options.

A number of observations follow from the three types of motives that shape decisionmakers' preferences to accept or avoid risks:

1. Risk-taking preferences will be strongest when all three types of motives work in the same direction. When, for example, judgment, prior disposition, and group dynamics reinforce risk-taking behavior, risk-averse behavior is unlikely.

2. Dispositional forces are more powerful than judgment or inertia, especially when information is ambiguous.

3. Most decisionmaking groups have a mix of motives. The balance of motives, the distribution of power among those with different motivations, and group dynamics will shape the the aggregate risk-taking disposition of the group.

4. In social contexts, those who have more forceful risk dispositions often become the more influential group opinion leaders and, if they are not challenged by other members, may well influence the group to go along with their personal risk preference.

5. Decisionmakers can have dissimilar data bases or even similar information that allows for dissimilar interpretations. In these cases, decisionmakers who have equally risk-averse or risk-acceptant dispositions will ex-

hibit different risk preferences because of their different definitions of the situation.

6. When decisionmakers' risk preferences are embedded in different cognitive attributes (beliefs and values) and different personality attributes, decisionmakers who have similar information are likely to have dissimilar risk preferences in the same situation.

7. The interactive nature of risk-assessment processes may result in a number of perceptual incongruencies and concomitant asymmetries that influence risk-taking preferences by the adversaries in a dispute: (a) asymmetry in the acuteness of risk awareness among the adversaries; (b) asymmetry among adversaries in the manner and openness of communicating their risk-taking propensities; (c) asymmetry in perceptions of adversaries' risk-taking propensities; and (d) incongruence in actors' assessment of how other actors interpret the level of risk perceived by the focal actor. Such incongruence may result either from self-inflicted error or from successful deception by the adversary.

8. Where taking a high risk involves a one-step decision, dispositional and judgmental factors are more likely to be the salient precipitators. But when risk taking is a process, when it emerges incrementally in an iterated fashion, social factors and, to a lesser degree, judgmental factors are more likely to be the main reasons the risk is taken.

The Nature of Political-Military Problems

Problems in the political domain, particularly in international politics, are frequently ill defined, meaning that they are characterized by open attributes. These attributes "include one or more parameters the values of which are left unspecified as the problem is given to the problem-solving system from outside or transmitted within the system over time" (Reitman, 1964: 292–93; see also Abelson and Levi, 1985; Milburn and Billings, 1976; Voss and Post, 1988). And because the foreign policy environment is very dynamic, this inconvenient state of affairs does not improve with time. The course of world events requires continuous updating of personal conceptual maps about the present and future; uncertainty, mostly of the unstructured type, is the decisionmaker's permanent companion (Downey and Slocum, 1975; Voss and Post, 1988: 265). This combination of dynamism, complexity, and uncertainty often makes it extremely difficult, if not impossible, to see a problem in focus and decide when a proposed solution is acceptable. As a result, problems that are ill defined are also poorly understood. Analysts have to impose on the problem a stable structure of their own choice and define the nature of risk within the context of that structure, which makes it difficult to take a detached and objective look at the risk assessment. The analysts become personally involved and develop strong feelings in the matter or even ideological dogmatism (Sjöberg, 1980; Strauch, 1980). With an ill-defined problem, determining which information is relevant becomes highly uncertain. Information that could prove relevant and important is ignored, or hyperattention and indiscriminate collection of large quantities of information overburden the information-processing system and harm its problem-solving capacity. Nor are decisionmakers always aware of the differences between well-defined and ill-defined problems; in fact, they often unconsciously transform ill-defined problems into well-defined ones by ignoring indeterminate attributes (Fischhoff et al., 1981: 9–14; Vertzberger, 1990: 54).

At the extreme, problems become wicked (Mason and Mitroff, 1981: 10–16). A wicked problem has no definitive formulation or solution. Alternative formulations co. espond to different and inconsistent solutions or policies. Each problem has enough unique features to make the value of learning from the past doubtful, and conceptual maps acquired with experience cannot be reused. The wicked problem has no clearly identifiable root causes, and particular symptoms have no exclusively identifying value but can be interpreted as symptoms of another problem. Hence, it is difficult to identify the boundaries of the problem and the level at which it has to be attacked. Because there is no immediate or ultimate test for evaluating solutions, it becomes impossible to tell when the problem has been resolved and whether an ending point has been reached.

Definition of a problem's boundaries affects risk estimates. The narrower the boundary definitions, the less the potential risk perceived. But when the boundaries of a problem are not easily recognizable, deciding which variables are most pertinent to problem solving becomes an arduous task. Similarly, setting the temporal boundaries of the problem—that is, determining the time span within which consequences of alternative solutions will materialize and within which consequences should be viewed and evaluated—is virtually impossible.

Because wicked problems have these attributes, a great deal of information and knowledge is required before a solution can be attempted. These attributes also impose the need to accept risk. Still, vital information is often unavailable, so the analysts have to use rough estimates. Such estimates induce doubts about validity and eventually reluctance to act on the available information. The decisionmakers may then ignore all or part of the information and either take a daredevil approach or choose the other extreme of doing nothing. When the decisionmakers eventually do gamble on a solution, the consequences are extremely difficult to become dissociated from or to undo, causing a rigid commitment to the solution and introducing a bias toward conservatism.

Intervention decisions, like most important political-military decisions, pose problems that are complex, ill defined, indeterminate, and sometimes wicked. The organized complexity characteristic of many such decisions means that the decision situation involves a web of problems and issues that tend to be highly interrelated and cannot easily be broken down and isolated for separate treatment. To map the large number of interactive relations relevant to the problem requires a degree of abstraction of which decisionmakers are often incapable. A solution for one problem does not necessarily reduce overall complexity, and often no single solution or policy will obviously yield the best outcomes (Downey and Slocum, 1975; Dror, 1986: 167–76; Elster, 1989: 7–17; Mason and Mitroff, 1981: 4–9). Policies implemented to solve one problem create unintended consequences that feed into other problems. The full range of the effects of alternative strategies becomes unpredictable, and the responses of adversaries and the subsequent implications of those responses are increasingly difficult to anticipate.

One way to deal with this state of affairs is to refine policy incrementally. This process allows a complex problem to be solved with limited computational resources—but without regard for the total environmental implications. An incremental approach tends to result in premature decisions made with incomplete and simplified information about the alternatives. When more detailed information becomes available, initial decisions affect how the information is interpreted and used (the anchoring heuristic). In high-risk decisions, this incremental process provides a false sense of confidence in the controllability of the situation and induces cognitive rigidity, which discourages adaptation and learning.

Problem Framing and Risk Estimation

Political-military problems do not often have single, obvious, self-explanatory boundaries, structures, and forms. It is the analyst who, in framing the problem, imposes boundaries, structures, and forms, which convey a particular truth but not the only truth. This process is succinctly described by Stern (1991):

> If all risk messages accurately reflected available knowledge, they would still engender intense conflict because there are different ways to tell the same truth that create different impressions on audiences. It is therefore impossible to determine which among the many possible accurate messages are unbiased. The problem lies in the fact that messages do more than simply convey information. They highlight whatever information they present, drawing attention away from other aspects of the issue at hand, and they give meaning to information. Because a risk message highlights information, it is less a mirror of scientific knowledge than a telescope: it makes the knowledge much clearer, but we see only part of it. Because it gives meaning, it provides less a photograph of reality than a portrait, or sometimes a caricature. (p. 107)

The framing of problems is therefore of great importance to their apprehension and to the search for solutions by decisionmakers, but it is especially critical in dealing with political-military problems that are ill defined, complex, indeterminate, and wicked. "Framing is controlled by the manner in which the choice problem is presented as well as by norms, habits, and expectations of the decisionmaker" (Tversky and Kahneman, 1988: 172), and it involves "the decision-maker's conception of the acts, outcomes and contingencies associated with a particular choice" (Tversky and Kahneman, 1981: 453). With problems that are complex, ill defined, indeterminate, and wicked, especially when time is short and does not allow for mulling over, competitive framings play a critical role in formulating the problem, because the attributes of the problem make any one self-evident and consensual comprehension impossible.

Because with these types of problems alternative framings are equally plausible, framing is also a functional and integral part of the argumentation stage of problem solving—the stage at which decisionmakers try to figure out what is at stake. Framing highlights and directs attention to particular dimensions of the problem and diverts attention from other dimensions. Focusing or diverting attention, in turn, affects estimates and judgments of probabilities, outcome values and their combination, and the validity attributed to the judgments.[14] But framing can also be misleading because "formulation effects can occur fortuitously, without anyone being aware of the impact of the frame on the ultimate decision. They can also be exploited deliberately to manipulate the relative attractiveness of options" (Kahneman and Tversky, 1984: 346).

Framing focuses on the *process* of decisionmaking or on the *outcomes* of decisions. In the first case, framing is centered on particular variables and specific causal sequences. In the latter case, framing calls attention to specific outcomes by amplifying them and attenuating the probability and import of every other outcome. Thus, policy change is sometimes preceded by a reframing of the problem to make salient different dimensions from those that influenced the formulation of the initial policy. In reframing the problem, the weight and range of value dimensions and probabilities could change to such a degree that risk acceptability is modified. Reframing can be either the cause of policy modification or a means to justify it to the public or other decisionmakers. In both cases, framing amplifies particular beliefs and values and triggers a selective search for frame-specific information.

Framing takes one or both of the following forms: it shapes how the risk dimensions of a problem are structured and presented, and it shapes how probability and value estimates are expressed. Decision problems with high stakes, such as foreign military intervention, tend to be unique, comparatively rare, and unfamiliar, and in many cases top decisionmakers have had no foreign policy experience before assuming office. When decisionmakers are faced with such problems, they are not always sure which of their values and beliefs are relevant or what the exact implications of those values and beliefs might be. They have simply never had an opportunity to think the matter through, and the contexts in which such problems arise often do not allow much time for contemplation. They will therefore search for some schema to help in interpreting the situation.[15] How the problem is framed and posed can thus have a strong, even decisive influence on which values and beliefs will be triggered and become reference points in the choice process, because people tend to evaluate options in relation to reference points suggested or implied by the statement of the problem.[16] When a decision problem is indeterminate, it will be reframed to appear determinate. For example, an option that stands out along one particular dimension can be made to seem the most attractive choice by framing the issue so that the dimension in question takes on decisive importance (Elster, 1989: 26; Slovic et al., 1982b; Tversky and Kahneman, 1981).

The threat-opportunity dichotomy determines the basic framing options for political-military issues. Decisionmakers have general schemata for threat and opportunity; that is, they identify threats and opportunities in their environment by comparing the characteristics of specific issues to their cognitive representations of threats and opportunities, which serve as anchoring points. Threats have clear negative connotations: they involve the high likelihood of loss without gain, feelings of low controllability, and a personal sense of being underqualified. Opportunities have positive connotations: a perceived high potential for gain without loss and for successful resolution of the problem, a sense of controllability, and a sense of being qualified, and having the freedom, to take effective action. But threats and opportunities also have characteristics in common: both involve high-priority issues, direct competition with others, and, often, short response time (Jackson and Dutton, 1988, Sitkin and Weingart, 1996). That threats and opportunities have both distinct and common characteristics underlines the importance of the initial framing of a high-value issue, for the framing will determine whether decisionmakers will treat the problem as a threat or as an opportunity.

Framing is bound to location. How a political-military problem is framed depends on where in the decisionmaking system the framing takes place. To illustrate: the risks of intervention decisions are assessed and then accepted or rejected at the political-military level, by the military high command, and by the field commander (Page, 1987). The framing of risk, and the assessment of the salient dimensions of risks, is quite different at these different levels. Whereas the goal of assured victory may sometimes seem a low-risk goal at the military-field level because of the decisive asymmetry in power between the intervener and the target, it may seem a high-risk goal at the political-military level, where victory can include not just battle outcomes but also the political consequences of using military force, such as international responses. Therefore, risks are assessed at the political-military level in more abstract terms and at the lower military-command levels in more concrete, easy-to-measure terms. The consequent differences in approach to risk raise doubts about how well the queries of the political elite about level of risk are understood at the lower levels of military command.

In intervention decisions it is not unusual for political objectives to be vague at the beginning—either because decisionmakers want to be opaque or because they are not sure what they want to achieve—and for the definition of objectives to evolve as intervention proceeds. This characteristic of political objectives makes it difficult for the military command to formulate an internally consistent and cohesive definition of military objectives. But even when the political objectives are understood, the compatible military objective that will make the political ones attainable is not always obvious or simple. Lacking such guidelines, military, strategic, and tactical decisions are made piecemeal, with each decision representing a narrow local view. Narrowness of view is bound to cause underestimation of the risks involved, because only a limited amount of information is used, rather than detailed, comprehensive information; in other words, factors not pertinent to the local decision but very significant for the problem as a whole are underweighed. In foreign military intervention, then, the lack of clearly set political objectives, and the difficulty of tailoring military operations to these objectives, seriously compounds the uncertainty involved in attempting to calculate cost effectiveness and the probability of success.

Consciously and unconsciously, those who frame a decision influence risk acceptability by manipulating the way adverse consequences are presented in relation to gains. The relative salience of perceived costs (or cost avoidance) compared with that of gains is an important consideration in decisions to take or avoid particular risks. Decisionmakers who avoid risk (e.g., oppose a decision of intervention) will emphasize the prospective losses, whereas those who accept risk (e.g., support a decision of intervention) will emphasize the prospective gains. Where the negative consequences (losses) cannot be ignored, the supporters of an intervention policy will take one of two approaches. They may legitimize the costly policy by emphasizing the costs of not adopting it rather than the gains from its implementation; such an approach would be in keeping with the negativity bias, according to which more weight is given to possible costs than to gains (Kanouse and Hanson, 1971: 56). Alternatively, they may frame the problem by emphasizing the probability of immediate gains and imply that the negative consequences are likely to be incurred only in the long run (Jones and Johnson, 1973; Levin et al., 1987). The threat to public-sanctioned legitimacy from negative outcomes is thus further away in time. This manner of argumentation takes advantage of some people's expectation that they will have more control over and be able to prevent delayed outcomes, and takes advantage, too, of the possibility of immediately using positive outcomes to legitimate the project in the eyes of the public. When intervention is framed as an opportunity to both make a gain and avoid a loss, but with an emphasis on the former, the decision is more likely to be in favor of intervention. When intervention fails and the choice between withdrawing and escalating is framed in terms of minimizing losses in the face of a threat of even greater losses, the decision is more likely to be risk averse, that is, to involve a retreat from intervention. But when the problem is framed to emphasize the losses entailed by retreat from intervention—for example, a threat to the credibility of the intervener—the decision is more likely to be escalation of intervention. These conclusions can be inferred from prospect theory, according to which the prospect of losing an amount of money weighs more heavily than the prospect of gaining the same amount (Kahneman and Tversky, 1979, 1982a; Tversky and Kahneman, 1992).[17]

Risk preferences are thus not revealed exclusively through the choices made but may be detected and exposed at the earlier stage of problem framing, which could bias choice and shape policy goals. A positive frame, one that emphasizes the probability of gains, is likely to lead to decisions aimed at seeking gains. A negative frame, one that emphasizes the probability of losses, is likely to result in decisions aimed at avoiding losses. But the process works both ways: decisions may also influence the frame into which they are structured. The preference for a particular frame may be motivated by the need to make the decision comprehensible and sensible to others (Heimer, 1988: 508; Levin et al., 1987: 53).

Risk labeling is another form of framing that plays an important role in establishing risk acceptability. The burden of responsibility for undesirable outcomes may be mitigated by labeling events as random and unforeseeable (Sjöberg, 1987). Calling them accidents could lower the threshold for risk taking by exonerating the decisionmaker of future responsibility for consequences in case of failure. Another salient labeling effect involves the distinctly different implications of framing negative outcomes as either losses or costs: "Losses are more aversive than costs" (Kahneman and Tversky, 1984: 348). The key decision point in intervention, when an intervening power has to decide whether to withdraw or to invest in it more heavily, can often be traced to the moment when further investments begin to be perceived as losses rather than costs.

Another important facet of framing concerns how probability and value estimates are expressed by decisionmakers and how they are understood by other members of the decisionmaking group. On various grounds, people resist expressing their opinions in numerical terms, preferring less precise formulations. With ill-defined problems involving risk, the related uncertainty is vague, not precise. Thus "people seldom know, describe, or utilize numerical values for these [various] likelihoods, and their intuitive knowledge of these values and their abilities to compare the risks of widely different hazards are often quite poor" (Kuipers et al., 1988: 180; see also Budescu and Wallsten, 1985; Clark, 1990). In using subjective probabilities, decisionmakers use symbolic descriptions of numbers (e.g., highly likely, probably) that express categorical and ordinal relations with certain focal or boundary values. But verbal probability expressions hinder communication because individuals interpret verbal probability terms differently. Variability in interpretation increases in contexts involving controversial or ambiguous topics. Most people are not aware of these facts (Beyth-Marom, 1982; Brun and Teigen, 1988; Yates and Stone, 1992a); consequently, they misjudge how well their views are communicated to and are understood by their audience. The use of verbalized impressionistic descriptions (e.g., likely, possible, probable) may also diminish or exaggerate some of the real risks, allowing a subjective risk perception that is incongruent with reality. It is, however, necessary to strike a balance between the need for precision in communication and the need to avoid imposing unwarranted precision on vague information (Wallsten, 1990)—something that is more easily said than done. One way to achieve this goal is by presenting precise numerical estimates along with information about how soft the estimates are and how much they may fluctuate.

Even when likelihood is described in numerical terms, it is treated as a landmark. The use of subjective probabilities that are essentially landmarks does not allow for arithmetic multiplication and addition or for measurement of errors and biases as distances from the correct values (Kuipers et al., 1988). In situations with more than two possible outcomes, a person asked to express the probability of an outcome verbally prefers to use expressions implying high-level probability for alternatives that actually have an objective probability of less than .50. Verbal exaggeration is most evident where all outcomes have equal chances of occurring.[18] "More generally, subjectively assessed probabilities seem to be regarded as properties of the particular outcome in question, with little or no attention paid to the total set of outcomes" (Teigen, 1988: 172).

In value estimates in which, for example, intervention decisions are concerned, framing plays three important roles. First, intervention decisions involve judgments of the salience of the national interests being threatened and the degree of commitment to defend these interests. Framing can affect both judgments. Placing the intervention decision in the context of broader historical trends tends to raise the perceived stakes significantly. By viewing Vietnam within the sequence of events leading from the Bay of Pigs to Vienna, Laos, and Berlin, decisionmakers gained a powerful sense that the line against communism had to be drawn somewhere, even if that meant entering the "wrong war."

Second, because the relative importance of various value dimensions is not fixed but, like the preference itself, depends on stimuli (Goldstein, 1990; Whipple, 1992), those value dimensions that were primed by the framing of the problem and that are readily available will be perceived as the more important ones and will dominate choice.

Third, decisionmakers must at some point operationalize cost assessments of intervention in terms of economic, political, and human-life values. Comparing the costs of alternative courses of action requires evaluations of trade-offs between each cost dimension and the other two. But decisionmakers, especially politicians, are reluctant to calculate trade-offs or to reason probabilistically about costs, particularly with respect to the cost of human life. They "consider it immoral and unacceptable to work with explicit values of human life in planning. . . . It is also likely that a politician would feel anxious if pressed to take a standpoint on the value of human life, among other things because such a standpoint could be easily criticized in the mass media. He would be seen in the eyes of the public opinion as a cold and cynical man" (Sjöberg, 1980: 311). One controversial and painful choice is how and whether to discount lives, that is, whether to give deaths far in the future less weight than present deaths. Consequently, problems are framed to underestimate loss of life or, if that is not possible, to avoid an explicit, precise statement of trade-offs. This evasion allows for misleading ambiguities.[19]

Framing plays an important role not only in the making of the decision but also in mid-term evaluation of the policy, when available feedbacks become a basis for deciding between continuing and changing the policy. Mid-term policy evaluations, especially when outcomes are ambiguous, are based on two considerations: decisionmakers' perceptions of actual policy outcomes and decisionmakers' perceptions of the outcomes that might have occurred if a different policy had been adopted. In the latter case, the decisionmakers compare actual results with counterfactual results to determine the value of the current policy relative to possible alternatives (Miller et al., 1990). Here, framing affects how the actual outcomes are presented—for example, whether the loss or gain dimensions are emphasized—and it influences how counterfactual results are presented. Because decisionmakers deal with events that did not occur, assessments of which are not given to validation or refutation, the room for manipulative framing is extensive. Framing can be used deliberately or even undeliberately to affect the comparison with the actual outcomes, considerably influencing the future of the policy under review. The actual outcomes can be made to look very poor or very positive in comparison with what might have been.[20] For example, actual losses might be presented as much lower than the losses that could have occurred had a different policy been adopted, thus leading to greater commitment to the policy that produced losses.

A final but important form of framing involves, not the content of the problem, but the style of framing itself. A particular framing may be presented as the only possible one, thereby discouraging the audience from searching for alternatives. Where access to information is unevenly distributed among members of a decisionmaking group, control over the sources of the most valid information ("Trust me, I know better") can become a powerful means for one member to dominate the framing process to the point of monopolizing it. This, of course, gives the monopolizer considerable leverage over choice.

In summary, and notwithstanding the specific forms that framing takes, to understand the essence of framing, it is useful to distinguish between *substantive framing*, which is oriented to problem solving, and *strategic framing*, which is instrumental in influencing others, in the elite or the public, to accept and support a particular policy. Substantive framing reflects decisionmakers' beliefs about the nature of the problem itself. Strategic framing, on the other hand, reflects decisionmakers' beliefs about the most effective way of convincing others to support a particular policy approach to a problem. The same frame can serve both the substantive and the strategic functions. In other instances, especially in coping with highly controversial and divisive issues, problem solving and influencing others involve two separate frames that may even be incompatible, requiring that the substantive frame guiding the decisionmakers be hidden from others, who are exposed only to a manipulative strategic frame. For obvious reasons, such a split between frames almost inevitably leads to miscommunication among different players in the decisionmaking process.

The Communication of Risk

Risk taking is not merely a means to achieve a goal; it is also a symbolic form of nonverbal communication. In relations among nations it is particularly important when, because of lack of trust, verbal communications are not regarded as fully credible. Risk-taking behaviors, however, are a credible way to signal seriousness of intention and depth of commitment (Triska, 1966: 11–12). In that sense, "intervention is in many ways an essential symbolic act, meant for a larger audience than the people of the country of intervention" (Brands, 1987–88: 623). Behaviors that convey commitment to risk taking are thus an important part of strategic interaction. Risk-taking behavior assumes knowledge, correct or incorrect, of the risk perceptions and preferences of the other side, knowledge that underlies a decisionmaker's own estimate of the likely reaction of the adversary.

The important role played by the communication of risk raises the question of whether risk perceptions and preferences should be kept secret from the adversary. Knowledge about how adverse an act of intervention will be considered by the other party may, for example, influence the focal actor's decision to intervene. At the same time, making the adversary guess what response can be expected to an act of intervention, thereby raising the adversary's level of uncertainty, could either deter or incite intervention, depending on how the adversary interprets the situation. The opposite option is for an actor to state risk perceptions and preferences or demonstrate them as a means of influencing the adversary's behavior. This option raises the question of credibility, because the adversary may suspect manipulation and react by distrusting and ignoring the risk perceptions and preferences communicated. A further complication is the effect of the domestic response in the focal state. Clearly stated risk perceptions and preferences could trigger enough domestic opposition to undermine the credibility of the statement. The communication of risk perceptions and preferences should therefore take into account that risk taking is a "two-level game" where the domestic and international consequences may work at cross-purposes and produce undesirable results. A third option is to create ambiguity with respect to risk perceptions and preferences. A communicator who prefers this choice either has not yet made up his or her own mind concerning the level of risk involved or is willing to take risks but does not want to admit it prematurely because of domestic or other constraints.

The communication of risk is problematic not just between nations but also within the national decisionmaking system. In collective decisionmaking contexts, intragroup, interorganizational, and intraorganizational discussion and communication of risk require that risk have a well-defined meaning that is shared by all the participants in the assessment; otherwise, the discussion becomes a dialogue among the deaf. In military risk analysis, the vertical structure of military command systems may have an adverse effect on the communication of risk by impeding the full sharing of knowledge about the problem among participants in the decision process. When several echelons of command are involved in risk assessment, each passes on its final judgment of the overall risk, rather than a detailed analysis of which factors should be considered and how they should be interpreted. Thus, each echelon in the hierarchy filters the information available to it and does little to increase the store of information available to the echelon above. This is so because command decisions are structured as unitary decisions and do not reflect dissent or reservations. When the same process repeats itself at each level of the chain of command, the dissenting views are not evident anywhere along the chain. Hence, dissenters can seldom form cross-echelon coalitions; they are not aware of one another's existence. At the same time, their commanders, not aware that there are dissenters in other echelons, consider the seeming lack of support for a dissenting view at their level of the echelon a further justification for overriding the particular dissent. Similarly, if information indicating high risk is not concentrated but scattered across the chain of command, so that information available at any specific level is insufficient to raise the alarm, it is not likely that information will accumulate and cause the risk assessment to be revised upward as it moves up the chain of command. This phenomenon is the dilution bias.[21]

Risk communication should thus be considered an integral part of risk management. The communication of risk is so important for gaining cooperation within the decisionmaking system and mobilizing public support for a policy that risky options which cannot be communicated effectively may have to be reformulated or completely rejected, even if they are the most effective options.[22] Successful risk communication does not necessarily lead to better decisions, because risk communication is just one component of risk management. Even successful communication of the desired message will not necessarily result in consensus about controversial issues nor in a uniform behavioral response. Where people do not have common interests, shared knowledge can lead to diverse responses (National Research Council, 1989: 27–28, 147–48). Effective risk communication within the decisionmaking system is a necessary but not sufficient condition for effective risk management.

Conclusions

I have made five main points about the risks associated with complex, ill-defined problems. These observations depart in important ways from the conventional wisdom on this subject.

First, risk is defined here in terms of three components: adverse utilities, probabilistic uncertainties, and the validity which the decisionmaker attributes to the utility and probability estimates.

Second, to make risk a useful construct in understanding choice, policy problems must be analyzed in relation to three types of risk: the real risk posed by the policy; the perceived risk, which is the interpretation of real risk by the decisionmaker; and the acceptable risk, which is the degree of risk that is acceptable to the decisionmaker. A policy perceived, correctly or incorrectly, as a source of risk will be rejected if the perceived risk is higher than the acceptable risk.

Third, people can be viewed as risk consumers; that is, in deciding whether to take or avoid a particular risk they do not consider only the price and degree of risk but take into account their taste in risk as well. Their taste, in turn, depends on the texture of the risk. People's attitudes toward problems with similar levels of risk will not be identical but will depend on the attributes of particular risks. People may prefer one risk texture over another, even if the levels of risk represented by both sets of risk attributes are comparable. In other words, from the point of view of the decisionmaker, risk is not a uniform, standard construct; rather, each risk has a distinctive texture that affects risk preference.

Fourth, the framing of a problem—the process by which it is represented and its risk dimensions are elaborated—is critical for risk assessment, risk acceptability, and, eventually, for risk-taking preferences, especially when problems are ill defined at first, which allows broad latitude in framing them.

Finally, because risk taking is a means of communicating with an adversary, decisionmakers should be aware of the negative implications of miscommunicating their risk-taking disposition. At the same time, they should realize that a risk-taking disposition may not always be well communicated even within their own decisionmaking system and could consequently have adverse effects on the quality of risk management.

CHAPTER 3

The Formation of Risk Judgments and Risk Preferences

A SOCIOCOGNITIVE APPROACH

Risk, like beauty, is in the eye of the beholder. This important fact raises two sequential questions, one cognitive and the other behavioral. First, what causes a situation, or an opportunity for decision, to be perceived as risky? And second, once a situation is defined as risky, how do decisionmakers cope with it? These questions involve risk judgment and risk acceptability, which are related but not always tightly interdependent; similar risk assessments by individual decisionmakers do not necessarily trigger identical behavioral responses.

A number of problems apply to all crucial risky decisions, including intervention decisions. First, the information for an astute decision is not always available and unambiguous; and when it is available, there are often serious doubts about its validity. Second, even when the information is available and is considered reasonably valid and reliable, decisionmakers' limited cognitive capabilities and bureau-organizational constraints may not allow them to process and integrate the information in a comprehensive, systematic, and coherent manner that will lead to accurate and goal-efficient calculations of the costs and benefits of a given policy. Third, even if optimal calculation is cognitively possible, decisionmakers and decisionmaking systems are not necessarily oriented toward, nor do they give the highest priority to, pursuing a comprehensive and systematic analysis of all the relevant considerations.

In the following discussion I identify the broad and complex set of factors that account for risk judgment and risk acceptability. With some modifications, the broad structure of the analysis follows the sociocognitive approach developed in an earlier book (Vertzberger, 1990). Accordingly, the factors impacting on risk judgment and risk acceptability are categorized into contextual, cultural, personality, group, and organizational variable-sets. The state of the art does not allow us to structure a tight theory in which the causal connection of each variable with every other variable or outcome is unequivocally established.[1] What is offered here is a looser theory. Metaphorically speaking, variables are treated as Lego blocks; they do not change color and form but can be arranged and rearranged to produce the different explanatory designs required by different situations. The theory identifies the relevant variables for diverse risk contingencies, together with a general set of causal relations among variable-sets. The case studies in Part II provide a systematic method to test the practical relevance of these variables and the validity of deductively inferred links among them.

A variable is considered relevant if it significantly affects one or more of the elements of risk judgment and risk acceptability: (1) anticipated outcome utilities; (2) assessment of outcome probabilities; (3) confidence in the validity of assessments of outcomes and probabilities; (4) risk preferences (avoidance versus acceptance). Many of the variables affect both judgment and preferences, and the multiple effects are sometimes causally related.

Risk analysis requires four related mental operations. One is recognition of the risk dimensions associated with an event or premeditated action. The second is causal attribution: people have explicit and implicit probabilistic theories of causality that relate a particular event or action to specific risky results. The third is judgment and estimation of the magnitude of the probabilities, and the utilities or disutilities of outcomes, together with the validity of these assessments. The fourth is the choice of how to respond to risk. Decisionmakers have different attitudes toward risk: "Risk avoiders do everything possible to play it safe. Risk takers are willing to take chances on occasions when the prize seems big enough. Risk makers aggressively seek out chancy situations which offer the possibility of a significant payoff" (Adler, 1980: 11; see also Machlis and Rosa, 1990). In the last, and probably least common, case a decisionmaker does not just respond to risk but creates opportunities that involve it, and fosters a decision culture in which subordinates are also encouraged to take chances and to view risk taking as paying off.

I do not take a position on the wisdom of either risk acceptance or risk avoidance.[2] In fact, the observation made more than three decades ago by Kogan and Wallach in their classic study on risk is still pertinent: "Risk-taking or conservatism may be equally disastrous, depending upon the particular circumstances and situation in question. We can, however, begin to make value judgments about behavior that is consistently risky or consistently conservative—in other words, behavior that relegates situational inducements and constraints to a secondary role" (1964: 211).

Contextual Influences

Risk judgment and preference formation are affected by three contextual influences: (1) the vividness and salience of each risk; (2) prior planning for contingencies that involve risk; and (3) existing commitments to a particular course of action. Each can have an important conscious or unconscious effect on the particular risks chosen for attention, on the relative weight assigned to each risk, and, concomitantly, on the determination of which risk will dominate calculations in the process of choice. These consequences result from the effects that the three contextual influences have on the anticipation of policy outcomes, on the assessment of their probable occurrence, and on decisionmakers' perceptions about the validity of these anticipations and assessments. These factors also affect risk preferences more directly by making certain risks more acceptable than other ones.

VIVIDNESS AND SALIENCE OF RISKS

The vividness of risk is associated with the inherent nature of the risk, and the salience of risk is related to the context in which a risky behavior, event, or development occurs (Fiske and Taylor, 1984: 185–94; Vertzberger, 1990: 56). In order for risks associated with a policy option to be assessed and compared to risks associated with alternative policy options, they first have to be noticed and marked for attention. Vividness and salience play a critical role in this process, because within the cluster of risk dimensions associated with each policy option, not all risks are equally vivid and salient, that is, not all risks will be noticed or get equal attention (Slovic and Lichtenstein, 1968). In comparisons across options, the preference for a particular option will be affected by which risk dimensions are noticed and how the risks are weighted. Where risks fail to get noticed or are given little weight, policy options closely associated with those risks will become more attractive because they will seem less hazardous.

In general, people are threat biased, that is, they are more sensitive to threat-consistent information than to opportunity-consistent information, and once they recognize a threat, they become less sensitive to information that is threat discrepant. Hence, decisionmakers "conclude threat (and not opportunity) is present when available information is ambiguous, and they do not conclude that threat is absent even when available information is clearly contrary to the presence of threat. In contrast, when available information is clearly contrary to the inference that opportunity is present, perceived opportunity decreases accordingly" (Jackson and Dutton, 1988: 384–85). This threat bias appears consistent with the finding that people value loss prevention over gain. Unless given a credible reason to deny the existence of threat completely, decisionmakers alerted to a high-priority issue are likely to see the threat dimensions of that issue more vividly than the opportunity dimensions.

What determines the vividness of a risk? First, outcome severity: the more severe the anticipated adverse consequences, and the more certain they are to materialize, the more vivid the risks. Second, outcome ambiguity: outcomes whose negative aspects are more concrete are more vivid. Historical lessons, particularly those that are part of the personal life experience of key decisionmakers, can act as both powerful deterrents and powerful incentives for risky decisions, such as foreign military intervention, by illuminating specific risks and making them more vivid. Objections within the U.S. military to intervention in Indochina in 1954 were no doubt stimulated by the still fresh and poignant memory of a bloody land war in Asia, the Korean War. The vividness of the Korean experience helped to highlight the costs that intervention would entail, rather than the gains that would ensue. Opposition to intervention was powerfully reinforced by one of the main participants in the decisionmaking group, Army Chief of Staff General Matthew Ridgway, who had experienced the Korean War firsthand as commander in chief of the U.N. forces in Korea. The vividness effect hence shaped risk perceptions and made intervention in Indochina less likely that it would have otherwise been.

The salience of a risk is the result of other risk attributes:

1. Risks about which information is detailed and definitive are more salient than lesser-known risks.

2. Risks believed likely to materialize in the near future are more salient than risks distant in time. "People seem unable to get involved in and evaluate future events (other than those which are very close in time and close to oneself). . . . Time gradually loses its realism the longer the time-perspective is. . . . This leads to the reflection that risk-factors—negative effects with low probability of occurrence—are underestimated" (Björkman, 1984: 37). Decisionmakers will thus be more sensitive to short-term risks than to long-term risks, even when the latter are more comprehensive in magnitude. The adverse consequences of a decision are likely to be more salient than the positive consequences when both are close at hand, but people are more likely to choose risky alternatives if the negative consequences are delayed (Björkman, 1984; Jones and Johnson, 1973)—all the more so if the positive consequences are near in time.[3] In comparing possible costs and gains of risky options, the salience of the near-in-time consequences dominates evaluation. People expect to have much more control over, or be able to prevent, delayed consequences. Also, positive outcomes can be utilized for mobilizing public legitimacy in the short term, when the threat to legitimacy by the negative outcomes is distant in time.

3. Events that can be defined as risky independent of contextual factors are more salient than events that are considered benign, neutral, or risky depending on the existence of certain contextual factors at the point when the focal outcomes occur.

4. Irreversible outcomes are more salient than reversible ones.

5. Adverse consequences for which decisionmakers believe they will be held personally accountable are more salient than adverse consequences for which they believe they can shift the blame to somebody else.

6. Risks that are familiar or are associated with issues uppermost in a decisionmaker's mind have a priming effect; they call attention to similar risk di-

mensions in the focal policy option that will then be readily noticed and heavily weighted, sometimes at the expense of other, even more important risk dimensions. For example, as I discuss in Chapter 6, after the bloody takeover of power in Grenada in 1983 by a radical Marxist faction, the perceived risk that hundreds of American students in the country might be taken hostage played an important role in President Reagan's decision to intervene. The taking of American hostages by the revolutionary Iranian government and the serious domestic repercussions were fresh in the president's mind. As a rule, the chance that people will associate particular risk dimensions with a specific policy option depends on the priming of these associations. Hence, risks supported by and consistent with prior theories, knowledge, and expectations are less likely to go unnoticed than unfamiliar risks (Vertzberger, 1990: 56–66).

7. Risks that emerge in an incremental, piecemeal fashion are more likely to go unnoticed and are more easily ignored than risks that appear all at once.

Vividness and salience not only are responsible for the cognizance of risk but also affect the judgment of risk prominence. An outcome that is perceived as highly vivid and salient triggers an unconscious bias toward exaggeration in judging its risk components. The adverse consequences are assessed as more adverse than they actually are; their occurrence is given a higher probability than is justified; and information about both adverse consequences and their probabilities is attributed much more validity than is warranted. In other words, vividness and salience strongly accentuate risk judgment.

The vividness and salience of risk often are more strongly affected by individuating episodic information than by base-rate (distributional) information (Bar-Hillel, 1980; Borgida and Brekke, 1981). The resulting risk assessments become resistant to disconfirming information (Anderson, 1983). As a result, cognitive conservatism sets in, and it leads to self-entrapment. Conservatism tends to persist when decisionmakers believe in their competence to cope effectively with risks if they materialize, or when they can maintain the illusion that the course of action can be reversed at any chosen point (Slovic and Lichtenstein, 1971).

Crisis situations are prone to produce faulty and incomplete risk assessments due to vividness and salience effects. The limited time for response and the high stakes involved can result in information overload, which hampers assessment. Limited time and high levels of stress may cause a narrowing of cognitive perspectives. Subsequent information processing is particularly vulnerable to the biasing effects of vividness and salience, resulting in serious impediments to the identification and systematic consideration and comparison of all the relevant consequences of the various policy alternatives. Furthermore, because short-term outcomes are more vivid under the situational constraints of crisis than long-term outcomes, the decision horizon of policymakers is shortened, and they tend to focus on immediate implications (Brecher with Geist, 1980; Holsti, 1989; Janis and Mann, 1977; Smart and Vertinsky, 1977). In the end, risk analysis is bound to be limited and selectively biased toward the most immediate, vivid, and salient risk dimensions.

These points in themselves do not necessarily imply an inclination toward the most risky choices in crisis situations. On the contrary, decisionmakers who make choices under pressure of time tend to take cautious risks, for manifold reasons. One is that time pressure adds to the anxiety induced by the potentially adverse consequences threatened by the situation. A cautious decision will reduce anxiety and contribute to feelings of safety and control. A further reason is that the association between unpleasant feelings of stress and the chance of adverse consequences may, under pressure of time, lead to an exaggeration of the estimated probability of losing. The salience of losses is reinforced by the fact that time pressure induces decisionmakers to process what is considered most important first—hence the emphasis on processing the loss dimensions, which results in a preference for less risky choices. As time pressure becomes less acute, additional dimensions are considered—especially the gain dimensions—thus riskier choices become more likely (Ben Zur and Breznitz, 1981). This argument implies that if a policy problem that has both loss and gain dimensions can be framed so that the gain dimensions are processed simultaneously with the loss dimensions or even first during situations requiring a rapid response, the result may be a preference for a riskier choice.

Familiarity, mentioned earlier as an important cause of salience, may also directly affect the disposition to take risks. Familiarity is a source of confidence; it gives decisionmakers assurance that there are no hidden costs of which they are unaware (Ridley et al., 1981; Sitkin and Pablo, 1992). Confidence results in reduced uncertainty and increased belief in one's ability to cope with risk, which may increase the propensity for risk taking as more time is spent working at a task (Streufert and Streufert, 1968, 1970).

An important source of familiarity is historical memory. In the case of foreign military intervention, a major factor affecting the predisposition to intervene is the success and failure of past interventions, especially when the past had a direct influence on the career development of the decisionmakers involved. Past failures are more salient than past successes. Successful cases of intervention generate much less inquiry into enabling causes; failures, on the other hand, because of their highly visible negative consequences, trigger an intensive search for causes and justifications and thus leave a strong memory imprint (Rosenau, 1981). Memories of failure tend to be highly salient, both to the general public and to decisionmakers, when a new intervention decision is considered. To the extent that decisionmaking is influenced by memories of past failures, which are perceived as relevant analogies to present and future situations, the perceived risk element of the policy is enhanced. But if a present policy can be viewed as clearly separate from past circumstances that resulted in failures, the risk element made salient by memories of past failures is explained away as currently irrelevant.[4]

PLANNING

Risky policies are often preceded by planning, which provides the broad framework for making individual decisions, although not all decisions associated with the policy are necessarily informed by and flow from the overall plan. Planning consists of formalized premises, beliefs, action directives, and expectations that organize and structure present and future realities to remove much of the inherent uncertainty and allow the framing of operational guidelines toward goal-directed outcomes.[5] Planning is supposed to anticipate risks and provide for their management in the least costly and most effective manner. Careful and meticulous planning is imperative for success when the problem is complex, uncertainty is high, and the stakes are considerable, as is the case with foreign military intervention. Yet planning is not a sufficient condition for success, although it is a necessary one; in fact, it introduces its own biases. The same attributes of planning that remove or reduce one type of threat may also inadvertently introduce other, different threats.

The most obvious psychological impact of careful planning is a sense of familiarity with and control over the future. Familiarity, whether it is justified or not, increases the propensity toward risk taking by reducing uncertainty about unforeseen consequences and apprehensions that they might happen. In other words, familiarity tends to increase the confidence in success, even when planners and decisionmakers are aware of the inherent uncertainties and ambiguities in the situation and the inadequacy of available information.

Overconfidence can be further traced to the simultaneous operation of three heuristics that affect judgment (discussed below). The plan becomes an anchoring point that resists adjustment to changing realities and instead mandates a reinterpretation of reality to fit biased perceptions of the effectiveness of the plan itself. The biasing effect of planning also involves the influence of availability and representativeness. Aspects of the environment that fit existing plans are more cognitively available and hence salient, and are less likely to be ignored, than environmental dimensions that were not anticipated or considered in the plan. Reality looks much more representative of the expectations of the planners than it actually is. The interaction among these heuristics makes the plan seem a more accurate reflection of present and future reality than is the case, which reinforces the tendency to avoid adjustment.

When decisionmakers evaluate the fallibility of complex plans, they concentrate on the most obvious aspects and are less likely to pay attention to what is not easily available. They may thus overestimate the completeness of their analysis. They often fail to consider imaginative ways in which human error can disarrange even the most carefully prepared plans. A well-prepared, tightly organized plan gives decisionmakers a sense of confidence in its success that reduces their sensitivity to what is risky and what could go wrong. Such a plan also reduces sensitivity to the degree to which expectations for the plan depend on the validity of the assumptions underlying it. Insensitivity in such matters may produce an enduring tendency to overlook changes in the environment and not make required adjustments in or reevaluations of the plan; such omissions have implications for the chances of the plan's overall success. In the process of evaluation and adjustment people are also often insensitive to the difference between assertions and inferences, so they fail to critically evaluate inferred knowledge (Fischhoff et al., 1977, 1978). In sum, a well-prepared plan absorbs risk awareness.

The biases of planning pose acute problems in cases of protracted intervention, where the combination of complexity and uncertainty inherent in intervention has its most serious effects. Yet decisionmakers are often unaware of the very serious effect that time has on the relevance and validity of existing plans. Indeed, because of the strength of the bias toward belief confirmation and cognitive conservatism (Anderson et al., 1980; Evans, 1989; Nisbett and Ross, 1980; Snyder and Swann, 1978), decisionmakers are likely to attempt to engineer the environment to fit their plans, rather than adapting their plans to a changing environment. The histories of U.S. intervention in Vietnam and Israeli intervention in Lebanon, discussed in Chapter 7, illustrate this behavior.

A pertinent and difficult task facing planners of intervention is the correct estimation of the time requirement for goal attainment. This is critical for anticipating the cost that intervention will entail and thus has serious implications for continued domestic support, as well as reactions by third parties. Two complementary determinants bias even experienced planners toward underestimating time requirements. One is the cognitive constraint that affects political-military plans with a high level of abstraction. Decisionmakers usually fail to enumerate all the time-consuming components of planned actions, which results in their systematically underestimating the time required to carry out the plan. Field commanders are left with the task of translating the plan into concrete operations, and they are not always aware of the time frame imagined by the political leadership; indeed, they often incorrectly assume that it is up to them to set the time frame. The other determinant of biased time estimates is motivational. As time pressures increase, the tendency to overestimate what can be accomplished within the available time increases, because people desire to accomplish all their tasks, their ability to do so being an indicator of competence and success. Cognitive and motivational factors may interact: as the task requirements increase in relation to the time available, the motivation to underestimate the time required for carrying out each task increases (Buehler et al., 1994; Hayes-Roth, 1980).

Furthermore, it is frequently the case that high-level civilian policymakers who make strategic decisions on intervention do not have the experience to understand the military modus operandi and relate it correctly to the time requirement of the operation at hand. On maps, distance and terrain may appear to them a less impressive hindrance than they actually are. Once faced with the politicians' expectations, generals are often reluctant, for self-serving and other reasons, to admit that they cannot carry out their tasks within the allotted time, especially when they support the policy of intervention. They are apprehensive that by raising doubts about time requirements they may either weaken the decisionmakers' will to intervene or place their own professional capabilities in question. Overall, the underestimation of time requirements increases the risks of military operations.

COMMITMENT AND ENTRAPMENT

The decision to choose a risky course of action, even when the risks are high and transparent, is often brought about by events that constrain decisionmakers' ability to choose otherwise. Two major sources of constraint are commitment and entrapment.

Commitments impose the implementation or persistence of risky policies and often preclude serious consideration of alternative courses of action. Commitments produce patterns of behavior that follow from "expectations about what we will do in the future. These expectations surround our behavior and constrain us to act within them. Commitments thus mold our attitudes and maintain our behavior even in the absence of positive reinforcements and tangible rewards" (Salancik, 1977: 63).

Commitments develop in two observable patterns. The holistic pattern is associated with formal, usually contractual agreements obligating one party to support a particular state of affairs (e.g., the status quo) or a particular course of action under given circumstances (e.g., deployment of military force in case of a threat to an ally). The piecemeal pattern is most likely to emerge where the commitment is formed tacitly through a sequence of decisions and actions, rather than through a premeditated single choice with a clear predetermined goal. Most cases of tacit commitment involve a commitment to a particular state of affairs, which is translated into a commitment to a particular course of action as the only one allowing the preservation or restoration of that state of affairs. Tacit commitments to intervention can, for example, flow from how an intervening actor conceives its international role (e.g., regional leader) even in the absence of formal agreements requiring intervention. In the case of holistic commitment, the nature and magnitude of the associated risks are more likely to be considered in a comprehensive rational-analytic manner before the commitment is made (Bilder, 1981). With piecemeal commitment, risks usually are not considered comprehensively but are often ignored or, at best, incorporated incrementally with each successive decision or action taken. When risks are considered incrementally, decisionmakers lose sight of the comprehensive risk effect, so they perceive the policy as less hazardous than it actually is. The incremental consideration of risk also provides an illusion of control that serves as an unconscious incentive for risk taking. Indeed, gradual commitment is incompatible with the principles of both substantive and procedural rationality, and to the degree that the risk element figures prominently in decisions, gradual commitment adversely affects their quality.[6] Finally, whereas a premeditated commitment to a high-risk decision may be assumed to reflect risk preferences, in cases of incremental commitment this is not the case. Even risk-averse decisionmakers may find themselves incrementally committed to a high-risk policy, having recognized too late that the decision to escalate or withdraw depends not only on their inherent risk preferences but on other contextual considerations beyond their control, such as the costs of withdrawal. It is not unusual in cases of intervention, especially protracted intervention, to find mixed types of commitment that contain elements of the two ideal-types described.

When formal commitments are supposed to deter an adversary from taking actions that will activate the commitment, the intervening actor is sometimes unprepared to carry the commitments out effectively. Having miscalculated the intentions and tolerance limits of the adversary, the actor may be forced to intervene; but once forced into intervention, the actor will stay the course in spite of the risks involved. The key here is the intervener's belief in the controllability of the commitment at each stage: first, that intervention can be avoided altogether and, later, that a limited show of determination will prevent escalation of the challenge. The reason for such optimism can be traced to the more general findings that under uncertainty decisionmakers are likely to ignore or to insufficiently consider the decisions of rivals, which results in overconfidence in their own judgments of the other party's response (Downey and Slocum, 1975; Griffin and Ross, 1991; Zajac and Bazerman, 1991). Specifically, an intervener may not appropriately weigh the importance attributed by the adversary to the adversary's interests, or the adversary's level of tolerance for the inflicted punishment. At the same time, the intervener overestimates the leverage maintained by the intervener over the local client and the ability to control the client's behavior and avoid being drawn into unwarranted escalation. Consequently, the risks of intervention seem limited and controllable, especially when compared with the risk of losing credibility due to lack of resolve and thus running the risk of inviting additional challenges and threats by the same or other adversaries or the risk of causing other allies to doubt the credibility of commitments made to them (Jentleson, 1987; Jervis, 1991; Lockhart, 1978; Schelling, 1966). This perception biases decisionmakers toward premature closure in the initial intervention decision and makes the intervention look less risky than alternative courses of action.

As the intervention develops into an established policy, additional factors become influential. A domestic coalition forms around that policy, so politicians, bureaucrats, and military leaders develop vested interests that make it difficult for another policy to replace it (Lockhart, 1978). Furthermore, "the more committed the decisionmaker is to the policy under challenge, the greater the degree of stress generated whenever he is tempted to change" (Janis and Mann, 1977: 280).

A policy of foreign military intervention represents a multidimensional high commitment. First, the investments involved are highly visible; a reversal of policy may thus have serious social-political costs for decisionmakers. Second, policy formulation and implementation follow from intensive cognitive and affective efforts invested in argumentation and the policy's justification. The policy becomes closely associated with the self-identity of decisionmakers, so policy reversal poses a threat to self-esteem. "High commitment makes for an initially high threshold of regret, which is manifested as a tendency to dismiss attacks on the current policy as relatively trivial and unimportant" (Janis and Mann, 1977: 286). Third, the nature of these political-security decisions is in many cases a reason for classifying, or provides an excuse to classify, feedback information as secret. Preventing access to it makes it unavailable for learning even within the intervener's decisionmaking system. Ignorance on the part of potential dissenters then becomes a factor in delaying policy reevaluation and amplifying policy rigidity.

Commitment implementation carries with it the possibility of commitment escalation and entrapment, to which interveners are vulnerable. Because decisions to escalate commitments involve high cognitive conflict, once made they affect information processing: there is a decrease in cognitive complexity. Decisionmakers become more vigilant for cues supporting their decisions and less vigilant, or completely obtuse to, cues that could raise doubts about the astuteness of their decision or that could raise accountability-related questions (Arkes and Blumer, 1985; Brockner and Rubin, 1985: 151–52; Schwenk, 1986; Tetlock, 1985, 1992). Risks, therefore, are taken for reasons that have little to do with the inherent risk disposition of decisionmakers, nor are they necessarily based on sound cost-benefit considerations.[7]

Escalation of commitment takes place in different contexts. With success-derived escalation the initial implementation of the commitment had positive consequences, inspiring a strong temptation to escalate commitment, at the expense of more prudent behavior, in order to further increase the expected payoffs; initial success generates a view of success as a one-way linear process. This view causes policymakers to ignore information that an overextension of resources and capabilities could easily turn military success into military disaster, or that third parties watch the expanding appetite of the intervener with growing alarm and may counterintervene. Indeed, throwing caution to the wind after an initial success is a well-documented historical phenomenon (Maoz, 1990b; Oren, 1982). The essence of this behavior is a diminished assessment of the risks of the policy, a sharpened anticipation of further gains, a reduced evaluation of the probability of loss, and increased confidence in the validity of these optimistic assessments.[8]

The second form of commitment escalation, entrapment, is counterintuitive. Entrapment is a process in which decisionmakers escalate their commitment to a previously chosen, albeit failing, course of action where there are opportunities to persist or withdraw and where the consequences of either are uncertain (Brockner and Rubin, 1985: 5; Drummond, 1995; Staw and Ross, 1987: 40). Entrapment is not driven by a single motive (e.g., Brockner, 1992). Different decisionmakers have dissimilar psychological needs, face different political contingencies, and deal with unique situational contexts. Entrapment occurs in both individual and group decisionmaking (e.g., Kameda and Sugimori, 1993; Whyte, 1991) and stems from the convergence of situational opportunities, motivated cost-benefit assessments, cognitive and attributional biases, and group dynamics.

The ambiguity inherent in most political-military problems allows manipulative access to knowledge and beliefs, and the resulting motivated reasoning leads to entrapment in a failing policy. This happens even though decisionmakers in positions of responsibility, when faced with high-stake decisions whose consequences could be costly, attempt to be rational and to present themselves as such to the public. They arrive at a desired conclusion only if they can generate evidence and arguments that will make it appear objective to a dispassionate observer (Kunda, 1990). Before escalating commitments, they try to prove that escalation will result in goal attainment at an acceptable cost. "In many ways, evidence that something is working is probably less important than the belief that it is working. For it is the belief that sustains activity, not the evidence" (Salancik, 1977: 72). Motivated reasoning is difficult to guard against, because people do not always realize that their reasoning is biased by their goals. They may process information in-depth, but because they access only a part of their relevant knowledge, the processing produces, not valid, but desired conclusions. Ironically, the in-depth processing provides a sense of confidence in the conclusions that is unwarranted, yet almost impossible to undermine.

Entrapment is easy because ill-defined problems do not have an obvious termination point where decisionmakers are forced to assess their performance. In most cases the criteria for success or failure are not all that clear. Sometimes decisionmakers purposely avoid defining the point at which the success of their policy should be evaluated. Their knowledge of the situation may be limited or opaque, they may believe that the uncertainty is instrumental in confusing and deterring their foreign adversary, or they may believe that by allowing ambiguity to persist, they can make it difficult for the domestic opposition to hold them accountable for a failing policy. This manipulative uncertainty may become a permanent feature of the problem, making it even easier to find excuses for escalation. The decisionmakers can claim that success will be achieved after turning the next corner, when the benefits of intervention will significantly outweigh the costs. The cost-benefit assessment leading to persistence in an entrapment situation involves the rewards that decisionmakers expect if they eventually manage to turn things around and attain their goals; the presumed increasing proximity of goal attainment; and the presumed costs of giving up on their prior investment.

Other powerful incentives relating to costs and rewards are primarily psychological; still others are social and political.

Where entrapment involves direct potential rewards, it is driven by the "gambling for resurrection" phenomenon. Individuals who engage in wagers with increasingly high payoffs and decreasingly low probabilities do so not with an expectation of gain but in order to salvage past losses (Downs, 1992; Highhouse and Yüce, 1996; Staw and Ross, 1989). Leaders who are faced with failing intervention policies that have already involved high costs, and who are aware that they will be held accountable, may choose escalation rather than changing course because it holds a glimmer of hope for success. They will consider escalation their last opportunity and therefore a sensible choice. This preference has two complementary explanations. According to prospect theory, people are more likely to take a chance on a larger, probable loss than to accept a smaller, certain loss. When loss seems certain and immediate, "the larger risks of escalation which are probable rather than certain may be preferable" (Stein, 1992: 221; Whyte, 1986). Even when probable losses become a reality, cognizance of the losses is slow to follow. Leaders are very reluctant to incorporate losses into a revised perception of the situation (Jervis, 1992: 200). Recognition that a probable loss has become a certain loss lags behind the materialization of the loss for both psychological and practical reasons, allowing commitment to a failing policy to grow.

The omission bias, too, has important implications for the persistence of failing policies. As a rule, people judge harmful commission to be worse than corresponding omission (Baron, 1994; Spranca et al., 1991). In high-risk environments where the implications of policy change are completely unknown or veiled in uncertainty, decisionmakers prefer risking losses—especially losses of lives—that stem from policies in place rather than chancing similar or even somewhat smaller losses that might accrue from implementing new policies. This preference, however, will prevail only up to the point where losses from the current policy are perceived as so costly and salient that even the implementation of a new policy that could result in further losses seems less threatening. When policy persistence becomes too costly, decisionmakers are ready to contemplate and implement change.

The psychological rewards of entrapment have to do with ego-defensive needs, which become critical when personal responsibility for prior losses is high. These needs have two motives: self-diagnosticity, which is the case when a commitment to escalation is made to maintain self-identity, and face saving, which is the case when a commitment to escalation is made to restore, maintain, or boost self-esteem. An attack on those commitments may cause a boomerang effect, strengthening the commitments and producing entrapment (Brockner and Rubin, 1985: 143; Kiesler, 1971: 65–89; Kogan and Wallach, 1964: 191–92). These psychological needs are most likely to be triggered when decisionmakers believe that their initial commitment to the failed policy was undertaken out of free choice (Brockner and Rubin,

1985; Kiesler, 1971: 176; Salancik, 1977; Staw and Ross, 1987). If, however, they believe that the commitment was imposed on them (e.g., inherited from the preceding administration), disentanglement from the commitment will be much less costly, and entrapment is more likely to be avoided.

All decisionmakers need to avoid losing face or credibility through the exposure of the policy as a costly failure, especially when there are no exonerating circumstances. Current costs of withdrawing from the policy become much more salient than future costs of failing with it (Akerlof, 1991; Brockner, 1992; Pallak et al., 1974). When the public identifies a policy as the responsibility of a particular individual or organization, the cost of withdrawal becomes so high that it deters those responsible from seriously considering that option. Ironically, the tactic of entitlement in the process of policy legitimation—the emphasis on the direct link between the decisionmaker and the policy and on the commendable aspects of the policy (Auerbach, 1989: 336)—becomes a major cause for the persistence of that policy when it fails.

The legitimation process required at the early stages of policy implementation hence becomes an obstacle to disengagement from the policy when circumstances change and policy revision is in order. When the policy is linked with rivalry or competition among participants in the decisionmaking system, as was the case with Soviet intervention in Afghanistan, withdrawal may be perceived as a personal defeat that will affect the decisionmakers' position and status in unacceptable ways. The tendency toward escalation rather than revision is reinforced by social norms that emphasize the importance of behavior consistency—that is, decisionmakers who persevere in a course of action are seen as better leaders than those who fluctuate from one policy to another (Staw, 1976, 1981; Staw and Ross, 1987; Whithey, 1962).

Where a policy reflects a broad political consensus that took time and other resources to build and maintain, its architects tend to have a positivity bias when evaluating its success. Foreign military intervention is usually such a policy. Especially in cases of protracted intervention, some policies are associated with ancillary interests that build up over time (e.g., a defense industry that depends on demand for its products) and make withdrawal more difficult. Another problem is administrative inertia, which intervenes between any decision to withdraw and subsequent actual behavior. Inertia becomes institutionalized when a particular policy becomes the essence of the organizations involved, making it unlikely that the policy will be phased out.

Escalation of commitment is even more likely to occur in cohesive decisionmaking groups than in fragmented ones. Two mutually reinforcing lines of research support this contention. One is research on the risky shift. Briefly, the phenomenon of group polarization suggests that groups consisting of individuals who lean toward escalation are even more disposed toward escalation after group discussion (Brockner, 1992; Dion et al., 1970). But even more revealing is the groupthink literature (Janis, 1982; Park, 1990). When groupthink prevails, members are reluctant to suggest a change of course de-

spite negative feedback about the current course because doing so would require the existing consensus to be abandoned. Two other phenomena associated with groupthink, the illusion of invulnerability and selective collective
information-processing and rationalization, also lead members to disregard
negative feedback. Thus, highly cohesive groups tend to escalate commitments more than individuals do.

This effect is stronger for high-responsibility groups than for low-re-
sponsibility groups, hence is pertinent to groups making high-stake political-
military decisions. Because groups want to avoid cognitive dissonance, high
responsibility induces a strong need to justify losses and, consequently, a
high commitment to the decision, coupled with the conviction that escalation will turn the situation around and prove that the decision was rational
(Bazerman et al., 1984). The incidence of group-driven escalation will be reduced if the need for self-justification declines in importance through the diffusion of responsibility for initiating the failing policy (Hartnett and Barber,
1974; Simonson and Staw, 1992; Whyte, 1991). But even this remedy has a
limited effect because, as indicated, entrapment may have multiple causes.

The causal attribution process is another source of entrapment. If decisionmakers believe that the causes of prior setbacks were situational and
temporary, they tend to upgrade estimates of future success and goal attainment and are more likely to become entrapped than if they believe that the
causes of failure are stable and unchanging. Similarly, if they believe that
prior failures can be traced to controllable rather than uncontrollable causes,
they are more likely to be entrapped because of their confidence that things
can be turned around and that all that is required is, for example, more determination or more material resources (Brockner and Rubin, 1985: 39–42;
Drummond, 1995; Vertzberger, 1990: 160–64). If the setback is judged to
result from temporary difficulties, which the decisionmakers believe can be
removed with further, moderate investment, there is an incentive for escalation. In this context, spreading costs across moderate increments and framing the costs as an investment (Davis and Bobko, 1986; Nincic and Nincic,
1995) may cause questionable returns or actual losses to be viewed as un-
alarming, which will reduce the incentives for withdrawal. Defining policy
costs as an investment biases decisionmakers to expect future gains, not
losses. These biasing effects are augmented when the policy is represented or
structured as a long-term investment with long-term payoffs, which allows
decisionmakers to ignore or explain away short-term losses, however salient.

In cases of protracted intervention, by the time decisionmakers are willing to recognize failure, the underlying causes of the policy failure lag so far
in time behind their effects that it is difficult to discern the connection and
becomes easy to bias the attribution of causality in favor of more immediate
causes (Tefft, 1990). These causes, because of proximity and the short time
perspective, are often considered mere human errors that can be rectified or
policy accidents that are merely short-term disruptions. Because the under-

lying causes are not identified or associated with the policy failure, expectations about developments are unrelated to the actual causes. Accordingly, risk assessments are unrealistically optimistic and encourage entrapment.

Another cause of entrapment can be found in burden shifting: another party is expected to prevent risk escalation. To the extent that risk taking is an interactive process that depends on an actor's expectations about the opponent's risk preferences, escalation may result unintentionally from misplaced expectations. Although these expectations are not necessarily lacking in common sense in and of themselves, they may prove wrong. First, the decisionmaker may believe that the rival understands the risks of escalation and will know when to quit. But the rival may not have the relevant information or may err in interpreting it or may be committed to the point of being unable to withdraw. Even if both adversaries share the goal of preventing escalation, it may still occur inadvertently and produce unanticipated risks (Maoz, 1990b: 103–34). Second, and related, is the expectation that the opponent does not want to face escalating risks and perceives the stakes and risks in much the same way the focal actor does. The focal actor assumes that both sides share an interest in preventing a risky situation from escalating beyond a certain tacitly understood threshold. From this point of view the opponent's objectives mirror the focal actor's objectives, but the opponent's actual goals and risk calculations may be quite different from the actor's own.

A third type of expectation comes from projecting onto the other party the focal decisionmaker's own risk-averse disposition. Projection takes place even when only a bare minimum is known about the personality attributes and preferences of the adversary's decisionmakers. It is grounded in the supposition that leaders share certain political and personality attributes (e.g., being responsible, being cautious, often engaging in verbal posturing but not meaning what they say). These images of leaders are often based on impressionistic common wisdom that draws on inferences from history or everyday life, rather than being based on valid case-specific reality-testing.

Two factors exacerbate the effects of these errors that lead to entrapment. More than one of these expectations may operate simultaneously, mutually reinforcing their effects. Also, these expectations may operate simultaneously in both sides, thus compounding and reinforcing each adversary's biased expectations and increasing the level of unanticipated risk.

To summarize: decisionmakers may get entrapped in success, in failure, or in ambiguity. Each of these situations has the potential to trigger judgment biases toward conservatism. In the case of initial success, decisionmakers may refuse to recognize that even a successful policy has its limits and that the context that made success possible in the past has changed. They tend to believe that past outcomes are a good predictor of the probability of future success. Consider the gambler who has been winning at the casino all evening: she could leave with a substantial reward but tries to double or triple her winnings and ends up losing everything. In the case of policy failure, the failure itself becomes a powerful motive for staying the course: success may be around the next corner. The underlying belief here is that past outcomes are poor predictors of the probability of future outcomes. Now consider the gambler who has been losing all evening: he continues to play, believing that his luck must soon turn around, but ends up doubling or tripling his losses. In the third case, if policy outcomes are ambiguous and it is unclear whether the trend is toward success or failure, a policymaker may prefer to stay the course rather than reevaluate the policy, believing that it is too soon to make a judgment and that the policy might take a little longer to bear fruit. Nothing inherent in policy outcomes, then, no matter what they are, would necessarily induce decisionmakers to reevaluate their past decisions.[9] Conservatism and a reluctance to admit poor judgment prevail over the rational monitoring of the causes of policy effectiveness and ineffectiveness.

Is policy revision likely in response to new information that raises doubts about the utility of the existing policy? Behavior revision by governments can take the form of adaptation or learning. With adaptation, behavior change involves adding new activities or dropping old ones without reexamining the theories and values underlying current policies. The emphasis is on altering means and not ends. Even if new ends are added, little attention is given to their congruence with existing ends; change tends to be incremental. The less frequent case is that of learning, where behavior change involves questioning the essence of policies. When questioning occurs, both means and ends are likely to be redefined so that ends have a common logic and coherence (Haas, 1990: 3). Intervention decisions and the high level of commitment that they entail make conservatism and policy persistence more likely than policy change. When change occurs, it is more likely to be a matter of adaptation than learning. Modification of the beliefs underlying the policy takes place only gradually, so costs continue to mount. Learning, which entails a comprehensive restructuring of the policy, will occur only after accumulated costs have become extremely high, causing public dissent that produces heavy pressure on the leadership to do more than merely adapt the policy.[10] Comprehensive policy change is usually associated with sweeping personnel changes in the decisionmaking structure. The years of painfully slow learning and the accompanying change in the holders of key decisionmaking positions—both of which characterized U.S. policies in Vietnam, Israeli policies in Lebanon, and Soviet policies in Afghanistan—convincingly demonstrate these points.

Three contextual factors—vividness and salience, planning, and commitment—were identified as sources of risk judgment and risk preferences. If dimensions of a situation are vivid and salient, higher values are attributed to outcomes associated with them, those outcomes are given a higher probability of occurrence, and the cause of vividness, when vividness results from familiarity, is a source of confidence in the payoff and probability assessments. The vividness and salience of risk lead to risk avoidance in many cases but in other cases, if framed properly, could result in risk acceptability.

The negative consequences of risk perceptions based on vividness and salience are of two kinds. Decisionmakers could focus on particular risks to the exclusion of other, no less important and relevant risks, or they could neglect to consider the cumulative effects of limited, secondary risks.

The second contextual factor, planning, produces familiarity, a sense of control, and perseverance in following the plan. These tend to bias assessments toward underestimation of negative outcomes, and confidence in the assessments will be relatively high. As a consequence, risk acceptability is likely to increase.

The third contextual factor, commitment, whether emerging incrementally or comprehensively, tends to result in underestimation of the probability of negative outcomes and to enhance risk acceptability. Underestimation could lead to entrapment in a failing policy; in that case, the prospect of salvaging past losses is overestimated, as are the costs of withdrawal, leading to an increased disposition toward accepting risk.

Although the three factors are not necessarily related, planning often produces commitment as well as vividness and salience; commitment often produces vividness and salience; and vividness and salience together with commitment often shape the targets of planning. Planning, in turn, often reinforces vividness, salience, and commitment.

Culture and Risk

Culture represents a unified ensemble of ideas that are shared by the members of a society. The ensemble can include sets of common premises, values, expectations, predispositions to action, and preferences for particular ways to organize social activities. Culture-based attitudes are more pertinent than any other attitude in unfamiliar, ill-defined, and ambiguous situations (Doty, 1986; Gaenslen, 1986), in which decisionmakers look for cues that will allow them to interpret and make sense of available information, that will provide guidelines for responding, and that will give them reasonable confidence in the chosen interpretation and response. Cultural biases thus permeate the perception of risk and the formation of risk preferences. Attitudes toward risk and risk taking are influenced by cultural identity. Decisionmakers from different cultures worry about different risks (Thompson et al., 1990; Wildavsky and Dake, 1990) because "in risk perception, humans act less as individuals and more as social beings who have internalized social pressures and delegated their decision-making processes to institutions. They manage as well as they do, without knowing the risks they face, by following social rules on what to ignore: institutions and their problem-simplifying devices" (Douglas and Wildavsky, 1982: 80).[11]

Three caveats, however, should be mentioned with respect to these assertions. First, cultural explanations cannot account for or predict by themselves, or even be the main explanatory variable for, risk-acceptance or risk-avoidance preferences in specific instances. In fact, it is extremely difficult to prove empirically and with certainty a causal relation between cultural identity and specific risk preferences.[12] Risk taking is too complex an activity. Situational, social, and personality variables have to be considered as well as cultural ones. Furthermore, an exclusively cultural explanation would have to be based on the assumption that individuals with identical cultural biases would have similar risk propensities, meaning that individuals from the same culture would respond identically to similar risks—an inference that is not empirically justified. It is more realistic to view culture as a source of guidelines for a sanctioned risk orientation. Whether decisionmakers follow cultural cues depends on the impact of other variables, such as the situational context, group dynamics, and organizational and personality attributes. Because of the broader range and context of explanatory variables, to say that socialization is based on culture does not necessarily imply that risk-taking dispositions will be identical across issues and circumstances, because of interaction effects among variables.

Culture plays a role in molding perceptions of risk and in defining risks as acceptable or not. "Culture would seem to be the coding principle by which hazards are recognized. The cultural standards of what constitute appropriate and improper risks emerge as part of this assignment of responsibility" (Douglas, 1985: 68). Each culture has a unique hierarchy of social values. The definition of costs and benefits, the notion of what constitutes an unacceptable cost, and the notion of what constitutes a threat to values differ from one culture to another. The search for information is accordingly biased toward information that promises knowledge about threats to high-priority values and toward salient cost dimensions as defined by the focal culture. This bias explains, at least partly, why in interactive situations decisionmakers from one culture often fail to communicate the extent of their preparedness to take risks in pursuing their goals to decisionmakers with a different cultural bias, resulting in such adverse effects as dialogue of the deaf and failure of deterrence (Cohen, 1990).

The interaction between culture and the shaping of risk perceptions requires brief elaboration. By providing filters through which people look at the world, culture may affect the assessment of all three components of risk. Whether particular consequences will be considered adverse or benign, and to what extent, is influenced by culture. For example, the value of human life varies across cultures. In cultures that put a high premium on life, decisions involving risk to lives produce a risk-averse propensity, whereas in cultures that put a low premium on life, similar decisions do not preclude a risk-acceptant propensity. Cultural values may also influence the judgment of outcome probabilities and the manner and precision with which probabilities are expressed, because of cultural variations in probabilistic thinking.

More specifically, decisionmakers from fate-oriented cultures are more disposed to think nonprobabilistically, when dealing with uncertainty, than are decisionmakers from other cultures. Asians, for example, adopt a less differentiated view of uncertainty, both numerically and verbally, than do the British. The implications for decisionmaking are important. Successful nonprobabilistic thinkers may remain flexible in response to an uncertain future if they are aware of their limited ability to think probabilistically; unsuccessful nonprobabilistic thinkers tend to make overconfident predictions about the future. Probabilistic thinkers predict the future in probabilistic terms, but errors may ensue from imperfect calibration. This cross-cultural variation has negative implications for the communication of uncertainty across cultural boundaries (Cohen, 1990; Phillips and Wright, 1977; Wright and Phillips, 1980).

Culture may foster expectations of optimism and thus provide an incentive for conservatism in the face of feedback information that the focal policy is not paying off. This is manifested in the availability of terms like "weathering the storm" and "staying the course." An exaggerated confidence that losing situations can be rectified is reinforced by culture-based social values that emphasize perseverance and consistency and that are associated with the image of strong leadership (Staw and Ross, 1987). Cultural biases, then, can contribute to entrapment in a failing policy. They also have an important effect on temporal orientation; short or long time horizons can mean very different risk preferences (Belbutowski, 1996: Douglas and Wildavsky, 1982: 3), as well as divergent assessments of outcome values (in gain-loss terms) and outcome probabilities. Here, again, probabilistic and nonprobabilistic thinkers differ; the latter focus on the immediate future because the distant future may seem too uncertain (Wright and Phillips, 1980: 254).

Finally, cultural values and customs can affect the criteria by which the validity of outcome-probability and outcome-value assessments are evaluated. In other words, the criteria for judging what is valid vary across cultures. Some cultures have strict rules of validation based on objectivity and other structural characteristics of available information; other cultures emphasize looser, more subjective criteria, such as the social position of the information source. Some emphasize analytic rules of evidence and impartiality; others tend to value and prefer reliance on intuition and experience in assessing the veridicality of judgment.

Not only is risk assessment culture-bound, but the preferred strategy of coping with risk sometimes depends on attitudes nested in cultural norms about the desirability of risk acceptance and risk aversion (e.g., Hong, 1978; Wallach and Wing, 1968). When cultural norms encourage risk acceptance, the costs of risk acceptance compared with those of risk aversion are perceived to be lower, and vice versa. In the military domain, the cultural construct that explains risk-taking preferences is strategic culture, which represents the accumulated effects of historical experience, political culture, and geopolitical imperatives on attitudes and beliefs about the role and efficacy of military power as applied to national defense and security policy. Taken as a whole, the set of long-lasting predispositions form the national strategic culture (Booth, 1979; Gray, 1986; Johnston, 1995; Klein, 1991). This becomes the prism through which all information pertaining to national security and defense is processed, and it shapes the emergence of a preferred strategy and operational style.

All else being equal, dissimilar strategic cultures should result in dissimilar strategic and operational dispositions. For example, preferences for a high-risk versus a low-risk military strategy and the estimate of the probability of success for each depend on the values attributed to risk acceptance (or even risk seeking) and risk aversion. Where risk acceptance is a dominant value, the probability of success and the payoffs of a high-risk strategy are estimated to be higher than those of a low-risk strategy. Probabilities and payoffs are inferred from culturally based norms, especially when other information about them is ambiguous or lacking.[13]

In the political-military domain, where decisionmakers face information overload and selective attention is common, social institutions selectively emphasize risks that threaten or reinforce the moral, ideational, and political order holding the social group together. Shared views of risk are one source of societal cohesion. They provide a sense of identity, "us" against "them." Consequently, certain dangers from particular sources are amplified; others are attenuated or even ignored. An important criterion for attention to particular risks is, therefore, that culture-based judgments sanctify those risks, so that major policy initiatives and consequences can be legitimized or delegitimized on cultural, moral, and ideational grounds (Dake, 1992; D'Andrade, 1990; Douglas, 1985: 60; Rayner, 1992: 87).

As a rule, to formulate an acceptable risky national policy and generate legitimacy for it, the private risk perceptions and risk-acceptability preferences of the decisionmaking group members must become broadly shared among members of their society. To cause that transformation from individual perceptions and preferences to shared cognitions requires a stimulus that triggers a convergence of cognitive and affective responses among the members and makes these reactions salient for forming shared risk perceptions and dispositions. The evoking stimuli can be any number of shared cultural symbols that are considered relevant to the policy problem at hand—such as an association of the problem with valued objectives—"defending the American way of life" or "making the world safe for democracy"—an association with affect-loaded analogous historical events that are embedded in the collective memory, for example, the appeasement of dictators in the 1930s. Although symbols may trigger shared justification for risk acceptance, culture-based legitimating principles are a necessary, but not sufficient, condition for both normative and cognitive policy legitimacy. To illustrate, the risks of U.S. intervention in Vietnam were mitigated by the moral argument of defending democracy—a preferred way of life—and those of Israel's intervention in Lebanon by the moral imperative of the national obligation, in light of the Jewish history of persecution, to defend a religious minority, the Lebanese Maronites, from mortal enemies. Cultural biases are built into the judgment process and are not easy to monitor—especially in the foreign and defense policy domains, which, more than other issue-areas, presumably manifest shared national beliefs and values rather than parochial ones. Thus, risk perceptions and risk-taking preferences are both products of worldviews and instruments for preserving and perpetuating them (e.g., through self-fulfilling prophecies)—in some cases well after these worldviews have become irrelevant and out of tune with reality.

That cultural norms shape risk assessment does not imply, however, that a person's idiosyncratic beliefs and values cannot dominate. Within a given culture, individuals vary in the extent to which they are good cultural representatives. Strong individuals with strongly held countercultural beliefs and values based on personal experiences or knowledge sometimes override cultural norms, especially when the situation and the information are ambiguous, and the ambiguity can be used for circumvention of cultural norms. But people would rather not bluntly diverge from cultural values; they fear that doing so would prove difficult to justify, particularly in case of failure.

"Serious risk analysis should also focus on the institutional forces of decisionmaking" (Douglas and Wildavsky, 1982: 81). That is because institutions play an important role in collecting, editing, storing, and communicating and acting as biased reminders of selective past experiences. Through this role, they affect judgments of present risks and the preferences for risk acceptance or avoidance in particular situations. "People who adhere to different forms of social organization are disposed to take (and avoid) different kinds of risk" (p. 9). By mandating specific norms and rules for appropriate social and related organizational behavior (e.g., conformity, hierarchy, incrementalism), cultural variables introduce biases toward specific types of risk preferences. This is best demonstrated by comparing cultures that stress individualism (e.g., the American culture) with ones that stress collectivism (e.g., the Japanese culture). In individualistic cultures, construal of self "derives from a belief in the wholeness and uniqueness of each person's configuration of internal attributes" (Markus and Kitayama, 1991: 226). The normative imperative of this cultural syndrome is to become independent of others and to express one's unique personal attributes.[14] In collectivist cultures, "the self becomes most meaningful and complete when it is cast in the appropriate social relationship" (Markus and Kitayama, 1991: 227; see also Triandis, 1993, 1995). The normative imperative is to maintain and enhance interdependence with others.

These two cultural syndromes suggest a number of practical deductions about decisionmaking behavior. The most obvious is that decisionmakers in collectivist cultures are more likely than those in individualistic cultures to expose themselves to peer-induced compliance. In fact, they are motivated to search actively for and value compliance with others in the group. Groupthink will therefore occur more frequently than in an individualistic culture and will be considered not pathological but desirable. Compliance by an individual with collectivist cultural norms is likely to differ from compliance by an individual with individualistic cultural norms. In the first case it will involve internalization of group attitudes and beliefs. In the latter case it is likely to involve a self-serving getting-on-the-bandwagon attitude. The implications of this difference for risk-taking behavior are explored in the discussion of group risk-taking later in the chapter.

In political systems nested in collectivist cultures where cultural norms demand that decisions be made by consensus, the openness to dissonant information once a decision has been reached is likely to be low, because participants will prefer not to take the political risk of attempting a time-consuming reformulation of the consensus. Decisions are also likely to be framed in vague terms that reflect a common denominator and that make it easy to manipulate risk assessments, thus allowing the bureaucracy a broad latitude in interpreting the decision, especially during the implementation phase. On the other hand, where cultural values emphasize strong centralized control, decisional inputs and feedbacks from subordinates are impeded. Subordinates are reluctant to question, or introduce information that questions, the quality of risk assessments made at levels above them. At the same time, authoritarian cultural values provide broad latitude for top decisionmakers who wish to introduce policy changes, for they know that they can count on the compliance of subordinates.

To illustrate: the liberal tradition in the United States is largely responsible for the need to conceal coercion behind consensus. The approach to the use of force in international affairs, especially large-scale force, is reticent. Hence, unless the vitality of the national interests involved is obvious, the use of coercive large-scale force is not likely to be the first policy preference or to be applied as a first step in response to threat. More likely is an incremental increase in the use of military force, even if that is less effective than the immediate application of large-scale force would have been. A second aspect of American political style is the engineering approach in coping with problems, which has its roots in the centrality of science in American conservatism. This approach is characterized by a drive toward certainty that leads to oversimplification and to rating purely technical but measurable elements of the problem higher than intangible, hard-to-measure ones. The engineering approach is typically piecemeal, too, with a focus on the issues at hand rather than on deeper structural causes and long-term effects (Hoffmann, 1962, 1968). These characteristics result in a short-term practical approach that enhances a tendency to oversimplify complex problems and a preference for incrementalism over comprehensiveness in dealing with risk.

In sum, cultural variables have subtle, hard-to-detect, but pertinent, direct and indirect influences on risk judgment and risk-taking disposition. Cultural cues are especially germane when situational ambiguity is high because they are readily available to decisionmakers and direct their attention to particular risks. In such situations cultural norms provide important cues for setting the time frame within which the outcomes are assessed, establishing the orientation toward optimism or pessimism, and, in general, suggesting whether outcomes should be perceived as benign or adverse. Culture affects the way probability judgments are expressed and communicated and the way probability assessments are calibrated. Culture also provides cues by which payoffs and probability assessments are validated and, in some cases, legitimates or delegitimates boldness or caution. Culture's indirect influences occur through the mediating effects of decisionmaking institutions and social processes sanctioned by cultural values that affect decisional outputs through group dynamics, conformity seeking and avoidance, and the likelihood for entrapment due to in-built culture-driven mechanisms for perseverance.

The Decisionmaker: Cognition, Motivation, and Personality

This section addresses three sets of factors driving a decisionmaker's risk judgments and preferences: cognition, motivation, and personality traits.

COGNITIVE FACTORS

As I pointed out earlier, risky problems confront decisionmakers with serious obstacles to coping because of the nature of knowledge in the political-military domain. These problems include ambiguity, incompleteness, lack of agreement about cause-effect relations, lack of a unifying theory that ties together isolated cause-effect links into a coherent knowledge structure, and lack of consensual knowledge on important issues. The last results in competing premises and theories about the nature of the political world. In such fields as physics or biology, in contrast, a substantial portion of the knowledge is consensual, even if there are areas of disagreement.

Faced with these obstacles to problem solving, the decisionmakers turn to their personal belief systems for guidance. Within the entire set of beliefs, the subset of operational-code beliefs (especially the philosophical part) contains basic images about political reality and plays a major role in political information processing (George, 1969, 1979b, 1980b; Holsti, 1977; Walker, 1990). As a diagnostic and prescriptive framework, the operational code is particularly important in one or more of the following situations: novel situations; highly uncertain situations in which information is scarce, contradictory, unreliable, or overwhelmingly abundant; and stressful situations involving surprise and emotional strain (Holsti, 1977: 16–18). Risky decisions are frequently taken when more than one of these descriptions apply.

The operational code of each decisionmaker contains a set of core beliefs that are grouped into two clusters. Each cluster consists of a combination of general and specific philosophical beliefs, together with their counterpart instrumental beliefs. The first cluster includes philosophical beliefs about the essentials of politics, the nature of the political universe (i.e., whether harmonious or conflictual), and the nature of political conflict. The cluster's instrumental counterpart is a set of beliefs about the best way to select political goals. Additional philosophical beliefs in the cluster concern the fundamental character of allies and opponents. Their instrumental counterparts prescribe how to deal with these actors.

For our purposes, the second cluster is more important. It consists of beliefs about the controllability and predictability of historical-political developments in terms of the onset, timing, pace, duration, and nature of events. Included are philosophical beliefs about how much control one can exert over historical-political developments. The instrumental counterparts are beliefs about how risk should be calculated and managed and how much risk should be taken. Other philosophical beliefs in the cluster involve the question of whether the future is predictable; the instrumental counterparts deal with the correct timing for action. Arising from these beliefs is optimism or pessimism, that is, whether one should be optimistic about the prospects for achieving one's ends. The related instrumental beliefs concern the best and most efficient means of doing so.

The first cluster of operational-code beliefs provides decisionmakers with general conceptions of the environment in which they operate. The second cluster provides insights into the odds they face and how best to cope with them. The specific content of beliefs about control over events and outcomes can reflect a number of alternative patterns: a generalized belief or disbelief that one has such control, or restricted and refined beliefs about the controllability of specific issues or actors—for example, beliefs that one has control over events and outcomes in specific domains but not in other domains, or that one can affect the behavior of particular actors (e.g., allies) but not other actors (e.g., the enemy), or, finally, that one has control over particular events and their outcomes. Beliefs about control are personal constructs but may be shared when they are embedded in culture.

Locus-of-control beliefs have information-processing implications that affect risk-taking preferences. In the first place, to the degree that people believe in an external locus of control, they devalue information because it is considered to be of low instrumental value for exerting control over events or other actors' behavior (Davis and Phares, 1967; Rotter, 1966; Trimpop, 1994: 161–79). The search for information is directed mostly toward information that will be relevant to coping with, rather than controlling, their social environment. The signals and communications of the adversary are consequently likely to go unattended and are accorded low relevance and importance at best. Comprehension of the motives of the other side is considered of merely academic value and hence of marginal utility. Attention and interest is given to information about one's ability to deal with the situation. When threatening situations or events materialize, decisionmakers who believe in the impossibility of controlling events experience increased stress and respond with a defensive avoidance of information about the threat, unless it also includes information about how to cope with it effectively and at a reasonable cost. On the other hand, belief in the controllability of events is likely to be associated with a general predisposition to take risks.

The difference in risk-taking dispositions insofar as it is related to beliefs in the controllability or uncontrollability of events can be traced to the dissimilar expectations that the opposed beliefs trigger with respect to outcomes. When a person believes in the controllability of a situation, he or she views its outcome as derived from skill. Someone who does not believe in controllability views the outcome as derived from chance. This dichotomy has two implications. First, "persons characterized by a belief in their ability to order their own destiny should be expected to become more involved in a task where success or failure is contingent only upon chance than individuals who believe their fate is not under their control" (Slovic, 1964: 229). Second, decisionmakers of the former type demonstrate more confidence in their success than do those of the latter type (Hale, 1987; Howell, 1971, 1972; Vlek and Stallen, 1981; Wehrung et al., 1989). The implications for risk taking are obvious: when uncertainty is perceived to be embedded in the external environment, low confidence in success will cause risk-averse behavior; but when uncertainty is construed as internal, the perceived probability of success increases, and so does the propensity to accept risks.

Controllability is thus a mental construct concerning the perceived relation between the task environment and a decisionmaker's attributes and skills. Decisionmakers' perception of self-efficacy will lead them to perceive risky situations as controllable and as affording opportunities, not threats. In group situations, members' mutual reinforcement of their confidence in the quality of their performance creates a shared sense of self-efficacy and group efficacy that disposes members to perceive risky situations as more controllable than they actually are; members will therefore prefer risk-taking options to more cautious ones. A similar effect results when sycophants reinforce the group leader's belief in his or her self-efficacy—even when others call attention to deficiencies in expected performance—because judgments of self-efficacy tend to be more influenced by positive than by negative information (Klein and Kunda, 1994; Krueger and Dickson, 1994; Stone, 1994).

By implication, decisionmakers who believe in controllability believe that they can not only anticipate and steer developments to serve their objectives but also reverse errors after a decision has been made (Shapira, 1995: 80–82). The more confidence that decisionmakers have in their skills, expertise, and experience, the surer they will be of their ability to fix initial errors, and the more tempted they will be to take risks. Some might even come to see risk taking as a welcome challenge to their skill in handling difficult situations and using them to their advantage. Belief in controllability will be enhanced by expectations of a second chance that induce decisionmakers to make risky decisions, even without systematic information processing and deliberation, confident that they will not have to bear the costs of risks nested in errors of judgment.

In military confrontations we can assume that both risk-acceptant and risk-averse decisionmakers want to win, or at least not to lose. The difference in their approaches has to do with their expectations. Risk takers believe that limited wars can be controlled and that escalation and de-escalation can be deliberately manipulated. To use other terms, risk takers hold an actor model of control and escalation; risk-averse decisionmakers, a phenomenal model. The actor model holds that escalation is an unilateral act by specifiable individuals and institutions, an independent and conscious decision to execute a certain action. The phenomenal model holds that escalation is a natural aspect of war, a process that gets started and keeps growing on its own, partly outside the control of any participant—that war "naturally" tends to expand (Smoke, 1977b: 10–11, 21).

From the actor model comes the belief that escalation is reciprocal. That is, the escalation sequence has a controllable, foreseeable two-step action-reaction structure. From the phenomenal model comes the belief that escalation is an unpredictable cycle of action and reaction with no clear and definitive end, so the situation cannot be fully planned in advance and is likely to involve unexpected costs (Smoke, 1977b: 26–30). Risk takers believe that escalation proceeds in identifiable and distinctive steps that can be monitored and hence prevented from getting out of control. Risk avoiders believe that escalation proceeds gradually in small steps that are not easily monitored; it can get out of hand and lead to a quagmire.

Risk takers view the probability of adverse outcomes as low and limited. Risk is expected to stay within recognized and acceptable limits, the danger of losing domestic legitimacy is expected to be negligible, and, most important, belief in the controllability of escalation reduces anticipation of postdecisional dissonance. Hence the reluctance to make risky decisions decreases. The expectation of and resultant preference for risk taking or risk aversion are often fed by one or both protagonists, leading to a "reverse mirror image" syndrome. Those who wish to take risks assume that they operate in an environment where ambiguities and uncertainties are few and controllable, whereas their adversary operates in an environment abounding with ambiguities and uncertainties whose effects are not easy to control. In contrast, risk-averse actors assume that there are many uncontrollable factors in their operational environment but few in their antagonist's.

Risk takers are therefore not necessarily bolder, more courageous, more irresponsible, or inherently less cautious than risk avoiders. They differ in their beliefs and theories about the world of people and nations and have different cause-effect premises, but they are not necessarily different in their personality attributes. The theories held by decisionmakers may or may not deviate from conventional wisdom or societal consensual knowledge.[15] But when the theories are shared by much of the public, confidence in their validity is high, making them resilient to dissonant information.

Most people assume that they know their propensity to take risks and the level of risk that they consider acceptable. There are motivational grounds for this belief, namely, the need to feel in control. The sense that one knows one's own preferences on important matters is reassuring. In fact, however, self-perceptions are often inaccurate. First, people tend to believe that their risk preferences are in line with the dominant social norms. Cautious people in a society that sanctions risky behavior are motivated to believe that they are more risk acceptant than they actually are. Similarly, risk-acceptant individuals in a society that values caution are motivated to consider themselves, and present themselves, as more cautious than they actually are. Second, people are unable to accurately assess in advance the effects of social influences, such as group pressure, on their individual risk preference. In individualistic cultures they tend to believe that they are not likely to be influenced by others, because they view such susceptibility as a form of weakness. The reality, however, is different. Third, in situations where risk taking is incremental, decisionmakers take more risks than they initially intended because the monitoring of an incremental process tends to be biased downward, creating a misleading impression at each stage regarding the degree of risk taken. Only at the end of the whole process is the picture clear enough to allow for a more precise estimate of the overall risk taken. In this way, decisionmakers often take more (or less) risk than they intended. Sometimes they simply do not know in advance how much risk they are willing to take in the pursuit of particular goals, and learn their risk propensity after the fact.[16] They will translate and internalize this into a general rule about to how much risk is acceptable to them in such circumstances.

If beliefs are a critical source of the disposition to take or avoid risks, then one way to change dispositions is to change the generating beliefs. Thus, no less important than the content of beliefs is the confidence with which they are held and their resilience to information that invalidates their underlying sources. As a rule, believers show overconfidence in their knowledge and reasoning (Fischhoff et al., 1977; Koriat et al., 1980; Otway, 1992) and fail to realize the impact that their beliefs and overconfidence have on information processing. They tend to think that information has only a single, self-evident interpretation, which cannot be reconciled with other, totally different beliefs. They see their interpretation of information as compelling and as independently confirming their beliefs and, by definition, disconfirming other beliefs (Jervis, 1976: 181–87). Their beliefs become powerful incentives for suppressing doubts; they encounter new, even highly discrepant information without altering their confidence in their theories to the extent required. Even if they later discover that the evidence on which their beliefs are based is false, the theories may still survive because of their extensive emotional investment in the beliefs. Emotional investment or not, people tend to seek out, recall, and interpret incoming evidence so that it sustains their beliefs, even when the evidence is weak (Koopman et al., 1995; Nisbett and Ross, 1980: 167–92; Slovic et al., 1980). The tendency to maintain core beliefs significantly reduces the frequency of change in the disposition to take risks, to the extent that risk disposition is derived from these beliefs.

Stereotypes are one type of belief that is held with great confidence. Rigid cognitive constructs that are extremely difficult to disconfirm (Hamilton, 1979; Snyder, 1981), stereotypes have a powerful influence because of their simplicity and their putative validity. In risky decisions, stereotypes provide an escape from the complexity of the issue. The stereotype of the adversary becomes a salient informational input into the decision to take or avoid risk and shapes expectations about the consequences of risk taking. A disparaging stereotype of an adversary can significantly downgrade the assessment of how adverse the consequences of taking a risk will be; the reverse can occur when the stereotype concerns potency. President Lyndon Johnson's initial optimism about the outcome of American intervention in Vietnam was undoubtedly based to some extent on his view of North Vietnam as "that raggedy-ass little fourth-rate country" (Tuchman, 1984: 321).

Decisionmakers' value systems affect risk preferences in two ways. Risk and caution are, in and by themselves, considered high-priority values in some general or specific circumstances. As a rule, "risk-averse people appear to be motivated by a desire for *security*, whereas risk-seeking people appear to be motivated by a desire for *potential*. The former motive values safety and the latter, opportunity" (Lopes, 1987: 275). But these value dispositions are sometimes modified by situational factors. More important, the value system provides a benchmark for assessing the stakes involved in a decision. If the stakes are related to one or more values, the weight of the stakes in the decisionmakers' calculus increases in proportion to the importance of the relevant values. Where risks are incurred in the pursuit of high-priority values or where the consequences of a risky policy choice are closely related to high-priority values, risk acceptance is more likely to be the case than when low-priority values are involved.

The set of coherent beliefs and values that form an ideology motivates decisionmakers to take risks when they believe that their commitment to the ideology requires it. Risk taking is justified by the stakes involved and the expectation of a favorable outcome. During 1981–83, for instance, President Ronald Reagan viewed U.S.-Soviet relations in Manichean terms. The West— the United States in particular—represented the values of freedom and democracy. It was locked in a struggle against aggressive communism, whose values were believed to be irreconcilable with freedom and democracy and whose adherents were believed to be determined to enslave all free and democratic nations. In Reagan's view, the United States had an obligation to face up to the communist threat and confront the Soviet Union wherever it attempted to expand its influence—regardless of the risks posed—because the very existence of democracy was at stake.

MOTIVATIONAL FACTORS

Incentives for taking or avoiding risks are not only cognitive but also motivational and affective (Brehmer, 1987).[17] Motivational factors are particularly influential when decisionmakers have no or little base-rate information or when decisionmakers choose to disregard available base-rate information. Without information, they can make overoptimistic judgments when they perceive events as controllable and have committed themselves to a particular outcome. Decisionmakers are then motivated to exaggerate the importance of facilitating actions and behaviors, an approach reinforced by selective memory of similar actions that enabled goal attainment. Thus, the more undesirable the outcome, the stronger the tendency to believe that one's own chances are less than average; the more desirable the outcome, the stronger the tendency in the other direction (Golec and Tamarkin, 1995; Weinstein, 1980).

Being in control is an important goal to most people, and they are frequently motivated to exaggerate the extent to which they have managed to reach it, as well as their degree of control over particular events that are actually chance driven. The illusion of control comes from cues in chance situations that are interpreted as suggesting that skill is involved. In fact, such cues distract attention from the actual situation (Fiske and Taylor, 1984: 128–29; Langer, 1975; Schwenk, 1984; Tyler and Hastie, 1991). The illusion of control encourages risk taking. Foreign military intervention decisions are prime candidates for triggering this optimistic bias.[18] The military capabilities of the intervener are superior to those of the target, and decisionmakers tend to identify control over power resources with control over events and outcomes. The magnitude of the costs that such decisions entail, and their high transparency, commits decisionmakers to their assessments of the anticipated outcomes of the decisions and contributes to the optimistic bias. Past failures in similar circumstances are explained away as resulting from inefficient or ill-advised behavior on the part of other decisionmakers or nations and are therefore considered irrelevant to learn from.[19] For example, reflecting on the disastrous French experience in Indochina, American decisionmakers concluded that, after all, "the French were French" (Kattenburg, 1980: 174–75)—a tacit statement that Americans would do better.

The power of motivational incentives—like those that generate the optimistic bias—to increase the tendency to take risks is not unconstrained. "People will come to believe what they want to believe only to the extent that reason permits" (Kunda, 1990: 483). Specifically, decisionmakers, particularly when faced with high-stake decisions, strive to be rational and to project themselves as such. But to arrive at a seemingly rational conclusion, they indulge in a biased memory search and access beliefs and rules that support the desired conclusion. On occasion, they may even combine newly accessed knowledge with creatively constructed new beliefs to buttress their position (Griffin and Ross, 1991; Kunda, 1990).

As a consequence, both extremely risky and extremely cautious decisions tend to be made with greater confidence than more moderate decisions are. Decisionmakers are not likely to take extreme risks unless they have convinced themselves that doing so will pay handsomely or that the alternatives are much worse. Confidence in their risky decision reduces the likelihood that they will pay attention to disconfirming information. Premature cognitive closure is the result. Similarly, when extreme caution provides a strong sense of control, increased confidence is the result, with the same consequent premature closure.

To avoid having to justify adverse consequences, people prefer to be involved in situations or dealing with issue-areas in which they consider themselves knowledgeable and competent. Risk taking is, therefore, often specific to a context or issue. It is not unusual for decisionmakers with expertise and experience mostly in domestic politics, for example, to be much more disposed toward risk acceptance in domestic policy issues than in foreign policy issues. They trust their intuitive judgment in familiar issue-areas even if the level of risk is high and distrust it when dealing with problems outside their experience, even tend to procrastinate or avoid decisions altogether. A good example is Lyndon Johnson. His expertise was in domestic politics, and he frequently demonstrated his ability to take risks in this area. But in foreign affairs he was extremely reluctant to trust his political intuition, preferring instead to defer to respected members of his administration who had an impressive background in defense and foreign affairs. Consequently, his initial risk-taking disposition in foreign policy reflected the risk preferences of his advisers rather than his own.

The explanation for such behavior is both cognitive and motivational. People may have learned from experience that they generally do better in situations they understand than in situations where they are less knowledgeable. The motivational explanation involves credit for success and blame for failure. When the decisionmaker has limited understanding of the problem, failure will be attributed to his or her ignorance, whereas success is likely to be attributed to chance. But if the decisionmaker is knowledgeable, success is likely to be attributed to his or her expertise, whereas failure will sometimes be attributed to chance (Heath and Tversky, 1991: 8). Many decisionmakers, therefore, do not regard a calculated risk in their area of competence as a gamble. Those who consider themselves competent military strategists and astute foreign policy experts will thus be less hesitant than others to choose risky options that involve the use of military force to attain political goals.

Risky decisions often cause decisionmakers intense inner conflict and doubt. From the moment they make a choice, it is subject to cognitive dissonance (Elliot and Devine, 1994; Festinger, 1957; Janis and Mann, 1977; Zeelenberg et al., 1996). Because they are regret averse, they prefer to be as sure as possible that their choices will be regret minimizing, which, depending on the situation, may entail either risk avoidance or risk taking. They are therefore more likely to take risky decisions when responsibility can be shared with others or attributed to external causes. In the event of adverse outcomes, they will argue that their decisions were not taken of their own free will or that the adverse consequences were unforeseeable (Collins and Hoyt, 1972; Cooper, 1971; Little, 1975; Sherman, 1970; Wicklund and Brehm, 1976: 51–71).

Once the adverse consequences become obvious, the decision to persist in the risky policy or to withdraw will be affected by the decisionmakers' response to the regret effect created by the situation. Decisionmakers who are psychologically committed to the decision and associate it with affirmation of their core values are more likely than others to get entrapped, persist, and rationalize in order to reduce or overcome regret (Abelson, 1959; Janis and Mann, 1977; Little, 1975: 136–55; Lyndon and Zanna, 1990), mostly by bolstering their decision with new arguments or linking it to high-priority values.

Doris Kearns, in describing the development of President Johnson's bolstering of the arguments concerning the stakes that the United States had in the Vietnam War as it continued to escalate, posits five stages. Johnson went from saying the war was "just a test case for wars of national liberation" and a "lesson for aggressors," to maintaining how necessary it was to "prevent the fall of Southern Asia" and moved toward the "containment of China," arriving finally at the ultimate justification: "America fought in Vietnam to prevent the otherwise inevitable onset of World War III" (1976: 270).

Decisionmakers also question the validity of the disconfirming information or its sources by claiming, for example, that it is premature to reach conclusions about the future of the intervention policy and that it will take time, patience, and sacrifice before gains materialize.

When decisionmakers are behaviorally, but not psychologically, committed to a risky policy, they may justify persistence by subordinating their concern to a higher value. Secretary of Defense Robert McNamara, once he had lost faith in the Vietnam War effort, solved his own cognitive dilemma by a form of transcendence. As the historian Barbara Tuchman describes it, McNamara remained "loyal to the government game" and continued to preside over a strategy that he himself believed to be futile and wrong. "To do otherwise . . . would be to show disbelief, giving comfort to the enemy" (1984: 346). He dealt with what must have been an extremely bothersome imbalance—between the emotional intensity of his now altered attitude toward the war and his strong belief in the value of loyalty to the president—by making loyalty superordinate to antiwar feelings. Decisionmakers who harbor serious doubts about a policy will, when a sense of cognitive dissonance is aroused, react with an attitude change and a shift of preferences toward more caution in dealing with, or even withdrawal from, the policy. It is, however, extremely important for them to be able to present their preference shift as a matter of statecraft, not opportunism, in order to protect their image and self-respect. When a dramatic change of attitude would be embarrassing, it may not take place at all or may take place slowly and incrementally, regardless of how badly the policy needs immediate adjustment or termination.

A major problem in some cases is to identify a success threshold that will justify the termination of a risky policy. Identification is hard when the goal whose achievement will qualify as success is vaguely defined. There is no obvious point at which the decisionmaker can decommit. To make matters worse, major policies have a life of their own and generate vested interests that help perpetuate them. When decisionmakers are sensitive to criticism that they are "chickening out" or "being spineless," especially when the sunk costs of a policy are high, they are motivated to avoid termination. They want to be able to confidently argue that their goals have been achieved. Thus the vagueness of the termination threshold may sometimes provide an excuse for decisionmakers who are risk averse and uncomfortable with the policy to decommit themselves—again by claiming that the policy goals were in fact accomplished.

Finally, two affective factors—stress and mood—occasionally have notable, even if secondary, effects on risk-taking behavior. Situations that require risky decisions and arouse inner conflict are stressful. The level of stress experienced, however, will vary with the propensity to take risks. As risk increases, risk takers are likely to experience only a moderate level of stress; risk-avoiders, rising stress and anxiety. Being more comfortable with stress, risk takers who experience just a moderate increase in stress may be insufficiently alerted thereby to warning signals; high risk, on the other hand, brings out the best in them, and they often cope most effectively in such situations— even better than with lower-risk situations that do not challenge their capabilities. In fact, sensation seekers find pleasure and have self-confidence in situations that most people would find frightening and anxiety inducing. The pleasure is in taking a risk, quite apart from the issue of winning (Horvath and Zuckerman, 1993; Klausner, 1968).

For the average person, mild levels of stress have constructive effects, encouraging increased vigilance and improved analysis. Stated generally, there is a curvilinear inverted-U relation between stress and the quality of problem solving (George, 1980b: 48–49). Still, even in risk-acceptant individuals, stress beyond a certain level can adversely affect the quality of judgment and decisionmaking. It narrows the attention span, reduces the number of cues attended to, and decreases the accuracy of cue-filtering procedures, so that pertinent dimensions of the situation are not considered. There is a propensity to focus on short-term consequences and ignore long-term ones.

Stress also induces intolerance of ambiguity; cognitive rigidity, which impairs the ability to deal with complexity; diminished creativity; and failure to adjust existing beliefs and theories to new information. Reliance on stereotypes, random rules, and simplistic single-criterion judgments increases, reducing the likelihood of complex multidimensional processing, which would produce competing interpretations and the need to choose among them. Stress fosters a tendency to seize on solutions before all alternatives have been considered and to scan the alternatives in a disorganized and unsystematic fashion. Decisionmakers magnify the benefits of their preferred option and the deficiencies of other options, thus making choice clear-cut. They tend to believe that a single option satisfies all or most value dimensions without requiring value trade-offs. After a decision is made, attention to negative feedback is limited (Holsti, 1989: 25–37; Holsti and George, 1975: 275–84; Keinan et al., 1987; Post, 1991), to avoid stress rearousal.

Yet this common view of the effects of stress needs to be refined by reference to specific persons and situations. With experienced decisionmakers, for instance, when time is moderately limited, sometimes higher stress

> will stimulate a more thorough search for information from multiple sources and a more careful assessment of its contents. In short, there will be a more active focus on the acquisition and evaluation of information in the crisis period. . . . Many of the laborious and time-consuming intermediary layers will be eliminated, with more information being elevated rapidly to the top of the organizational pyramid. The result is that senior decision-makers' perceptions under high stress will be formed largely from their direct access to information relevant to a crisis. (Brecher, 1993: 378)[20]

It should be noted, however, that even these positive effects of stress are conditional and do not necessarily improve the overall quality of information processing, especially as time constraints become more acute. True, decisionmakers are exposed directly to information from multiple sources, but they often do not have the resources (e.g., time, skills) to take advantage of the information and are likely to deal with it in a perfunctory manner. Furthermore, being directly involved increases the level of confidence in the decision and commitment to it, which makes cognitive closure to feedback information from its implementation a real possibility.

In brief, stress has a dual effect on risk taking. If it distorts decisionmakers' judgment, making them emphasize the low probability of adverse consequences and highlight opportunities, the result is a tendency to take risks. But if stress biases judgment toward overestimation of adverse consequences and their probabilities, then it contributes to risk avoidance.

Mood is the other affective factor whose role in shaping risk preferences cannot be overlooked. Although moods are rarely decisive, they often indirectly facilitate risk-preference formation by influencing judgment. "We tend to make judgments that are compatible with our current mood, even when the subject matter is unrelated to the cause of that mood" (Johnson and Tversky, 1983: 30). Yet people making judgments are apparently unaware of this fact. When the risk is ambiguous or low and known, people in a positive mood are more likely to take the risk because their mood produces optimism about the outcome. But when the risk is high, they are more likely to be conservative and risk averse to protect their mood (Arkes et al., 1988; Isen and Geva, 1987; Isen and Patrick, 1983). In opportunity situations, then, positive moods facilitate risk-acceptant behavior, but in high-risk situations it encourages risk avoidance.

Affect influences not only the content of judgment but also the process of judgment and problem solving, particularly in instances requiring constructive processing and substantial transformation of information. People in positive affective states tend to reduce the complexity of a situation by using satisficing strategies, rather than attempting to optimize, and they tend to do little or no rechecking of information, hypotheses, and tentative conclusions.

Effective performance may result only when the problem is routine and intuitive first-guess solutions are relevant (Forgas, 1989, 1995; Isen et al., 1982; Mann, 1992). But impaired performance may be the result when the problem requires systematic analysis of its complexity, as is the case with foreign military intervention decisions. The tendency to reduce complexity may result in increased risk acceptance when the short-term policy outcomes are positive and the long-term negative consequences are ignored.

This effect of affect explains why foreign policy decisionmakers react differently to victory. Victory produces elation and satisfaction (positive affect). Some decisionmakers may perceive victory as opening new opportunities, which may be exploited as long as the risks seem low or ambiguous. In these situations they prefer risky choices and may eventually overextend themselves militarily. Other decisionmakers, more concerned with losses that could be incurred if they reach for more than they have already achieved, tend to be cautious and risk averse. Thus, at times of victory the dual influence of problem framing (in terms of opportunity or threat) and positive affect explains differences in the preference for rashness or prudence. Positive affect and a threat frame will induce prudence, whereas positive affect and an opportunity frame will induce daring and risk taking.

PERSONALITY FACTORS

Certain individuals, because of innate personality traits, may be disposed to develop beliefs and motivations that encourage a consistent preference for risk taking across situations. Although research on this question is less revealing than could be expected, it has produced some interesting insights. Individuals with great autonomy and endurance—who are not so concerned with the admiration of others and who are capable of sustained waiting for relatively small gratification—tend to choose what they perceive as low-risk options with high cumulative value (Cameron and Myers, 1966). These personality attributes sometimes result from experience. Top-level decisionmakers who reached their position after a lifetime in government bureaucracies become socialized to adopting a cautious decisionmaking style. President Bush is an example. On the other hand, those who are exhibitionist, aggressive, or dominant pursue admiration by others and tend to behave in a manner calculated to draw attention to themselves, as in the case of Hitler. They will prefer high-risk strategies, hoping to elicit others' response.

In general, risk takers are persistent, confident, outgoing, aggressive, domineering, manipulative, opportunistic in dealing with others, and needful of achievement (Kowert and Hermann, 1995; Plax and Rosenfeld, 1976; Scodel et al., 1959).[21] Many of these traits characterize narcissistic individuals, whose certainty of ultimate success "leads to both a sense of omnipotence and a feeling of invulnerability that they cannot go wrong. This underlies the narcissist's capacity for risk-taking" (Post, 1993: 103). But behind this facade, which is self-attributed, the narcissist is psychologically vulnerable and cannot admit mistakes, cannot afford to acknowledge ignorance, and cannot believe that any aspect of his or her knowledge can be faulted. Because these attributes make learning from mistakes difficult (Post, 1993), the narcissist is a likely candidate for entrapment in failing policies.

Personality factors are most salient in politics when risk taking is a compensatory behavior or an instrument in image building and image amplification. In the first case, risk taking can be triggered by deep psychic needs to compensate for feelings of shame and humiliation (Steinberg, 1991, 1996). Political leaders are vulnerable to threats and injuries to their self-esteem because of their need to project a public image of integrity and wholeness. Narcissistic injury and failings in self-esteem that result in feelings of shame and humiliation, especially under conditions of high stress, are therefore likely to induce aggressive responses for the purpose of alleviating these feelings and restoring a sense of self-worth. The responses motivate risk-taking actions— for example, from leaders who are generally not predisposed to take risks and even from leaders who are as a rule extremely cautious. A case in point is President Bush's intervention in Panama (the "wimp factor").

Another instance when a national leader takes risks is to create a heroic image, a key constituent of charisma. A leader becomes a hero by daring to achieve a difficult and dangerous task against all odds and by accomplishing it successfully when the policy or action is perceived by the public as beneficial for the nation. High-risk foreign policies can therefore be instrumental in shaping charismatic leadership (Willner, 1984: 90–92). Charismatic leaders gain a national legitimacy based on almost blind confidence, and their supporters' trust makes it easy for them to pursue risky foreign policies. Charisma may even stimulate the pursuit of risky policies, for risky policies enhance the charismatic image. If the policies fail, charisma could decline over time as the costs to followers mount.

The relation between policy failure and the decline of charisma is neither linear nor immediate. Even dramatic and costly failures will not cause the immediate decline of charisma. In some cases, costly failures can even be manipulated to enhance the leader's charisma. The leaders may identify their position with impossible odds, unprecedented adversarial forces, and the threat of national disaster if someone weaker than themselves gains power. By projecting an image of strength in the face of adversity, charismatic leaders become figures to rally around, able to pursue risky failing policies and, up to a point, retain the support of followers.

To summarize, in this section I analyzed the implications of three individual-level causations that affect risk judgment and risk-taking preferences: cognitive, motivational, and personality. Of the three cognition-based causations—beliefs, stereotypes, and values—beliefs about the controllability and predictability of events are probably the most important. Beliefs play an important role in information processing and in encouraging or discouraging acceptance of risk. Because central beliefs change slowly, belief-related risk assessments and choices tend to persist, depending on the degree of confidence with which the beliefs are held. Stereotypes are so conclusive that they add rigidity to risk estimates and preferences. Values may encourage or discourage risk taking if boldness or caution are salient values or if policy outcomes can be linked to important values.

Motivational factors, especially the illusion of control, may result in an optimistic bias, with a consequent increase in the acceptability of risk. Frequently the stress generated by the risky situation affects the quality of risk-related information processing. Whatever the causes of risky choices, decisionmakers anticipate postdecisional regret. Confidence in the ability to manage and contain that regret therefore fosters risk taking. Some personality attributes dispose decisionmakers to accept risk, whereas others encourage them to avoid it. Charismatic leaders are risk acceptant and even risk seeking, but at the same time they also make premeditated functional use of risk taking to build and sustain their image. In most cases, personality-based causations are of secondary importance by themselves but are effective in enhancing trends set by cognitive and motivational sources of causations.

Beliefs, values, motivations, and personality attributes do not necessarily operate independently. They can be related in one of two ways. First, personality attributes dispose individuals to adopt beliefs, values, and motivations that provide cognitive and affective justifications for personality-based drives and allow them to take situational variability into account every time personality attributes are triggered by a problem. For example, narcissists are very likely to adopt beliefs in the controllability of events, which will justify their sense of omnipotence. Similarly, preferences for particular beliefs and values are sometimes driven by underlying motivations. The second way beliefs, values, motivations, and personality attributes are related is when they synergistically enhance each other's effects on whether risk is accepted or avoided.

The Decisionmaker: Heuristics and Biases

As emerged from the discussion of the definition and nature of risk in Chapter 2, before a risky policy option is chosen, decisionmakers have to anticipate policy outcomes, evaluate their utilities, estimate the probability of each outcome, then make a meta-judgment on the validity of the estimates. The task is highly demanding and people, being poor intuitive statisticians, frequently perform poorly. Their judgment of probabilistic phenomena is often biased and in violation of fundamental normative rules (Abelson and Levi, 1985; Kahneman et al., 1982; Nisbett and Ross, 1980; Reason, 1990; Vertzberger, 1990). They make probability judgments and assess utilities without the training, the tools, or even, in many cases, the data necessary to do so.[22] They tend to apply simple trial-and-error learning schemes, ignore uncertainty, and rely on habit or simple deterministic rules.[23] In considering what could go wrong, they ignore pertinent risk dimensions because of forgetfulness or lack of imagination, hence are likely to seriously underestimate the chances and costs of failure. Bias and ignorance, however, do not deter them from confidently announcing judgments, estimates, and inferences without investing much thought in them.

Decisionmakers are able to make snap judgments by applying several common heuristics in preference to using algorithms.[24] Heuristics are general principles for reducing complex judgment tasks to simpler mental operations by emphasizing certain properties of the data and ignoring others. They are informal intuitive procedures that are applied both consciously and unconsciously. Faced with the need to figure out the outcomes of competing policy options and their probabilities, people act as cognitive misers, either because they do not have the time, energy, or intellectual resources to analyze the problem systematically or because they do not want to bother to do so.

Four heuristics are available: representativeness, availability, simulation, and anchoring and adjustment (Kahneman et al., 1982; Tversky and Kahneman, 1974). Because their application involves selective use of information, they are a source of judgment biases—deviations in judgment that are consistent and predictable. Biases may result in either risk-acceptant or risk-averse policies, depending on the direction and content of the bias. Heuristics, therefore, do not directly affect risk acceptance. Their effect on risk disposition is through their influence on the judgment of probabilities and payoffs and through the confidence that they inspire in the validity and completeness of the estimates. Thus, heuristics become influential cognitive mechanisms that in conjunction with other cognitive and noncognitive factors help bias decisionmakers toward risk acceptance or risk avoidance.

The *representativeness* heuristic is applied in assessing the probability that Object A belongs to Class B or that Event A originates from Process B. The essential features of the sample events or sample statistics are matched to the parent population or the generating process. Representativeness is applied in judgment tasks because it is readily accessible; it correlates with probability, but people tend to overestimate the correlation. Although similarity is sometimes a useful and valid cue in judging likelihood, other factors unrelated to similarity are frequently critical in determining it. In judging by representativeness, sample size is not considered and base-rate information is generally ignored. Consequently, the gambler's fallacy, which involves the belief that deviating outcomes balance out, is recurrent. People predict outcomes by determining what these outcomes should be to balance out any present or past deviations from expected values (Tversky and Kahneman, 1971).

If representativeness is high, people tend not to scrutinize the reliability and validity of evidence. Hence, even where representativeness is a normatively legitimate rule of judgment, it can lead to errors in likelihood estimates; and neglecting to take the reliability of the information into account causes overconfidence in one's judgment (Cross and Guyer, 1980; Slovic et al., 1977), which results in premature cognitive closure. A high degree of representativeness leads to a high level of faith in predictions, even when there is an awareness of factors limiting the accuracy of the prediction. Representativeness becomes so central in the mental process that factors limiting accuracy, such as insufficient or imprecise information or the unique aspects of a situation, are discounted, seen as marginal, or underestimated.

Representativeness also affects causal judgment. In making attributions, people exaggerate the role and weight of representative causes in accounting for outcomes. They tend to reason from causes to consequences, not the other way around, and are more confident in their judgments when they reason forward than when they reason backward. Outcomes are attributed to the success or failure of a specific preceding behavior on the assumption that if Event A preceded Event B, then A caused B. Representativeness is expressed in the linear time sequence between the events. The fallacy lies in identifying the sequential order with a causal one, a conclusion that has no necessary objective justification.

Finally, representativeness may cause a monotonicity bias, expressed in the unconditional judgment that "if some is good then more is better, and if some is bad then more is worse" (Kanwisher, 1989: 670). This bias occurs, for example, in the behavior of military commanders in intervention operations who always request more troops and ask for more firepower, even when it becomes counterproductive. The indiscriminate use of firepower in Vietnam only made the local population more alienated. Likewise, the use of a large number of troops ("to be on the safe side") in the early stages of Israel's invasion of Lebanon in 1982 only led to an unnecessary increase in the logistic burden and blocked main roads leading to the battleground with military units that did not participate in the fighting.

The representativeness heuristic provides a sense of overconfidence and inevitability regarding policies that should in fact be constantly and critically evaluated. Feedback information indicating unacceptable negative consequences is ignored, although it should signal the need for policy reevaluation and revision.

The second common heuristic is *availability*. In judging the frequency of a category of events or attributes, the likelihood of an event, or the frequency of the co-occurrence of events, people often use as a criterion the ease with which they can bring specific instances to mind (Tversky and Kahneman, 1973, 1974). Thus, for example, an estimate that a certain policy is doomed to fail may depend on the ease and speed with which a decisionmaker can imagine difficulties to be encountered or recall a similar event that failed. Although availability can be useful for taking shortcuts to reasonable judgments of frequency and probability, it can also be highly misleading, because frequency and probability often are predicated on factors other than availability.

Availability operates, with potentially detrimental consequences, both when decisionmakers have experience and when the situation is novel for them. In the first case, availability amplifies the role of experience and dominates other decisional inputs, even if other inputs are as important as or more important than experience-based knowledge. In cases where decisionmakers lack experience, especially when time is at a premium, availability causes the decisionmakers to use the information at hand without critically scrutinizing its validity. The sample of readily available information, used in judgment, is not necessarily random or unbiased. Because people tend to search for positive, confirming instances of the hypotheses that they hold, they fail to make adequate use of disconfirming evidence when they encounter it. Thus, disconfirming information is less readily available, even if decisionmakers are aware of its existence. Vivid and salient data are readily available, too, especially in judgments involving ambiguous events, although vividness and salience may have little or nothing to do with frequency or likelihood. Nonsalient data are less available than salient data, even if they are highly diagnostic. Thus, nonoccurrences are used less in judgments than occurrences are because they are not salient, although they may have an important diagnostic value (Einhorn and Hogarth, 1978; Nisbett and Ross, 1980: 113–38; Taylor, 1982).

People making risky decisions have specific beliefs about the relative importance of the components and dimensions of risk—for example, beliefs about the relative importance of probabilities and payoffs in a gamble. These beliefs derive from previous experience, from logical analysis, or sometimes from irrational fears and prejudices. The risk dimensions that people consider more important dominate analysis to the exclusion of other dimensions; limited information-processing capabilities reinforce this tendency (Slovic and Lichtenstein, 1968). These points are particularly relevant to highly complex decisions, when a decisionmaker is expected to integrate a large number of risk dimensions into a single risk assessment. The complexity of the assessment is reduced by focusing the assessment on one cognitively available risk dimension.

For the same reason, people focus on either the payoff or the probability dimensions of risky decisions, whichever seems to them more salient. When the probabilities have a small distribution range, they are undervalued, except when large losses are involved. People do not combine probabilities multiplicatively with payoffs, as normative theory requires (Schoemaker, 1989). Similarly, an increase in the complexity of gambles leads to an attempt to simplify (Payne and Braunstein, 1978). In a choice among equally valued alternatives, decisionmakers select the one that is superior on the most important dimensions, even if it is inferior in terms of overall utilities (Mintz, 1993; Slovic, 1975). Thus, in choice problems people often note the most important factor and choose the option that maximizes that factor; they may overweigh a single important factor and underweigh other relevant factors. For instance, in making intervention decisions American decisionmakers have often ranked U.S. security high while ranking other factors—cost in human lives, economic costs, and human rights—unrealistically low (Kanwisher, 1989: 666–70). Similarly, they tended to evaluate the uncertain outcome of U.S. intervention in Vietnam by using the more readily available quantitative indicators—firepower and body count—and to ignore factors that were harder to measure, like morale, determination, and quality of leadership.

Easily available theories or causes are likely to be considered in making attributions, and the availability produced by mentally associating events can produce illusory correlations (Chapman and Chapman, 1969; Tversky and Kahneman, 1973). Yet the theories on which associative correlations rely are rarely scientifically tested. They often have little or no factual basis and may have their sources in superficial impressions. To compound the problem, an available associative covariation between two events leads to the conclusion that they covaried more often than they actually did, and from there to the inference that the two events do indeed have a basic substantive connection. Such misperceptions withstand even disconfirming evidence.

An example is the misplaced correlation between military superiority and the inevitability of victory (see Chapter 4), which has been proved incorrect so often and has beverled so many military commanders. The domino theory, another central organizing concept in U.S. containment policy, was based on an easily visualized metaphor that made it readily available and meant that it was thoughtlessly applied. Decisionmakers who accepted the domino theory predicted a very high correlation between reneging on even minor commitments, the lessons that the adversary would infer from any such revocation, and the threat to the overall strategic posture of the United States because of inevitable challenges to other commitments. The validity of these assertions was neither based on fact nor embedded in irrefutable logic (Jervis, 1991; Snyder, 1991a).

Availability can bias judgments of the likelihood of dangers. Certain risky outcomes are highlighted, making them easier to imagine or recall and therefore likely to be given more weight than they deserve in the overall calculus of a decision. The availability heuristic accounts for the "unsqueaky-wheel trap" (Janis, 1989: 194–96), according to which decisionmakers focus their information search, evaluation, and emergency planning entirely on the elements known from the outset to be fraught with dangers while they totally overlook subtle risks embedded in the routine steps of implementing the plan. The error is not in paying more attention to riskier steps, which is logical, but in disregarding risks that should be considered even if their probability is low.

To the extent that causes or reasons for a predicted outcome are readily available, decisionmakers tend to increase the estimate of that event's probability (Levi and Pryor, 1987). For example, most American decisionmakers could at one time explain why the United States, with its decisive technological superiority, could not possibly lose the war in Vietnam once it was fully involved militarily. The choice of a particular theory for anticipating the results of a policy is, then, strongly influenced by its availability, rather than by a meticulous comparison of its relevance and validity with alternative theories. A decisionmaker's search terminates with a satisficing theory, rather than continuing until the best-founded, most valid theory is discovered. Nevertheless, people are overconfident in the explanations that they generate in this questionable manner. They bring to bear theories that often are founded on culture-based stereotypes of other nations; on knowledge that has not passed validity tests; on hearsay, movies, newspaper stories, novels, or their own limited and unsystematic acquaintance with history. They use and abuse a sample of one or a few highly available historical cases as a basis for confident and highly generalized inferences and predictions by analogy (Jervis, 1976; Khong, 1992; Neustadt and May, 1986; Vertzberger, 1990).

A readily available historical experience that focuses attention on the unacceptability of particular risks, such as those associated with foreign military intervention, and makes these risks more vivid than any possible benefits from intervention, is likely to result in risk aversion. The debate over U.S. intervention in Bosnia, for example, was dominated by the Vietnam analogy. Furthermore, the effect on risk disposition could depend on whether the lessons of history are used to make the risks associated with a particular option, and not with other options, seem familiar. In this case, decisionmakers with a preference for known risks over unknown risks are more likely to choose the option with the known risks.

Lack of ability to imagine certain threats may therefore reduce the subjective probabilities attributed to their occurrence, even when higher probabilities should be accorded to these outcomes. And because it is easier to envisage a desired than an undesired future, people tend to forecast the desired future, even if there are dubious objective grounds for doing so. No wonder decisionmakers seem to be more sensitive to the risks of change than to the risks of prevailing policies. New policies trigger images of unexpected pitfalls, not images of improvements. Existing policies, because they have already been experienced, bring to mind their beneficial effects or useful and proven means of coping with what might go wrong.

The third often-used heuristic is *simulation*. When people have to predict a future event, estimate the probability of an event, make a counterfactual judgment, or assess causality, they run a mental simulation of the event in question. The ease with which any outcome can be simulated becomes a basis for judging its likelihood (Kahneman and Tversky, 1982b). The simulation heuristic is also applied when judging the plausibility of both positive and adverse outcomes. Decisionmakers construct scenarios that consist of causal chains, depicting the consequences of not intervening compared to the consequences of intervening. Here the simulation heuristic clarifies to the decisionmaker the relative advantages of intervention versus nonintervention, as well as convinces others, at the argumentation stage, why they should support one or the other strategy. Scenarios are also used to assess the probability of events. Vivid, plausible scenarios are attributed a high subjective probability and dominate the process of policy choice.

Where the simulation heuristic is part of the planning process, it causes an overoptimistic assessment of the plan's success, owing to the underestimation of the probability of unanticipated events. Scenarios that explain events that have already occurred are perceived as highly plausible and even inevitable. The envisioned causal chain is perceived as a reconstruction of the decision process; in fact, however, it is contaminated by knowledge of the outcome, so that evaluation of the decision process is confused with evaluation of the outcome, even when there is little ground to assume high correlation between decision and outcome (Baron and Hershey, 1988; Bukszar and Connolly, 1988; Fischhoff, 1975, 1982; Fischhoff and Beyth, 1975). Hence, in hindsight people assume that the outcome could have been foreseen and judge harshly those who did not anticipate it; worse, they remain overoptimistic about their own ability to anticipate and understand future outcomes.[25] On the other hand, possible future events for which no elaborate simulated scenarios are easily available are deemed highly improbable.

In estimating the overall probability of entire scenarios, decisionmakers tend to attribute a higher probability to a scenario that consists of causally linked and representative components(those typical of observed events or behavior) than to individual components. The more detailed the scenario, the more probable it is judged to be. Even weak links buried in a mass of details are given credence that they do not deserve. In 1982, for example, Israeli Defense Minister Ariel Sharon was able to overwhelm the cabinet with details and convince the government of the high probability of success if Israel intervened in Lebanon. Decisionmakers allowed their intuition to overpower their logic—which should have warned them that the more detailed the causal sequence envisaged, the less probable the outcome was (Kanwisher, 1989; Slovic et al., 1976; Tversky and Kahneman, 1983).

This conjunction fallacy is particularly evident when the argumentation or cognitive processes involve counterfactuals. These are used in reevaluating a policy once it has been implemented. Decisionmakers compare the known sequence of events to what might have happened had they chosen a different course of action (Miller et al., 1990).[26] The counterfactual scenario then competes with the known sequence of events, and for lack of any other means of estimating counterfactual probabilities, the representativeness of the scenario becomes the main source of judgments about probability. Representativeness in turn competes with the overpowering availability effects of the known sequence of events that actually happened. In hindsight, and reinforced by the conjunction bias, actual sequence-dependent outcomes seem more deterministic than they were, hence more vivid than counterfactual outcomes (Fischhoff, 1975). The vividness of the actual outcomes acts as confirmatory feedback information that reinforces the belief that the policy and its consequences could not have been avoided, thus making reconsideration of the continued relevance of the policy less likely than if counterfactuals had not been invoked.

The fourth heuristic is *anchoring and adjustment*. Beginning with a base value, decisionmakers adjust estimates as more information becomes available (Tversky and Kahneman, 1974). The point of departure can be an ideology or any theory that prescribes expectations and rules of behavior. The degree to which the point of departure is a suitable base depends on its continuous updating as circumstances change and as information becomes available. But this updating often does not take place. Likelihood estimates are strongly susceptible to anchoring, and the anchoring effect has been found to be robust (Plous, 1989). Where the point of departure is a master plan, its predictions of success are not systematically adjusted. Decisionmakers are anchored to the original expectations of success, because the plan provides a sense, often not justified, of power and control over the situation. Moreover, the likelihood of success, which depends on the execution of a sequence of actions, is seen as higher than is warranted because decisionmakers are likely to overestimate the probability of the occurrence of conjunctive events. People do not adjust the end-state probability to the decline in the probability of interdependent outcomes as one moves along the chain of conditional links.

Anchoring causes the neglect of information that would require adjustments in expectations about policy outcomes.[27] The failure to learn, even when a policy is clearly failing, correlates with the ability to erect effective cognitive barriers to the lessons of failure. One barrier is severance of the causal link between the decisions leading to the failing policy and the negative consequences themselves. By delinking these two, the negative consequences can be attributed to causes that do not necessitate major adjustment in the policy.

Another barrier to change is personal responsibility and accountability for the initial decisions that led to failure. A feeling of responsibility stimulates ego-defensive mechanisms that prevent learning in order to curtail the emotional and practical costs of admitting responsibility for failure. Faced with unexpectedly adverse outcomes, people may exaggerate the degree to which they did not expect the outcomes to occur (Mazursky and Ofir, 1990). The biased reconstruction of past expectations ("I never expected it to happen") justifies not having taken measures to reduce the costs of the adverse outcomes (e.g., a stock market crash, escalation of conflict to war). Furthermore, it reduces learning from mistakes by amplifying the surprise effect (e.g., outcomes could not be expected). The involved individual is kept from learning to take precautions, is desensitized to potential, even if remote, risks in the future, because he or she views the case as unique.

Policy change is more likely to take place if decisionmakers can dissociate themselves from the decisions that led to failure (e.g., "I inherited this mess from my predecessor," "I was presented with a fait accompli," "The situation forced my hand; there was nothing else I could do"). These excuses do not have to be true; those using them just want to plausibly deny responsibility and thereby detach themselves from the policy and change course.

The use of heuristics is most likely to take place when decisions are of low importance and errors are not costly or irreversible. Heuristics are, however, also used when the stakes are high, time is at a premium, and decisionmakers are confronted with information overload. In these cases they use heuristics as a shortcut to judgment. An individual's level of expertise and cognitive sophistication is another factor. When decisionmakers are unsophisticated and untrained, they prefer to simplify coping with a complex world by relying on heuristics. But when decisionmakers are highly experienced and sophisticated, they may also use heuristics that encapsulate expertise and experience. In either case, the heuristics are usually applied thoughtlessly to save time and effort (Chaiken et al., 1989: 226–27; Sherman and Corty, 1984).

The use of heuristics has important consequences for estimates and judgments of utilities and probabilities: overconfidence in their validity and overestimation of the completeness of the analysis, which reinforces overconfidence (Fischhoff, 1991: 144–45; Fischhoff et al., 1978; Slovic et al., 1976, 1980, 1982a). The combination of overconfidence and overestimation prevents critical evaluation of estimates and judgments, precludes questions about the validity of underlying assumptions, and reduces the motivation to search for and be alert to additional information.[28] Even when the use of heuristics and the particular heuristic chosen are normatively justified, the fact that judgment was based on heuristics tends to be forgotten with time, and resultant judgments and inferences become inputs for assessing new information. If new information allows reassessment of heuristics-derived prior judgments without resorting to heuristics, that does not have any effect; previous judgments prevail and are not reexamined.

Risk-Taking Behavior in Small Groups: Causes and Processes

Some decisions are made by individuals, but more often decisionmaking is a collective enterprise carried out by organizations and small groups. It is therefore critical to understand whether and how risk aggregation affects risk-taking dispositions in collective decisionmaking contexts. In the case of foreign military intervention decisions, the action involves, by definition, convention-breaking behavior, so it is unlikely that final decisions are made by way of SOPs. The decision on whether to intervene is made by top-level decisionmakers, either a single individual or a small group mostly composed of leaders of important organizations. Organizational inputs are, therefore, crucial. Organizations collect data and disseminate them, as well as their interpretations of them, to individual decisionmakers through a highly politicized distribution chain. They also define the problem and the relevant risk dimensions as seen from their perspective, creating in the process a pluralistic, competitive, complex, possibly enlightened, but also confusing, decisionmaking environment. The aggregation of these multiple inputs and the resulting decisions take place in the small group.

How do the different types of risk—real, perceived, and acceptable—relate to group decisionmaking? Faced with an opportunity for decision, decisionmakers will encounter varying real risks, depending on the subsequent personal and political implications of the policy. That is, the consequences could, for example, mean demotion and loss of prestige for one member, or promotion and increase in influence for another, with an important impact on the balance of power in the group. Perceived risk may or may not vary among decisionmakers depending on how they interpret the situation. The extent to which group members share perceptions of risk could affect their cooperation or the lack of it and thus their sense of shared responsibility for outcomes. In other words, the extent to which risk perceptions are shared affects group dynamics and risk preferences.

But even when risk is identically perceived, the level of risk that is acceptable to different members of the group is likely to be dissimilar. Identifying collective attitudes toward risk requires the aggregation of individuals' attitudes into a single collective, coherent judgment of risk and a single collective preference for risk taking.[29] The difficulty of identifying aggregate risk assessments and risk-taking preferences stems from the fact that group attitudes toward risk are not a simple average of the attitudes of all group members, nor do group attitudes necessarily reflect the position of the majority. Group preferences, especially in tasks involving judgment rather than knowledge, may be shaped by a minority and result in a position quite different from that of either the average aggregate position of all members or the position of the majority.[30]

A balanced view of how group risk judgments evolve and how risk preferences are shaped is possible only if we favor an approach that takes into account a broad range of causes and processes, one that is not dominated exclusively by social interaction variables and that is contingent on situational attributes and the composition of the group. We thus avoid overemphasizing social interaction at the expense of other important forces that determine group information-processing and choice outcomes. This approach better reflects the essence of real-life situations.

The expectation that social interaction within a group will be the main agent shaping members' views, perceptions, and preferences is often unrealistic, especially in political decisionmaking groups. What must be taken into account is that group membership is not necessarily a matter of free choice. Constitutional requirements, institutional arrangements, or the force of circumstances (e.g., election outcomes that mandate a coalition government) could make a group of a particular composition emerge as the legitimate decisionmaking body. If that is the case, social interaction will not necessarily be sought after, nor will individuals readily abandon their views for shared or group beliefs. It is in factions where social interactions are likely to play the dominant role in shaping judgments and preferences, not in the group as a whole, because faction membership is voluntary, a matter of choice. A faction represents explicit or implicit commonalties: intellectual (common beliefs), functional (common goals), or emotional (mutual respect between members or shared admiration for a leader). Faction members choose to share physical and social space with other members and are therefore open to influence from one another.

Policy groups do not always meet frequently enough or long enough for social interaction to have preponderance over all other influences. Members of policy groups are subject to and exposed to the pull of contradicting social-interaction influences because they are simultaneously members of more than one peer group and have to be attuned to persuasion by different reference groups, which may substantially reduce the influence of interaction in any single group. Nor are all group members equally influenced, for some are more amenable to persuasion. Group composition and the consequent distribution of personality attributes in the group help determine how important social interaction is likely to be in shaping risk preferences.[31] Finally, group members are not required to reach a meeting of minds to act. Circumstances often impose a need for agreed-on stopgap decisions that do not reflect a commonality of beliefs and assessments but are merely the most practical response to the situation at hand. This operational consensus is sometimes incorrectly interpreted as evidence of consensus reached through social interaction.

This analysis, therefore, is based on the assumption that the underlying motivations for collective behavior are found neither in the simple aggregation of the attributes of the members of the group nor in the group as an autonomous entity, but rather in the confluence of the two. Both former theoretical approaches have some validity; individuals matter even when they are part of a group, and the group itself can shape individual preferences. The interaction between the two reflects a nexus more generally known as the agent-structure relationship. I argue that it is this interaction that results in two clusters of factors that shape risk preferences in collective decisionmaking. One cluster concerns group dynamics. The other cluster reflects a range of group attributes, including the distribution of personality attributes; procedural arrangements that allow interventionary manipulation; the distribution of power and coalition politics; and shared group beliefs.

GROUP DYNAMICS

How can differences between aggregate risk dispositions and average individual risk dispositions be accounted for? One explanation must be sought in the dissimilarity between individual and collective decisionmaking—that is, in the strong impact of social influences that are grounded in both motivated and unmotivated reasoning. Decisionmakers are likely to be exposed to and receptive to social pressures when they are faced with problems whose consequences are potentially serious, and this influence is likely to increase as the ambiguity, complexity, and uncertainty of the situation increase. Decisionmakers need to validate their judgments and choice preferences by comparing them with those of other group members.[32] Their need for security and reassurance when the costs of error can be high are satisfied if they conform and adapt their preferences to those prevailing in the group. Conforming with other group members becomes even more tempting when, in addition to external uncertainty, people have doubts about their abilities that are embedded in low self-esteem, the risk and complexity of decisions, or moral dilemmas raised by the issues (Moscovici, 1976: 25–31; Steiner, 1989).

For a long time, conventional wisdom has held that groups make riskier decisions than individuals do. The "risky shift" was explained in terms of motivation, cognition, social interaction, and statistical aggregation of risk preferences. More specifically, it was attributed to responsibility diffusion, persuasion by more risk-acceptant group members, familiarization with the problem, and the cultural salience of the value of the risk (Davis, 1992; Dion et al., 1970; Vinokur, 1971). In spite of its widespread acceptance, the risky-shift argument is neither convincing nor supported by observation, at least in the field of foreign policy. Groups can be both risk averse (e.g., Kennedy's cabinet and advisers in the Cuban missile crisis) and risk acceptant (e.g., Johnson's cabinet during the early years of intervention in Vietnam). These examples suggest that groups as such are not consistently and automatically risk acceptant and that group choice is influenced by the content of the problem and the initial distribution of individual preferences within the group (Cartwright, 1971; Gologor, 1977), as well as other factors that shall be discussed later in the chapter. It should be noted that not all significant decisions are perceived as primarily risk dominated; some important decisions are perceived to be opportunity dominated and thus induce less cautious and more risk-acceptant preferences.

Later findings on group behavior are more sensible. According to these findings, the generic process that takes place in groups is polarization, which may result in either risky or cautious shifts. Polarization indicates that social interaction within a small group tends to accentuate individual predispositions (Minix, 1982; Myers, 1982; Myers and Lamm, 1976). In other words, group decisionmaking contexts produce a pull toward an extreme, whether risk or caution, and not exclusively toward risk itself.[33] Specifically, group discussions tend to amplify the initially dominant point of view in the group.

This conclusion implies a more extreme risk-averse (cautious-shift) or risk-acceptant (risky-shift) preference for a group than for an individual under similar circumstances. A risky shift in a small group could have been preceded by judgments that de-emphasized warning information or emphasized coping and risk-management skills. A cautious shift, on the other hand, could be associated with judgments that emphasized warning information or de-emphasized coping and risk-management skills. The underlying social-interaction determinants of polarization and its direction can be found in a modified version, presented below, of any one or a combination of the explanations that originated in the risky-shift literature.

Social interaction affects both risk judgment and risk-taking preference. The influence of social interaction on risk preference occurs either directly or through the mediation of risk judgment, whereby risk-taking preference is determined by the evaluation of the risks involved. But how is the impact of social interaction generated? And what makes it effective? In brief, social influence works through four processes: responsibility diffusion, persuasion, familiarization, and value illumination and amplification.[34] These processes affect the judgment of risk and shape the preference for risk acceptance or risk avoidance by satisfying some basic social and psychological needs.

According to the *responsibility-diffusion* explanation, the most commonsensical of the four listed, in group decisionmaking contexts responsibility and accountability for consequences are diffused among group members. This reduces fear of failure, so decisionmakers are readier than before to make risky decisions.[35] Foreign policy decisions are susceptible to the consequences of risk sharing because they are made for others (the nation or a group within it). When choosing for others, people tend to prefer more cautious decisions than when they make decisions for themselves. Failure in the first case would be accompanied by self-blame and a need to justify the choice to others. But when decisions for others are made by a group, the tendency to avoid risk is less pronounced because failure can be shared with others and anticipated personal responsibility can thereby be reduced (Zaleska and Kogan, 1971).

The diffusion of responsibility in group contexts also encourages risk acceptance because postdecisional dissonance—which depends on perception of responsibility for outcomes—is anticipated to be low. That is only the case, however, when individuals are convinced either that other group members are willing to share responsibility or that responsibility sharing can somehow be imposed on them. Group discussions will be a way for each member to find out whether others are willing to commit themselves to share responsibility. This information will help decisionmakers in the group make up their minds about how much risk should be taken. Still, the judgment of whether others are willing to share responsibility can be biased. People tend to see their own judgment as appropriate for the circumstances and shared by others (Alicke and Largo, 1995; Ross et al., 1977). The false-consensus bias can generate a misleading sense of confidence, which stems from misplaced perceptions of social support in group decisionmaking contexts and which encourages risky or cautious shifts based on false assumptions of shared views and responsibility. This phenomenon is robust when information about other group members' views is ambiguous and members are reluctant to reveal their true positions, as is particularly common in new groups, but it is not exclusively limited to these groups.

The effect of responsibility sharing is more robust to the extent that a decision cannot be attributed to any single member but only to the group as an entity. Anonymity encourages a sense of group-shared responsibility as opposed to individual accountability and enhances the incentive to take risks. The actual probability of adverse consequences from the decision (policy risk) is not reduced, but the anticipated adverse consequences to the decisionmaker (political risk) do diminish. Where personal accountability is un-evenly distributed, those members of the group who do not feel strongly about the problem (e.g., are less responsible for outcomes) become subject to persuasion by members who do feel strongly (e.g., are more responsible). The preferences of the more ardent, responsible members set the direction of polarization for either risk taking or risk aversion.

Responsibility sharing works both ways. Where a decisionmaker prefers a cautious but unpopular option, one not in line with the prevailing mood in society at large, that decisionmaker, in a group where others have a similar preference, could be encouraged not to give in to the public mood. Moral support comes from similar-minded individuals in the group; important as that support may be, even more important is reduced risk. Holding onto a deviating preference that could have very unpleasant personal consequences, especially when it proves to be an error of judgment, is less risky when re-sponsibility is shared.

In determining what causes responsibility sharing, the formal institu-tional context in which decisions are made is often less important than the actual diffusion of responsibility within the group. When the members of a small group (e.g., cabinet) are known to share power and influence, group de-cisions may indeed be riskier than individual ones. But when power and influ-ence are not equally distributed, and that fact is transparent and publicly known, the group setting does not by itself necessarily produce a risky shift, because leading members know that they will bear most of the responsibility and the burden of accountability in case of loss.[36]

Hence, in analyzing group decision settings attention must be paid to the following: the formal procedures for decisionmaking (e.g., majority rule, unanimity rule); the actual, rather than formal, diffusion of power and influ-ence within the group; the public view of who is in charge (in democracies the public decides who to hold responsible for losses); and individual decision-makers' value systems (certain individuals may feel more responsible for losses than others do). The combined impact of these factors will decide the extent of responsibility-sharing expectations triggered by decisions made in a group setting, if any, and the resultant polarization effect. This set of consid-erations implies that not every group context produces responsibility-sharing effects; other influences will then be required to produce polarization.

According to the *persuasion* explanation, persuasion does not necessar-ily result from a logical and informative argument but is largely a matter of social self-presentation. Thus, people in a group who have more radically polarized judgments and preferences have a stronger incentive to invest more resources in attempts to lead and influence others (Rim, 1964, 1966).[37] The more self-confident and assertive members of the group are often capa-ble of communicating expectations that their position will prevail. Thus, the assertive members present themselves as winners, creating the expectation in other members that eventually the majority will support them. These expectations may motivate undecided members of the group to support the assertive members' positions. Correctly or incorrectly, members who have been waiting to see which way the wind is blowing interpret assertiveness as a cue and throw in their support. Expecting that side to win, they make winning possible, resulting in a majority for the more polarized position.

Risk-acceptant members tend to be strongly committed to their positions, and their arguments are forceful and influential because a risky choice requires deliberative efforts. But this is not always the case. When the cautious members are more committed than the risk takers, then a cautious shift is more likely than a risky shift (Burnstein and Katz, 1971; Stroebe and Fraser, 1971). More frequently, however, risky choices elicit greater confidence and commitment from supporters than cautious choices do. Risk-acceptant group members, because they are deeply committed, tend to ignore initial negative feedback from their decisions; they will, therefore, increase their commitment to the risky choice, making it even more difficult for them to back off. As their persistent commitment makes them more influential, they are likely to carry the whole group toward risk escalation and even entrapment.

An influential and assertive member of the group, often the most influential and assertive member, is its leader. Therefore, an important dimension of the persuasion process concerns the leader-follower relationship. It is to a large extent shaped by the type of leadership involved: transactional or transformational. With the more common, transactional type, leader-follower relationships are a matter of valued transactions, implicitly or explicitly anticipated (Bass, 1985; Hollander, 1986). The power of a leader to influence risk-acceptant preferences of other group members depends on what desired rewards he or she can offer to followers in the group. That, however, is not the case with transformational leadership, of which charismatic leadership is the most salient manifestation.

Charismatic leaders influence their followers to accept risks through two separate but interrelated processes. They inspire and arouse deep emotional attachment in their followers; they instill in them a sense of faith and trust in themselves that translates into confidence that risks are worth taking in pursuit of superordinate goals.

At the same time, leaders affect their followers through the lateral relationships that develop among the followers themselves (Bass, 1985; Meindle, 1990). The leader's charismatic effect is sustained and enhanced by the collective, interactive admiration and romanticization of the leader by the followers. In these cases, we often witness anticipatory compliance—group members try to second-guess the leader's position. Members may go to either extreme, risk aversion or risk acceptance, depending on their judgment of their leader's preferences. They may even indulge in competitive anticipatory compliance, trying to outdo each other in conforming with a position that has not even been explicitly expressed. They not only passively conform but try to identify, locate, and bring to the leader's attention information that supports the position in question. Policy revision becomes less likely. The latent process of influencing is very real here, despite the absence of any observable attempt by the leader to direct members' preferences.

Persuasion, however, is not just induced by the influence of an assertive individual. No less important is the social atmosphere in the group, which fosters self-imposed, sometimes anticipatory conformity with the dominant opinions in the group.[38] Self-imposed conformity is most evident where there are no conformity-constraining factors, such as prior commitment to the position of a reference group that is different from the dominant view in the decisionmaking peer group (Asch, 1958; Lamm and Kogan, 1970; Myers, 1982; Vertzberger, 1990: 236–37), or normative commitment to multiple advocacy.

Conformity takes two forms. In the predecision stage it entails accepting the prevailing position if that position conflicts with one's own judgment. In the postdecisional stage it entails continued support for the decision even if one concludes that it was the wrong decision and should be reversed. Entrapment results from the latter type of conformity. Group members become entrapped by the justifications, especially public justifications, that they provide for their initial decisions and actions. Forcefully made justifications gain a life of their own and are extremely difficult to refute at an acceptable cost.[39] Justifications that are necessary to ensure the acceptability of a decision can thus become roadblocks to change. Initial justifications prevent deviation from a position that prevails in the group, even when it is recognized as incorrect.

The most extreme manifestation of group-induced conformity can be found in groupthink situations, where all group members think as one. The occurrence of groupthink depends on a number of antecedent conditions (Hensley and Griffin, 1986; Janis, 1982: 242–56; Steiner, 1982; 't Hart, 1990). To start with, the group is highly cohesive or at least group members desire acceptance and approval by others. Cohesiveness can be caused or reinforced by the homogeneity of members' social background and ideology. Cohesiveness reinforces group unity and confidence in group skills and creates an illusion of invulnerability.[40] The defective performance of a cohesive group is most pronounced in situations of high stress from external or internal sources where there seems to be little hope of finding a better solution than the one advocated; such a situation increases members' dependence on the group and the likelihood of groupthink.

Another condition is insulation from outside influences, which works to establish the value and credibility of information coming from internal sources, especially when the group lacks explicit norms requiring methodical procedures for search and assessment. But the most potent condition comes from leadership practices that are directive assertive and are not modified by a tradition of impartiality. They may sometimes be disguised by a false facade of openness: the voicing of dissonant views is encouraged, but only so they can be shot down.[41] Thus, group members learn that deviationists pay a price for their nonconformity and stand only a slight chance of improving the accuracy of the group's consensual conceptions. This belief, combined with the perceived material and emotional benefits of not causing disturbances, produces a tendency to rationalize the group consensus as accurately reflecting situational reality.

Groupthink will be associated with either overoptimism or overpessimism. The group that becomes overoptimistic lacks the motivation to search for or attend to all available information, becomes convinced of its invincibility and morality, and thus experiences a risky shift. In other cases, groupthink produces collective agonizing, magnifies doubt by distributing it, and thus generates excessive pessimism, which results in procrastination, inactivity, and a cautious shift.[42]

Peer pressure to conform, which is common in group decisionmaking and characteristic of groupthink, may result in two broad kinds of conformity. One entails true conversion to the group position and internalization of group attitudes. The other is to comply with group pressure for conformity while privately doubting the group's views and course of action (McCauley, 1989; Vertzberger, 1990: 234–41). Does the nature of conformity matter? In the short run, the reason for conformity does not make a difference. The consequences are the same whether it stems from compliance or internalization. But in the long run, the two types of conformity reflect different levels of long-term commitment to the group's attitudes and policies. If feedback information from the policy threatens the group's cohesiveness, and the policies become an issue of contention within the group, those who merely complied are more likely to distance themselves from those policies or challenge them. Those who internalized group attitudes are less likely to be attentive to dissonant information and reluctant to substantially reform the failing policies, because converts are usually the staunchest believers owing to their investment of emotional and cognitive energy in internalizing the new attitudes.

At the core of a group, particularly a group whose membership represents diversity, there are commonly one or more factions, each composed of a temporary alignment of a few individuals. Factions generally do not have a formally structured hierarchy or a set of rules and procedures; rules are often fuzzy and tacit, although they may become formalized over time. The two group phenomena, polarization and groupthink, are more likely to occur in the faction than in the larger group, because faction members' beliefs, attitudes, and schemata are more similar and more specific than in the group as a whole; they reinforce one another and will therefore be more extreme. Groupthink is more likely because the valence and initial sharing of beliefs, attitudes, and schemata create a climate for the emergence of all the other symptoms of groupthink.[43]

When faction members perceive themselves to be facing competition from other factions in the group and view their membership as a common shield against it, their information processing is likely to be biased toward maintaining and accentuating the psychological distance between in-group and out-group. That will result in more information sharing within the faction and significantly less information sharing with those who are not members of the faction. Those faction members who by their extreme positions best represent the in-group–out-group dichotomy are likely to become influential and persuasive, and faction members will move psychologically closer to other in-group members, enhancing the possibility of factional groupthink (Jackson et al., 1995; 't Hart, 1990: 108–11; Vertzberger, 1984; Wilder and Allen, 1978). Factional polarization and groupthink have a spillover effect on the group as a whole when the coercive power that a faction wields can be translated into the ability to affect the risk preferences of the group or when a well-defined position presented with confidence and persistence by the faction carries the rest of the group with it.

According to the *familiarization* explanation, the group is a social arena where the diversity of specialized information and opinions among the members is expressed and exchanged. Group discussion increases familiarization with a problem by making available to the members information that was unknown or unattended to before (Myers, 1982; Sniezek, 1992). As familiarization increases, both caution and boldness tend to increase. When people are unsure, they usually prefer to avoid an extreme position. But as individuals are exposed to new information and argumentation through group interaction, ambiguities and inconsistencies are clarified, and the participating decisionmakers are assured that there are no hidden costs of which they are not aware (Marquis and Reitz, 1969; Ridley et al., 1981). The broader and more diverse are the views expressed, the more effective familiarization is, resulting in reduced apprehensions about making incorrect assessments of utility and probability and strengthened confidence in the validity of estimates.[44] Having fewer doubts, group decisionmakers are disposed to be uninhibited about being more cautious or more risk acceptant than are average individual decisionmakers facing similar problems.

In assessing polarization caused by familiarization, an observer needs to take into account a number of important factors affecting the actual practices and utility of information sharing. The notion that group members exchange specialized information, that some members have and others do not have certain information, and that sharing it will make for better-informed decisions is based on three underlying assumptions. The first is that group members have a common understanding about problem definition and preferred type of solution and therefore know which information is relevant to share with others. The second assumption is that group members have a common goal of solving the problem and therefore have no hidden, self-serving agendas but a shared interest in a successful solution. The third assumption is that solving the problem will not adversely affect some members and benefit others by redistributing power and influence.

These assumptions do not hold true for groups that are tasked with solving many important political-military problems. The issues at stake are often deeply contested and divisive, and information is neither equally accessible nor readily shared (Halperin et al., 1974; Metselaar and Verbeek, 1997;

Preston, 1997; Stasser, 1992). In foreign policy areas where secrecy is considered critical, some members of a decisionmaking group may be cut off from crucial sources of information. Scarcity of information increases the chances that in assessing the actual level of risk, group members will make lower risk estimates than are called for. Sometimes information is not shared because the holder does not understand its relevance. In other cases, a successful solution would reward certain members and the organizations that they represent with improved status and influence, which would give them long-term advantages in competing for access to societal resources. Those whom a successful solution will not favor may obstruct a successful outcome by declining to share essential information. Knowledge is power, in any case, and hoarding information is sometimes perceived as a wise general strategy in anticipation of some rewarding but still unspecified use for this knowledge in the future.

The preceding discussion points to a potentially serious bias that is unique to group decisionmaking: an illusion among participants that they have canvassed the whole range of relevant information and are aware of all policy implications. The reality, however, is that there are barriers to information sharing that reduce exposure to information about policy costs that might substantially affect risk assessments. The illusion of comprehensiveness and the reality of information hoarding result in a lower-than-appropriate risk judgment, which could create a predisposition to espouse policies that are riskier than they seem, along with a predisposition to uphold the policies with greater confidence than is warranted.

Two little-noted consequences of familiarization should be mentioned. In decisionmaking groups afflicted by bureau-organizational parochialism, where members are mostly concerned with the risks affecting their organization, the interactive effects of multiple risks will be ignored, even if the risks are familiar. Nobody will monitor and advertise these risks, even though interactive effects are difficult to control. The consequence of narrowly focused attention to risk and neglect of interactive effects is the underestimation of risk assessment by each member. When underestimation is shared, it has the cumulative effect of lowering group perceptions of risk and increasing confidence in the controllability of risks. This may well encourage a risky shift in the group.

The other little-noted consequence of familiarization is that with tasks for which a correct solution is not self-evident or remains uncertain, even if information is shared to the fullest extent, group decisionmaking is inferior to individual decisionmaking because of the aggregation of errors and biases. "When individuals are error prone and functioning under a majority-type process, groups would be expected to make more errors (or more extreme errors) than individuals" (Tindale, 1993: 111). Many important foreign policy decisions, including decisions concerning the use of military force, belong in this task category.

According to the fourth explanation for polarization, illumination of *appropriate cultural values*, risky decision opportunities elicit one of two core cultural values: risk acceptance or caution. Cultural is used here in a broad sense and could refer to societal, political, or organizational culture. Once a cultural value becomes salient, a couple of mechanisms cause a shift in preference in a direction consistent with the revealed value (Brown, 1965; Hong, 1978).[45] First, the salient value affects information flow so that it generates more verbal arguments that support it than those that oppose it. Second, group discussion reveals to each participant the distribution of risk-taking preferences within the group and indicates which decisions are consistent with the prevailing group value. Each member can thus choose to appear bold or cautious with respect to others (Kogan and Wallach, 1967: 20; Myers and Arenson, 1972; Tullar and Johnson, 1973; Wallach and Wing, 1968; Witte and Arez, 1974).

Group decisions are more likely to reflect cultural values than individual decisions. Generally, in a risk-oriented culture, decisions made by a group are more likely to reflect the norm favoring risk and to be riskier than decisions made by individuals. A social environment that supports and encourages risk taking and associates it with superior ability (Jellison and Riskind, 1971) enhances the tendency to take risks. For similar reasons, in a culture where caution is valued, group decisions are likely to be more cautious than individual decisions are. Specifically, a political culture that emphasizes failure avoidance for the sake of political survival induces risk-averse policy preferences, especially when decisionmakers believe that the potential political costs of failure outweigh possible benefits from policy success. On the other hand, when the costs of inaction are believed to be high, even moderate stakes may trigger risk taking.[46]

The four explanations for polarization, which invoke different causal mechanisms, are not mutually exclusive.[47] Very often more than one factor produces the risky or cautious shift. The relevance and relative weight of each explanation depends on context—for example, issue-areas or attributes of group members (e.g., cognitive style, personality, status in the group). Stated more generally, the relative importance of the two generic modes by which groups exert influence on individuals—the normative mode, which is based on the desire to conform with the expectations of others (persuasion, responsibility defusion, cultural values), and the informational mode, which is based on the acceptance of information from others (familiarization)—may vary by situational context and by issue.[48] In a group that develops a siege mentality owing to real or imagined threats to its position, members will be exposed to normative influences, so they will forge a consensus in order to face the threats. When positions have to be stated publicly, normative influence is also more likely to be salient than information influence. When positions can be kept private, however, informational influence is more likely to dominate. With intellective issues, when correct, fact-based solutions are considered attainable, informational influences are more likely to dominate.

With issues that involve behavioral, political, or ethical judgments, when there are no demonstrably correct answers, the group will attempt to produce a consensual best approximation to the "right" answer through persuasion. In all these instances, the unanimity decision rule enhances the effect of the dominant influence mode—normative or informational—associated with the situation or issue-area in question (Kaplan, 1987; Kaplan and Miller, 1987).[49]

These generalizations should be qualified by three observations. First, politically contested issues are not easily defined within one issue category. In fact, by framing an issue as intellective or judgmental some group members may acquire more influence than others in shaping the group's position. Those who have access to information supporting their position will prefer to define the issue as intellective. But those whose main resource is political manipulation of group dynamics will want to describe the issue as judgmental in order to weaken the weight of arguments based on information. Second, in a cohesive group there is likely to be a tendency for normative rather than informational influences, even with intellective issues. This is not the case, however, if norms of professional competence and epistemic validity are embedded in group culture. Third, personality differences between group members can be important. Cognitively complex individuals are mostly influenced by information sharing in the group; extroverts, by the need to be accepted and to comply with majority views.[50] Most experimental work does not control for these and other important contextual sources of variability in the group environment.

In brief, decisions involving risk are taken or avoided based on two general considerations. First, risks are taken when they are perceived to be below a predetermined nonacceptability threshold. Second, risk-taking preferences increase or decrease if the acceptability of risk increases or decreases. The four causal explanations of the effect of group contexts on risk taking imply one or both of these two situations. The responsibility-diffusion and familiarization explanations imply processes that result in change in individual risk acceptability and risk perception, mainly the latter. The persuasion and cultural-value explanations imply processes that result in group-induced change in individual risk acceptability.

GROUP ATTRIBUTES

Variable forms of group dynamics affect in multiple ways the risk judgments and preferences of group members and their aggregation into a group judgment and preference. Still, group dynamics is not an exclusive cause. Four additional factors shape risk behavior in a small group. They are embedded in group attributes, but whether some or all are synergistic depends on the situation.

Social interactions do not have similar effects on all individuals; certain people are more likely than others to be influenced or influential in group contexts. People who are low in self-esteem have a strong need for social approval and tend to be highly responsive to all forms of actual or even anticipated social pressure. They tend to fear failure and be keen to share responsibility for the possible failure of a risky course of action; group discussion therefore has a powerful influence in disposing them to take greater risks or avoid them (Kogan and Wallach, 1967: 265; Wood and Stagner, 1994). People whose cognitive style is dominated by affect are likewise more influenced by the group context and the possibility of sharing responsibility (Henderson and Nutt, 1980). In addition, there are issue-bound considerations. Group members who are high on evaluative-cognitive consistency (the consistency between attitude and supporting beliefs) and know a great deal about the issue under discussion are much less susceptible to influence than others are (Wood and Stagner, 1994). Thus, susceptibility to influence emerges from both the kind of person one is and the kind of situation or issue one is facing.

Group composition is an important variable, then, because as the number of independent-minded individuals increases, so does the chance for diverse opinions and eschewal of compliance with peer expectations. Leading decisionmakers recognize these tendencies and do not trust group processes to overcome disagreement and produce a communality of views. In many cases, they attempt to make sure that the decisionmaking group is composed of like-minded individuals or at least those who are not likely to be too independent. In these cases, the direction of polarization can be predicted from a group's composition. In a group whose membership is determined by loyalty to the group leader, the direction of polarization is determined by the position of the group leader. In a group for which entry is decided by shared views and like-mindedness, discussions are dominated by the shared views, and choices shift in the direction indicated by those views (Levine and Moreland, 1990; Moreland and Levine, 1992).

Change in the balance of personalities can consequently influence the risk-taking preferences of the group.[51] The balance of personalities is affected when new people with different beliefs, values, and personality attributes are brought in or when a single individual, a new leader, joins the group. This second form can have a substantial impact on group processes even if all other members of the decisionmaking group remain the same. The new leader's style—including accessibility, expressed tolerance for dissident views, preference for the advice of particular group members, use of consultation, and an interest in extensive searches for information—can affect group preferences through its impact on decisionmaking and on the balance of influence among group members. Even when a new leader does not immediately change the composition of the group, group members who cannot adjust to the change in leadership style leave and are replaced by others, thus continuing the change in the balance of personalities that was triggered by the change in leadership.

Group risk-taking preferences can also be shaped by decisionmaking procedures and norms and their purposeful manipulation. There are obvious differences between majority and unanimity rules. Majority rules require less effort and less time to drive a group to accept or support risky policies. The quality of risky decisions produced by the rules are consequently different. Unanimity rules motivate a group to process information systematically— as opposed to heuristically—to pay attention to minority viewpoints, and thereby to avoid tunnel vision. The investment of time and cognitive effort contribute to the group's commitment to a decision and to the belief in its inevitable correctness.[52] Commitment can backfire by leading to perseverance in the face of disconfirming information and to eventual entrapment.

Decision procedures and norms may be manipulated by individual group members intending to affect the group's risk-taking preferences and choices. Manipulators can make sure that those who might oppose particular policies will be conveniently unavailable to attend important policy meetings, and they have other options as well (Halperin et al., 1974; Hoyt and Garrison, 1997; Maoz, 1990a; Raven, 1990). First, they can attempt to set the agenda and determine the procedures by which group decisions are made, influencing who participates, what is discussed, and, consequently, what choices are arrived at. Second, when a majority is coalescing around an alternative, they can introduce a new alternative to split the supporters of the previous alternative, resulting in either a majority for the new alternative or a majority for what was the previous minority alternative. Third, they can frame choice in a manner that affects perceived trade-offs to enhance the inclination for risk or caution. Fourth, they can use a salami tactic, breaking down a high-risk course of action into gradual policy steps. Each new policy deviates only marginally from the prior one and at the same time sets the stage for the next decision in the series, so that the manipulated members are unaware that they are cooperating in producing a much more risky policy than the one that they initially supported.[53]

The distribution of power between more risk-acceptant and more risk-averse members is another factor that affects a group's risk-taking preference. In a group with both risk-acceptant and risk-averse members, where the balance of power between the two subgroups does not allow either side to impose its views on the other and where both subgroups are equally committed to their risk-taking preference, the amount of risky or cautious shift reflects the results of bargaining and compromise.[54] Over time, the changing balance of influence among group members allows for particular policy choices. But when the choices are implemented, they enhance the power of those who supported them, putting the supporters in an even stronger position to ensure persistence of the policy.

With foreign policy decisionmaking groups the composition of the group and the power distribution within it reflect the institutional and political contexts in which the group is embedded. That is, the group may be a coalition of individuals representing government organizations or political parties. Risk-taking preferences and the behavior of individual members are consequently not driven by social processes alone but are also driven by political imperatives (Lamborn, 1985, 1991; Preston, 1997). Key members of the decisionmaking group may have different views of which suboptimal policy will best serve their parochial interests. Intense factional politics in a competitive organizational environment means that risk-taking preferences will not necessarily be a manifestation of polarization. Instead, when coalition maintenance dominates calculations within the decisionmaking group, the main interests and preferences of the participants must be at least minimally satisfied. In this case, fundamental policy changes are eschewed; decisionmaking is a matter of muddling through. The approach to decisions, then, is frequently cautious, and behavior is risk averse. Under these circumstances, members averse to running political risks (e.g., assaulting vested interests in their constituency) may decide to run external risks, but only if their internal position either depends on it or is not threatened by it. As a rule, decisionmakers will choose a suboptimal policy with expected minimal adverse power-redistributive effects over an optimal policy choice, risky or not, that poses a serious threat of adverse power redistribution.

Risk-taking preferences of group members also vary depending on the anticipated level of control over policy choices and implementation. The smaller the group, the more control each member thinks he or she has. Therefore, small decisionmaking groups are more likely to prefer strong policies with high-risk elements than large groups are. The less control group members expect to have over a policy and its implementation, the more indifference they feel toward accepting a risky policy choice, as long as it does not generate any threat to their power position in the short term. But when group members have little control over policy choice and anticipate adverse redistribution effects in case of policy failure, they tend to be strongly risk averse.

Finally, group members' beliefs about and aspirations for the group as a whole and themselves within the group are an important input into risk acceptance or avoidance. A group that believes in its invulnerability to mistakes is more likely to make risky decisions than a group that believes in its vulnerability ('t Hart, 1990: 81–85; Thompson and Carsrud, 1976). Groupthink is only one source of a socially reinforced belief in invulnerability (Janis, 1982). Similarly, decisionmaking groups that have a record of successful decisions, or believe in the superior capabilities of their members because of past performance and achievements ("the best and the brightest"), or have faith that they are being guided by some divine power are more prone than other decisionmaking groups to take risky decisions. Arguably, this is an extreme case of the more general phenomenon of the illusion of control.

Shared beliefs may become group beliefs, that is, "convictions the group members (a) are aware that they share and (b) consider as defining their 'groupness' " (Bar Tal, 1990: 36). When beliefs acquire the status of group beliefs, their centrality increases, and they are easily available and therefore accessed frequently when the group deals with judgment and choice tasks. As individual members begin to share core beliefs and acquire a sense of common identity, the group's success becomes more closely associated with their self-identity and positive self-perception, and they become more motivated to favorably assess the group's past performance. The overestimation of past success then becomes the anchoring belief from which members infer their own expectations of self-efficacy, which are therefore overoptimistic (Cervone and Peake, 1986; Lant, 1992; March et al., 1991). The more exaggerated the positive perceptions of past performance, the greater the optimism about the group's ability to succeed in the future. This often unjustified optimism encourages members to take risks and reduces their propensity to pay attention to information that casts doubt on or contradicts their optimism, which enhances their overconfidence and raises the attention threshold for dissonant information through group-serving attributions.[55]

Evidently, a group-initiated risky policy could also result from shared group cognitions that would trigger a collective assessment by the group of being in the loss domain. These cognitions are less likely to be challenged in the group context than in an individual decisionmaking context, because of the added costs entailed in establishing a collectively shared view of probabilities, outcomes, and risky policy remedies.[56] Group cognitions are held with great confidence and tend to persist even in the face of discrediting information. The likelihood of reversing the resulting policy declines in proportion to the antecedent invested efforts. Feedback from decisions that threatens the validity of group beliefs may go unattended and contribute to policy entrapment. Accountability is a possible remedy (Kroon et al., 1992; Kroon et al., 1991; Tetlock, 1985; 't Hart, 1991: 84–85). It induces cognitive complexity and systematic attendance to relevant information and improves the quality of deliberations at both the individual and group level. In the latter case, the collective and especially the individual accountability of group members reduces the chances of thoughtless concurrence-seeking and thereby the likelihood that the group will display groupthink patterns and the related proneness to take risks. It also reduces the effect of deindividuation, a symptom associated with groupthink, which reduces concern about long-term consequences of the groups' action and desensitizes members to a whole range of risks that flow from these consequences. Yet the remedial power of accountability is limited because the cognitive complexity that it induces occasionally exacerbates biases in judgment (Simonson and Nye, 1992; Tetlock, 1992: 344–46, 358–59) and because certain organizational imperatives (to be elaborated later) neutralize the effects of accountability by reinforcing group-caused entrapment. More remarkably, the power of accountability to enhance cognitive vigilance is derived from decisionmakers' increased awareness of the fact that they will have to pay the price of failure. This implies that accountability will have only limited effectiveness where fear of failure is low. It could be expected that decisionmakers who have high confidence in their problem-solving skills in general, or in their ability to handle successfully a particular problem, will pay less attention to the im-

plications of accountability in the case of failure, because they will anticipate the risks of failure to be minimal. Lack of accountability-induced vigilance will be even more pronounced when decisionmakers believe that even
if they fail, there will still be an opportunity for rebounding, especially if
they have prior experience in getting out of tough spots and thus perceive
themselves to be survivors no matter how tough the situation.

To summarize, group contexts affect risk-taking preferences of individual members primarily in two ways. One is through a variety of social influences that are relatively independent of substantive change in the knowledge
base of individual members. Thus, the composition of the group, its social
atmosphere, and the decisionmaking procedures can be transformed to manipulate risk-taking preferences. The other way the group influences members involves the significant exposure of the knowledge base of individual
members to new information that causes substantive changes in their beliefs
and values, leading in turn to a change in their evaluation of utilities and
disutilities and in their judgment of probabilities. Successful manipulation
would have to be knowledge based rather than social or procedural.

The group itself is affected by a broad set of factors. The balance among
the factors determines the output of the group decisionmaking process. The
influences of social interaction on group members and group judgments and
choices are primary and important, but not exclusive. Social interactions operate through a number of channels. Responsibility diffusion may encourage
the acceptability of risk by reducing accountability. Persuasion, which is often but not always the tool of risk-acceptant members, and peer pressure toward compliance influences members either to accept risk or to avoid it entirely. Familiarization increases the amount of available information and
could result in polarization in either direction—risk acceptance or avoidance. Group discussion reveals and makes salient the main cultural values
and, depending on these values, also tends to induce polarization in the direction of risk acceptance or avoidance.

Still, the effectiveness of social influences and their scope depend on group
attributes. The distribution of personality traits within the group, which depends on its composition, determines how many members, if any, are more
or less likely to be amenable to influence. Group preferences can also be affected by political manipulation of the institutional arrangement or be shaped
by political needs, such as coalition maintenance. Finally, the group response
to risk is not only the result of aggregating individual preferences; it also has
to do with the sense the group has of itself as an entity with a distinctive identity, including a well-defined set of group goals and beliefs and a sense of
competence. Ironically, and for diametrically opposed reasons, both more
self-confident and less self-confident groups are likely to be inclined to take
risks—the former because they are confident that they can control process
and outcomes, and the latter because desperation has driven them to gamble
on boldness, although each individual expects to be sheltered from negative
consequences by the shared responsibility.

The Organizational Setting

Collective decisionmaking does not begin or end in the small group. Bureaucratic organizations play a role both before and after a choice among competing risk-taking preferences is made in the small group. As opposed to groups—whose main impetus is personal relationships, which shape cognition and motivation and which could be agents of change through the process of reaching agreement and a common ground—organizations help to buttress the status quo by imposing compliance with preset rules on their members (Burns, 1978: 295–302; March, 1994). Not that bureaucratic organizations are risk neutral. Rather, members are aware that it is their task to implement policy and that the choice of risky or cautious policies has important and diverse implications for various organizations, their own among them, in terms of material resource allocation and the chances for success or failure in carrying out tasks. Material resources and success or failure affect status and influence, the distribution of power among rival organizations, and future access to resources. Therefore, organization members view risky decisions, especially when risks are severe, as impinging on their vested interests, and organization members are aware of the critical importance of influencing the definition of a problem to serve their organization's interests. This is to a large degree why collective decisionmaking is first, though not solely, about choice among competitive framings of the problem. Different institutions advertise distinct risks and promote the taking of particular risks from among those advertised. Risks that do not have an institutional mentor are least likely to be attended to sufficiently, if at all.

In a competitive organizational environment, organizations have a vested interest in advocating policies and actions derived from their experience and repertoire of schemata (Anderson, 1987; Harris, 1994; March and Olsen, 1989). Knowing in advance which policies and actions serve their goals and suit their capabilities biases the acceptability of risk even before a risk estimation is reached. Determining what level of risk is acceptable beforehand could then have a backward biasing effect on risk assessment. "One could argue that what institutions do is to provide us with a series of vivid experiences that then, through the availability heuristic, make us more likely to overestimate some risks and to underestimate others. Institutions might similarly supply stereotypes that make some cases seem more representative than others, frame choices in characteristic ways, and suggest reference points" (Heimer, 1988: 499).

Factors that induce risk aversion in individuals may be even more prominent in the organizational context. The multiple layers of accountability and personal responsibility increase the tendency to avert loss. Organizations tend to evaluate the performance of employees at short intervals. This encourages taking decisions incrementally, one at a time, rather than treating them as instances of a broader category; and future decision opportunities are neglected, enhancing a built-in aversion to taking risks. Many organizational decisions are perceived as both unique and major at the level within the organization at which they are initially made, thus inducing caution even if they could be considered small and recurring for the organization as a whole. The organizational structure that reflects multiple priorities and preferences and has established formal and informal processes for coordinating and aggregating these preferences is likely to mellow the preference for risky decisions at the top of the organization. In essence, complex organizations are formed to prevent undesirable and unpredictable outcomes. Risk taking could involve both (Demchak, 1991: 20–27, 31). Yet organizations differ in their degree of caution, just as they differ in organizational culture, that is, in their system of shared meaning. Common beliefs and norms about the optimal means to achieve goals and, in the case of military organizations, the beliefs and norms that constitute a shared paradigm about the optimal way to fight and win a war, affect the degree of risk aversion (Legro, 1994; Pfeffer, 1981). At different times the same organization could demonstrate different preferences for taking risks, depending on the prevailing shared paradigm.

The reality of organizational behavior is therefore complex. In spite of an inherent disposition toward risk aversion, organizations are liable to be Janus faced and display risk-acceptant behavior. Such behavior has intraorganizational and extraorganizational motivations. The former are associated with the triggering of an optimistic bias (Kahneman and Lovallo, 1993). In the intraorganizational debate over competitive options the probability of accepting one option and rejecting its competitor is higher if its promoter forecasts highly favorable outcomes and underestimates risks of adverse outcomes. The forecasts often become statements of commitment, which induce motivated optimistic assessments of future accomplishments. To express pessimism about the future performance of the organization can be readily interpreted as disloyalty. Avoiding that opprobrious label causes self-censorship, so pessimistic views are suppressed.

Similarly, organizations, as policy followers, are induced to behave riskily after the country's decisionmakers have made a risky decision and it has become the task of the organizations to implement the resultant policies.[57] At this point, an organization may find the policies useful and even worthy of expansion, if expansion increases its influence and the resources that it commands. Indeed, interorganizational rivalry and competition in implementation can produce an addiction to action that becomes a source of pressure to expand even risky programs. With expansion, as the stakes increase, program reversal becomes less likely. At this point, the common attributes of organizations contribute to and reinforce policy persistence and escalation even in the face of negative feedback information. Decisionmakers in organizations are caught in situations involving a string of choices among disutilities. "Each decision actually makes the situation worse; each becomes a factor in the propagation of worsening utilities surrounding further choice. The actors in these cases have become caught up in the peculiar 'logic' of what might be termed a 'decisional trap' " (Schulman, 1989: 33–34). The problem is that the decisionmakers engage in a spiral of progressively greater risks on behalf of what is essentially the same utility. "As probabilities of obtaining the initial utility decline over each decision, the 'expected value' of each subsequent decision must decline over its predecessor" (p. 36).

A number of reasons account for this peculiar behavior. The most obvious is the routinized nature of organizational behavior. Once established, organizations carry out their missions according to their standard operating procedures (SOPs). To transform the set behavior sequence is very difficult; SOPs are not only guides to action but also constraints on unhindered cognitive operations. In fact, they are a source of anchoring and lack of adjustment; they provide ready-made responses to internal or external stimuli, so that the stimuli can be interpreted and responded to with little reflection. In this sense, SOPs are rational only if they hold a degree of built-in flexibility that allows adaptive responses. But SOPs are usually inflexible. Paradoxically, the more complex and uncertain the environment for which the SOPs are programed, the less likely they are to be flexible. Such environments generate SOPs that represent a large investment of time and intraorganizational bargaining. Once an SOP is established and routinely utilized, organizations become committed to it and avoid making necessary adjustments. It is also true that the more complex and ambiguous the environment, the higher the perceived probability of failure. Because substituting an untested mode of operation for an established and familiar routine—which has already gained acceptance and has some record of past success—is perceived to be a gamble, conservatism prevails.

Governmental nonmarket organizations tend to reward appropriate decisionmaking processes.[58] Market organizations, in contrast, reward outcomes, placing substantially less emphasis on how decisions were made and basing rewards on how well the decision turns out.[59] Emphasis on process control reduces the perception of risk for the individual decisionmaker (Sitkin and Pablo, 1992). There is less uncertainty and more predictability ("follow the correct process and you will be safe"). Therefore, individuals following SOPs may lead organizations into risky enterprises because they are not intimidated by unanticipated outcomes and are confident that as long as they follow the appropriate procedure, they are on safe ground, even if the outcome is unfavorable.

In general, organizations are slow to adapt cognitively and operationally to environmental cues; they demonstrate defensive conservatism, acting in the belief that the future is only marginally different from the past and present. Information that indicates a need for nonroutinized change and reorientation is ignored, belittled, or explained away. Preset programs and SOPs construct reality to fit the prevailing assumptions through their impact on the processing of information. Built-in inflexibility is part of organizational culture.

Another cause of organizational insensitivity to the need for behavior change is the lack of a point of comparison that would allow the recognition of policy failures. Unlike market organizations, which are oriented to profit making, nonmarket organizations do not have clearly defined and measurable goals that provide an obvious consensual standard by which their performance can be judged. To construct agreed-on measures of success or failure at each stage of program implementation is difficult or impossible. Failure or success, then, has to be dramatic to be conspicuous. Routine feedback from performance, because it is ambiguous and because an organization is inertial, is likely to cause little or no adjustment; indeed, feedback is manipulated to serve organizational inertia. The ambiguity of feedback information about foreign policy performance makes that information particularly susceptible to manipulation, so that failure is recognized only when the time to adjust effectively or without great costs has already passed. In addition, measuring how policies have fared poses a severe problem because each organizational unit measures performance according to what it believes to be important; these measurements are seldom integrated into a uniform evaluative system. The more numerous the government agencies that must agree on the appropriate policy for goal attainment, as is the case with most ill-defined problems, the more likely it is that shared expectations about policy performance will be general, nonspecific, inconsistent, and given to ambiguity. Feedback becomes fragmented and manipulable, and the actual failure to achieve significant goals goes unobserved (Christensen, 1985; Hermann, 1990; Vertzberger, 1990: 205–16). In these circumstances, the recognition of the need for change will be slow to penetrate, making entrapment and escalation more likely than withdrawal.

A known commitment to a particular policy at the highest levels in an organization acts as a disincentive for lower levels to pose vigorous challenges to the policy. Information collection and dissemination within the organization will be selective, reinforcing commitment to the policy and diverting attention from disconfirming information. Again, the result is entrapment. For example, the commitment of the Bush administration to pursue friendship with Saddam Hussein may have discouraged vigorous and probing questions by intelligence and policy specialists regarding the continued supply of considerable economic and indirect military aid to Iraq (George, 1993: 42).

Even clear negative feedback information is not always effective in instigating policy reappraisal, because information that indicates decline in organizational performance induces conservative, status quo–maintaining interpretations and decisions (Levitt and March, 1988; March and Olsen, 1989: 60). By implication, when risky decisions taken in the past turn out badly, organizational response is likely to be, not innovation and reevaluation of current policies, but entrenchment and the normalization of deviance. Moreover, as conditions of adversity trigger restrictions on information flow and processing, decisionmakers will be partly or fully unaware of negative feedback and hence unlikely to consider change in current policies. The tendency is reinforced by policy centralization, which is encouraged by poor performance (Singh, 1986; Vaughan, 1996). The combined effects of centralization, restriction on information processing, and conservatism produce a powerful incentive for inertia. When a current policy involves risk taking, as with military intervention, organizations are likely to respond first with escalation of the unsuccessful policies, rather than with innovation, which could lead to de-escalation.[60]

In fact, organizational behavior inadvertently reinforces other causes of individual and group-caused entrapment. The slow process of learning and adjustment to negative feedback creates a self-defeating momentum. As time passes and the old policies lead political decisionmakers to invest more and more resources in solving a problem, the decisionmakers may ignore suggestions to adjust. The cost of admitting responsibility for expensive mistakes has by then become too high to allow it, either practically or psychologically.

To summarize, organizations may attempt through their representatives in a decisionmaking group to direct the group toward a particular assessment of and disposition for risk that is congruent with organizational goals. Their ability and motivation to act this way stem from the facts that organizations are a source of information to the group and that once a decision has been made, organizations are responsible for its implementation. The nature of organizations and organizational information processing and action tends to introduce selective attention to particular risks and make an organization prefer responses that are part of its repertoire, as well as prefer risk avoidance. Once a policy is implemented, policy persistence is a common preference even in the face of negative feedback and increased risk, raising the prospect of entrapment in a failing policy.

Conclusions

I depart sharply from the rational choice approach to risk. The decisionmakers who emerge in rational choice theory—the proverbial Superman type—rarely err in complex computations and, if they do, learn quickly and in time to adjust. They assess outcomes, judge probabilities, are confident of their estimates, calculate expected utilities, make the choice that maximizes expected utilities, and are ready in an instant to take up the next mission.

In contrast, the decisionmakers that emerge from this study, based on the sociocognitive approach, are not sure what their goals are; they are biased, lack confidence in their judgment, are short on time and long on doubt, pulled in different directions by cognitions, hidden motivations, personality needs, cultural socialization, the organizational setting in which they operate and influenced by the attributes of and interactions in the small decision group of which they are a part. Consequently, they agonize over decisions. They search for ways to deal with complexities and come up with solutions that are occasionally effective but that, in many cases, have distorting effects on the rationality of information processing and choice.

Decisionmaking is messy, not linear, and risks are not taken in a single step

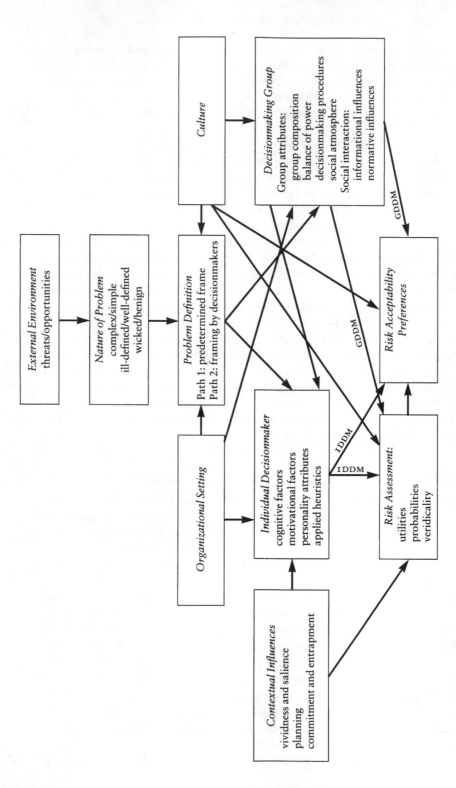

Fig. 2. Risk Assessment and the Formation of Risk Preferences. (IDDM = individual-dominated decisionmaking process; GDDM = group-dominated decisionmaking process.)

but in an incremental process. Decisionmakers change their minds a number of times before eventually deciding to take risks. Both assessing risks and choosing risks involve inner conflict, but conflict is more severe in the latter case because action implies greater commitment than judgment does. The relative influence of group, cognitive, and other variables often fluctuates during this predecision stage. Rarely do policymakers make an immediate intuitive decision and stay with it. Figure 2 presents the process by which risks are assessed and a preference for risk taking is formed, but the causal variables there do not necessarily operate in chronological or any other sequential order. They may in some instances, but in many cases, some of the variables operate simultaneously and others operate sequentially. The graphic presentation is simply a heuristic way to clarify the interaction patterns among variables and does not necessarily imply a highly structured process.

Risk taking is the result of either an individual-dominated decisionmaking process (IDDM)—as in, for example, the United States in the Grenada intervention—or a group-dominated decisionmaking process (GDDM), as in, for example, the Soviet Union in the Czechoslovakia intervention. The same variable-sets push and pull groups and individuals toward specific risk assessments and risk-taking preferences, though not necessarily in a linear process. The end result, the sum of the push and pull of the various influences, is situation specific. In this chapter I have mapped these influences and their direction under different circumstances, thus providing a loose but realistic contingent theory of how risks are assessed and how the preference for risk taking is formed. I believe that this is the only practical approach.

It should be obvious by now that the description and explanation of risk assessment and the formation of a risk-taking preference defy parsimony. Severe risks, such as those posed by foreign military intervention, are extremely complex and hard to model and quantify, and their time boundaries are ill defined. No single cause effectively explains either risk assessment or the formation of a risk-taking preference; instead, explanations require a broad sociocognitive approach, namely, integration of a number of interacting clusters of variables: problem framing; contextual effects; underlying cultural attributes; decisionmakers' cognitive, motivational, and personality traits; attributes of the decisionmaking groups and; organizational setting. These sets of causal variables shape the content and precision of risk assessments. Risk-taking preferences are either formed in response to risk assessments (deliberative risk taking) or shaped by cultural, individual, and social attributes (dispositional and social risk taking).

The product of this process, a risk-taking preference, is not necessarily enduring (March, 1988), even if it does not change easily. There are two main ways in which preferences for risk acceptance or risk aversion might change. One occurs when decisionmakers receive and attend to new information that increases or decreases their perception of risk, thus making a policy based on their perceptions more or less attractive. The other way occurs when the taste for risk taking changes. Sources for change in the acceptability of risk can be intrinsic or extrinsic. There can be a change in attitude toward risk taking in a particular context, such as an ego-defensive response to criticism, or a change in the social and political environment, so that it is more or less tolerant of risk taking. After the Vietnam War, for example, the level of acceptable risk for foreign military intervention was substantially diminished, because decisionmakers were aware that the American public would not tolerate the costs of another such intervention.

People can adapt and learn, but doing so is not a linear process nor a systematic, well-organized one. Even when the required information is available and the motivation for learning is there, learning should not be taken for granted and will still be much less than optimal. In a perfect world populated by ideal decisionmakers, the decisionmakers will improve their performance by learning the right lessons about degree and timing. But in real life, people tend to learn too much or too little, either overlearning and performing tasks thoughtlessly, without considering the uniqueness of a situation and a task, or ignoring important lessons from experience, thus missing opportunities to perform better in the future or to adjust their performance in ongoing tasks. With overlearning, they will take or avoid risks according to inflexible rules applied without careful consideration of the circumstances; with underlearning, they will fail to adjust the level of risks that they take in spite of warning signals.

People also learn too quickly or too slowly. Intuitively it may seem that fast learning is always an advantage and slow learning always a disadvantage; but that is not so, because what one learns depends on when one learns. Individuals and organizations extract lessons before receiving feedback information on the results of their decisions (March et al., 1991). They may repeat a decision and perpetuate a policy because they learn to enjoy the thrill of taking action and the sense of control and power that it gives them and because they have overoptimistic expectations about outcomes. Similarly, they may learn to avoid further decisions or to withdraw from a given course of action, not because it failed but because they have found that risky decisions produce stress, inner conflict, and a sense of commitment. They may also reach conclusions without waiting to see the long-term consequences. Premature learning can therefore be extremely counterproductive. But the opposite, learning too late, also has adverse consequences. By waiting for all the feedback information to become available before learning and applying operational lessons, opportunities for adjustment are often missed, and costs accumulate. Optimal learning should be an ongoing process in which feedback information is continually monitored and lessons are continually adapted, incorporated, and revised. In other words, decisionmakers must learn how to learn if they want to avoid the adverse consequences of either cognitive and operational conservatism or trial-and-error incrementalism.

CHAPTER 4

Foreign Military Intervention

NATIONAL CAPABILITIES AND CONSTRAINTS

Large-scale foreign military interventions are risky and costly ventures, at times much more so than decisionmakers originally anticipated. Because they test the limits of national capabilities, astute decisions should be made with three sets of national capabilities in mind: military, economic, and political. Decisionmakers analyzing these capabilities should avoid the temptation of making crude unidimensional quantitative comparisons of power discrepancies between intervener and target. Planning intervention requires not only a detailed analysis of the adversary but also a candid and penetrating assessment of the weaknesses of the intervener's own resources, society, and institutions. Each new stage in climbing the ladder of escalation should be preceded by the same comprehensive political-strategic planning. Only then can intervention serve as an effective instrument of statecraft.

The capabilities required for successful intervention are those that allow for effective and simultaneous management of both the domestic and external environments. Decisionmakers must recognize that these environments are interdependent, as are the various dimensions of capability. A decline in management effectiveness vis-à-vis the external environment will directly affect the ability to carry out the intervention policy successfully; failure will feed back into the domestic system and exacerbate the problem of domestic management, and vice versa. The management of both environments requires the employment of military, economic, and political capabilities. Below I discuss the nature of the risks and constraints associated with each.

I begin by defining foreign military intervention, then analyze how the assessment of capabilities should be incorporated into the calculus of intervention decisions by taking into account both external and domestic considerations. Next I analyze the antinomic effects that time has on outcomes of intervention policies; the reality and illusion of power seen from the intervener's point of view; and the resultant difficulties in establishing and maintaining an optimal level of intervention. The chapter concludes with some cautionary policy-relevant advice to decisionmakers contemplating intervention as an instrument of statecraft.

A Conceptualization of Foreign Military Intervention

The term *intervention* refers generically to a broad range of activities that encompass many, if not most, of the activities directed by one state toward another. The activities can be political, diplomatic, economic, or military; they can have various levels of intensity and scope, and they represent a balance between the intervener's interests, power, and opportunities and the structural vulnerabilities of the target state (e.g., a weak but domestically stable state offers fewer opportunities for intervention than an unstable one does) and its determination to bear the costs of resisting the intervener. The balance among all these factors determines the initial calculations of comparative risk or limited liability—that is, judgments based on cost-benefit analysis related to the likelihood of success and its cost, which in part determine the willingness to intervene, as in the subjective expected-utility (SEU) approach (Beloff, 1970). Yet intervention decisions involve a much more complicated set of considerations and are often not the result of rational calculations that the expected utility approach posits for decisionmakers.[1]

Before discussing the calculations involved in intervention, we need to narrow our focus to a well-defined type of intervention; the broad term can be equated with practically all forms of international conflict. Without a precise definition, analyzing intervention as a distinctive category of behavior is of questionable value (Rosenau, 1968). Let us focus, then, on large-scale overt foreign military intervention.[2]

Conceptually defined, foreign military intervention is coercive state-organized and state-controlled, convention-breaking, goal-oriented activities by one sovereign state in the territory of another, activities directed at its political authority structure with the purpose of preserving or changing that structure, affecting thereby its domestic political process, and/or certain of its foreign policies by usurping its autonomous decisionmaking authority through the use of extensive military force. Operationally defined, foreign military intervention involves the following indicators: the direct, overt commitment of uniformed, combat-ready military formations, including ground forces (i.e., battalions, brigades, and divisions rather than military advisers or irregular forces used in covert operations), to conduct, when necessary, conventional ground warfare operations that are, in expressed purpose, continuous but limited in time.[3]

Intervention may be hostile, as when opposing a target government or aiding rebels (e.g., Soviet intervention in Czechoslovakia); friendly, as when supporting a government or opposing rebels (e.g., U.S. intervention in Korea); or neutral, as when influencing particular target-government policies, but without any intention to support or oppose a particular regime (e.g., Israel's initial intervention in Lebanon).[4] On the part of the intervening state, such acts involve nonroutine decisions and entail multiple objectives that are both concrete and abstract. Overt foreign military intervention decisions and their implementation, which have highly visible consequences, are made by decisionmakers at the highest level, and they impose on those decisionmakers a substantial commitment owing to the high stakes and risks involved and the transparent lines of responsibility and accountability. Throughout this study the term *intervention* refers only to overt large-scale foreign military intervention; all other forms of intervention will be referred to as "involvement."

The risk attributes of foreign military intervention affect the interests that trigger this type of statecraft. Obviously, the interests involved are perceived as vital national interests, because of either their substantial value or their reputational implications (e.g., as a symbol of national resolve or credibility). Rarely is intervention triggered by a single interest; it is likely to be triggered and then justified by multiple interests in order to make the stakes compatible with the high risks involved and enable decisionmakers to form a domestic consensus in favor of the risky venture. Consensual support will reduce their direct responsibility for failure if it occurs.

Intervention, then, is likely to be triggered and justified by some combination of the following interests: territorial acquisition, protection of political, religious, or ethnic groups, protection of economic interests, protection of diplomatic interests (embassies, diplomats), protection of military-strategic interests (military bases, access to important air, sea, and land lines of communication), preservation of or change in the regional power balance, and ideological motivations (Pearson, 1974a). In reality, to draw a clear line between these types of interests is not always analytically easy, for they are often intermingled. Protection of an ethnic or religious group, for example, can be related to an ideological commitment or can be perceived to serve political, economic, and strategic interests, or both (e.g., Israel's intervention in Lebanon in 1982).

Another problem associated with the definition of the national interest is to operationalize interests in terms of specific objectives. The objectives will then dictate the rules of engagement. In the U.S. intervention in Somalia in 1992, the dilemma was whether to attempt to disarm all Somali factions or to focus on delivery of food. In the end, the decision was to do the latter and seize arms in a very limited and selective way (Haass, 1994; Mazarr, 1993). Furthermore, the interests of interveners and their clients' are not necessarily congruent but are likely to be complementary, so, most of the time, there will be short-term coalitions of convenience rather than long-term alliances. Thus, even a successful intervention can turn sour. Success on the battlefield does not guarantee the intervener favorable treatment in the long run.

The implications of these attributes of large-scale foreign military intervention decisions for the process by which such decisions are made and the manner in which decisionmakers deal with feedback information were discussed in preceding chapters. In the comprehensive analysis of specific cases of intervention provided in later chapters, we will look at constraints and influences on intervention decisions at all three levels: the international system; the state (intervening and target); and the decisionmakers.

Military Capability

Military capability is the core resource for overt foreign military intervention. It has complementary dimensions: combat capability and logistic capability. The first subdivides into considerations of (1) force structure, (2) hardware, and (3) strategy.

Decisions on intervention necessitate careful considerations of the size and composition of the force structure. Specifically, the intervening country must determine whether it has a sufficient number of soldiers under arms to sustain intervention as well as all its other commitments, and whether it will be able to accomplish its goals by depending solely on regular troops or will have to call up reservists. Heavy reliance on reservists may pose problems regarding the availability of adequate numbers of highly qualified and trained soldiers.

> It is inevitable that most [U.S.] army reserve units would perform their missions less well, and at a greater cost in casualties (some resulting from sheer physical stress) than the active duty units might. Small wars frequently entail fast deployments to extreme climates (e.g., the Persian Gulf), and it is absurd to think that reservists can be as physically hardened to such challenges as their active duty counterparts, or that they can be equally practiced in their individual skills. (Cohen, 1986: 299–300)

The scarcity of regular troops may constrain both the ability to get involved in large-scale overt military intervention and, once involved, to escalate the intervention. An intervention that requires mobilization of reservists raises the risk of unwelcome social effects—first among them, the arousal of public sensitivity to the problem. Public scrutiny of the policy may become a direct constraint on the decisionmakers' flexibility and freedom to escalate, and it may have further indirect effects due to the increased importance of domestic legitimacy for intervention.

The use of reservists has two additional social implications, namely, social expectations and economic costs. The first is less tangible but is nevertheless very real. Soldiers, especially professional soldiers, are expected to be prepared to be casualties; such losses are usually considered (within limits, of course) to be within the realm of acceptable risk. Not surprisingly, "polling data indicate that until 1967 the Vietnam war was not particularly unpopular. This was particularly true in the early 1960s, when the troops stationed there were primarily volunteers and officers" (Cohen, 1985: 109). But when intervention leads to high rates of casualties among civilians called up for military duty, the level of sensitivity concerning casualties becomes very high, and the realm of acceptable risk quite limited.

Between the two extremes of an all-volunteer professional military force (e.g., U.S. intervention in Lebanon in 1958 and French interventions in Chad in the 1980s) and an intervention force depending largely on conscripts and reservists (e.g., Israel's intervention in Lebanon in 1982) lies a third option—that of a military force based on a mix of volunteer career soldiers and conscripts (e.g., Soviet intervention in Afghanistan and U.S. intervention in Vietnam).[5] Here, the level of sensitivity depends largely on the degree of domestic legitimacy for the intervention and the ratio of career soldiers to conscripts in the intervention force. It is also affected, as in the case of a reservist-based force, by perceptions of the fairness of burden sharing. If the burden is perceived to be equally shared by all sectors of the society, the tolerance for costs is high, particularly in open societies, where burden sharing is transparent. It is no wonder, then, that "a system of selection which offered a good chance of escaping military service [as was the case in the Vietnam War] fostered more opposition [in the United States] than an all-inclusive draft [as was the case in the Korean War]" (Cohen, 1986: 283).

Interventions that require large-scale use of reservists have direct, disruptive effects on the economy of the intervening country. In the longer run, they may also have the indirect effect of depleting economic resources; an economic slowdown due to mobilization of skilled workers may eventually reduce the ability to sustain a protracted intervention. The economic hardships may, in turn, be translated into domestic disaffection with the intervention and increased opposition to the intervention policy.

When a government is, for domestic political reasons, reluctant to call up reservists, its ability and incentive for extreme escalation are restricted, necessitating speed in achieving its objectives to avoid an extended intervention that will require larger numbers of troops. At the same time, if the target realizes the intervener's domestic political vulnerability, it will have a higher tolerance for the cost of resistance and a tendency to be intransigent, making goal achievement by the intervener less likely within a short period and further increasing the risks and potential costs of the intervention.

The second component of combat readiness is hardware. Decisionmakers must determine whether the intervening military force has the quantities and types of hardware required by the task and by the geographic features of the target country (terrain and climate). Availability of military hardware depends on both demand-side and supply-side conditions. The demand-side conditions are a reflection of the military doctrine and strategy that the armed forces have adopted. Doctrine and strategy take into account the types of military operation that military planners anticipate. Limits on available troops and financial resources make it impractical to acquire specialized hardware and to train the military to operate in all possible contingencies and geographic locations, just those anticipated as the most likely. If the military considers a particular geographic deployment and type of intervention less likely, it will accordingly be less prepared for it.

Modern armies involved in intervention operations in the Third World may find themselves at a disadvantage. "The texture of these landscapes hampers mechanized operations and often provides a multiplicity of niches for guerrillas, requiring a great number of regulars to carry out successful sweeping actions" (O'Sullivan, 1987: 34; see also Quinlivan, 1995–96). This lesson was learned by India in its intervention in Sri Lanka in 1987. In such cases, the cost of intervention is high in terms of casualties and lost hardware during the learning and adjustment stage. The increased amount of time and other resources required, but not necessarily available, negatively affect domestic support for the policy.

The supply-side conditions for hardware availability apply mainly to countries that do not produce all their own military hardware, depending instead on what is available either in the arms market or through military assistance programs. When available hardware is inadequate for an intervention, adjustment entails growing dependence on external sources of supply, notwithstanding the increased financial costs of such an adjustment. The intervening country must pay the political costs of increased dependence, as well as satisfy the demands of external legitimacy for the intervention imposed by allies and other military hardware suppliers as a condition for assuring a steady flow of hardware. Building up a large inventory is one way to avoid dependence on foreign suppliers. Even then, when intervention exceeds the limits of initial plans, the accumulated attrition of hardware, including ammunition and spare parts, can severely limit the intervening government's freedom of choice. Thus, assessing and matching hardware with the scope and time frame of a planned intervention is an important and difficult task. Decision-makers cannot afford to be fundamentally wrong about either the rate of hardware attrition or the length of time that the military operation will require.

As a rule, preparedness for a particular intervention depends on its predictability. To the degree that a country has a well-thought-out political-military doctrine and strategy and a well-defined set of goals and priorities, it is better situated to anticipate likely types and locations of intervention and thus make appropriate investments in hardware. Adjustment of hardware to the intervention requirements can be time-consuming. Even if a willing supplier is available, the hardware cannot usually be bought off the shelf, and there will be some delay until production gets under way and the hardware is supplied. It also takes time to train military personnel in the effective use of the new hardware. In other cases the situation requires new technological solutions that take time to develop and implement. However, military hardware is often multipurpose and can be used for different military objectives. Even if the location and type of intervention have not been anticipated, intervention will not necessarily require a major hardware adjustment. Generally, then, where hardware acquisition is time-consuming and requires readily available and steady suppliers, the economic and political costs of intervention increase.

The third element of combat capability, strategy, binds the first two—force structure and hardware—in a goal-directed and cost-effective manner.

Strategy is the relation between political ends and military means. It is a set of prescriptions on how military forces should be structured and employed to respond to recognized threats and opportunities in order to attain predetermined goals. Intervention strategy demands not only foreseeing the contingencies for which military force may be required and used but also foreseeing correctly the overall strategic setting and the duration of military deployment. For instance, it has been assumed that the conditions under which the U.S. Rapid Deployment Joint Task Force (RDJTF) can be deployed range "all the way from a relatively stable situation in all geographical areas except the Persian Gulf to one wherein global tensions are high and hostilities involving the United States appear imminent in, for example, NATO, Europe, Korea and the Persian Gulf" (Fabyanic, 1981: 351).

Strategic inadequacy can occur for a number of reasons. The most likely cause is that the intervention and the types of military operation involved were not anticipated as a probable contingency, so no forethought was given to building up an appropriate force structure. In the 1960s, for example, having focused on preparations for conventional and nuclear wars, the American military found it difficult to adjust to fighting a subconventional war, the kind usually associated with intervention. The military was not psychologically ready or trained for intervention, nor was the American force structure appropriate for carrying out a combination of military, civilian, and police operations in target countries. Because of the standardization of military training, most officers had little training in fighting counterinsurgency wars, which required both special military and social engineering skills. The military is also sometimes reluctant to get involved in small wars that absorb resources that it believes to be best invested in what it regards as its main task — winning major wars. Not infrequently there is a discrepancy between plans for small wars, imposed on the military by the political leadership, and military allocation of available resources — the "strategy-resources mismatch" (Dagget, 1989).

Another problem relating to the efficiency of strategy has to do with the consequences of self-imposed constraints on the use of particular weapon systems and their geographic deployment — constraints intended to curtail the risks of escalation and third-party intervention. One example is to refrain from attacking the territory of a third party even when it provides a sanctuary to hostile military forces. Closely related are moral, value-based, self-imposed restraints. In an international system emphasizing the norm of sovereignty, intervention is often counternormative, and an intervening power can therefore only rarely unleash its full military power for fear of being denounced as an aggressor and possibly losing external legitimacy for the intervention policy. Thus, its strategy is often based on the use of limited force at the cost of relative inefficiency.

Moreover, unlike wars, which are usually carried out in well-defined military zones, intervention operations are executed in a mostly civilian environment. The lines between civil and combat zones, where different rules of conduct apply, are often fuzzy or even nonexistent. The question Who is the enemy and who is a noncombatant? often has no clear answer. The moral dilemma raised is very serious, especially for democratic liberal societies. The ambiguity of where the rules of war apply and where civilian rules of conduct apply imposes significant restrictions on the way military power can be deployed and on the rules of engagement (cf. Hosmer, 1987: 57–64). An army untrained and unprepared to deal with problems of this kind will find itself denied the fullest use of its skills and resources to reach its objectives, and yet it will be expected to accomplish its assigned mission. At the same time, it could face growing external and internal criticism that undermines morale, further eroding its ability to carry out its mission effectively and within an acceptable time frame.

This moral dilemma has far graver implications when there is an asymmetry in normative rules of conduct between the intervener and the local adversaries. An intervener that cares about human rights and therefore, with the local allies, observes strict humanitarian rules of conduct in dealing with the local population will suffer significant comparative disadvantages if the adversaries do not share similar concerns. The local population might be terrorized into supporting the less humanitarian side and thereby contribute to its victory (Tullock, 1974: 56–58). In fact, for the intervener the choice may be between, on the one hand, self-imposed moral restraints, which bring the benefit of support from the domestic public, and, on the other hand, unrestrained behavior, which risks losing support at home but compels the population in the target state to support the intervention. The assumption with the latter choice is that if the policy brings success, criticism at home can be contained. But cases of protracted intervention pose the threat of the moral corruption of the military, even where the intervener has clear self-imposed normative constraints for military operations. The stress of operating for long periods of time under life-threatening conditions in an often hostile society could cause even well-trained and disciplined military units to behave in a manner that stands in total contradiction to the moral values and ethics of their own society. This was evidenced at Mai-lai in Vietnam and at Sabra and Shatilla in Lebanon. Such moral breaches undermine both domestic and external legitimacy for the policy of intervention.

Technological innovation can have important implications for reducing some political risks associated with intervention. Greater precision with firepower, which can be exploited to minimize civilian casualties, especially in a friendly population, could remove or reduce a main cause of domestic and foreign criticism of intervention. In the U.S. interventions in Panama and the Persian Gulf in 1989 and 1991, respectively, this was demonstrated by the use of Stealth bombers and Apache helicopters. Stealth technology also improves the survivability of the most expensive and technologically advanced weapon platforms, such as fighters and bombers, thus reducing the economic costs of limited military operations while increasing their effectiveness. Increased survivability also reduces military casualties. The combination of a diminished risk to civilian populations, reduced economic costs, and a lowered risk of military casualties reduces the inhibition to intervene.

A critical aspect of limited military operations is their control and manageability by the central political authority. Intervention, being an instance of coercive diplomacy, "requires centralized management of intervention instruments and signals of intent. It calls for speedy and accurate communication vertically through the institutional hierarchy and a high degree of compliance with central directives. Communication channels must be available and signals must be clear" (Bobrow, 1977: 103; see also Fabyanic, 1981). These conditions allow for the optimal combination of force and diplomacy, prevent uncontrolled escalation, reduce the probability of intervention by a third party, and limit the risks of delegitimization of the policy at home and abroad.

To achieve a controlled and quickly adaptable military environment that is highly sensitive to political needs and changing developments, decision-makers require an effective system of command, control, communications, and intelligence (C^3I). When this is lacking, it is much easier for field commanders to misunderstand the goals and directives of the political echelon or even to overzealously implement or misinterpret orders for the purpose of serving personal and organizational interests and ambitions; unplanned escalation may well result. Yet a very tight C^3I system in the turbulent environment of the battlefield suppresses boldness and initiative on the part of field commanders, which often make the difference between success and failure (Lanir et al., 1988).

When two or more interveners have the same or complementary goals or assist the same party, some form of military cooperation between the interveners tends to emerge, as in Britain, France, and Israel in Suez in 1956 (Dunér, 1983a, 1985). To be effective, that too requires a robust C^3I capability to assure not just vertical control (within one's own military forces) but also horizontal control—that is, coordination of the goals, strategies, and operations of the military forces from different countries. Differences in doctrine, command structure, technology, and language complicate the problems of C^3I significantly.

Even a formally established unified command does not fully obviate the problem, because national sensitivities may limit the level of actual coordination. A unified command is usually very cumbersome, slow to produce decisions and even slower in implementing them. Because the unified command is often improvised and untried—that is, the organizational structure is a quick-fix response to an acutely pressing need—it is rarely highly effective, especially in the initial phases, and may have to go through a long trial-and-error learning process before it becomes effective. The quality of vertical and horizontal C^3I capabilities, both technological and organizational, is, then, an important element in assessing the risks of unplanned and uncontrolled escalation.

The complementary element of military combat capability, one that also affects estimates of the economic costs of intervention, is logistics. The logistics of intervention concern the ability to transport soldiers and matériel into battle and then to support and sustain combat operations as long as necessary. The crucial issues in logistics are distance from the theater of operations and technological ability to provide an efficient, fast, and uninterrupted flow of personnel and matériel. There is a trade-off between distance and technology: the closer the intervention target, the less advanced the transportation technology and capacity have to be. But technology can compensate for distance, so long distances, though a hindrance, need not entail a combat disadvantage.

A realistic assessment of the logistic burden of distance has to take into account four factors: (1) distance in kilometers, (2) the physical characteristics of the terrain and the constraints that they impose on mobility of supplies and troops, (3) the technology and transportation capabilities available to the adversary, and (4) vulnerability to interdiction by the adversary's combat forces. If there are more than one adversary, the factors should be analyzed for each adversary in order to uncover comparative advantages and disadvantages. American assessment of USSR capabilities for intervention in Iran in the 1980s indicated that the Soviets would face serious problems if they tried to invade Iran because there were only three axes of movement from the Soviet Union into Iran by land, because the transportation route along each axis was very limited, and because each route—road or rail— had to pass over difficult mountain terrain, making each route vulnerable to blockage (Epstein, 1987; Wohlstetter, 1968).

If the surprise and shock effects of military intervention are to be amplified and used advantageously, the intervener must commit adequate forces on short notice. A long distance implies that these forces must, at the early stages, either be largely self-contained or supportable by air, for time will probably be insufficient to establish sea or land supply lines. An airlift is difficult because it "requires elaborate and extensive aerial port facilities—runways, loading and unloading equipment, and refueling capabilities [e.g., staging facilities en route], including extensive fuel reserves" (Pickett, 1977: 143). Often political circumstances quickly narrow the choice of en route staging facilities to one. Depending on a single facility and on very vulnerable cargo planes, which are exposed to attack by practically any jet fighter, no matter how technologically backward, and by portable surface-to-air missile systems during takeoff, landing, and low-altitude flight, introduces an additional risk factor that jeopardizes the achievement of the intervention objectives at an acceptable cost.

A possible trade-off for distance is the pre-positioning of military hardware in a country bordering the anticipated target or in the territory of the target itself. Alternatively, an intervention force may secure access to permanent bases at mid-distance points or even in the target state. For example, American military bases in Europe have made the logistic tasks of supporting intervention in the Middle East much easier. Yet if the country in which the bases are located fears provoking its neighbors, third parties, or its own domestic opposition (McNaugher, 1987), the use of the facilities may be fully or partly denied to the intervener in times of dire need. If the bases are used against the host country's wishes, long-term prospects for access to the host country may be jeopardized, thus raising both the long- and short-term costs of intervention. There is an advantage, then, in having bases for which sovereignty is not at issue (such as the U.S. base on the island of Diego Garcia) or, alternatively, in pre-positioning soldiers and supplies on ships at sea, or even implementing the technologically feasible idea of an artificial mobile base at sea (Vlahos, 1987).

Another important aspect of distal intervention has to do with the combined effects of distance and the force structures of the intervener and the adversary. Rapid deployment to distant regions can be achieved only by air-mobile forces; these are light forces that after landing are mostly foot mobile and do not possess great firepower but are charged with the difficult tasks of forcing entry and establishing a defensible perimeter. If the forces in opposition are equipped—as even many in Third World countries are—with tanks, artillery, and armored personnel carriers, the qualitative and quantitative superiority of the local forces gives them a decisive edge. In other words, the requirements of rapid deployment to the battlefield contradict the requirements of staying on it. But this asymmetrical balance of disadvantages could change in favor of the intervener if intervention is protracted. The intervener then has time to deploy sea-mobile forces and, in general, bring its superior capabilities to bear.

Hence, paradoxically, the most appropriate forces for intervention in terms of mobility are often the least appropriate in terms of combat capability.[6] The alternative, a strategy of preemptive deployment to a crisis spot before hostilities begin, requires that the political decisionmakers be willing to commit themselves to a policy that will not be easy to justify at home and abroad because the immediacy of the crisis and the necessity for deployment of forces may still be ambiguous at the time of deployment. Preemptive deployment could even cause the escalation that it was designed to deter (Record, 1981: 20).

High costs and the difficulty of maintaining domestic support together make distant intervention relatively rare and most likely to be the choice of a major power, whose public is socialized to view such intervention as an integral part of their country's international role. In the rarer cases when it does involve a small country (e.g., Cuba's intervention in Angola in 1975), the small country acts as a proxy or ally for a great power, and the intervention almost always involves authoritarian regimes, which are much less accountable to their citizens.[7]

As a rule, perceptions of distance affect the public's view of whether the costs involved are justified and whether the intervention itself is vital to the national interest. The more distant the target, the less justified intervention seems. Therefore, the more distant the intervention, the clearer, less contested, and less abstract its main objectives must be in order to mobilize public support for the intervention. Major military-strategic and economic interests are likely to be the main triggers of distant interventions. Such interests seem easy to explain and justify, and they are associated with specific objectives, which are thus perceived as attainable.

There seems to be considerable reluctance "to oppose distant governments when the possibility of prolonged (most domestic dispute interventions lasted longer than six months), costly involvement in targets' domestic conflict existed" (Pearson, 1974b: 447). At the same time, "interveners may be willing to pay the cost [of distant intervention] if they act on behalf of a target government in a situation unlikely to produce effective governmental resistance to intervention" (p. 456). Therefore, when distant interventions occur, they are usually of the friendly-intervention type because of the relative ease of establishing policy legitimacy in these cases. Sometimes, however, perceptions of political and cultural proximity to the target can compensate for geographic distance. The intervening government can utilize these perceptions at home and abroad to justify intervention.[8]

Access and isolation (Connaughton, 1992; Woolley, 1991) are additional geographic factors, apart from distance, that are important in determining risks and costs of intervention. Geographic conditions that determine physical accessibility, or lack of it, have apparent importance for the costs and sustainability of both combat operations and logistic support for intervention. There is, however, a trade-off between distance and accessibility. Even a short-distance intervention could be a costly operation if the target area is not easily accessible. This was, for example, an important consideration in the U.S. decision not to intervene in Peru or Colombia in 1989 in its war on drugs. The other factor that must be considered is whether the conflict can be isolated from direct intervention or other forms of involvement by a third party. If the topography of the target area does not allow isolation of the battlefield, the opposing forces may be in a position to expand the scope and intensity of the conflict with the assistance of a third party and deny the intervener a quick victory. As a consequence, the costs of the operation are likely to escalate, and domestic support for the intervention will be difficult to maintain.

Economic Capability

The deployment and maintenance of a sizable military force beyond national borders can turn out to be a significant economic burden, with both direct and indirect costs. In cases of protracted intervention, the costs of hardware, spare parts, and ammunition can become a decisive factor in determining the intervener's capability to maintain the level of military activity required for success (Neuman, 1986: 90–106). When intervention requires large-scale reserve mobilization, there are, in addition to the direct costs, indirect costs entailed by the disruptions to various economic sectors. The costs can be underwritten by external powers willing to carry part or all of the burden by supplying economic aid or military hardware and ammunition, or both. If external sources of aid are unavailable, protracted intervention can have profound structural effects on the intervener's economy. The accumulation of a growing governmental budget deficit will fuel domestic inflationary pressures, which will have adverse domestic social consequences. In export-dependent countries inflation reduces their competitive edge in external markets and negatively affects their balance of payments—that was the effect of the Vietnam War on the U.S. economy.

When the intervener is economically or militarily dependent on a third power or multilateral organization for economic aid, or is heavily dependent on access to international markets and credits, decisionmakers have to take into account the potential costs of economic sanctions by the third party or multilateral organization (e.g., the Asian Development Bank stopped aid to Vietnam following its invasion of Cambodia in 1978; the United States threatened to cease supporting the pound sterling in 1956 if Britain did not end its military operations against Egypt). The more protracted the intervention is expected to be, the higher the likelihood that economic sanctions of some kind or another will be threatened or imposed by the international community, either collectively or by some members individually. The least vulnerable are superpowers and major economic powers, in particular the United States; smaller states have varying degrees of vulnerability to sanctions. In general, the less dependent the intervener's economy is on world markets, the less vulnerable the country is. With the exception of the United States, open economies are, as a rule, more vulnerable than closed economies.

The economic costs of intervention may be offset when intervention provides direct economic gains, such as access to natural resources or control of an international waterway. Alternatively, intervention is sometimes used to prevent economic loss, such as the disruption of production or loss of access to vital natural resources (e.g., oil), or to defend capital investments from nationalization. In all these cases, the economic costs of intervention have to be weighed against the economic benefits that may accrue to the intervener. If the economic costs are higher than the opportunity costs of nonintervention, there is a net economic loss from intervention, and vice versa.

In sum, the salience of the interests believed to be threatened usually makes economic costs a matter of secondary importance in the initial intervention-decision calculus. As intervention becomes protracted, however, and economic costs escalate, the consequent social side-effects will increasingly influence the decisionmakers in the direction of terminating the intervention.

Political Capability

Political capabilities and constraints represent the combined costs imposed on decisionmakers by the structure of the domestic environment, the institutional framework, and the processes by which choices among policy alternatives are made. They set the degree of freedom that decisionmakers have in making, implementing, and changing their decisions. Whereas the military and economic capabilities are relatively easy to characterize and, in some cases, to measure quite accurately, that is not always the case with assessing political capabilities and their cost implications. The difficulty of quantification makes the task of estimating risks and costs much less manageable. Unlike military and economic capabilities, which primarily concern the constraints that determine the probability of success in achieving the objectives of intervention (policy risks), analysis of political capabilities focuses primarily on the interaction, conflicting views, and bargaining of decision-making individuals, factions, and institutions which recognize the possibility that the policy will negatively affect their political position (political risks).

The higher the political risks associated with the policy, the greater the incentive to change either the policy choices or the strategies and tactics identified as necessary to achieve the policy objectives. Where consideration of political risk dominates the formation of policy preferences, one of two outcomes may result. Either a policy with a higher-than-optimal risk component will be chosen, or a policy with a high probability of success will not be sustained long enough to maximize its intrinsic worth. In either case, the outcome is the result of political expediency (Lamborn, 1991: 58–60). If decisionmakers are confident that they can deal with domestic politics, they will focus more of their attention on policy risks. The less confident they are, the more attention they will give to the domestic arena and not the policy risks.

Freedom of decision is determined by available tangible and intangible assets. The size of the parliamentary majority is a tangible asset. Examples of intangible assets are rules and laws that set the limits of authority of decisionmakers and decisionmaking organizations, and informal, incidental phenomena that are specific to time and person (e.g., a particular person's charisma, authority, and experience in manipulating the system).[9] The combination of tangible and intangible assets determines not only what types of decisions can be made and at what cost but also whether decisions to intervene can be reversed and undone at an acceptable cost once an intervention policy is recognized as inappropriate and requires either extreme escalation or de-escalation and withdrawal.

> Once pressure has been put upon the political authorities, either from within the formal decisionmaking structure, or externally by groups within the political system, a whole dynamic process might be set in motion that would mesh with the appeal for aid from one party to the civil strife. Once a government were committed to the first, apparently minor step, others might follow because of domestic pressures which make it difficult for decisionmaking elites to retreat "down the ladder" away from further entanglements. Under certain circumstances, the whole process may possess a horrid dynamic of its own, firstly because resources had previously been formally committed and expended, and decisions taken in official circles, where there would be an unwillingness to re-open bygone and dis-

credited arguments; secondly, because planning and prediction in official and rival government agencies had been involved; and thirdly, because official prestige was at stake, either domestically or internationally, through commitments having been publicly declaimed as policy, and reputations staked on the policy's successful conclusion. The last factor might be exceedingly potent when the political reputation of a key political figure is associated with a policy, and when an elite's retention of political authority within the system is perceived as being seriously affected by failure of that policy. (Mitchell, 1970: 174–75)

When the reputation of the decisionmakers who initiated the intervention is at stake, a change of leadership becomes a prerequisite for the termination of intervention.[10] Yet a sweeping leadership change may take a long time to occur, causing protracted intervention and resulting in mounting costs that could have been avoided through an earlier policy change.

Three overarching issues are affected by the institutional framework within which decisions are made: (1) resource mobilization (extraction and distribution), (2) the imposition of losses on powerful constituencies,[11] and (3) postdecisional accountability. All three influence government capability, government incentive, and the will to take risks associated with foreign military intervention. The more open the institutionalized political-bureaucratic system is to the pressures of outsiders' demands, the more dependent decisionmakers are on the consent of, and the more accountable they are to, a wide circle of individuals and organizations. They must take a broad view of where their own interests lie (Adomeit, 1982: 34–38; Allison and Halperin, 1972; Hermann, 1990; Hermann et al., 1987; Ostrom and Job, 1986; Putnam, 1988; Risse-Kappen, 1991). In other words, decisionmakers have to include a range of considerations in calculating the merits of intervention, not all of which are necessarily relevant to the decision.

This requirement is compounded by the complexity of the authority structure, expressed in the ratio of consent and approval foci (veto points) relative to the number of decisionmaking foci.[12] In some governmental structures responsibility is dispersed, requiring extensive interagency consensus, bargaining, and compromise; in others, responsibility is concentrated, and lines of authority are clear. The political difficulty of reaching a bureaucratic consensus on intervention and the substantial investment of resources required to do so encourages inertia once a decision has been made. Decisionmakers are highly reluctant to risk the erosion of an authoritative internal consensus reached in a time-consuming and resource-consuming bargaining process (Brands, 1987–88; Kissinger, 1969; Lockhart, 1978). Hence, intervention decisions are often difficult to make but, once taken, are equally or even more difficult to reverse. The more numerous the government agencies associated with the formulation and implementation of the intervention policy, the greater will be the resistance to acknowledging the failure of the policy and to effecting necessary adjustments, especially its abandonment.[13]

A government structure with a small, autonomous, highly professionalized and centralized bureaucracy is likely to produce daredevil irrationality, with the leaders making dramatic and rash decisions because of the absence of checks and balances (Mandel, 1984). Such a propensity for risk taking increases the likelihood of a careless and undifferentiating use of the instrument of intervention. In most ways, authoritarian political systems have a greater capability for imposing losses on their constituencies than democratic systems have, and are therefore less constrained in undertaking and escalating interventions. Yet even authoritarian regimes must take into account the human and material costs of protracted, escalating interventions. Societal unrest—caused perhaps by mounting casualties or defense expenditures that affect levels of social welfare—has to be considered, because it can undermine the power ba. of the decisionmakers, especially if the issue becomes a cause of contention between the ruling elites or between competing organizations and bureaucrats (e.g., Soviet intervention in Afghanistan, 1979–89).[14]

In institutional frameworks where the locus of accountability is centralized and easy to identify, decisionmakers are likely to be averse to policy risks because they anticipate more difficulty in avoiding accountability in case of failure than do decisionmakers in political systems where the locus of accountability is diffused, for buck passing lowers perceptions of political risk. Hence, the important distinction between parliamentary and presidential systems is in the locus of accountability. Owing to the clear hierarchy of accountability in parliamentary systems, the governing party or parties and their leaders are held accountable for government decisions and actions. In a presidential system, like the American one, power is shared and decisions are negotiated between branches, so accountability is more diffused, which encourages buck passing and decreases the incentive for risk avoidance (Weaver and Rockman, 1993a: 14–15). Parliamentary governments, with the exception of minority governments, are in general more able to impose losses on powerful interest groups and constituencies than are presidential systems, which have multiple veto points. But government leaders in parliamentary systems often lack the political will to confront powerful interest groups because the centralized locus of accountability exposes them to political retribution in case of failure (Weaver and Rockman, 1993a: 17, 26–27; Weaver and Rockman, 1993b); accountability raises their risk aversion and provides an incentive to learn from feedback.

Anticipation of the irrevocability of a decision causes decisionmakers to view it as risky and to approach it with caution and a reluctance to commit themselves. But once the decision is made, even if it produces negative results, decisionmakers are locked into it and find it difficult to adjust without paying a very high personal price. In some cases, escalation can result simply for lack of ability to disengage from a policy. Decisionmakers are forced to take risks that are much higher than they initially anticipated, and they become increasingly entrapped in a failing policy.

Prior awareness that decisions, once set in motion, may become practically irreversible leads decisionmakers to choose one of four different behavior modes for coping with the risk that the intervention decision may turn out to be a mistake. The first two behavior modes tend to reduce the probability that the adverse causative factors will occur. The latter two modes are directed toward minimizing exposure to adverse consequences once they materialize. The four modes are:

1. Setting a high threshold for intervention decisions, which reflects a reluctance to intervene unless it is perceived as absolutely necessary because of the extremely high costs of nonintervention.

2. Launching a bold large-scale intervention from the very beginning to reduce the chances of having to pay the high costs of protracted intervention by taking quick advantage of overwhelming military superiority to achieve rapid and complete victory. This approach characterized Soviet interventions in eastern Europe (Tatu, 1981).

3. Engaging in cautious incremental intervention to avoid fully committing oneself. This involves starting at a very low level of overt military intervention and carefully monitoring the feedback before deciding whether to continue. Decisionmakers thus leave themselves a way of paying low costs for errors in case intervention does not seem to work. The danger, however, is that negative feedback can indicate the inadequate initial level of military force rather than the inappropriateness of intervention itself. Incremental application of military force may also allow the target to prepare, and thus the intervener loses the benefits of the multiplier effect of surprise, ultimately raising the level of force and the costs above those required in a situation in which larger-scale forces were used from the start.

4. Arranging for intervention by a loyal surrogate nation, so that the leadership of the surrogate pays the price in case of failure. Where major powers are in competition, using a surrogate reduces significantly the risk of direct counterintervention by another major power; it also reduces the risk of domestic dissent, foreign criticism, and possible friction with the local client in case of friendly intervention.

Most of the benign results from the use of a proxy can accrue even if the surrogate's relationship with the patron is transparent, as is evident, for example, in the manner in which the Soviets have used surrogates, especially in Africa (Adomeit, 1986; Bar-Siman-Tov, 1984; David, 1987; Dunér, 1981, 1987; Hosmer and Wolfe, 1983; Ra'anan, 1979). In democratic societies, however, the public may not condone even indirect intervention through proxies, especially when the proxy has a transparent relationship with the patron and its intervention behavior clearly violates core norms of the patron's society, or when the public opposes the quid pro quo granted to the proxy in return for its intervention because the quid pro quo is considered exaggerated, morally repulsive, or harmful to the patron's long-term interests.

Patrons should be aware, too, of the limits of their leverage over their proxies and the risk of their getting out of hand—a risk that is more acute in protracted interventions. In a protracted intervention the proxy's insufficient, limited military capability increases the possibility that the patron will inadvertently get trapped into intervening directly under probably unfavorable circumstances (cf. Foster, 1983; Klare, 1989). The proxy may or may not intend to drag its patron into the conflict. The patron that keeps its distance to avoid entrapment may seem ready to abandon its proxy, thus making the proxy unwilling to continue the fight or perhaps willing to surrender to or reach an accommodation with the adversary; the proxy may, if it has the opportunity, even align itself with another patron.[15]

Choosing among these four behavior modes often depends on a number of considerations: available options, the personal preferences of key decisionmakers and their proneness to take or avoid risks, and the national style or cultural orientation of the decisionmaking system. The pragmatic strain in the American national style, for example, has caused interventions to be undertaken in a piecemeal, technical fashion (Rosenau, 1981: 138–39).

Legitimacy is also vital for the success of most major policies and, as such, is extremely important in assessing political capabilities, especially when the costs of the policy are high, its normative justification is disputed or not transparent, and the benefits are not immediately obvious or realizable. Legitimacy for foreign policy has two dimensions, cognitive and normative, and two markets, or targets—domestic and external.

Normative legitimacy establishes the desirability of a policy in terms of its consistency with fundamental national or international values; cognitive legitimacy requires that the leadership prove that its policies are feasible and suited to goal attainment (George, 1980a; Trout, 1975). Decisionmakers' awareness of the need for legitimacy is demonstrated in the semantically loaded titles given to intervention operations: Israel's intervention in Lebanon was named Operation Peace for Galilee; the U.S. intervention in Panama, Operation Just Cause.[16]

Cases of protracted intervention pose a particularly serious problem of risk communication for decisionmaking individuals and institutions, one that affects the ability to sustain legitimacy. Interventions are intended to be short and in their initial stages are presented and justified to the domestic public as short-term necessities. As intervention becomes protracted, decisionmakers associated with the intervention policy lose credibility and trustworthiness, and that undermines their effectiveness in communicating risk assessments for subsequent policies. The public feels cheated, and the leaders encounter growing domestic doubt and suspicion. They will find it difficult to gain domestic legitimacy for follow-up policies.[17]

The external sources of legitimacy are foreign governments and the more elusive "world public opinion." An intervening government attempts to convince third parties abroad, both allies and foes, that it knows what it is doing and that its policies will produce the desired results. At the same time it will cloak the intervention in legitimacy by attempting to persuade the external actors that the policy is in line with the norms, ethics, and laws of international society—for example, by quoting an invitation to intervene given by the government of the target country (cf. Brands, 1987–88; Little, 1975; Schraeder, 1989b).

If allies are convinced that the policy of intervention is normatively appropriate and will succeed, they may support it either actively (e.g., by sending a symbolic contingent of their military forces or by allowing the use of bases in their territory) or diplomatically when the policy is challenged in international forums.[18] Allies will then not feel that their relation with the intervener, or the latter's ability to carry out other commitments, is at stake, thus reducing the risk that they will seek alternative alliances or de-emphasize their links with the intervener. At the same time, the intervener's rivals may find it diplomatically and morally difficult to engage in counterintervention and thus be deterred from it, although they may engage in lower-intensity involvement (e.g., covert operations). These observations are particularly valid when the third party is an open society whose citizens are well aware of events occurring beyond the national borders, and where public opinion is a significant input into government decisions.[19]

Intervention is one of the foreign policy issues most likely to become a subject of domestic discourse. Domestic legitimacy, like external legitimacy, involves cognitive and normative dimensions. Belief by the public that their government knows what it is doing and that the policy will succeed makes costs and sacrifices acceptable even when they are high. But when the public doubt the efficiency and morality of the policy, even low costs will be considered a burden not worth bearing. Domestic opposition to a policy, however passive, undermines the probability of its success and, particularly in democratic societies, undermines decisionmakers' freedom of action—specifically, the freedom to escalate or de-escalate the intervention. The combination of possible domestic disaffection and failure of a high-stakes policy not only raises the perceived risks to the national interest but is often perceived as threatening decisionmakers' personal interests, such as survival in office or being judged positively by history.

The public is not always fully aware of the complexity and uncertainty involved in intervention decisions. Hence, it is much easier to mobilize public support for intervention when a threat to national interests is clear and sharply defined than it is to maintain an intervention once it begins to unfold and hidden risk dimensions come to light with mounting costs. Yet decisionmakers underestimate the difficulties of sustaining public support because of the ease with which it was initially mobilized. Their confidence in the assumptions that underlie the intervention decision and their lack of foresight regarding the resources that will have to be committed—both of these together complement and reinforce their confidence in their ability to maintain public support.

When intervention drags on, the national interests involved and their importance become less obvious to those groups whose interests are adversely affected. Because the war was defined as limited and national survival is not at issue, the troops are also likely to grow less convinced that risking their life is worthwhile or necessary. Morale is likely to suffer, and the public will question and even oppose the goals and deny the moral justification for the means used against the adversary. The division within the intervener's society and the growing alienation of military organizations from the society will eventually result in constraints on the intervener's ability to mobilize its full material resources and bring to bear its superior capability. The asymmetry in power between intervener and target thus becomes less and less relevant as the conflict escalates.[20]

With the intensification of domestic opposition, third parties, even if militarily inferior, may perceive the risk of involvement as worth taking. They become willing to challenge the intervening power, either on their own or in tacit or explicit coalition with other actors who feel threatened by the intervener's power projection and goals and who assess the situation as an opportunity to change the balance of power in their favor (Maoz, 1989; Maoz, 1990b: 214–50). They perceive the situation as an opportunity to increase pressure on what would otherwise be a much more intimidating adversary, and are ready to run risks they would normally avoid. These developments offer the intervener's domestic opposition another reason to question the cognitive legitimacy of the policy and to demand withdrawal, claiming that the cost-benefit balance has tilted in the direction of costs.

As a rule, domestic legitimacy for foreign military intervention is based on four elements: goals, costs, probability of success, and means. Each requires the establishment and maintenance of cognitive and normative legitimacy. "Those who oppose the goal would have a cost-tolerance of zero. Those who supported it could select a cost ceiling from a range of given costs, or they could support the goal at all given costs" (Mlotek and Rosen, 1974: 107). Or, people might value the goals and be willing to pay the costs, but at the same time doubt that the goals can be successfully attained at acceptable costs, if at all (p. 116). Even if the public supports the goals, is willing to pay the costs, and accepts that the goals can be achieved, legitimacy may be withheld if the means (e.g., saturation bombing of civilian targets) are unacceptable. The lack of legitimacy (cognitive or normative) in relation to even one of the four components may undermine the structure of legitimacy for the policy as a whole. Indeed, legitimacy is uncertain even if it is temporarily established.

External and domestic legitimacy are thus not unrelated. When a policy loses domestic legitimacy, the effect is to erode its external legitimacy as well. Similarly, a decline in external legitimacy leads the public at home to question the validity of the policy. Still, for the decline of external legitimacy to affect domestic legitimacy, the domestic public must be aware of the attitudes of foreign governments and the world public. In other words, the linkage between external and internal legitimacy is more likely to occur when the intervener's society allows the free flow of information. Reflecting on U.S. policy in the 1990 Gulf crisis, former Secretary of State James Baker recognizes this linkage: "For better or worse, there was a synergistic relationship between the international coalition and domestic support. The stronger the coalition the easier it was to generate consensus at home. . . . The reverse was also apparent—if the coalition began to fray, congressional and domestic support would be undercut" (Baker with DeFrank, 1995: 333).

Success and failure have direct and immediate impacts on both external and domestic legitimacy. The requirements of accountability are likely to be much less severe in cases of quick, low-cost policy successes than in cases of high-cost policy successes or transparent policy failures, especially if the failures are expensive. With quick, low-cost successes, the fine points of legal and normative propriety are unlikely to trigger significant negative reactions from allies or other international actors and are lost on the domestic public. The permissive domestic response can be attributed mostly to the "rally 'round the flag" syndrome. The effect of this substitute legitimacy is strongest with tough actions, such as military intervention, but is usually measured in months, not years (Lamare, 1991; Russett, 1990: 34–35).

International criticism of an intervening authoritarian government cannot affect a domestic public debate over intervention policy when there is no debate or when the critic is denied access to the target's society. But the leadership of the authoritarian society is often aware of and sensitive to international criticism, because of its implications for relations with foreign governments or because competing factions may use international criticism to strengthen their power position and challenge the leadership. Not all international critics, however, will have the same impact. The intervener's response will be affected by its perceptions of the critic's partiality, the critic's influence in the international community, and the power of the message to attract followers and create a bandwagon effect. [21]

Ideology is another generator of political incentives and a source of post facto legitimization for intervention. It defines the scope of the stakes in specific geographic regions, countries, and situations and consequently affects assessments of the risks worth taking to shape developments abroad. American ideology in the twentieth century, which equates an active foreign policy with the enhancement of liberty at home and abroad, is informed by a belief in Anglo-Saxon racial supremacy, as well as by a distrust of revolutionary change and a preference for political change that occurs incrementally and within the existing system. This ideology provided a strong incentive for intervention against revolutionary change in both Latin America and Asia. After 1945 perceived moral obligations and practical interests have been interpreted as requiring that the United States live up to its historic duty of rescuing people from the perils of communist oppression, thereby containing communist expansion (Hunt, 1987).

Ideology thus affects both the motive for taking risks and the assessment of risk.[22] An ideology that justifies or advocates intervention as a legitimate and necessary instrument of statecraft or as a way to achieve specific goals and national interests—for example, spreading the faith, freeing oppressed people (see, e.g., Hamburg, 1977; Hunt, 1987; Young, 1974)—encourages intervention, notwithstanding the risk. In this way, ideology can assure decisionmakers in the intervening state that intervention will bring the desired results, for example, by reassuring them that it will provide the final incentive for domestic opposition in the target state to rise up and bring down the regime (not resist and close ranks against the intervener).

The resulting confidence, generated by ideology, that intervention is going to work and that it is justified at virtually any cost makes ideology a powerful instrument for building cognitive and normative legitimacy for intervention. To the degree that it increases confidence that the goals will be achieved, it also acts as a barrier to learning from negative feedback; initial negative feedback can be ignored or discounted as marking temporary setbacks. This reduces the chance that necessary policy adjustments will be made.

Target-State Attributes

Even when the balance of capabilities and international systemic incentives favors the intervening actor, some attributes of the targeted state can have important, even decisive influence over the short-term and long-term costs and outcomes of intervention. When the target's society is socially cohesive, has a popularly supported government, and is administratively and militarily competent and when the goal of intervention is limited to stabilizing the regime, rather than establishing and imposing a government of one's choice, the likelihood of the success of friendly intervention is high (Engelhardt, 1989; MacFarlane, 1985: 37–39; Jentleson et al., 1992: 309).

The reasons are obvious. First, a capable client state can use external support effectively; an unreliable client state will become a burden on the intervening power. Second, when the client is an effective ally and the intervener's goals are limited, the costs of intervention are low, the benefits are often immediate and transparent, and burden sharing between patron and client is fairer than otherwise. In these circumstances, the risk of domestic opposition to the policy of intervention will decline significantly in the intervening power. Third, when a majority of the population in the target state supports the intervention, external and domestic legitimacy for intervention are fairly easy to establish and maintain. Finally, a third party is not too inclined to counterintervene when its own prospective client in the target state is weaker than the intervener's client. In general, decisionmakers should scrutinize carefully and assess realistically the attributes of the potential client in the target state before making a decision to intervene.

Intervening states tend to overvalue the stability and reliability of local allies, assuming that the chief threat to the ally is a military one, and once it is neutralized, the ally will be able to consolidate its political position (Kupchan, 1992: 256). This view fails to take into account the local political-social context and to recognize that the ally's vulnerability has deeper political and socioeconomic causes than being in a temporary militarily disadvantaged position. The biased assessment is heavily influenced by the local ally. The intervener's governmental agencies responsible for the assessment of the situation usually have a close working relationship with the leadership of the local ally and a history of commitment to it that results in a vested interest in supporting the ally (Kupchan, 1992: 260). The foreign intelligence agency of Israel (Mossad) has had such a relation with its Maronite Christian clients in Lebanon since the mid-1970s. The biased assessment of the client's position reduces the perceived risks of intervention on its behalf and feeds unrealistic expectations of gains from a policy of intervention.

The Antinomies of Time and Policy Effectiveness

Time is a particularly salient element in cases of protracted intervention, but it plays a role in the decisionmaking process for short-term interventions as well. The amount of time available to decisionmakers can be crucial in determining the nature as well as quality of the decisionmaking debate. The shorter the time for deliberation, the more likely decisions are to be guided by a few core values and beliefs, schemata, stereotypes, historical analogies, organizational SOPs, and cultural biases. Compared with decisions made with more deliberation, quick decisions are less likely to be based on a systematic analytic evaluation of alternative options and their short-term and long-term consequences. Commitment to the policy of intervention will thus have to be made faster and at a relatively early stage, before all appropriate information becomes available. The effect of feedback will be limited, and rigidities will set in prematurely and dominate policies, precluding necessary adjustments unless they are marginal. These constraints on the quality of decisions are augmented by the stress and anxiety created by a combination of the short time and high stakes involved (George, 1986; Holsti and George, 1975; Vertzberger, 1990).[23]

In cases of protracted intervention, time may have a positive effect on testing the assumptions that guide the policy of intervention by providing perspective and the opportunity to learn from experience. Time may help in the reduction of uncertainties and ignorance both about the intervener (e.g., whether the domestic public is willing to carry the costs of intervention for an extended period of time), the target (e.g., how much the target actor is willing to sacrifice in order to deny success to the intervener), and the international environment (e.g., whether other actors will recognize and legitimize a fait accompli even if it is counternormative). Time is thus a necessary, but not a sufficient, condition for learning and adjustment.

Although time is an important factor in learning, its effect on the parties to a protracted conflict is likely to be asymmetrical. With foreign military intervention the asymmetry in learning can be traced to the incentives for learning. In most cases, the intervener has a clear and decisive military advantage over the target. This results in overconfidence and arrogance, which—when coupled with the short-term and medium-term capabilities of the intervener to absorb the negative consequences of a failure to learn and adapt— act as barriers to learning. The weaker party, however, faced with the intervener's overwhelming military power, has no choice but to use the passing time to learn how to effectively cope with and compensate for the power imbalance. If intervention is protracted, these uneven motives for learning result in the erosion of the power imbalance as the target actor learns to make the most of its limited resources and raise the costs to the intervener by exploiting its weaknesses.

When time does not seem to reduce uncertainties but rather to enhance them or add new ones, and when a clear-cut resolution of the uncertainties does not seem likely, an intervener may face a growing problem in maintaining domestic and international support for the policy of intervention. The more uncertain the outcomes seem, the more reluctant are decisionmakers to invest resources and increase the long-term risks associated with intervention.[24] They also require a larger margin of safety.

Time poses a special dilemma for countries engaged in friendly interventions. Decisionmakers cannot be sure that the positive attitude of their clients will persist. This problem is serious in cases of protracted intervention; intervening powers, viewed as saviors, come to be resented as occupying powers once they have overstayed their welcome (e.g., India in Bangladesh following the latter's liberation in 1971). The intervener might even end up fighting local forces that, only shortly before, welcomed the intervention.

The passage of time may also become a cause of physical, economic, psychological, and moral attrition in the intervener's society, and attrition could result in domestic pressure for a change of policy or change of government, or both. Concomitantly, it could make the intervening government vulnerable to external pressures as it faces rising costs and a decline in available resources and in domestic legitimacy. The end result is very likely to be withdrawal from the policy of intervention without achieving stated goals. On the other hand, as time goes by and the accumulated costs of intervention grow, decisionmakers become identified with and committed to the intervention policy. With their reputations at stake, they may become reluctant to recognize in a timely way the need to disengage from the policy completely, even when its disastrous implications have become obvious. When they recognize the ineffectiveness of the policy, they may be unable to make a clean break; and if they disengage incrementally, they may increase costs unnecessarily.[25] When decisionmakers are incapable of either cutting their losses or escalating to win, the resultant compromises could make intervention a policy quagmire (e.g., Israel in Lebanon, 1982–83).

It is not unusual to mistime a decision to take risky measures in dealing with a problem. When the status quo is satisfactory, there is a built-in reluctance to take risks, so decisionmakers may dismiss such questions as whether to intervene or escalate or terminate an intervention, or they may treat them as constraints on options, not as problems that must be addressed separately. This tendency is noteworthy when the national interest is not clearly defined;

lack of sharp definition combined with military conservatism allows the avoidance of risk (Davis and Arquilla, 1991a). When a problem becomes inescapable, the solution may have become so risky that it poses immensely greater difficulties than if the problem had been tackled earlier.

The amount of time that decisionmakers hold office helps determine the style and nature of intervention. In some cases, having a short time left in office may lead to risk avoidance. Or, when elections for another term are imminent, decisionmakers may choose to play it safe, even in a risk-acceptant administration. Or, having a short time left in office may encourage decisionmakers to take unwarranted risks, perhaps advocating massive escalation, in the hope of getting the job done as fast as possible. By committing the nation (e.g., its prestige and credibility) to the policy in question, they present the next administration with a fait accompli, making sure that the policy they support will persist even when they are no longer in power.

In sum, time can lead either to risk-acceptant or risk-averse policies, creating (1) congruence between optimal and actual intervention, (2) actual intervention that exceeds optimal intervention, or (3) actual intervention that is less than optimal intervention. These situations will be discussed later.

A policy-relevant product of ambiguity is the measure-of-effectiveness problem. The intervening forces, in assessing how well they are doing, are often misled "simply because no clear measure, no line on the map, tells one how the war is going" (Cohen, 1992: 267). This state of affairs has a number of important implications. In the first place, it allows interested decisionmakers to manipulate information to prove that intervention was successful. Second, it produces unrealistic assessments of how long the intervention will last. Third, it makes recognition of an inappropriate policy slow to emerge. By the time the problem is recognized, decisionmakers may be so committed to the policy that entrapment becomes almost unavoidable.

But even if the measure-of-effectiveness problem is resolved, decisionmakers assessing the cost-effectiveness of the intervention face incongruence between short-term and long-term consequences.[26] Even when a decisive military victory is achieved at an acceptable cost, the long-term results may have a destabilizing effect on the target state or the region.

> In the longer term, the record would suggest that intervention is likely to be unrewarding. The military instrument may perhaps buy time, but it is too blunt to resolve the political and social conflicts which provoked the intrusion. In fact, in that it may encourage local clients to believe that they need not address the grievances and aspirations of their rivals, it may impede the target state's political integration. Furthermore, even where intervention succeeds in establishing a local client in power or perpetuating his rule, the cases of Angola and Ethiopia suggest that the influence gained thereby is by no means absolute and permanent. The local actor's debt to his external benefactor does not prevent him from pursuing policies inimical to the latter. In general, the strength of nationalism throughout the Third World suggests that influence, however gained, is likely to be circumscribed and short-lived, particularly where the external actor attempts to exercise it in a manner inconsistent with his client's aspirations. (MacFarlane, 1985: 55)

Yet for reasons that were elaborated in the preceding chapter, decision-makers are likely to accord much more weight and consideration to short-term consequences than to long-term consequences. Among other reasons, it is politically expedient to exploit an immediate, vivid success, especially when the costs (negative long-term consequences) can be transferred to the subsequent administration. But there are also unmotivational cognitive reasons for this bias. The long-term negative results are ambiguous and uncertain; being hard to envisage, their representativeness and availability for the calculation of costs are low. The opposite holds true, of course, for short-term results of intervention.

Capabilities: Illusions and Realities

The keys to successful foreign military intervention are the ability to make decisions that do not overstretch available capabilities and the ability to implement these decisions effectively. This translates into the ability to specify, employ, and sustain the level of force necessary to achieve the intervention goals at an acceptable cost. The problems inherent in carrying out these decisionmaking tasks are often misjudged by decisionmakers owing to the underlying structure of intervention situations and the implications for risk assessment. In most intervention situations there is an adversarial relationship and a substantial imbalance in military power between the intervener and the target of intervention. Because of the imbalance the intervening power implicitly poses the threat of invasion and occupation, but the target does not pose a reciprocal threat. "The asymmetry in conventional military capabilities is so great and the confidence that military might will prevail is so pervasive that expectation of victory is one of the hallmarks of the initial endeavor" (Mack, 1975: 181–82). Decisionmakers in the intervening state focus on their military superiority and tend to exaggerate their resolve and competence in using it. They believe that they will either achieve a quick victory or isolate the theater of war from their citizens' everyday life and win by attrition because of their superior resources and resolve.[27]

The lessons from foreign military interventions demonstrate that military defeat of the intervening power is not unlikely. Where complete victory is not attained, however, a second factor comes into play. As the costs of intervention mount, the theater of war extends well beyond the battlefield to encompass the polity and social institutions of the intervening power. A stalemate or even success on the battlefield that is short of total victory can be transformed into political defeat, eventually manifested in the exhaustion of the intervening power's will to continue the struggle. The intervener's superior military capability becomes largely irrelevant to deciding the actual outcome of intervention. Although this phenomenon is historically well recorded, decisionmakers often blunder into intervention operations by misjudging the complexity of the situation, making crude comparisons of military power and resolve, and expecting a quick and easy success.[28] When success does not materialize, decisionmakers are slow to adjust the policy or, worse, obstinately persist in it or even increase their commitment to it.[29]

In general, the sources of this inevitable-win bias can be traced to the availability, simulation, and representativeness heuristics (see Chapter 3). The first two make scenarios of winning much more available and credible than they realistically are and cause underestimation of the risks of a divisive domestic conflict in the intervener's society. The representativeness heuristic leads to the misidentification of control over resources with control over actors and outcomes.

Specifically, the biased judgments that often produce failures in military intervention are embedded in a number of common misjudgments and misperceptions. One misperception is that control over resources is fungible with power and influence over actors' behaviors and policy outcomes. A second is the failure to realize the complex and multidimensional capabilities required for successful intervention, to consider factors besides quantifiable military capabilities (e.g., number of troops, tanks, aircraft)—which are themselves grossly misleading indicators of overall military power. Third, an intervening power sees a war of intervention as limited. Hence "the prosecution of the war does not take automatic primacy over other goals pursued by factions within the government, or bureaucracies or other groups pursuing interests which compete for state resources" (Mack, 1975: 184; see also Jentleson et al., 1992: 305). Fourth, decisionmakers are not cognizant of the invisible impact of time on the effectiveness of the state, however powerful, in producing successful intervention. In other words, decisionmakers often initiate intervention policies because of excessive confidence in their success, exaggerating assessments of their military capabilities and underestimating the adversary's tolerance of costs.[30]

Optimal and Actual Intervention

The discussion of capabilities and the dynamics of intervention over time indicates that astute policymakers must avoid overcommitting national resources and must instead think in terms of an optimal level of intervention determined by stakes, resources, and constraints. This optimal level has to be constantly reassessed and adjusted. How to take into account the sociocognitive factors operating on judgment and choice (Chapters 2 and 3) in a real-life situation is neither obvious to decisionmakers nor necessarily easy to accomplish in a timely fashion.

When intervention is a plausible option, decisionmakers face two interrelated decisions. They have to decide whether to intervene and, if they choose to intervene, on the scope and intensity of intervention. Analytically, it is possible to compare optimal intervention (I_O) and actual level of intervention (I_A). Both can range from zero (nonintervention) to very high levels involving a substantial share of the intervener's military, political, and economic resources. Ideally, as in market economics, the intervener should strive to reach the point of metaphoric equilibrium between optimal and actual intervention levels. Optimal intervention would be the level of intervention that allows the intervener to achieve the main objectives at the lowest acceptable cost.

An equilibrium point is operationally hard to recognize; it is inherently unstable owing to the constantly changing external and internal strategic political landscape, as well as the intentional attempts of adversaries to interfere and undermine the equilibrium. Still, a comparison of the optimal and actual levels of intervention is heuristically useful. This comparison yields three logically exhaustive and mutually exclusive situations.

1. Actual intervention equals and is congruent with what should be optimal intervention ($I_O = I_A$). In this case the cost-benefit ratio should be optimal as well. But as I shall argue below, this equilibrium is bound to be inherently unstable and is not self-perpetuating.

2. Actual intervention is greater than optimal intervention ($I_O < I_A$). In this case the cost-benefit ratio is tilted in favor of costs. Goal attainment cannot justify the enormous costs of the process, and the intervener is overextended. Where major powers are involved, the drain on resources may even lead to a direct confrontation between them, as in 1962, when Soviet intervention in Cuba led to the Cuban missile crisis.

3. Actual intervention is smaller than optimal intervention ($I_O > I_A$). By implication, goals are not likely to be attained at all, or at best only partially attained. Another consequence is that intervention is likely to extend over a longer period of time than was anticipated, resulting in unanticipated costs. Incremental intervention is sometimes an instance of this general situation. Lack of success in goal attainment by the intervener may tempt a third party to intrude either by direct counterintervention or by provision of military and economic assistance to the target of intervention.

Naturally, an intervening power would like to be in the first situation, where optimal and actual intervention are congruent. Yet it is more likely to end up in the second or third situation, each of which represents a state of incongruence between optimal and actual intervention. Identifying the level of optimal intervention in advance is extremely difficult because of the complexity of the computation, which requires combining and integrating relevant variables—some measurable, others not—into a coherent single optimal assessment. Sometimes the only way to do this is by trial and error.

Even if the optimal level of intervention is clearly identified, decisionmakers cannot always implement the desired policy, either because of political-domestic or external constraints or resource constraints. The constraints do not necessarily generate a level of actual intervention that is lower than the level of optimal intervention; political-domestic or external constraints may impose on decisionmakers the necessity of risking a level of intervention that is higher than optimal.

Even if an equilibrium is reached, it is likely to be unstable. Equilibrium means an optimal balance of costs and benefits (in terms of goal attainment) for the intervener. It may be in the interests of a third party or the target actor to raise the stakes and thus change the level of optimal intervention and unbalance the relation between actual and optimal intervention.

Because optimal intervention is dynamic, changing constantly in response to changes in resources and political circumstances for the intervener as well as for the target actor or third parties, it requires an ongoing assessment of intervention requirements and a very flexible and adjustable decisionmaking system, especially in cases of long-term interventions (e.g., the United States in Vietnam). Yet we know that decisionmaking systems have built-in rigidities (e.g., Etheredge, 1985; Evans, 1989; Steinbruner, 1974; Vertzberger, 1990). The problem is much less acute in cases of short-term intervention (e.g., the United States in Grenada), when most of the operating variables can be assumed to stay constant. When the level of optimal intervention is dynamic, adjustment must be farsighted and preemptive, starting even while the equilibrium holds—even when there are no obvious immediate pressures for policy change—in order to prevent a lag at a later stage. To accomplish this requires that future needs be predictable and anticipated. Decisionmaking systems do not commonly operate in this optimal manner.

The probability of achieving an equilibrium between optimal and actual intervention depends, then, on both the complexity of the environment in which decisions have to be made and the adjustability of the relevant variables. The other deciding factor is the quality of the information-processing system whereby the intervention situation is defined, evaluated, monitored, and updated.

Conclusions

The vast complexity and uncertainty attending foreign military intervention means that to make a rational, balanced decision, the decisionmaker must aggregate the factors discussed in this chapter into a coherent calculus in order, first, to assess the wisdom of using intervention as an instrument of statecraft given the military, economic, and political capabilities and constraints. Second, the decisionmaker must determine a level of actual intervention that is congruent with the optimal level of intervention and the degree of acceptable risk. In particular, risks to be taken must be well understood and considered prior to making a commitment to a policy of foreign military intervention, not only because of the costs of initiating intervention policies but also because of the political costs of terminating them. The performance of either task could result in errors and biases in risk assessment (discussed in Chapter 3); these errors and biases are embedded in the very difficult process of aggregating multiple risk dimensions into a coherent risk calculus.

What complicates these tasks is the inherent tension between the communicative role of intervention and its substantive goal of pursuing vital national interests. Intervention as a signal of intentions and determination requires taking low risks and committing low levels of military resources. Yet if the target is undeterred by the implied threat or chooses to disregard it, low-level intervention may merely warn the opponent, allowing the opponent time to prepare and thus requiring much higher levels of subsequent intervention than if surprise had been the decisive consideration in the first place.

Another important problem in calculating the cost-effectiveness of an intervention policy is the tension that is likely to arise between short-term and long-term consequences. Even when a decisive short-term military victory is achieved at an acceptable cost, the long-term results may be unfavorable and have a destabilizing effect on the target state.

Finally, decisionmakers must be well aware of the differences between the variables affecting intervention. Some variables are nonmanipulable and must be taken as parameters—for example, the attributes of the international system (that will be discussed in the next chapter). Others, such as technology, are manipulable in the long run but not in the short run; and only a few are given to rapid adjustment and flexible manipulation. In any case, the combination of the variables analyzed forms the basis for determining the level of actual risk that the decision involves. As we have seen, intervention decisions are highly complex and risky, and making successful ones requires foresight, sophistication, and cognitive complexity. The mental computational skills required place on decisionmakers the heavy demands of integrating quantitative and qualitative variables, domestic and external considerations, and military, economic, and political factors in an orderly and analytic manner. Complexity, together with the high costs of failure, suggests the need for constant monitoring of and sensitivity to feedback, along with quick, timely judgments of required policy adjustments and their implementation. Intervention situations often involve extremely difficult choices between admitting defeat, on the one hand, and committing substantially more resources and possibly placing decisionmakers' reputations and careers on the line, on the other. Paradoxically, the nature of such decisions and the demands that they pose may too often result in the temptation to avoid, rather than confront, the burdens of complex calculation by vulgarizing the analysis and the judgment of risks.

CHAPTER 5

The International Milieu and Foreign Military Intervention

WHEN AND HOW MUCH DOES THE MILIEU MATTER?

In many or most cases of foreign military intervention the intervener's assessment of the threat and the balance of capabilities between it and its target provides the main, if not the only, incentive for intervention. Yet, as is obvious from the lessons of history, the causal relation of raw power (im)balances between adversaries and intervention outcomes is neither linear nor simple. The main reason is that other factors, aside from the balance of capabilities, have important mediating effects on the structuring of intervention outcomes, particularly in cases of protracted intervention.

In this chapter I argue that some prominent international factors should be incorporated into the calculus of intervention decisions in order to avoid misjudgment of the risks involved and unrealistic expectations of success. The international environment is the generator of "conditions of possibility for state action" (Wendt, 1987: 342) that shape opportunities, provide incentives, or act as constraints on the use of this particular form of statecraft. More specifically, the international setting affects the payoff structure, duration, targets, location, frequency, and unilateral or multilateral format of intervention. At any point in time, particular actors have fewer or greater opportunities and incentives to intervene, depending on the prevailing external conditions.

Three central issues are addressed: (1) the systemic enabling conditions and constraints on making and successfully implementing foreign military intervention decisions; (2) the roles played by third parties and how these roles affect the payoff structure of such decisions; and (3) the salience of capability-related variables compared with international variables in forming decisionmakers' views. In the following sections I analyze alternative systemic structures and the effects of structural and nonstructural attributes, then discuss the various forms that third-party responses might take and their implications. From the discussion of system and state-level constraints attention is directed to the mutually reinforcing cognitive and motivational explanations that account for the selectivity bias—that is, the relative marginal weight of external constraints in the initial stages of planning and choice of intervention policies compared with the weight given to threat and capability variables. The analysis of these issues indicates that systemic structural attributes are much less influential than nonstructural factors in determining the power of the incentives for intervention or nonintervention. Even more important than systemic factors are state-level third-party responses, both by allies or friendly nations and by adversaries. Finally, in spite of the importance of international factors for success or failure of intervention, motivational and cognitive causes may bias decisionmakers toward paying less attention to international factors in making the decision to intervene than these factors merit.

Systemic Constraints and Imperatives for Intervention

Intervention decisions are taken within a specific international setting, and their outcomes are affected by the attributes of the international system within which they are made and implemented. Comprehending the implications of system attributes for intervention provides decisionmakers with improved probabilistic assessments of whether the policy will be acceptable to the international community, how and which third parties are likely to react, and what the potential costs of intervention and the constraints on its effective use are.

The systemic attributes that affect the opportunities and incentives for intervention are of two kinds: structural and nonstructural. Different types of systems have distinct norms and rules about the use of large-scale military coercive intervention as an instrument of foreign policy. Differences among types of international systems with regard to the diffusion, suppression, and isolation of internal wars directly affect opportunities for intervention. Whether the norms, rules, driving logic, interaction patterns, and political climate of a system are favorable to insurgents or incumbent governments indirectly affects the incentives for intervention (Modelski, 1964). These factors in turn influence the reactions of third parties.

Third-party responses became a particularly salient factor in 1945–90, when superpower competition added the danger of nuclear escalation to the traditional risk calculus of foreign military intervention. For the superpowers it was more essential than ever before to define, tacitly or explicitly, and reach a mutual understanding about their exclusive spheres of influence. During the Cold War direct intervention by one superpower in the domain of the other was assumed to require the other superpower's military response, significantly increasing the probability of nuclear escalation. Nonetheless, most regions in the Third World and some in Europe (e.g., Austria, Yugoslavia) were in a gray area where explicit rules did not exist, tacit rules were fuzzy, and uncertainty prevailed regarding the reaction of one superpower to military intervention by the other.[1]

The lack of a clear commitment by the superpowers to use nuclear weapons in the Third World rather than accept the subjugation of local allies removed an important deterrent to military intervention. It was astutely observed at the time that "the greatest danger lies in areas where (1) the potential for serious instability is high; (2) both superpowers perceive vital interest; (3) neither recognizes that the other's perceived interest or commitment is as great as its own; (4) both have the capability to inject conventional forces; and (5) neither has willing proxies capable of settling the situation" (Betts, 1988: 91). This is in sharp contrast to the situation in Europe or Japan, where the threat of immediate, massive military retaliation to intervention from outside in a superpower's sphere of influence seemed clear and resolute.

The following analysis does not adhere to the strict systemic structural realism of the Waltzian approach (Waltz, 1979, 1986). K. N. Waltz admits that "structures never tell us all that we want to know. Instead they tell us a small number of big and important things" (1986: 329). For our purposes, structural variables provide neither accurate nor specific-enough inferences and predictions regarding the location and frequency of interventions, nor do they suggest policy-relevant prescriptions regarding the contextual, temporal, and spatial conditions under which an intervention policy is most cost-effective.[2] A richer systemic analysis is preferred, one that takes into account not only the structure of the system but also nonstructural variables—aggregate unit-level processes, interactions, and patterned behaviors.[3] Structural and nonstructural variables are presented separately only for the sake of conceptual clarity.

What distinguishes nonstructural variables from state-level attributes and makes them system attributes is the systemwide effects they have: the ways they affect the system are "qualitatively different from the way the particular attributes of particular states affect their interaction with other individual states" (Buzan et al., 1993: 70). Consequently, I discuss the strict quasi-deterministic consequences of the structural attributes of a system, which are conditioned and modified by nonstructural variables that cut across different system types and reduce some of the anticipated behavioral diversity among state actors in different system-types. I suggest that there is more continuity between systems than is inferred by focusing exclusively on structural attributes. By exploring the interaction of structural and nonstructural variables and arguing that systemic effects are codetermined by both, we gain precise and insightful answers to questions about foreign military intervention: Who? Why? Where? and When?

Structural Attributes of International Systems

To explore opportunities for and constraints on intervention that are generated by the system, let us start by analyzing the implications of the structural attributes of the international system.[4] In the twentieth century these have included the balance-of-power, tight and loose bipolar, and emergent post–Cold War centro-polyarchic systems.[5] As a general rule, in all these systems "the more extensive the disparities in power, the greater the opportunities for intervention" (Young, 1968: 180), because the weaker actors are unable to fend off external intervention. It is tempting, therefore, to use intervention "when distribution of power is changing more rapidly than the distribution of other values" (p. 183), sometimes resulting in competitive interventions by more than one external actor. Revisionist states may attempt intervention to bring about a redistribution of values; satisfied states may use intervention to stem the tide of revolutionary change.

A functioning *balance-of-power system* should limit both the opportunities and incentives for intervention, though not prevent intervention altogether. This is the case for three reasons. The power distribution in the system is supposed to protect the independence of weaker states by deterring interventions of stronger states. The main balancing mechanism—the shifting of alliances in face of a threat to the balance—requires that shifts be independent of the nature of the regime and domestic order in the countries involved. Therefore, the more dispersed the power structure of the international system, the less likely that radical change will result from a single development in the system. Occasional domestic upheavals, which under different systems may provoke intervention, are not perceived as threatening enough to trigger intervention in response.

Some analysts maintain that in a balance-of-power system there is a built-in incentive to insulate internal political instability from external intervention.[6] Yet this statement should be modified to pertain particularly to states essential to the balance and to circumstances under which intervention might undermine the balance and give a decisive edge to the intervening power, either because the target states are essential to the balance, being important powers, or because they have other characteristics, such as geostrategic salience, as with a buffer state, that gives them a role in reducing friction between major powers.[7] There is, however, an incentive to apply the norm of nonintervention as uniformly as possible (even to actors nonessential to the preservation of the balance) to avoid setting an unwanted precedent.

Intervention in a marginal state may erode the principle of nonintervention, especially in cases of "twilight-zone violations" (Kaplan, 1964: 102). The fear of precedents sometimes makes it possible to mobilize action by the major powers against unilateral intervention, even when it could not destabilize the balance of power, and inhibits armed intervention. Nevertheless, the possibility that a window of opportunity could provide an incentive for collective intervention or collusion among some of the great powers to divide the spoils among them is not precluded. Similarly, an interventionist alliance among major powers can prevent the development of competitive unilateral interventions and their escalation into a confrontation between rival major powers, as in the joint U.S.-Japanese intervention in Siberia in 1917. Interventionist alliances may also support the established government of a member of the system against a potentially disruptive domestic group, as with the Boxer uprising in China in 1900, when Western intervention broke the rebels' siege of Beijing.

In reality, the constraints on intervention in the balance-of-power system sometimes do not work because alignments against the aggressor seem too costly. On other occasions, intervention is not perceived as a threat to the interests of major actors because they are indifferent to the nature of the domestic regime that would be set in place by a particular intervention. The interventions that sometimes take place in civil-war situations could affect "elements of the international environment that are of interest to major powers, but do not bring into question either the underlying structure of the system or the ordering of the major powers within it" (Forman, 1972: 1127). Transparently obvious limited implications of the intervention sometimes allow intervention even at the risk of creating a precedent. In principle, then, intervention is not condoned in a balance-of-power system, but exceptions are allowed if they do not affect the overall stability of the balance of power. These exceptions are most likely to take place when the relations among the major powers are characterized more by trust than by suspicion and involve shared perceptions that all the major powers are, for reasons of self-interest, reasonably satisfied with the status quo and that none have revisionist designs.

The situation is substantially different in a bipolar system, either loose or tight. In a *tight bipolar system* military intervention is used mainly as an instrument of the bloc leader to maintain hegemonic order (Weede and Mannheim, 1978) or by smaller states when they act as proxies for the bloc leader. Bloc leaders disallow not only external intervention in their bloc but also intrabloc intervention by secondary members of their bloc that they did not authorize. They are also unlikely to allow independent external intervention by a member of their bloc in order to avoid being drawn into an unwanted armed confrontation with the rival bloc leader. Because in a tight bipolar system state actors tend to be closely identified with one bloc or the other, interventions tend to be mainly of the intrabloc type. Intervention in a country belonging to another tight bloc is very rare because of the high probability of resulting interbloc conflict.[8]

In a *loose bipolar system* the control of bloc leaders over secondary bloc members is limited; weaker states may sometimes pursue their own interests without the blessing of or even in defiance of the leaders. Even smaller states may use the instrument of military intervention to pursue their national interests, not only to serve major powers as proxies; examples are Israel's and Syria's interventions in Lebanon in the 1970s and 1980s. Because bloc leaders have fewer means to influence the domestic stability of bloc members, the probability of unchecked domestic instability is higher than in the balance-of-power system. The increased number of cases of domestic instability and the looseness of the system provide more opportunities for both extrabloc and intrabloc interventions (interventions not exclusively by the bloc leader). Intervention depends, however, on such additional factors as the intervener's capabilities, experience with interventions in the past, estimates of the risks involved, and the risk-taking preferences of its leadership.

As a rule, intervention is more widespread in a bipolar system, especially a loose bipolar system, than in a balance-of-power system for reasons related to both available opportunities and incentives. During the Cold War years, this trend has been enhanced by three additional factors. First, the underlying ideological dimension of bloc conflict that placed heavy emphasis on competing conceptions of the domestic and international orders injected a sense of zero-sumness into superpower relations in periods of tight bipolarity, encouraging intrabloc interventions to prevent possible defections from one bloc to the rival bloc. This was the case with Soviet intervention in Hungary in 1956. The more tightly the international system is structured, the more threatening is the potential shift of allegiance by any one state. The domino theory dominates strategic beliefs. It is expected that if one domino falls, others will fall too, owing to increased boldness on the adversary's part and a bandwagon effect, which will cause defections by allies. In this context, the reputational interests of the superpowers are at least as important as their intrinsic interests (Jervis, 1991). Therefore, domestic upheaval in a bloc state that would destabilize the status quo induces immediate intervention. Each bloc leader has an interest in preventing change in the domestic political systems of its bloc members and at the same time has an interest in supporting change in the political systems of the adversary bloc members. These interests generate incentives for intervention.

Second, the nuclear balance has had a dialectic effect, presenting decisionmakers with difficult decisions about getting involved in direct counterintervention once intervention by another nuclear power has taken place. In fact, for a competing superpower there has been an obvious advantage in preemptive intervention, that is, intervention in a weaker state when the risk of superpower confrontation is assumed to be minor; this was the case with Soviet intervention in Czechoslovakia in 1968 (Adomeit, 1986; Dawisha, 1984; Kaw, 1989; Valenta, 1991). Once a nuclear superpower intervenes, a rival superpower is not likely to deploy troops in a counterintervention, although it may invoke other measures, like extensive economic and military assistance to an ally; Chinese and Soviet economic and military aid to North Vietnam was such a response to American troop deployment in South Vietnam.

Third, the many newly independent states since 1945 have presented targets of opportunity; the weak new states facing serious internal rifts and tensions seemed to constitute low-risk targets of intervention. Some actions were moderated by the nuclear standoff, which forced the superpowers to limit their goals and to aspire to change the international environment through incremental measures rather than by attempting step-level changes through foreign military intervention.[9] At other times, the standoff between the superpowers allowed the emergence of a group of nonaligned states. Manipulation of the superpower competition allowed these states access to modern weapon systems, and they built up their military power to a level that tempted them to use armed intervention to coerce neighbors, as with Libya's interventions in Chad since 1971.

The power structure of the post–Cold War international system is more fractured than in the bipolar system that preceded it, but there is still one preeminent center of military and economic power coexisting with multiple centers of autonomous decision authority. This *centro-polyarchic system* has multiple autonomous decisionmaking units and one key central actor, the United States, which is the only remaining superpower militarily and even economically (in spite of its economic decline). A central player has more influence than an ordinary major power, but not enough to veto or impose all decisions that are desirable to it. This is why the system is polyarchic and not hierarchical. This definition does not preclude the possibility that one or more of the other major powers will also rise in the future to a position of central player within an essentially polyarchic structure.

The centro-polyarchic system maintains some of the key attributes of the loose bipolar system (e.g., Brecher et al., 1988: 13). But although some of the main trends of the loose bipolar system still prevail, there are also a number of important discontinuities. Specifically, with the demise of the Cold War and the diminution of the ideological and strategic competitive relations between the superpowers, the sphere-of-influence notion has lost most of its significance. Major powers are likely to undertake foreign military interventions only when they perceive a threat to their vital interests or when important interests are at stake but the risks of intervention are perceived as low to moderate. This was the case when the United States intervened in Panama and the countries of the Persian Gulf. Regional clients or would-be clients may consequently face growing difficulties in mobilizing proactive intervention or even defensive counterintervention by a major power patron unless the threat to the patron's interests is extreme or the risk of intervention low to moderate. Thus, we are likely to witness more interventions by regional powers and fewer by the superpowers.

Stated more generally, in a decentralized global system with one predominant military power, opportunities for intervention that are not matched with extremely strong incentives are not likely to be unilaterally exploited by the superpower. The United States is a highly risk-averse superpower, whose policy choices are constrained by domestic institutions, ideas, and norms very dissimilar to those of major powers in the past, which were driven by traditional hegemonic imperial motives. The development of democratic institutions that are in principle committed to transparency and accountability, the formal procedural and informal constraints on the executive, and the lessons drawn from experience—all these have significantly constrained the behavior of U.S. leaders in using military force as often and as widely as they might have done in earlier times. This constraint is particularly evident where there is a risk of combat casualties, even moderate ones, and considerable economic costs.

The domestic legitimacy of intervention is the key consideration, and public support is not easy to build and is even more difficult to sustain over protracted periods of time (Chapter 4; see also Jentleson, 1992). Collective intervention is therefore more palatable than unilateral intervention. Collective intervention provides a sense that military risks and economic costs are being shared, thus encouraging risk taking. Collective intervention also makes it easier to muster international support for intervention, which can then be used as an argument for mobilizing and sustaining domestic support. For industrialized democracies, multilateral intervention in the current international system, unlike intervention in the bipolar system, is in most cases no longer merely a convenience but a strong political preference.

Multilateral intervention may occur in three different power-structure contexts. The first is when a hegemonic power imposes on its allies symbolic or substantial participation in an intervention in order to share costs and risks or provide a facade of international legitimacy for the act of intervention. Second, when a hegemonic power does not have the capability or will to impose participation on its allies but can offer significant payoffs to those willing to join in an intervention, some other nations will join in. A possible bandwagoning effect could cause even more nations to participate for fear of being left out when the spoils are divided. Third, if the most powerful nation in a polyarchic system, unwilling to take on the responsibilities associated with its position of power, prefers that system governance takes the form of a concert of powers, multilateral intervention will result from a collectively perceived threat to the international order. In this case, the threat of multilateral foreign military intervention by the major powers deters nations from defying the international order and from attempting to break key rules.[10]

The present centro-polyarchic system is still in a formative stage. It is not unlikely that, with a further decline of the U.S. hegemonic position and the rise of China, Japan, and Germany as true global powers, a variant on the balance-of-power system may emerge. Alternatively, it is not unthinkable that a variant on the loose bipolar system will reappear. Such a development may occur, for example, if the fundamentalist Islamic states controlling oil-based wealth and nuclear weapons are grouped against a bloc of democratic industrialized states. A third bloc of moderate Islamic states and other developing countries could cooperate within an institutional framework of loose association and take a nonaligned position.

Notwithstanding the continued potential relevance of the historical and contemporary types of systems for the analysis of current and future developments, structural variables produce only very general and indeterminate conclusions, because a structural analysis in and by itself is insufficiently nuanced to capture the driving forces behind intervention policies. These policies are only moderately determined by the structure in which they are formed; the structure could remain unchanged while the pattern of behavior shifts dramatically. To understand why, we must look at the effects of nonstructural factors.

Nonstructural Attributes of International Systems

Four types of nonstructural causes have affected the preferences of major and lesser powers to utilize foreign military intervention to advance their national interests: (1) changes in the essence of relations between the super-powers that did not result from and could not be explained by the power distribution between the superpowers, (2) incentives or self-imposed constraints for would-be interveners that are embedded in changes in domestic societal attitudes or changes in international norms regarding the use of military force abroad, (3) an increase in the political bargaining power of small states, and (4) an increase in the military denial and independent action capabilities of small states. These sets of nonstructural causes have substantially affected the perceived cost-effectiveness of intervention and the probability of sustainable success at an acceptable cost. In many cases, the relation among nonstructural factors has been synergistic.

First are *changes in the relation between the superpowers*. The concern that an outbreak of violence between the superpowers might escalate into nuclear war has reduced the incentive by either of the superpowers to intervene in the other superpower's sphere of influence, even when the opportunity to intervene had been there; the United States, for example, did not intervene in Hungary in 1956. At the same time, because of the reduced risk of third-party intervention in a superpower's sphere of influence, the hegemonic power in that sphere has been more inclined to use intervention within its bloc as an instrument of statecraft, all else being equal.

But this simple equation has changed over time. The nuclear balance between the two superpowers, which was one of the pillars of the bipolar system after the Second World War, implied a danger of inadvertent escalation to nuclear war resulting from a runaway local conventional conflict involving the two superpowers.[11] The process could start with foreign military intervention by one superpower in a nonbloc state, leading to a perceived threat to the other's vital interest, resulting in counterintervention and uncontrolled escalation. This scenario, if carried to its logical conclusion, suggests that there would be no direct extrabloc foreign military interventions in a bipolar system, because whatever the expected benefits, the potential costs might outweigh them. Yet we know that the superpowers undertook foreign military interventions.

The explanation lies with one of the dominant processes that modified the consequences of structure: learning to cooperate in a relationship that was structurally adversarial. As the two superpowers came to realize the dire consequences of competition, they learned to establish explicit and tacit norms and rules of behavior that included monitoring and recognizing the limits of the adversary's tolerance, understanding how to accept faits accomplis, and learning to use multiple channels of communication and signals to prevent inadvertent escalation in highly adversarial situations (Kanet and Kolodziej, 1991; Midlarsky, 1991). Once the code of collective caution was in place, it was constantly refined to further reduce the risk of inadvertent nuclear escalation. The code provided confidence in the overall stability of the system that allowed foreign military intervention in the Third World within mutually understood and acceptable limits. During the Cold War, then, the superpowers had a mutual grudging acceptance of each other's foreign military interventions, so there were more interventions than an exclusive structural analysis of a bipolar system would suggest.

The code not only prevented counterintervention when one superpower's ground troops had already intervened, for fear of escalation (Kaw, 1989; Tillema, 1973), but also moderated and placed substantial self-imposed constraints on the intervener's military-strategic operational options, thereby reducing the effectiveness of military power. The lesson of the Cuban missile crisis was constantly on the minds of decisionmakers on both sides. The intervener's political leadership perceived moderation as necessary to avoid challenging the substantial and reputational interests of the rival superpower to such a degree that it would feel compelled to respond by direct counterintervention, starting an escalatory process that might lead to a nuclear showdown. In other words, the nuclear balance and mutual deterrence favored the power that intervened first and, at the same time, induced the intervener to be cautious, accepting strict, self-imposed constraints on the manner in which the intervention was executed in order to avoid provoking counterintervention, which would ultimately make the benefits of intervention less certain and more costly.

The Korean and Vietnam wars illustrate the imposition of constraints by the superpowers. In both wars the Soviet Union was reluctant to send supporting combat troops. During the Korean War the United States avoided attacking China directly, and in the Vietnam War it carefully averted a decision to send ground troops into North Vietnam. The U.S. offensive against the North Koreans seemed to threaten China's security and elicited counterintervention by the Chinese in November 1950. But President Harry Truman rejected the option of responding to the Chinese intervention by attacking across the Yalu River—among other reasons, for fear of involving China's ally the Soviet Union. An acute and growing shared recognition of the need to avoid any kind of direct military confrontation between the superpowers and a possible nuclear showdown had emerged. The incentive to take advantage of intervention opportunities in gray areas was consequently greatly moderated by the process of learning that reflected this recognition.[12]

The second nonstructural attribute of systems is *the effect of societal attitudes and international norms*. Since 1945 learning from one's own and others' experiences has brought home the potential high costs of intervention and the prospects of failure. The lessons of Vietnam, Afghanistan, and Cambodia have been illuminating. As a rule, modern democratic industrial societies have become more inward looking as citizens become better educated and more affluent. Domestic constituencies have become more interested in the low politics of the welfare state and more reluctant to bear the hardships and sacrifices of military conflict, especially as the risks associated with intervention have become more salient. In turn, the leadership of these states have found it increasingly difficult to build and sustain the necessary domestic support for large-scale, protracted military intervention (Morgan and

Campbell, 1991: Tarr, 1981: 51–52), thus raising the threshold for choosing this policy option.[13]

A major cause for this anticipation of increased intervention costs was the perception of the rising capabilities and willingness of the Soviet Union to project power (Rubinstein, 1988), which increased Western reluctance to risk intervention for fear of counterintervention. These considerations were particularly salient when the target country was close to Soviet borders or when the Soviets were perceived to have prominent interests in the target country. This reluctance and the decline of Western, especially American, determination to use superior technological and logistic capabilities to intervene were very evident immediately after the costly interventions in Korea and Vietnam (Blechman and Kaplan, 1978: 26–28). The considerably diminished risk of Western direct counterintervention increased the Soviet incentive to intervene.[14] The relatively low-cost Soviet intervention experiences (e.g., Hungary, 1956; Czechoslovakia, 1968) enhanced the Soviet incentive to intervene until the painful failure in Afghanistan in the 1980s taught otherwise.

Ironically, a weak Soviet/Russian threat, like a powerful one, has made major Western powers reluctant to intervene militarily. With a powerful Soviet Union, the potential risks reduced both opportunities and incentives for intervention. When the Soviet threat declined, mobilizing domestic support for the costs of unilateral foreign military interventions has become extremely difficult (see Chapter 4). In the first case, unilateral intervention was not out of the question if the Soviet threat was perceived as high enough. In the latter case, because the traditional threat to national security interests has been greatly diminished with the breaking up of the Soviet empire, even a high threat to universally accepted sociopolitical values, such as grave violations of human rights, is usually insufficient to bring about unilateral intervention; even if it does, the intervention will be practically impossible to sustain over a protracted period of time, as with U.S. intervention in Somalia.

In addition to attitudes based on cost aversion, explicit international norms, such as self-determination and sovereignty, have made foreign military intervention morally and legally difficult to justify in some instances but easier to justify in others.[15] These norms, especially the sovereignty norm, were codified and institutionalized in a number of international legal documents.[16] Establishing and sustaining either international or domestic legitimacy for intervention has thus become a difficult and uncertain task, contributing to the reluctance of the Western powers to intervene forcibly, directly, and openly, especially where intervention requires the use of large-scale military force for extended periods of time (Bull, 1986; Jentleson, 1992).

Some analysts (e.g., Lyons and Mastanduno, 1993) argue that current trends favor a shift in the direction of a more permissive attitude toward intervention in cases where universal values are threatened and members of the international community reach a consensus on the need for intervention. They do not seem to realize that this trend increases the opportunities but not necessarily the incentives for intervention. Countries may avoid intervention for domestic or other reasons, even if they can normatively justify it.

The rising tide of ethnic subnationalism that emphasizes the norm of self-determination does provide more opportunities for intervention in Third World countries by other Third World countries and, on a smaller scale, in eastern Europe. Most Third World countries are multiethnic societies where there has been only partial success in creating a universal sense of loyalty to the institutions of the state. In many cases, ethnic groups are divided by international borders (e.g., the Pathans in Afghanistan and Pakistan and the Kurds in Turkey, Iran, Iraq, and Syria). A discontented minority ethnic group in conflict with a home government is likely to look outside national borders to mobilize military support in the name of self-determination. A neighboring country where the same ethnic group resides and can pressure its home government into intervening on behalf of their kin is the obvious choice. For example, the Tamils of south India forced Delhi to deal with the ethnic conflict in Sri Lanka, eventually intervening militarily in 1987. The opportunities and incentives for intervention driven by domestic political imperatives of irredentism are likely to be even more salient as the pillars of the Westphalian system of state sovereignty decay (Zacher, 1992) and the emphasis on self-determination gains ground. The current polyarchic power structure lacks the interest, capability, and commitment of the hegemonic type to preserve stability and subdue secession conflicts on the periphery unless the conflict threatens the vital interests of the core.

The third nonstructural attribute of systems is the *increase in the political bargaining power of small states*. During the Cold War the competitive relation between the two superpowers turned out to be a blessing for Third World countries, who used it to deter intervention by pointing out the costs that the intervening power might incur in the competition for their hearts and minds. But interventions by regional powers in other Third World countries often gained support from superpower patrons, thus making such interventions a more attractive policy option, even as major powers' incentive for intervention declined.

The fourth attribute is the *increase in the military denial and independent action capabilities of small states*. The competition between the superpowers led to a series of alliances that forced both East and West to increase the capabilities of allies and nonaligned countries to deter external intervention or resist it successfully, preferably without the need for direct involvement by the patron power. Yet this increase in supposedly defensive capabilities produces two spillover effects. First, the buildup of the capabilities of lesser powers increased the temptation for them to undertake interventions of their own in neighboring countries, even when doing so was in conflict with the patron's interests.[17]

Further, when changed circumstances (such as a revolutionary change of regime) require the original patron to intervene in a former client country, the patron may find that its past efforts to build up the political, military, and economic resilience of its client now make the costs of its own intervention unacceptably high.[18] The demise of the Cold War did not resolve that problem because resulting cuts in the military budgets of the major industrialized powers reduced the demand for weapons. Their arms industries have an immense production surplus capacity, and laying off of the many workers may become a serious domestic sociopolitical problem for the arms-producing countries.[19] The arms bazaar has been turned into a buyer's market; willing and able-to-pay Third World clients have access to substantial quantities of powerful modern weapon systems with few strings attached.

The dramatic shift in the location of arms production and the pattern of the international arms trade has resulted in a decrease in Third World dependence on the major powers and a decrease, too, in the leverage that those powers have. There has been the emergence and steady growth of military sales by "second and third-tier" suppliers since the 1970s (e.g., Italy, Spain, Israel, Brazil, North Korea, China).[20] A number of Third World countries have also become self-sufficient in the production of certain weapon systems, such as small arms, artillery, and most types of ammunition—those weapon systems necessary for sustaining limited military interventions in smaller and weaker neighbors. Nonetheless, these countries have continued to rely on the major industrial powers for sophisticated weapon systems, such as high-performance jet aircraft. Another and much less important source of arms is the black market, including illegal sales of military hardware stolen or misappropriated from the stockpiles of major industrial governments (Karp, 1994; Laurance, 1992).

With the emergence and proliferation of second-tier and third-tier arms producers and suppliers, the major powers have lost their monopoly over the arms market. Their ability to manage and control belligerents has consequently declined.[21] The paradoxical logic inherent in this development dictates that because of the availability of alternative suppliers, in situations where the major powers might have considered an arms embargo beneficial to their interests, they are now likely to find themselves acting in contradiction to their best long-term interests. They know that there is a willing alternative supplier waiting to take their place, and they may lose both the market and their influence without having significantly influenced their client's policy. The availability of alternative suppliers, the transformation of the arms market into a buyer's market, and the competition among the major powers prompted the relaxation of arms export restrictions by the major industrial producers and raised the technological sophistication of the weapon systems transferred.[22] The diversification of the sources of arms for Third World countries has resulted in increased political autonomy for buyers and decreased control by suppliers over how the clients use the acquired weapon systems. The clients have thus gained increased opportunities and means to initiated interventions.[23]

But access to an uninterrupted flow of modern weapon systems has also significantly increased the military denial capabilities of Third World countries. Even when indigenous military forces are unable to prevent the presence of foreign military forces on national territory, they can threaten to raise significantly the material and human costs of intervention to the intervening state, thus causing a decrease in incentive to intervene. In fact, many of the weapons systems that could raise the cost of intervention to the intervening power if used effectively by a weak local adversary (e.g., small arms, mines, personal anti-aircraft and antitank missiles, as were used in Afghanistan and Somalia) are easy to transfer, simple to operate, and relatively inexpensive. At the same time, an external intervening power requires heavy investments in logistics and depends on sophisticated heavy weaponry (artillery, tanks, and armored vehicles), as well as combat and logistic naval and air support. This asymmetry, combined with the fact that the stakes for the target are in many cases higher than for the intervener, tilts the balance in favor of the target as the conflict becomes more a battle of wills than a matter decided by the asymmetry of raw military power. In a protracted intervention, the target may actually achieve its goals not by winning but by not losing; the intervener cannot attain its goals without decisively winning and preferably completely eliminating the local adversary's military capability.

This ability of a local target to access external weapon suppliers creates a major military problem for the intervening country, as is indicated by the success of the resistance in Afghanistan and Cambodia—due largely, in both cases, to the secured supply of Western and Chinese weapons. This state of affairs will become more acute as Third World countries gain ever more access to exotic weapon systems, such as ballistic missiles and chemical and biological weapons, that either increase their reach or raise their capability to inflict heavy losses on an intervener (David, 1989: 69–72; Krause, 1992: 187–92).

In sum, the proliferation of the arms industry, as well as the lack of control by a hegemonic power over the trade in arms, even when it adversely affected the bloc leaders' interests, started with the loosening of the blocs, notably the Western bloc, and the split in the communist bloc, which allowed China, France, Israel, South Africa, and others to pursue commercial and political interests through arms sales in defiance of the interests and wishes of both the Soviet Union and the United States. The emergence of a more permissive and less regulated environment for the development of independent defense industries further loosened the bipolar structure by allowing the smaller powers to decrease their dependence on either one of the two superpowers and has given them the means to pursue their national interests aggressively, sometimes to the detriment of one or both superpowers. The minimally regulated arms market in the post–Cold War period and initiatives taken by major powers to build up the military capabilities of their proxies together encouraged regional arms races. The expanded and upgraded military capabilities of Third World countries has made them formidable adversaries in case of armed conflict or intervention by a major power and in the post–Cold War era gives them leverage in dealing with the major powers by making hegemonic imposition through foreign military intervention an un-attractive policy.

Implications of the Systemic Analysis

These trends are likely to expand in scope and intensity, especially as for-mer bloc countries of eastern Europe earn needed cash by producing ad-vanced weapon systems and selling large quantities to whomever is able and willing to pay. Nuclear proliferation in the Third World will make interven-tion an even less attractive option for the major powers and their local non-nuclear allies, which might be held hostage. The current polyarchic structure is likely to become even more fractured and less regulated, thus allowing a great degree of freedom to decisionmakers on matters of national security and other issues. The United States and other major industrialized powers will show little enthusiasm for unilateral intervention in the Third World or, for that matter, in eastern Europe unless they perceive that their vital interests are threatened. With few exceptions intervention will take place mostly in their backyard rather than globally.[24] This decrease in the number of unilat-eral interventions by major industrialized powers must be juxtaposed with an increase in the use of intervention by emerging regional Third World mili-tary powers, which are less risk averse and less inhibited by domestic con-straints.[25] Whether the net results will be more or fewer interventions cannot be predicted.

A possible, if not highly probable, remedy for the increase in unilateral interventions by Third World countries is the emergence of an effective col-lective security system that could (1) strictly regulate the arms trade and the proliferation of nonconventional weapons; (2) create an effective multilat-eral military force to act forcefully when a threat to norms of sovereignty and self-determination is posed by illegal and unauthorized acts of foreign mili-tary intervention; and (3) authorize the multilateral force, through appropri-ate international institutions, to intervene in the domestic affairs of particu-lar states, for example, in cases of serious violations of human rights. These conditions would largely deter individual states from engaging in foreign mil-itary interventions; interventions would be mostly multilateral. However, the chances of these conditions materializing in the foreseeable future appear very slim.

This set of system-related arguments leads to an obvious conclusion: that the opportunities for foreign military intervention since 1945 have increas-ingly been in the Third World; only to a very limited degree have there been such opportunities in Europe, and these resulted in exclusively intrabloc in-terventions. This state of affairs has changed somewhat with the breakup of the communist bloc and the emergence of ethnonationalism in eastern Eu-rope and central Asia. Incentives to intervene have declined for the major powers but increased for lesser powers, especially Third World regional powers. Thus, interventions have mostly taken place in the Third World, have been undertaken to a growing degree by Third World states, and have been directed at Third World states.[26]

Could a case be made that, in the absence of a hegemonic threat, the opportunities to project core values, such as those related to democracy and human rights, would lead to more cases of intervention in order to advance these values? Probably not. Although there is broad support for advancing these values, support is not likely to be translated into a willingness to make the necessary sacrifices. As a senior American official commented in an interview with Thomas Friedman of the *New York Times*: "Do you really think the American people want to spill their blood for Bosnia?" (May 31, 1992). The same attitude was later manifested in the U.S. failure to act to prevent the unfolding Bosnian tragedy. In fact, "resource constraints would seem to demand idealism on the cheap" (Deibel, 1992: 106; see also Van Evera, 1990). This observation is true not only for the United States but for the European nations and Japan as well.

In the absence of a perceived hegemonic threat and with growing domestic competition for inadequate resources, multiple pressures are likely to combine against costly foreign adventures. If the reduction of military forces proceeds among industrialized countries as currently envisioned, it is highly likely that any major foreign military intervention in the future will require the forging of a coalition in order to reach necessary force levels—and not only for the purpose of sharing risks. Western industrialized democracies are therefore going to be even less anxious to take advantage of opportunities for unilateral or even multilateral intervention unless the threat is of a major strategic nature, involving access to critical resources or the threat of nuclear proliferation to nations perceived as irresponsible. Even then, resorting to intervention, whether unilateral or multilateral, is likely to come only after all other possibilities have been exhausted. It might well be, however, that procrastination before resorting to force will make intervention either irrelevant or so costly that accepting the unacceptable will seem preferable to challenging it.

The Response Modes of Third Parties

Focusing the discussion on the system level has provided useful but only general observations of the international environment in which states operate. Within this environment, specific identifiable actors are likely to play a critical role in the intervener's calculations. Intervention is rarely likely to remain a bilateral intervener-target affair. Other actors—the intervener's adversaries or allies—will be brought in at the target's request or under pressure from the intervener or even at their own initiative, when they perceive their substantive or reputational interests threatened. Let us address three questions: What forms might third-party responses take? What are the implications for the intervener of each response option? How can the intervener prevent or restrain interference by third parties?

The strategic relationship between intervener and an adversarial third party spans the whole gamut from compellence, to deterrence, to reassurance. On the adversarial third party's side the response to intervention is in essence a strategy of *compellence*, with four main response options applied singly or in combination. Each option carries a set of potential risks with varying levels of gravity for the intervening power.

1. *Direct counterintervention* involves the deployment of combat military forces by a third party—whether superpower, regional power, or even extraregional small power. In the last case, the third party may act as a proxy for a superpower or as its ally when intervention serves both its own and the superpower's interests. The patron superpower's direct involvement in logistic support for the intervention enables the third party to overcome technological and other resource limits on power projection. To illustrate: in the civil war in Angola, Cuban intervention in 1975 came in response to Zairian and South African interventions. The Soviet Union then assumed responsibility for transporting Cuban troops and Soviet equipment by sea and air, and it also assumed the financial and logistic burden of providing arms (Ebinger, 1976; Klinghoffer, 1980: 110–14; Valenta, 1978). In extreme cases, counterintervention, especially during the Cold War years, risked triggering a global war (e.g., an American counterintervention in Hungary or a Soviet counterintervention in Korea or Vietnam). It also significantly raised the economic and human-life costs of achieving the goals set for the intervention by requiring the deployment of more combat troops than was originally anticipated or raising the assessment of anticipated casualties for the troops deployed.

2. *Paramilitary involvement* entails sending military advisers, supplying intelligence, and, most important, providing substantial military aid to the client in the target state, but refraining from sending in combat troops. This was the case, for example, when the United States and China supplied arms to the insurgents in Afghanistan and Cambodia, respectively (e.g., Klare, 1989; Simon, 1975). Paramilitary involvement raises the costs of goal achievement for the intervener by making local resistance to the intervention more resilient and effective, as is indicated by the casualties inflicted on Soviet and Vietnamese forces in Afghanistan and Cambodia. In both the counterintervention and paramilitary response options, recognition by the target of the serious detrimental consequences to the intervener of external support for the target boosts the morale and confidence of the target and its willingness to resist; the target is thus less likely to give up or to make quick and easy concessions to the intervening power.

3. *Nonmilitary involvement* includes providing or increasing economic aid and political-diplomatic support for the target of intervention or imposing sanctions on the intervener, such as suspending aid.[27] A third party's

nonmilitary involvement raises the risks of intervention by undermining international legitimacy, among other things, and increases the target's capability and will to resist longer, which increases the economic and military costs to the intervener beyond those originally anticipated.

4. *Horizontal escalation* implies direct intervention, paramilitary involvement, and nonmilitary involvement by a third party against targets in regions of vital interest to the intervening actor but outside the region where the current intervention is taking place.[28] For example, the Reagan administration contemplated responding to any potential large-scale Soviet military intervention in the Persian Gulf by hitting back in Cuba, Libya, Vietnam, or even the Asian land mass of the Soviet Union (Epstein, 1987: 30–31). Horizontal escalation raises the costs for the intervening power by applying the first, second, and third options mentioned above in regions where the intervener has vital interests but power projection capabilities that are limited either because of distance and technological constraints or because of prior military commitments elsewhere, which leave it with inadequate resources for a concurrent counterintervention. Such considerations are of particular importance to a major power that has extensive, multiple worldwide commitments, for when faced with simultaneous demands on its resources, it will have to default on at least some commitments.[29] The possibility that the intervening power will be unable to effectively deter horizontal escalation in another region reduces its credibility with allies in those regions. Thus, foreign military intervention, which is often considered the strongest signal of credibility, may actually have mixed effects. By escalating commitments to intervention in one trouble spot, the intervening power signals its vulnerability in other potential trouble spots. Allies in those other trouble spots may then be tempted to reach an accommodation with the third party so that it will refrain from horizontal escalation in their region. That accommodation will not necessarily take into account the focal power's interests in that region.

Each of these options not only increases the direct costs of intervention but has additional, indirect spillover risk effects. Briefly, as direct costs increase owing to a third-party response, the domestic public of the intervening power is likely to become more sensitive to costs and more reluctant to go on supporting the intervention or its escalation, even if escalation offers the possibility of making the intervention successful in attaining its goals. As the obstacles to successful intervention mount, more time is required by the intervening armed forces to achieve a level of acceptable success. A protracted, inconclusive intervention increases the opportunity for the target state to mobilize external support while further straining the limits on domestic support for the intervention policy. Thus, the intervening power may lose its overwhelming edge over time.

The risks associated with a hostile third-party response make it advisable to limit intervention except in regions (1) where the intervening power has easy access and overwhelming military superiority, so that the objectives of the intervention can be accomplished quickly; (2) whose geography makes it possible to isolate the target from interference by a third party (see Chapter 4); (3) for which there is uncontested tacit or explicit recognition that the region falls within the intervener's sphere of influence; and (4) where other major contenders have interests of only secondary importance to them. In the vivid cases of foreign military intervention failure—the United States in Vietnam, the Soviet Union in Afghanistan, and Israel in Lebanon—two or three of these conditions were not fulfilled.

What affects a third party's response choice? Its preference is first and foremost driven by its assessment, often based on a misplaced worst-case threat analysis, of the potential strategic implications of the situation created by the intervention. The third party may consider a forceful military response necessary in the face of what it regards as a critical challenge to its substantial interests, deterrence credibility, and international status.

In contemplating the consequences of a U.N. victory in Korea and the emergence of a united Korea, China, for example, perceived several threats, among them, the emergence of a U.S.-Japan-Korea alignment against China; the threat to China's role as an Asian power; and the severe repercussions on internal Chinese politics of what seemed a threat to the newly established, vulnerable communist regime (Whiting, 1960: 156–60).

Besides the perceived vitality of the interests threatened, other considerations are the third party's capabilities and resources, assessment of the risks associated with each of the four possible third-party responses, and the related perceived benefits to the third party from employing one or more of these responses. Very few third parties have the resources to choose any or all four options. Only major global powers or sometimes major regional powers have that luxury. In most cases, therefore, a third-party response takes the form of paramilitary or nonmilitary involvement.

The prospect of third-party interference calls attention to an inbuilt policy dilemma. In many cases intervention is used to signal credibly to third parties how far the intervening actor is willing to go in defense of its interests. In some cases preemptive escalation through intervention is intended to prevent worse types of escalation. This was, for example, an argument used by Presidents Truman and Johnson to rationalize intervention in Korea and Vietnam (Hosmer, 1987; Kearns, 1976: 270; Truman, 1956: 336–39). In other cases, intervention is necessary to deter a third party from employing compellence measures against the target state. But the irony is that intervention may trigger the very same third party behavior that it was intended to deter.

The intervener can take either of two approaches in dealing with the possibility of third-party interference. The first is a straightforward strategy of *deterrence* that combines threats of a punitive response by the intervener and a show of determination to proceed with intervention no matter what. If, however, the intervener believes that a third party will perceive intervention as a challenge, a deterrent posture could result in amplification of the challenge to the third party, so that the actor that sees its reputational interests challenged believes that it must respond at any cost. In these circumstances, an intervener eager to avoid intervention or involvement by a third party should explicitly build into the intervention policy prior to its implementation a strategy of reassurance.

Reassurance requires explicitly or tacitly negotiated arrangements with the third party.[30] This approach works by moderating the third party's perception of an emerging threat to reduce its incentive to interfere. One way to reduce threat perception is by establishing a mutual understanding between the intervener and the third party about the purposes and limits of intervention, thus quieting the third party's fears in advance. Syria's intervention in Lebanon in 1975 made it necessary for Syria and Israel to establish a mutually understood (through a combination of public statements and diplomatic exchanges) set of constraints on Syrian military activities in Lebanon. These included limiting the presence of Syrian military units to the area at least forty kilometers north of the Israel-Lebanon border; refraining from positioning SAM anti-aircraft missiles on Lebanese territory; agreeing not to use the Syrian air force in Lebanon, particularly against the Christian forces; and not interfering with the freedom of operation of the Israeli air force in Lebanon (Bar, 1990: 127–31). These "red lines" stabilized the strategic relationship between Syria and Israel and prevented unwarranted escalation, allowing both sides to pursue their interests.

In such a context, a reassurance strategy has two advantages over a deterrence strategy. One is that a deterrent posture in fact challenges the third party, increasing the threat to its reputational interests provoking it into taking chances where the costs are significant. Second, a reassurance strategy allows for a more gradual escalation in the intervener's dealings with the third party than does a deterrence strategy adopted at the start. In some cases, however, an effective reassurance strategy is not a viable option.

Intervening powers have to take into account not only responses by adversaries but also the views of friendly nations and allies. For major powers and even for a superpower, support by friendly states is a source of external and internal legitimacy for a policy of intervention; losing such support may cause erosion of legitimacy. For lesser powers the consequences are even more severe: intervention may become unsustainable if allies withdraw their support, leaving the intervening power overextended politically, militarily, and economically. This is what happened in the case of French-British intervention in Egypt in 1956. The British decision to withdraw from the operation that they and the French had initiated in response to the nationalization of the Suez Canal was the result of active American opposition to the operation. The United States made it clear that it would not intervene financially nor make available the funds necessary to stop a run on sterling. The other lever used by the United States was the dependence on oil of its European allies: the Americans indicated that they would withhold oil supplies until the

British and French complied with the U.N. resolution on a cease-fire and withdrawal (Finer, 1964; Neff, 1981).

The responses of adversarial and friendly third parties are not unrelated. Successful deterrence of adversarial third-party interference, especially direct counterintervention, makes friendly third-party support for the intervener more likely simply because it has become less risky. Failure to deter adversarial third-party interference makes support by allies of the intervener less likely because of the increased risk. As a rule, awareness that the intervener's allies are not supporting the intervention provides an incentive for an adversarial third party to challenge the intervener because of the decline in potential risks. On the other hand, realizing that the intervention is supported by the intervener's allies provides a disincentive for interference by the third party because of the implied increased risk.

The unfolding events in the Suez crisis of 1956 illustrate these points. In the first days of the military operation the Soviets confined their response largely to public denunciation of Britain, France, and Israel and to taking hostile positions against them in the United Nations. The Soviets were still uncertain where the United States stood. By early November, when it became clear that the Eisenhower administration did not support the operation, the Soviets adopted a much more aggressive stance. On November 5 they threatened the use of force and made what could be taken as a threat to use nuclear weapons against Britain, France, and Israel. This state of affairs was a chief cause of the urgent American application of pressure on its allies to reach an early cease-fire and forestall a potential Soviet military intervention (Campbell, 1989; Finer, 1964: 416–17).

The exception to this rule occurs when a third party interprets the support of the intervener's allies as a qualitative increase in the threat against it. When intervention is viewed in this broader strategic perspective, support by allies provokes interference by the third party rather than deterring it. When Vietnam intervened in Cambodia, China's perception of the magnitude of the threat was substantially influenced by the Soviet-Vietnamese alliance. The resultant massive military attack on Vietnam by China (horizontal escalation) "was designed not only to blunt Vietnamese pride and restore China's credibility but also to force Vietnam to withdraw some of its units from Cambodia. It was also to be a lesson for the whole Western world, which appeared to China as paralyzed in the face of Soviet adventurism" (Chanda, 1986: 358–59).

The Salience of Relative Capability

International factors are instrumental in shaping the opportunities and payoff structure of interventions, and these factors, at both the system and state levels, should play a prominent role in facilitating or inhibiting the use of intervention as an instrument of statecraft. Why, then, are external factors given a relatively marginal weight in many intervention decisions, compared with decisionmakers' intense focus on the balance of military power between intervener and target? To be sure, some decisionmakers are better informed than others and do pay proper attention to the international environment. The following explanation pertains to those who fail to attend to external factors properly.

It can plausibly be speculated that a narrow capability-focused view can be traced to three common heuristic-related judgmental biases and one important motivational bias (Kahneman et al., 1982; Tversky and Kahneman, 1974; Vertzberger, 1990: 116, 144–56). First, the availability heuristic biases decisionmakers to think of military power as the decisive causal agent because military power is so salient compared with the other factors. The notion of an international system is too abstract to be grasped in operational terms, and a third party's response is a matter of speculation, whereas information on the balance of capabilities with the adversary is concrete and seemingly of immediately most relevant diagnostic value.

Second, the representativeness heuristic causes decisionmakers to directly and, in many cases, incorrectly correlate power over resources with power over outcomes. They, therefore, correlate their superior capabilities with positive policy outcomes.

Third, the simulation heuristic biases decisionmakers, who readily evoke scenarios in which overwhelmingly superior military power resolves difficult issues neatly and quickly, as unilateral interventions are initially structured and designed to do. Decisionmakers who commit themselves to a policy of intervention assume quick successes resulting in a fait accompli, which would substantially reduce or eliminate the risk of detrimental effects from international factors.[31] Decisionmakers therefore dismiss international factors and come to think of military power as the proverbial sword that cuts the Gordian knot.

Fourth, these cognitive biases are compounded by a motivational bias that is related to the need for, and perception of, control in threatening situations where the stakes are high. Decisionmakers who face risky decisions and have an acute sense of accountability are more likely to pay attention to variables that provide a sense of control over outcomes than to variables over which they believe they have little or no control.

National capabilities and their application to the problem at hand seem, correctly or not, to be policy instruments that decisionmakers control and that can be used at will to deal with and produce desired outcomes. This illusion of control is missing in considerations of third-party responses. Systemic attributes are perceived as even more imponderable and beyond decisionmakers' control. Capability variables therefore become the focus of attention in policy formulation. Once a tight cause-and-effect link is identified between capabilities and outcomes, decisionmakers—being the overburdened satisficers and cognitive misers that they are—are not very keen to reconsider, update, or expand their cause-and-effect attributional theories, especially because some of the additional possible causes are abstract constructs that are not easy to grasp, such as international systems. The mutually reinforcing effects of these four biases can produce dangerous misperceptions with regard to the ultimate efficacy of the use of state power. The international context, in which the intervening state has to operate, is de-emphasized, and quick success is expected, but in many cases, it does not materialize.

The policy consequence of expecting success and slighting international constraints is self-entrapment at home and abroad. Leaders of the potential intervening country become preoccupied with the reputational consequences of not demonstrating their power and resolve through military intervention. At the same time, self-perceptions of power and resolve ripple through the intervener's society and induce an attitude of overconfidence among important sections of the public. These sections of the public and their political representatives exert political pressure on the national leadership to use force in the service of the national interest. The leaders may perceive that not doing so will have serious domestic political repercussions.

This double entrapment is echoed in McGeorge Bundy's advice to President Lyndon Johnson on the eve of the crucial escalation decisions on Vietnam in 1965. He told the president that "even if it fails, the policy will be worth it. At a minimum it will damp down the charge that we did not do all that we could have done, and this charge will be important in many countries, including our own." In another assessment Bundy made the point that the "Goldwater crowd," who would attack Johnson if he pulled out, were "more numerous, more powerful and more dangerous than the flea-bite professors" who would attack him if he stayed in (Snyder, 1991b: 300).

Decisions to intervene are thus usually informed and driven by a blend of perceived threats to tangible and reputational national interests or, in some cases, a powerful sense of a threat to decisionmakers' political positions, which is translated into the perception of a threat to national interests; the perception of threats to national interests is accompanied by a strong conviction that foreign military intervention will be a straightforward or even once-and-for-all answer to the threats. These expectations are embedded not only in simplistic assessments of the balance of capabilities but also in misperceptions of the balance of resolve. Interveners believe that the balance of resolve favors them mainly because their goals are limited to the neutralization rather than the destruction of the adversary (Jentleson et al., 1993; Kupchan, 1992: 256, 258). That and the misinterpretation of the implications of the immense power discrepancy between the initiator of intervention and the target produce an "inevitable-win" bias (Vertzberger, 1992). Still, the inevitable-win bias does not automatically result in a pervasive preference for intervention over other options. Decisionmakers may believe in winning but still be reluctant to pay the costs of winning, even if they are sure that they will eventually prevail.

For both the decisionmakers and public the inevitable-win bias is strengthened by a highly positive, patriotic perception of their nation. This self-perception has two primary effects. Information consistent with the national image is better attended to and remembered than information inconsistent with it. Inconsistent information is distorted and recalled as if it were consistent. This kind of patriotism accounts for the popular support of American military interventions when they are represented as ways to protect freedom and democracy, two attributes closely associated with the patriotic national image. To the extent that intervention has negative aspects, such as the bombing of civilians, they are explained as unavoidable and beyond control—"there was no other choice"—and, in any event, as being ultimately for the victims' own good (Hirshberg, 1993). The perseverance of the positive national self-image elicits legitimacy for the policy of intervention, reduces the effect of critical inconsistent information, and enhances a self-righteous sense that state power is being used unselfishly in a good cause.

It takes a special cognitive effort on the part of decisionmakers to weigh properly the vivid inferences about favorable outcomes flowing from perceptions of disparities in resolve and capabilities between intervener and target against the implications of the much less vivid international factors, which usually suggest caution. It requires even more of an effort to analyze systematically and sophisticatedly the actual limits on power projection and the nonlinear relation between control over resources and control over outcomes. Therefore, when a potential intervener finds resolve and power disparities extremely meaningful, the temptation to resort to intervention is such that otherwise astute leaders may fall into the trap of the inevitable-win bias, fail to scrutinize closely the effects of international constraints, and remain unaware of the limits on the use and effectiveness of coercive state power.

When they decide to intervene, then, the international factors are marginalized or at best related to the possible adverse reputational consequences of not demonstrating resolve and not using their superior capabilities.[32] A perceived threat to the reputational credibility of an actor can be a powerful incentive to risk intervention. The domino theory is invoked; that is, once a challenge to the position of a major power fails to generate a forceful response, that power is likely to be challenged and compromised in other ways.

When international factors are influential in decisionmaking, the intervention option *is more likely to be rejected*. The Eisenhower administration considered the international context when deciding not to intervene in Indochina in 1954, but the Johnson administration did not consider it a salient factor when it decided to send troops to Vietnam in 1965.[33] The same pattern is demonstrated in the more recent Panama intervention. President George Bush rejected the possibility of sending U.S. troops for combat duty in Central America during his first year in office, emphasizing the sensitivity of Latin American nations to U.S. intervention. Yet when he eventually changed his position in late 1989, an informed administration source is quoted as saying: "[The Panama intervention] was a military decision to do it, not a political one. George Bush made the decision regardless of the risks on the political side" (*Washington Post*, December 24, 1989).

I do not argue, however, that decisionmakers are incognizant of international constraints, only that international constraints are often not given their due weight in the decision to intervene. In fact, participants in the decisionmaking process recognize intervention as likely to involve considerable investments of resources—material, human, and reputational—with the potential for very significant personal and societal consequences. The decisions are preceded by intense interelite debate. Decisionmakers are therefore probably aware of the range of the issues involved (Kupchan, 1992: 255–56), but they are also likely to misperceive the relative importance of international factors.[34] Although international factors play only a secondary role in the deliberations over intervention decisions, they can affect how the intervention policy is implemented by eliciting self-imposed restraints (e.g., the intervener can restrict its military forces to particular areas or not use certain weapon systems) and the accompanying justifications and legitimization strategy for the intervention policy.

Conclusions

The overall trend in international politics is toward reduced use of unilateral foreign military intervention; intervention is not as widespread as it could have been, given the great inequalities among nations in power distribution and the number of available opportunities. Many opportunities for intervention have not resulted in intervention in spite of substantial power disparities between the potential intervener and the potential target. But intervention is still, and is likely to remain, a common and important policy option. When it occurs, its consequences are frequently far-reaching, sometimes disastrous, and often unanticipated. Intervention by one power could become an incentive for counterintervention by another power, which in general makes intervention normatively more tolerable than otherwise, even if it is not legally condoned (Moore, 1969: 238–39).[35] Intervention often alters the regional balance of power by weakening some of the actors and may consequently create opportunities and incentives for future interventions that were unanticipated when the original intervention was contemplated. In the long run, there are rarely winners, and the question to be asked is not who won but who lost less. This is not to imply that intervention should never be used under any circumstances, but rather that in high-risk cases intervention is a minimax strategy that should be used wisely.

Briefly, in this chapter I analyze the effects of external factors, namely, systemic attributes and third-party responses and explain the cognitive and motivational biases that cause misperceptions in judging the importance of international factors in policy formulation. More specifically, it is argued that the analysis of systemic structural attributes that reflect the power distribution within a system result in general insights regarding the broad constraints and incentives for intervention, even if it does not provide specific policy-relevant knowledge or determinate predictions. There are quite a few exceptions to the general rules, and in comparisons across systems the most we can hope for are broad ordinal observations identifying the types of systems likely to have more (or fewer) interventions, everything else being equal. Adding nonstructural variables to the analysis results in greater detail and specificity. These variables explain changes in incentives and constraints on intervention *within the same system over time*, as well as similarities *across systems* with different power distributions. Thus, for example, system norms can explain why a preponderant power may be reluctant to impose its will in defense of its national interests through direct military intervention, as might be expected from a purely structural point of view. As we saw, nonstructural attributes of the system are also useful in identifying the regions where interventions are most likely to take place, as well as the type of regional actors that are most likely to become interveners or targets and the type of intervention they prefer (unilateral versus multilateral).

The actor-specific variable, third party's anticipated and actual behavior, is critical for determining an intervener's behavior. It is salient because the behavior of third parties directly affects the risks and costs of intervention. The potential third parties are specific, identifiable actors, and their behavior takes predictable forms. Taken together, the analysis of systemic structural and nonstructural attributes and third parties' behavior will provide decisionmakers with a map of the external risks and opportunities for using foreign military intervention as an instrument of statecraft. Misjudgment of the relevance and importance of the international setting will result in a faulty and probably costly decision. The reason is obvious. Where the potential intervener perceives capabilities and resolve as clearly in its favor, only external constraints can moderate the inevitable-win bias. Underestimating the role of external constraints enhances the salience of capability-based expectations of success; these expectations are not necessarily warranted and act as an incentive for intervention, where caution would have been advisable.

PART TWO

The Case Studies

A Comparative Analysis

CHAPTER 6

Foreign Military Interventions with Low to Moderate Risks

GRENADA, PANAMA, AND CZECHOSLOVAKIA

Two of the three analyses of low- to moderate-risk foreign military intervention concern cases of U.S. intervention in the Western Hemisphere but not at its immediate borders. The third case concerns Soviet intervention in eastern Europe. The interventions, Grenada in October 1983, Panama in December 1989, and Czechoslovakia in August 1968, were all short-term, involved low to moderate risk, and exemplify hostile interventions, that is, military operations directed against the governments in power with the immediate objective of changing the regimes or governments. Yet these interventions had dissimilar motivating causes, they were executed by two diametrically different regimes and types of decisionmaking systems, and they were legitimized, domestically and internationally, in different ways. The Grenada intervention had mostly external causes related to global conflict between the superpowers and their respective proxies and clients. The causes of the Panama intervention can be traced mainly to U.S. domestic politics—the Bush administration's effort to fight the drug war and bolster a declining domestic reputation. These primary causes were enhanced by a set of secondary causes. The Czechoslovakia intervention was triggered by a perceived threat to the Soviet-dominated domestic political order in eastern Europe and, by implication, to Soviet security.

By examining the military, economic, and political (domestic and international) risk dimensions of these three interventions and assessing how they affected risk preferences and risk acceptability within the administrations of Ronald Reagan and George Bush and the Politburo of Leonid Brezhnev, and by discussion both the pre-intervention anticipations of the decisionmakers and the realities of the actual interventions, I hope to illuminate discrepancies between the quality of planning and accuracy of expectations on the one hand and actual outcomes on the other hand. This approach also allows us to evaluate to what degree these discrepancies were significant in affecting the achievement of objectives and the costs. As we shall see, in spite of major differences in regime type the two interveners, the United States and the Soviet Union, had much in common. At the same time the cases illuminate differences between situations that in essence involve decisions by single leaders and situations that involve group decisionmaking.

The Grenada Case

AN OVERVIEW

The Caribbean island of Grenada lies 100 miles north of the Venezuelan coast. The most densely populated island in the Lesser Antilles, it has a population of 100,000. In March 1967, Grenada became a self-governing state in association with Great Britain. It was granted full autonomy over its internal affairs, but Great Britain remained responsible for external affairs and defense. In the first general election, in August 1967, the Grenada United Labor Party (GULP) won the elections, and its leader, Eric Gairy, became prime minister. His main opposition was young, overqualified, unemployed middle-class students educated mostly in the United States and Great Britain who had returned in the late 1960s. Ideologically they were inspired by the Black Power Movement of the time, the Cuban Revolution, and Tanzanian *ujamaa* village participatory democracy. Politically they were radicalized by the increasingly authoritarian behavior of Gairy's regime. Gairy responded to their radicalism with oppressive measures that were carried out by the secret police. When oppression accompanied by corruption and inefficiency caused further deterioration of the economy, dissatisfied young professionals and intellectuals founded two movements: the Joint Endeavor for Welfare, Education, and Liberation (JEWEL), which was led by the economist Unison Whiteman and the sugar factory clerk Selwyn Strachan, and the Movement for Assemblies of the People (MAP), which was founded by two young lawyers, Maurice Bishop and Kendrick Radix. In March 1973 the two groups merged to form the New Jewel Movement (NJM), which espoused pragmatic nationalism and saw itself as a radical populist alternative to Gairy. The NJM shifted to the left during the 1970s, but there was ambivalence in its leftist leaning that became a source of contention between its more radical and more moderate factions.

In 1974 the island won full independence from Great Britain against a background of growing labor unrest and increasing oppression by Gairy's secret police. When the political and economic situation worsened, the various opposition groups formed a coalition, the United People's Party. GULP won only a bare majority in the general election in 1976, and Maurice Bishop became leader of the parliamentary opposition. Increasing corruption and economic decline prompted much of Gairy's remaining support to shift to the NJM. In March 1979, during Gairy's absence on a trip to the United Nations, the NJM seized power in a move welcomed by most Grenadians. The NJM turned Grenada into a one-party system, modeling the structure of the party on that of the Soviet Communist Party, suspending the democratic constitution, and banning opposition. Soon enough, the People's Revolutionary Government (PRG) found itself accused of the same human rights violations as its predecessor. The economy, owing to effective management by Minister of Finance Bernard Coard, improved meanwhile, leading to a budget surplus and economic growth. With economic growth the confrontation between the pragmatists and the Leninists in the party intensified. The latter were led by Coard, an uncompromising believer in "scientific socialism," who claimed that the revolution was moving too slowly. Coard's radical faction challenged Bishop's more pragmatic faction, increasing its attacks in 1983, accusing Bishop of turning the NJM into a social democratic party and failing to get closer to the World Socialist Movement (Cuba, the USSR, and the GDR). A compromise between the two factions, reached with Cuban encouragement, called for joint Coard-Bishop leadership of the government. The agreement did not last long. While Bishop was on a trip to eastern Europe and Cuba, the Central Committee ordered the army not to take any more orders from Bishop, only from the Central Committee. Shortly following his return to the island, Bishop was put under house arrest, stripped of his positions, and expelled from the NJM. This provoked a public outcry that forced Coard to resign as deputy prime minister. On October 19 a crowd freed the popular Bishop from house arrest. Weapons were distributed to Bishop's supporters, but soldiers, on the orders of the commander of Grenada's military forces, General Hudson Austin, turned their guns on the crowd. Bishop told his supporters to surrender, whereupon he and his senior colleagues were detained. Later, Bishop and seven of his colleagues, including cabinet ministers and trade union leaders, were executed, and 60 or so of his supporters were shot down by the People's Revolutionary Army (PRA). The next day it was announced by Radio Free Grenada that the Revolutionary Military Council (RMC) headed by General Austin had taken power in Grenada.

The U.S. response to the 1979 revolution in Grenada, when the NJM seized power, and to the events culminating in the American intervention in 1983 must be viewed in the context of the extensive U.S. concern with the entire Caribbean region, specifically, its geostrategic perception of the Caribbean Basin as a region of exclusive U.S. dominance and potential vulnerability. American decisionmakers recognized U.S. interests in the region to be multifaceted—strategic, economic, and reputational—and therefore were likely to respond to a perceived serious threat to these interests with a strongly activist reaction, as was the case with U.S. policies toward Cuba and the Dominican Republic in the 1960s. They were alarmed by the attitudes and policies of the NJM because the party espoused a foreign policy intended to challenge U.S. hegemony through solidarity with the nonaligned movement and Soviet bloc countries, particularly Cuba, thus ensuring material and moral assistance. Castro, on his part, saw the NJM challenge as an opportunity to demonstrate what Cuban aid could accomplish. Under President Jimmy

Carter the American response was to warn the PRG against closer ties with Cuba. At the same time, the Americans attempted to avoid a direct confrontation by treating Grenada as a regional issue (Thorndike 1989: 255), preferring to avoid unilateral initiatives and to follow the lead of the Caribbean Community (CARICOM).

U.S. policy changed drastically with the Reagan administration, which was determined to adopt a "get tough" policy in the Caribbean Basin. A firm Cuba-containment policy became a top priority, for Cuba was seen as a Soviet surrogate. The Reagan administration regarded Grenada as a launching point for Cuban-Soviet operations against the rest of South America and paid particular attention to the strategic threat implications of a huge $71 million international airport project in Grenada that was largely funded and constructed by Cuba. U.S. policy consisted of applying multiple pressures on the PRG: restriction of diplomatic contacts, suspension of economic aid and active pressure on allies to do likewise, a sustained barrage of propaganda that affected tourism, and a series of military maneuvers and exercises that signaled possible military intervention. In October 1983 the bloody power struggle that erupted within the NJM provoked the Reagan administration to intervene directly. The United States cited the need to rescue 1,000 U.S. citizens, mainly students enrolled at St. Georges Medical School, and legitimized the action by mentioning an invitation to intervene under the umbrella of the Organization of Eastern Caribbean States (OECS) charter. The intervention was justified on the grounds that the military regime in Grenada had become a threat to Grenada's five eastern Caribbean neighbors.

The U.S. decision to intervene was made on October 22 and implemented on October 25. Operation Urgent Fury lasted a week. The bulk of the military forces were Americans. In addition to U.S. military units, forces from Jamaica and Barbados and police contingents from Antigua, Dominica, St. Kitts-Nevis, St. Lucia, and St. Vincent constituted the Caribbean Peace-Keeping Force. Part of the forces remained in Grenada until September 1985 to stabilize the new government and to preserve law and order.[1] The Grenada operation is, therefore, a typical case of hostile intervention as defined in this study. It was a short-lived military action that involved American and some Caribbean uniformed military units, including ground forces, operating in combat formations to overthrow the Grenada regime.

U.S. INTERESTS

The U.S. assessment of the significance of the revolution in Grenada and its aftermath was shaped by the perceived threat to its interests, which were viewed within the broader perspective of U.S. concerns for its position in the Caribbean Basin. Although the United States has had multiple interests in the region, the most salient ones pertaining to the decision to intervene militarily were the strategic and the related reputational interests that could be affected if decisive forceful measures were not taken. The intrinsic value of these interests was strongly influenced not only by the self-declared leftist anti-American nature of the revolution in Grenada but also by other events in the international system that took place about the same time. President Reagan's worldview endowed the events in the Caribbean with an extremely threatening content because he interpreted them as part of a global pattern, rather than as merely regional or local disturbances. The risks of inaction were thus perceived to be much higher than the risks of a limited operation, even if the intervention would be the largest combat deployment of U.S. troops since the end of the Vietnam War.

It was during the Reagan administration that U.S. officials once again emphasized the view of the Caribbean as a U.S. lake. The origins of this geo-political image can be traced to the mid-nineteenth century and the Clayton-Bulmer Treaty of 1850 between Great Britain and the United States restricting the former's influence in Central America. The idea acquired momentum during the Second World War and persisted when the underlying geopolitical logic was formalized in the cartographic concept of the Caribbean Basin. The concept implied not only the strategic unity of the islands and the Central American rimland but also stressed the strategic vitality for U.S. interests of a region in the U.S. "backyard." Consequently, American foreign policy became more sensitized to the vulnerability of the southern borders. This perception of threat derived from three historical experiences: German U-boat activity in the Caribbean from 1940 to 1944; the Cuban Revolution and the American failure to uproot the Castro regime; and the Cuban missile crisis of 1962, which led to U.S. assurances that it would not use force against Cuba. Together, these made U.S. foreign policy decisionmakers particularly cognizant of the strategic threat implications of the permanent presence of a foreign power in the region. This cognizance was expressed in the Johnson administration's intervention in the Dominican Republic in 1965. But the lessons of Vietnam and the growing preoccupation with other, more acute flash points, such as the Middle East and Europe, where the United States and its main protagonist, the Soviet Union, faced each other directly, deferred the resolve for direct military intervention in the Caribbean region. Indirect means of intervention were preferred—specifically, the use of economic destabilization against unfriendly governments, such as Michael Manley's pro-Cuban regime in Jamaica between 1974 and 1980.

The Reagan administration presented a Caribbean version of the domino theory, that is, the notion "that each island offers a potential beachhead for projecting power, presumably hostile, toward the United States" (Thorndike, 1989: 251). Grenada was perceived as a potential base for Soviet and Cuban long-range bombers, reconnaissance planes, or fighter aircraft operating against land or maritime targets and as a possible support base for insurgencies in the region (Zakheim, 1986: 178). Such concerns were rooted mostly in the Reagan administration's worldview, in the historical experiences mentioned above, and in U.S. contemporary commercial and economic interests.

A substantial share of U.S. oil imports either originate in or pass through the Caribbean. The Caribbean sea-lanes are important for maritime traffic to and from the Panama Canal, including a large share of U.S. seaborne commerce, especially oil imports handled by American ports on the Gulf of Mexico.[2] Caribbean countries are the most important source of bauxite and aluminum for the United States; they are also a major tourist destination for Americans and a source of legal and illegal immigrants. With the growing U.S. concern about drug trafficking, the fact that the region is both a producer and an important transit area for Latin American narcotics shipments into the United States became another reason for attention to the region. But these issues by themselves were not perceived as important enough to provide a justifiable incentive to use foreign military intervention as an instrument of statecraft.

What transformed the Grenada situation into a major issue of American concern and triggered foreign military intervention was the way Washington officials perceived and interpreted developments in Grenada in the broader global context. The core of President Reagan's foreign policy was to halt the decline of American power and influence in the world. This meant a reconstruction of American hegemonic influence and an active foreign policy involving the exercise of power as and where necessary. However, the policy required broad domestic support. Its success would be measured by containment of the main adversary—the Soviet Union—and its proxies, reinvigoration of U.S. alliances that fostered support for American foreign policy objectives, and, finally, renewed respect for the United States in the Third World (Haftendorn, 1988; Zakheim, 1986). Grenada was a test case where all these points could be brought home and was therefore a test of the validity of Reagan's foreign policy. In October 1983 the eruption of bloody factional rivalry and the alleged threat to 1,000 American students and citizens on the island created a window of opportunity for military action. The Reagan administration realized that because of its dependence on public opinion, it had only a short time in which to act.

Reagan harbored an intense ideological aversion to the left. This position stemmed from what seemed at the time to be a set of unchecked advances by the Soviet bloc in the face of a declining, weak American superpower that was preoccupied with secondary concerns, such as human rights and democracy, while neglecting its vital interest of containing communism. The bold challenges of the Soviet Union—the "evil empire," according to Reagan— took the form of direct military intervention in Afghanistan, support of Vietnam's intervention in Cambodia, and promotion of leftist movements both directly and through proxies like Cuba. The countries of the Caribbean, Central America, and Africa, according to this view, demonstrated the severity of the unprecedented Soviet threat. Events in Afghanistan, Cambodia, Angola, Nicaragua, El Salvador, and Grenada seemed to represent the pattern of Soviet behavior and that of its allies and, as well, a test of U.S. resolve. Failure on the part of the United States could result in a chain of Marxist revolutions that would turn the Caribbean Basin into a hostile region under Cuban influence. In this context, the revolution in Grenada was not a mere change of regime in a small, poor, unimportant island state but a salient link in a chain of events that, if allowed to continue, could threaten vital American interests. The Caribbean became a new front line in containing world communism, and Grenada became an example of what was increasingly known as the Reagan Doctrine (Meese, 1992: 222; Reagan, 1990: 455–56; Weinberger, 1990: 105–6).

The perception was that if the United States could not roll back communism in its own backyard, it would lose its position and credibility as a world power, invite more aggression and subversion, alienate its traditional friends, and prove that the Reagan Doctrine was nothing but empty words. Responding to the urgent request of OECS members on October 22, the president rhetorically speculated on what kind of country the United States would be if it refused to help small neighboring democracies defend themselves against tyranny and lawlessness. Secretary of State George Shultz expressed this view clearly: "If we want the role and influence of a great power, then we have to accept the responsibilities of a great power" (quoted in Burrowes, 1988: 74; Reagan, 1990: 451; Shultz, 1993: 324, 336). Confirmation of the reputational consequences of inaction came from Eugenia Charles, head of the OECS and prime minister of Dominica, whom Secretary of Defense Caspar Weinberger described as "extraordinarily eloquent and persuasive" (Weinberger, 1990: 121). During her visit to Washington on October 25, 1983, she firmly argued that inaction would make military action more difficult later and "would contribute in the meantime to the impression that the United States had not the resolve, and perhaps not even the strength, to take decisive action when it was required" (Weinberger, 1990: 121–22). Her words impressed President Reagan and the members of the National Security Council.[3] They not only seemed to justify earlier views held by the president but confirmed the need for action in Grenada at the same time that reports on the high loss of life among U.S. Marines killed by Muslim terrorists in the attack on the Beirut barracks became known.

From this global perspective, Americans could be expected to support a national-interest-based incentive for using military force. What was missing was opportunity, which the events of October 1983 provided. The Reagan administration assessed military intervention in Grenada as a low-risk occasion to demonstrate resolve without the danger of a serious confrontation with the Soviet Union; the United States could simultaneously gain domestic and international legitimacy for its role as a superpower and thereby for the operation. Administration officials knew, however, that Reagan's worldview and strategic analysis were contested among broad sectors in American society. They also realized that such an operation required broad bipartisan support from the public and the Congress to overcome anticipated opposition that might urge activation of the War Powers Resolution to interfere with the deployment of extensive military force.[4] Consequently, American interests had to be explained not merely in somewhat abstract geostrategic terms but also in terms of the specific, immediate obligation of a state to defend its citizens who face a life-threatening situation, even if the reality was quite different.

> In fact U.S. citizens were in no danger until the operation was launched. The assault on the island could easily have precipitated the taking of hostages by desperate men driven to desperate means to save their own lives. But rescuing students was really a smoke screen to conceal the real motive: the seizing of an unprecedented opportunity to rid the Caribbean of an expanding communist threat and at the same time permit the military to regain credibility in a situation in which they could not lose. (Adkin, 1989: 263–64)[5]

Yet the planned operation, whatever the motive, did contain some risks, which are discussed below.

NATIONAL CAPABILITIES

For a country as powerful and large as the United States to invade a weak, small country like Grenada, an island that can be easily isolated from external third parties' support, should have been no major military problem. Grenada offered a target of opportunity that "was a military planner's dream" (Burrowes, 1988: 134). Not only was it small and isolated but the population was hostile to the regime. The operation to be executed by Joint Task Force 120 seemed easy, and the original plan was accordingly simple. U.S. Marines were to take the northern part of the island, including Pearl Airport. The Army Rangers had the task of taking the southern part of the island, including the medical school campus and the new airport, still under construction, at Point Salines. The sea, air and land teams (SEALS) and special forces were supposed to ensure the safety of Governor-General Sir Paul Scoon, capture the radio station, and free the political prisoners at Richmond Hill prison. The 82nd Airborne was then to take over and engage in mopping-up operations to secure all points for the 350-strong Caribbean Peace-Keeping Force that was to follow. The assumptions were that the main part of the operation would be over within a day and that Grenadian resistance would be weak.

Yet there were risks: military risks, measured by casualties for both intervener and target, and political risks, both domestic and international, measured by support for the operation at home and abroad and its sustainability. The political risks made a military operation unthinkable without a legitimizing event to justify and trigger intervention. Dependence on such a trigger, whose timing could not be anticipated, created two interrelated problems. First, the window of opportunity might close, but fast action did not allow for proper planning and intelligence gathering. Still, according to Defense Secretary Weinberger, attention to Grenada predated the intervention by a good bit. "I had begun receiving regular intelligence briefings on Grenada early in the new administration" (Weinberger, 1990: 105). Yet attention did not translate into operational planning and systematic operations—

relevant intelligence gathering. Furthermore, when the opportunity for using military intervention is constrained by time, it is not the overall size of the intervener's force but the immediately available deployable forces that matter, giving a less-than-optimal structure of forces for the operation and therefore an increased risk of casualties. This is particularly true where the target country is distant and force deployment requires time. The combination of a narrow window of opportunity and a limited time for deployment makes only a fraction of the overall military strength of the intervener relevant, especially in short-term interventions like Operation Urgent Fury.

U.S. military planners had to contend with a small, poorly trained Grenadian military force, which consisted of 1,500 soldiers in the regular army supplemented by a militia estimated to be 2,000 strong (Weinberger, 1990: 131). These forces would be reinforced by a few hundred Cubans, including 40 advisers and 650 armed construction workers (Adkin, 1989: 159). Because of inadequate intelligence, there was uncertainty about the exact number of Cubans and the quality of their military training. It was assumed, however, that they must be well-trained soldiers who would put up a fierce fight, which could raise the number of casualties among the attackers. Although these numbers in themselves did not pose a substantial problem, given the huge asymmetry between U.S. and Grenada's military capability, constraints were posed by the narrow window of opportunity, the limited immediate availability of troops (because of distance from the theater of operations), and requirements for troop deployment to regions outside Grenada. These constraints were made more salient by Fidel Castro's ability to airlift perhaps 5,000 to 10,000 Cuban troops to Grenada within days, making secrecy and quick action even more essential (Beck, 1993: 105; Menges, 1988: 69). But assembling the intervention force presented problems for the United States about the same time that it faced a crisis in Lebanon. There was the imminent problem of assembling the required number of troops to handle both situations simultaneously given the time, distance, and transportation constraints. Secretary Weinberger summarized the dilemma: "We had to take the forces that were available and close enough to land quickly, and forces that came specially trained and that we could equip quickly to carry out a very difficult and potentially very dangerous mission of rescuing our students, without time for either the planning or the pre-landing reconnaissance that any military commander would want to have" (Weinberger, 1990: 125). The immediately available combat-ready force was the marine replacement group that was then under way for normal rotation with the units in the multinational force in Lebanon. The Middle East mission required only lightly armed forces, so the marines had to be reequipped for the new mission. Nor was there time to rehearse for the Grenada operation (Weinberger, 1990: 111).

An additional, psychological factor affected the number of troops considered necessary for the operation: the vivid memory of the failed rescue operation in Iran. Weinberger was determined to avoid a similar failure, so he insisted on at least doubling any recommendation by the Joint Chiefs of Staff (JCS) as to required force size, "since I always had in mind that one of the major problems with our attempt to rescue our hostages in Iran in 1979 was that we sent too few helicopters" (Weinberger, 1990: 111n3). This was one of the main reasons why he rejected suggestions to let the marines do it all themselves and insisted on sending Army Rangers along, reminding the JCS again and again to double whatever forces the field commander said he would need.[6] Weinberger also authorized General John Vessey, the chairman of JCS, to use the 82nd Airborne Division as a backup in case the opposition was stronger than expected (Weinberger, 1990: 112–13). Over 6,000 U.S. combat troops were eventually involved in an operation that started on October 25 and lasted for seven days. About 1,500 troops stayed until December 15, 1983, and by June 1984 almost all of them had been pulled out (Thorndike, 1989: 258).

The concern about casualties bore heavily on the mind of the secretary of defense because of the heavy loss of life two days earlier in Beirut, where 241 marines were killed, as well as because of limited information. The Beirut incident demonstrated once again that an adversary could inflict heavy casualties even on a military force with superior training and equipment. In fact, American casualties in Grenada were light: 18 killed, 93 wounded, and 16 missing.[7]

Planning suffered from confusion and last-minute changes in operational tasks given to the participating units. The changes were due to technical malfunctions in some of the transportation aircraft, lack of coordination, miscommunication, and late-breaking intelligence. These caused delays in the schedules, turning the operation from a nighttime to a daylight operation (Adkin, 1989: 201–4). "The Rangers lacked basic information on which to plan effectively; they were compelled to accept unrealistic timings for H-hour, which prevented surprise. The plan they were forced to adopt took no account of the inevitable difficulties that occur in every military operation" (pp. 211–12). Fortunately, not a single Ranger was killed by enemy fire during the drop at Point Salines, mainly because of Castro's order to the Cubans not to engage in combat unless first attacked, which only happened later (p. 212).

Later in the operation when there were three task forces on the ground, the Rangers, the marines, and the 82nd Airborne, there was no ground force commander as such. The absence of coordination increased the probability of fire from friendly forces; it also meant that commanders learned of changes in interforce boundaries only after delays. Although a linkup radio frequency had been designated, call signs had not been distributed, and joint fire-control measures had not been established. As far as communications were concerned, the navy did not have interoperability with the army and the air force. The radio frequencies given to marine pilots were incorrect, preventing them from making radio contact with the air force AC-130s or with ground units. When contact was established, it turned out that the marines and the Rangers were using different maps, which made target identification by marine pilots practically impossible. Similarly, the Air Naval Gunfire Liaison Company (ANGLICO), which was responsible for controlling naval gunfire and aircraft operating in support of ground forces, could not do its job for lack of necessary radio codes, call signs, and frequencies to communicate with the supporting arms coordination center on the flagship USS *Guam* (Adkin, 1989: 216, 289, 285).

One reason cited for the lack of proper training for the "come as you are" situation in Grenada was that standard military exercises were overprepared and thus unrealistic when notice was short (Metcalf, 1986: 279, 295). The problem of vertical coordination was severe because of the decision that all services and the special forces had to participate. The major participants had different operational styles, but the intervention force was composed of "units and staff that did not know each other, had never trained together, and were forced to plan in isolation unaware of what others were doing" (Adkin, 1989: 131). Responsibility for planning was given to the Atlantic Command, rather than to the Caribbean Command. The Atlantic Command was under Admiral Wesley McDonald, who was a naval aviator by training. "The headquarters at Norfolk lacked the intelligence and communications capacity to handle the situation. Neither did the staff have planning expertise for a large-scale ground operation" (p. 126). The staff had a great majority of naval officers because the primary role of the command was to win naval engagements in the Atlantic in a world war and were inexperienced in planning a ground-force operation. A last-minute attempt to deal with this deficiency was met by sending an army general, Norman Schwarzkopf, to advise Rear Admiral Joseph Metcalf. The navy resented his presence, and, in any case, he joined the command group too late to have any substantial effect on planning (Schwarzkopf with Petre, 1992: 246).

Although U.S. intelligence on Grenada was sorely deficient, information about Grenada was plentiful in Barbados. This information, including an assessment of the strength and deployment of PRA, was provided to Washington, but it does not appear to have had an effect on planning. There was a complete absence of horizontal coordination between U.S. and Caribbean forces. Also, the role of the Caribbean contingents was not clear to either U.S. or Caribbean commanders (Adkin, 1989: 130–31). Both secrecy and insufficient preparation time adversely affected the quality of planning, but no less serious was the intelligence failure.[8]

The Grenada intervention policy has been aptly described as "policy without intelligence" (Hopple and Gilley, 1985). U.S. decisionmakers were faced with "violent uncertainty," to use Secretary of State George Shultz's definition of the situation. A host of deficiencies in military intelligence can be identified: the nature and scope of the Cuban presence; the Soviet role; the military capability and intentions of Grenada; tactical information about the defensive positions of the PRA; Grenadian society, its regime, and its leadership, especially interpersonal relationships among the leaders (Shultz,

1993: 342). Prior to intervention, the United States lacked hard political intelligence or evidence regarding the long-term objectives of the NJM and the Grenadian leadership. Intentions and potential threats were inferred from the ideological orientation expressed in public statements by the leaders, Grenadian voting behavior in the United Nations and the Socialist International, and known agreements with Soviet bloc member countries about economic and military aid.[9] In fact, American policy objectives were given priority over intelligence inputs in providing the main guides to action.

Even though U.S. intelligence gatherers had focused on Grenada since the 1979 revolution, the planners of the military intervention had no detailed maps. The Ranger and marine units with the key tasks of seizing the airfield under construction at Port Salines in the south and Pearl Airport in the north had no hard intelligence on enemy locations, strength, and intentions.[10] The maps were inadequate, and the unit commanders had to use black and white photocopies of dated 1:50,000 British tourist maps. Similar problems arose when one of the Ranger units was given the task of attacking the main base camp of the PRA (Adkin, 1989: 194–95). Ultimately, this task could not be accomplished as planned by the understrength Ranger battalion. Because the operation was not on schedule and the Cubans and the PRA resisted much more fiercely than anticipated, no fewer than six battalions of the 82nd were deployed by October 27 (Adkin, 1989: 224; Metcalf, 1986: 286). Lacking intelligence about the PRA, the SEAL teams that captured the transmission station and Government House came unequipped with rocket launchers and were almost defenseless against counterattacks by armored vehicles. This led in the first case to a SEAL team's retreat and in the latter case to a team's being trapped for almost 24 hours after capturing Government House. Inaccurate intelligence about the Richmond Hill prison and the terrain around it likewise led to incorrect anticipation of a docile, sleepy enemy; the plan to land the Delta and Ranger teams alongside the prison proved completely unrealistic, and the operation failed. And because navy pilots went into combat on the first day without prior coordination with the Rangers and special forces operations, air support could not be called in (Adkin, 1989: 181–91). In short, a large number of the military casualties were unnecessary.

Intelligence about the whereabouts of the students whose safety was the declared purpose of the operation was grossly insufficient. Combat forces had identified one campus. Although the Defense Intelligence Agency (DIA) knew of two campuses by October 21 and passed the information to Admiral McDonald's staff, planning was based on the need to secure just one campus. Nor was it known that 200 students were scattered in houses between the two campuses (Adkin, 1989: 140–41, 270–71).

Logistics posed another problem. The Ranger battalions were reduced to half or less of their normal strength owing to a lack of air force crews trained for night operations (Adkin, 1989: 194). The long flight to Grenada required inflight refueling of C-130 planes en route by K-10 tankers and posed the risk of interception by Cuban Mig-23s. Those handling the airlift did have the advantage of using Barbados, only 120 miles northeast of Grenada, as a forward administrative base. Nonetheless, the combination of an inadequate number of airplanes for transportation, meager intelligence about the terrain and the enemy, and frequent changes of roles and objectives given to the participating forces during the planning and early stages of the operation created uncertainty for commanders, who responded by overloading their soldiers just in case. Lack of vehicles for the first few days, the overloading of the troops, who were not acclimatized to the heat and wore uniforms unsuitable to the tropics, and the difficult terrain significantly reduced the troops' effectiveness and caused many cases of heat exhaustion (pp. 290–91). The inattention to the effects of climate is surprising in light of the lengthy experience with tropical combat conditions in Vietnam. In general, this oversight, as well as other logistic problems, can be attributed to the JCS decision not to discuss the operation with the head of logistics; the senior logistics planner of the Atlantic Command was told about the operation only 22 hours before H-hour owing to secrecy considerations (p. 132).

Finally, to avoid civilian casualties, there was no gunfire, and only occasional fixed-wing aircraft from the carrier *Independence* were called up to deal with situations in which such support could have saved casualties and time, as in the attack on the Revolutionary Military Council headquarters at Fort Frederick. As a rule, the commander of the naval task force, Rear Admiral Metcalf, allowed commanders to use maximum firepower only on a case-by-case basis (Adkin, 1989: 242–43; Metcalf, 1986: 289). These self-imposed constraints increased the vulnerability of helicopters in spite of light and disorganized defenses and account for the destruction of 9 percent of the helicopters committed to Grenada and the damage to several more (Cypher, 1985b: 106; Illingworth, 1985: 140–41).

Economic costs were not a factor in the preoperation considerations; indeed, they posed a very limited burden. The overall estimated cost was $134.4 million: army, $74.9 million; navy, $46.8 million; and air force, $12.7 million (U.S. Senate, 1984: 15). These are very marginal costs compared with the size of the U.S. defense budget or considered as a percentage of the gross domestic product.

Even though Grenada posed a very low risk target for the U.S. military, the situation was somewhat different from the domestic political perspective. Reagan faced a Democratic majority in the House of Representatives that was largely unsympathetic to his active foreign policy agenda (his "cowboy mentality") on both ideological and practical grounds. Grenada was the largest combat deployment of American troops on an overseas mission since Vietnam, and it revived legal issues about presidential authority to take military action. In the short run, a determined president could use the constitutional powers invested in his office, and in his position as commander in chief, to initiate a military operation without prior consent by the Congress or other agencies and individuals. In the words of one observer, Reagan "exploited its monarchical potential to the full" (Hill and Williams, 1990: 5).

The president is authorized to undertake unilateral military action to protect the lives of Americans abroad without congressional authorization. The president is also empowered to make unilateral executive agreements with foreign countries—in this case, to grant emergency assistance to the OECS. A breakdown of law and order, unless it could be related to an imminent national security threat, does not legally justify the use of force. In a meeting of the Special Situation Group (SSG) convened by Vice President Bush on October 22, the implications of the War Powers Resolution were briefly discussed. The view of the participants was that it did not pose any serious political risks because the mission would be completed long before Congress could raise the issue (Beck, 1993: 134). That was actually the case. Still, to comply with the War Powers Resolution, the president informed five congressional leaders of his decision on October 24, when the operation was already under way, and gave them an opportunity for questions and comments. The two Republican leaders present supported the president, although Senator Howard Baker thought the intervention was "bad politics." The Democrats were cool, and House Speaker Tip O'Neill said, "Mr. President, I have been informed but not consulted" (Beck, 1993: 164; Shultz, 1993: 335). O'Neill later described the briefing of the House and Senate leaders in the White House: "I had some serious reservations, and I'm sure that my Democratic colleagues did as well, but I'd be damned if I was going to voice my criticism while our boys were out there" (O'Neill with Novak, 1987: 365). He believed that the Grenada operation was intended to distract attention from the Lebanon fiasco. The quick, successful completion of the operation prevented this latent resentment for the White House policy from becoming a confrontation between the administration and the Congress.

Reagan also complied with the other condition of reporting to Congress within 48 hours of the commencement of hostilities by submitting a report on October 25. The Senate voted 64 to 20 on a rider attached to an unrelated bill that never passed the Senate, and the House voted 403 to 23 to apply the 60-day time limit on the use of U.S. troops abroad; the War Powers Resolution specifies that time limit unless Congress gives its approval for continued deployment. These actions were routine assertions of congressional privilege vis-à-vis the president rather than expressions of opposition to the operation. Congress thus failed to invoke the War Powers Resolution by not passing a resolution that would start the 60-day clock by vote. In the event, U.S. combat troops were withdrawn on December 15, 1983, less than 60 days after the report was submitted (Hall, 1991: 188–200; Rubner, 1985–86). Congressional criticism later focused on specific aspects of the operation, such as the failure of U.S. intelligence, the restrictions on the media, and the mistaken bombing of a mental hospital (Payne et al., 1984: 164–65), but not on its legality.

The presidential power permitting unilateral military action allowed for secrecy and speed in planning and execution of the operation. There would have been serious opposition in Congress, especially in the House, on ideological and practical grounds had the preparation for the operation become public knowledge long enough beforehand to allow for public debate. The administration was keenly aware of potential opposition, and that was one reason for the elaborate veil of secrecy. "Frankly, there was another reason I [Reagan] wanted secrecy. It was what I call the 'post-Vietnam syndrome,' the resistance of many in Congress to the use of military force abroad for any reason, because of our nation's experience in Vietnam" (Reagan, 1990: 451). A swift, successful operation that would take advantage of the short-term "rally 'round the flag" effect was a political imperative. Once Reagan made up his mind, he was determined to carry out his decision. In spite of desperate attempts by the Grenadian and Cuban governments to assure U.S. officials that the threatened American students were safe and to halt the impending operation—and in spite of the tragic events unfolding in Beirut—President Reagan resolved to proceed with the military attack (Adkin, 1989: 102, 120–21).[11]

The question of legitimacy at home and abroad was very much on the minds of President Reagan and the civilian and military participants in the decisionmaking. Domestic cognitive and normative legitimacy required an easily justifiable reason to intervene once the operation started and became public knowledge. The rescue of American students, in light of the vivid memories of the Teheran hostage situation and the humiliation that it had entailed, seemed a valid objective. The operation was anticipated to be short and not costly, to have immediately transparent consequences in resolving a potential hostage situation, and to remove a potential long-term strategic risk in a region close and vital to U.S. interests—all of which made achievement of cognitive legitimacy seem assured. The OECS leaders assured Reagan's special emissary Ambassador Frank McNeil and his team that the Grenadians would welcome American intervention (which later proved true), that intervention would not become a matter of popular dissent, as was the case in Lebanon. West Indians are English speaking, and many have relatives in the United States, so the majority were pro-American and disliked the NJM regime (Lewis, 1990: 261; Sandford and Vigilante, 1984: 7). Thus, the intervention could gain cognitive legitimacy from the argument that the use of force not only had a justified objective but was also the only practical way to deal with a potential hostage situation and the long-term strategic threat and, at the same time, to free Grenadians from an oppressive regime.[12]

It was realized, however, that establishing normative legitimacy would pose a more difficult problem. The military, sensitive to the power of the press to undermine public legitimacy for military operations—a lesson learned from the Vietnam experience—and impressed with Britain's tight control over press coverage in the Falklands War of 1982, prevailed upon the secretary of defense to exclude journalists from Grenada during the first few days of the invasion (Burrowes, 1988: 89–90; Quester, 1985). It was clear, too, that practically occupying a foreign country would raise an international outcry, because it defied the universal norms of sovereignty and nonintervention; it would also be viewed as a violation of U.N., OAS, and even OECS charters, not to mention international law.[13] The U.S. invasion might even invoke comparisons with the Soviet invasion of Afghanistan. To have as strong a case as possible in international forums as well as with the American public, the Reagan administration relied on a number of arguments and measures. It stressed the obligation of a government to ensure the security of its own citizens, emphasized the illegality and the oppressive and brutal nature of Grenadian government, tied the U.S. military operation to the invitation of a regional organization (OECS) operating within the scope of Article 52 of the U.N. Charter, which permits "actions to restore order and self-determination in a setting of breakdown of authority" (quoted in Waters, 1986: 236), including seeking help from extraregional powers, and emphasized the request for assistance to restore order made by Sir Paul Scoon, the governor-general, which was supposedly the only link to legality in the absence of a legally elected government.[14]

These measures and arguments had little direct influence on international responses to the operation or on the American press and were only partially effective with Congress. But they were acceptable to Americans generally, who were willing to give their president the benefit of the doubt. The warm welcome accorded to U.S. troops, who were received as liberators, enhanced the public's perception of the moral correctness of the intervention. According to a CBS poll conducted immediately after the intervention, 91 percent of Grenadians supported the intervention (Sandford and Vigilante, 1984: 16). In a Gallup poll, 59 percent of the Americans surveyed approved of the way the president handled the situation, and only 32 percent disapproved, with the rest expressing no opinion. In fact, within a month of the operation Reagan's job-performance rating jumped to 53 percent from an average of 44 percent in the months before the operation (Adkin, 1989: 318–21; Burrowes, 1988: 90; Lamare, 1991: 16–17; Payne et al., 1984: 164–66). The support of the American public influenced the media and critics in Congress to change their positions from "snide, scathing, and condemnatory" (Shultz, 1993: 339) to grudgingly supportive.[15]

INTERNATIONAL FACTORS: INCENTIVES, CONSTRAINTS, AND JUSTIFICATIONS

As the discussion of U.S. interests indicated, international factors played a key role in providing incentives for military intervention. Reagan's commitment to reconstructing the leadership role of the United States in international politics was unwavering. He believed that the legacy of Vietnam had burdened Americans for too long and that they should prove their resolve to leave Vietnam behind and take the necessary steps to contain the Soviets and their allies. In the absence of a determined forward-looking American foreign policy, the Soviets had significantly increased their power projection and intervention capabilities and had shown that they were willing to take advantage of available opportunities to advance their position and acquire new assets at the expense of the United States.

The drive to shed the Vietnam legacy was associated with a more immediate concern that was embedded in another historical failure: the emergence and persistence of the Marxist regime in Cuba. Avoiding a "second Cuba" has instructed U.S. foreign policy since the 1960s. Castro and his regime survived all U.S. attempts at destabilization and became a permanent concern of U.S. foreign policy. Cuba's willingness to challenge the United States and take risks associated with its active foreign policy was perceived by American decisionmakers as posing a two-edged threat, with Cuba acting both on its own behalf and as a proxy of the Soviet Union, as in Central America and southern Africa, notably Angola. Grenada came to be perceived as a potential second Cuba after the 1979 revolution. "As Cuba sits astride the northern sea lanes to the region, so Grenada sits athwart the southernmost" (Meese, 1992: 214). It was common knowledge that Grenada had become the most thoroughly militarized island in the Caribbean.[16] It was becoming significantly closer to Cuba. Not only was Prime Minister Bishop a personal friend of Fidel Castro's, but Cuba was helping with the construction of a large international airport at Point Salines. The construction of the airport could be explained by the requirements of Grenada's tourism industry, but that fact that Cuba was providing the bulk of the materials, plans, and labor force, as well as the fact that the airport was so large, made the project suspect. From the U.S. point of view, as expressed by the president and secretary of defense in 1983, the island was becoming a site for projecting Soviet and Cuban power into Latin America and Africa (Burrowes, 1988: 98; Illingworth, 1985: 131); Grenada might also export revolution to other Caribbean islands. These apprehensions were reinforced by the anti-American foreign policy adopted by the PRG. In the United Nations, Grenada systematically voted in support of the USSR and its invasion of Afghanistan (Adkin, 1989: 109–12; Cypher, 1985a; Payne, 1990; Shultz, 1993). The perception of an immediate threat in an area within the American sphere of influence— a threat linked to a broader threat to American supremacy given the continuing retreat of American power in the bipolar international system—led Reagan to believe that he had to act to preserve the balance between the superpowers and their blocs. He reasoned that the longer-term steps already taken to increase the defense budget, which produced the largest military buildup in peacetime in American history, would—if combined with one daring act of limited risk—not only demonstrate American resolve to halt the retrenchment of U.S. power but also signal that the United States had left Vietnam behind and was again willing to act like the great power it was (Burrowes, 1988: 72–73, 142).

Urgent Fury, the largest foreign military operation by the United States since the end of the Vietnam War, was unlikely to take place without the involvement of third parties: allies, adversaries, and international organizations. To preemptively blunt unavoidable international criticism and Third World antagonism, the United States wanted the operation to be sanctioned by other Caribbean nations. There were no illusions as to the response of the United Nations, an organization that the Reagan administration came to view as inherently hostile to the United States. Although the operation itself did not depend on either an invitation to intervene or an endorsement by other Caribbean nations, such support was considered extremely useful in containing the diplomatic fallout from the operation, especially accusations of "Yankee Big Brotherism" (Weinberger, 1990: 121). Another reason for rallying Caribbean support was domestic. The Reagan administration could convincingly explain why the United Nations would not condone a military operation, but it would have found it extremely difficult to convince Congress, especially congressional Democrats, and the American public of the need for intervention if the immediate neighbors of Grenada did not feel threatened. In building bipartisan domestic legitimacy and support for a military operation, it was imperative, then, that the Caribbean nations invite U.S. support and intervention—especially in light of recent opposition to U.S. intervention in Lebanon and involvement in Central America (Reagan, 1990: 451; Weinberger, 1990: 122, 128). A third but marginal reason was the need to appease Prime Minister Margaret Thatcher, who was not consulted about the operation, even though Grenada was part of the British Commonwealth, because of the need to avoid leaks (Reagan, 1990: 454). The United States would argue with Britain that it had responded to OECS requests.

The situation was complicated by the composition of and decisionmaking rules in the two relevant organizations: the seven-member OECS, chaired by Eugenia Charles, prime minister of Dominica, and the thirteen-member CARICOM, chaired by Prime Minister George Chambers of Trinidad and Tobago.[17] The treaties of both organizations did not permit action on a majority vote, and it was not likely that a unanimous decision would be reached, especially because Guyana had consistently supported the revolution in Grenada. To avoid this problem, Tom Adams, prime minister of Barbados, decided to work through the OECS. He was told by U.S. Ambassador Milan Bish that a written request from Caribbean nations would be required for the United States to intervene. On October 21 the OECS agreed to invoke Article 8 of its Treaty of Association regarding collective defense. The decision spoke of "requesting assistance from friendly countries to provide transport, logistic support, and additional military personnel to assist the efforts of the OECS to stabilize this most grave situation within the Eastern Caribbean" (quoted in Adkin, 1989: 98). The countries to be invited to assist OECS were the United States, the United Kingdom, and Canada. The same day the prime ministers of Dominica, Jamaica, and Barbados presented the decision to Ambassador Bish. On Sunday, October 23, President Reagan sent a special envoy to Barbados carrying draft letters prepared by the State Department requesting U.S. intervention. These draft letters were finalized in the discussion between the American delegation and Prime Ministers Adams, Seaga, and Charles. The Caribbean leaders did not know that SEALS were already on their way (Adkin, 1989: 99–100)—proving that the operation did not depend on Caribbean endorsement.[18] The legal ground was now set for justifying U.S. intervention, yet there was concern among American decisionmakers. The State Department representative at the final planning conference of the Atlantic Command on October 24 responded to a request to postpone the operation by 24 hours by saying: "I would not recommend that we delay. The Organization of Eastern Caribbean States, which asked us to intervene, is a shaky coalition at best. There's no telling how long it's going to support this thing" (Schwarzkopf with Petre, 1992: 248).

Unlike the OECS members, allies in the North Atlantic Treaty Organization (NATO)—West Germany, France, Italy, and Canada—were surprisingly severe in their criticism of the operation, viewing it as a deviation from international law and a dangerous precedent (Burrowes, 1988: 90–91). But most damaging was the response of Britain, which had traditionally maintained a particularly close relation with Washington. Prime Minister Thatcher, a close ally of President Reagan's who shared many of his views, felt rebuffed when Reagan apparently did not take her reservations into account. The British government was unprepared for the unilateral action taken by the United States against a Commonwealth country. Being uninformed or misled by American officials embarrassed the government domestically and exposed it to trenchant attacks by the opposition in Parliament (Burrowes, 1988: 117–25; Shultz, 1993: 340–41). This unanticipated repudiation could, it was thought, spill over into a broader controversy endangering the deployment of U.S.-built Pershing II and cruise missiles in Europe scheduled to begin a few months later. Fearing the implications of a public backlash, the European allies eventually moderated their criticism, especially after the operation turned out to be short and successful and won overwhelming support among the people of Grenada and the other Caribbean states.

The strongest repudiation came from Latin American countries, whose traditional fear of U.S. intervention was confirmed by the Grenada intervention. Their views were expressed in meetings of both the United Nations and the Organization of American States (OAS). No condemnation resolution was passed in the OAS because it would not have been supported by Caribbean and Central American allies of the United States. But the delegates from the largest and most influential countries (Brazil, Mexico, Venezuela, Colombia) condemned the intervention (Payne et al., 1984: 174–75). In the United Nations the United States could veto any condemnation resolution in the Security Council, which it eventually did, but it could not expect to prevent such a resolution in the General Assembly. Because Reagan persisted in considering the United Nations an instrument of Soviet and other radical Third World states, the organization was marginalized in U.S. foreign policy calculations.[19]

One concern of decisionmakers in the United States and the Caribbean was the possibility of Cuban preemptive measures, with or without Soviet support, that would send additional Cuban troops to Grenada in anticipation of an American-OECS intervention. The pre-landing reconnaissance desired by the military was therefore eliminated to avoid alerting the Cubans and Soviets to the operation (Reagan, 1990: 450; Weinberger, 1990: 114). In fact, the Revolutionary Military Council (RMC) of Grenada requested Cuban reinforcement, placing Castro in a difficult position. On the one hand, Castro resented the murder of his friend Bishop, but, on the other hand, he had been committed to the revolution in Grenada since 1979 and had provided substantial military and economic aid. There were already hundreds of Cubans on the island, and sending in more troops might provoke a direct U.S. military response, which could escalate to an armed confrontation between the United States and Cuba. Consequently, Castro informed the RMC that "Cuba cannot send reinforcements, not only because it is materially impossible in the face of overwhelming U.S. air and naval superiority in the area, but also because politically, if this were to be merely a struggle among Caribbeans, it should not do so in order not to justify U.S. intervention" (Adkin, 1989: 161). Castro did, however, instruct the Cubans already in Grenada to fight, but only if attacked first by the Americans; when that happened, they fought under a Cuban commander, not under the RMC.

To summarize, the international repudiation of the intervention was broader than anticipated. The preemptive legitimizing measures taken by the Reagan administration were relatively ineffective, mainly because they lacked credibility and were incongruent with the norm of nonintervention. Yet the quick success of the operation and the broad support by the people of Grenada and the neighboring Caribbean island states, combined with the powerful U.S. position in the international system, rendered this criticism inconsequential and short-lived.

THE EVOLUTION OF AN INTUITIVE DECISION

In spite of what looks like a routinized procedure to evaluate policy options, in the final account the decision to invade Grenada was an individual, intuitive one that was not preceded by a systematic and careful consideration of the risks involved. In fact, it could be convincingly argued that the role of the standard procedural framework for dealing with emergencies was marginal.

Based on a suggestion by Ambassador Bish, contingency evacuation plans and possible rescue plans were discussed as early as October 13. On October 14 the Atlantic Command was alerted to start planning the noncombat evacuation of American citizens. This course was approved by the president on October 17. Formal planning for some type of involvement by the United States in Grenada was begun on October 17 by the Restricted Interagency Group (RIG), which included middle-level staff from the National Security Council (NSC), the Departments of State and Defense, the Central

Intelligence Agency (CIA), and the JCS; it was headed by the Assistant Secretary of State for Inter-American Affairs, Langhorne Motley. This group was concerned that the United States might be caught in a protracted hostage situation, like Reagan's predecessor, President Carter. The Department of Defense supported an Entebbe-like rescue operation. It was already preoccupied with the peacekeeping operation in Lebanon and refused to get involved in anything but a swift, limited rescue operation. The State Department held a hawkish view, emphasizing the reputational consequences to the United States if nothing was done (Beck, 1993: 106). The Restricted Interagency Group met again on October 19; after the meeting, Secretary of State Shultz briefed the president on the prevailing opinions within the National Security Council (NSC).

October 19, the day that Prime Minister Bishop was murdered, marked the beginning of planning for nonpermissive evacuation, which would require the use of force. The Atlantic Command introduced six different options the next day, three of which assumed a hostile operational environment. The Crisis Pre-Planning Group (CPPG) assembled in the White House on October 20 at a meeting chaired by Rear Admiral John Poindexter, the deputy to National Security Adviser Robert McFarlane, and discussed full-scale military intervention. In the late afternoon, after receiving a message from Prime Minister Adams urging U.S. military intervention, Vice President George Bush convened the National Security Council Planning Group (NSPG), and the participants decided to keep preparations secret and to divert the Lebanon-headed amphibious task force to Grenada. The emphasis was still on evacuation, though more extreme options were also considered. After the meeting McFarlane briefed the president. On Friday, October 21, the president left for a golfing weekend in Augusta, Georgia, with Shultz and McFarlane; the CPPG agreed that the JCS should plan for a full-scale military operation. The next day, October 22, the president was informed of the OECS invitation, and he said that the United States should respond positively and quickly. At this point, Reagan had already made his mind up to use military force. The OECS heads of government urged that the operation begin the next day. The president called Vice President Bush and instructed him to set up a NSPG meeting later that Saturday morning. At the meeting, representatives of the National Security Council and the State Department supported a military operation because of the threat to U.S. reputational interests. The Pentagon and the Department of Defense shared a more cautious position: they were worried about an expected new wave of anti-Americanism in Latin America and were opposed to a hasty military operation (Brands, 1987–88: 617–18). The president, who participated by phone, made his views clear after listening to alternative options. He said: "Well, if we've got to go there, we might as well do all that needs to be done" (Adkin, 1989: 120). On October 23, Reagan signed the National Security Decision Directive (NSDD) to authorize Operation Urgent Fury. During the next 24

hours he could have aborted the operation, but he did not, nor did he interfere with the military preparations (Adkin, 1989: 115–21; Burrowes, 1988:
136–40). He met again with the JCS on October 24 and posed some questions. Each military leader agreed that the plan would work, but they com-
plained about poor intelligence and inadequate rehearsal because of time
pressure. But the implied risks did not change Reagan's decision to move
ahead as planned (Weinberger, 1990: 113).

Essentially, then, the decision to invade Grenada was first and foremost a
decision by one person: President Reagan.[20] Other participants in the decisionmaking process had the task of justifying, explaining, and implementing
his decision, which was not guided by a systematic analysis but, as we shall
see, was driven by his core beliefs and values and informed by his selective
life experiences in dealing with communists (or perceived communists) in
Hollywood and by vivid historical analogies that provided the main source
of informational inputs. In other words, "the body of information and
knowledge upon which Reagan relied tended to be impressionistic and anec-
dotal" (Waldstein, 1990: 55; also Shimko, 1991: 120). Both the definition of
the problem and the identification of specific risks evolved from simplistic
analogies. An emphasis was placed on the risk of Americans becoming
hostages, even though the logic of the argument was flawed and had little
factual support. Yet the memory of the hostage situation in Iran was so vivid
for Reagan, who became president the day the hostages were released, that
this threat overshadowed other analytic considerations.[21] Similarly, he repeatedly referred to the Vietnam syndrome as a reason for intervention.

Although an analysis of risk taking should focus on the president, it should
also take into account the supporting roles of others. Planning activities for a
possible intervention operation took place routinely, but Reagan reached a
decision on his own and before the planning groups made their recommendations (Reagan, 1990: 450–51; Shultz, 1993: 329, 344; Weinberger, 1990:
113). Reagan had great confidence in his risk assessment. This confidence
was bolstered by his strong ideological convictions, which made alternative
courses of inaction seem much more risky than a quick military operation.
The effect of his beliefs was reinforced by his being in Georgia, where he was
insulated from Pentagon and Department of Defense reluctance and was in
the company of George Shultz and Robert McFarlane, both strong supporters of the military option (Shultz, 1993: 344).[22]

Reagan's decisionmaking derived from a belief system that revolved
around a few fundamental axioms; these beliefs, which were held with deep
conviction, were unaffected by the uncertainties associated with a more complex, informed, and nuanced worldview. President Reagan was interested
only in a small number of issues. And even with these, he set only the broad
policy guidelines, giving little attention to details. Still, his vision and ideological commitment were tempered by pragmatism. He perceived the world as a
place where two sets of values, translated into diametrically opposed systems
of social and political organization, were locked in an uncompromising strug-

gle. One represented freedom, democracy, and capitalism; the other delin-
eated various forms of authoritarianism that enslaved body and mind. The
ultimate manifestation of authoritarianism was the communist ideology
propagated by the Soviet superpower and its agents. A confrontational approach—backed by political resolve, superior military power, and a strong
network of like-minded allies—was deemed the best way to deal with this
threat.[23] Reagan believed, too, that the United States had an obligation to
prevent smaller nations from being preyed on and victimized by communist
authoritarian states, and when necessary, the United States had to be ready to
use force to deter and roll back communist advances.

His worldview was underlaid by a determined optimism, expressed in his
belief in the inevitable victory of the free world over communist aggression.
Equally important was his belief in the possibility of controlling confrontational situations to prevent them from escalating into nuclear exchanges.[24]
Therefore, he asserted that risks could and should be taken in assisting and
supporting allies willing to participate in the struggle against communism,
even if they did not necessarily uphold liberal democratic values. The United
States had to overcome the Vietnam legacy and be ready to intervene militarily. Reagan was confident that this was the only way to regenerate American
pride and confidence. But his optimism was not limited to the controllability
of limited militarized conflicts involving the superpowers and their allies; he
also assumed that the United States could and should maintain a military-
technological edge by using its superior economic system and organization
to advantage and by outspending the Soviet Union. The combination of
moral correctness, superior resources, and resolve would inevitably lead to the
victory of good (Western democracy) over evil (communism). Given Reagan's
self-contained and tightly structured operational code belief system, risk taking became both efficient and morally correct. He used his communication
skills to convince the American public of the necessity of an active and assertive policy abroad and an increased defense budget at home, which he presented as a highly cost-effective way to use taxpayers' money (Williams,
1990: 201).[25] His success in communicating this message to the public was to
a large degree because it reflected conventional wisdom about how the world
works and what the appropriate U.S. role was.

An important part of Reagan's method of decisionmaking was the manner in which he used subordinates, especially in those cases in which he had a
direct interest. He relied on a small group who were chosen on grounds of
personal loyalty and ideological affinity and who were expected to provide
detailed programs and take care of the everyday business of government. He
understood his role as that of chairman of the board and believed in providing strong leadership to set the tone; particular arrangements for generating
information or advice were of secondary importance (Spear and Williams,
1988: 204).[26] The homogeneity within his close circle of advisers reinforced
their tendency to search selectively for information to support the president's expressed or assumed preferences. Because Reagan had a limited ca-

pacity for critical thinking, once he had adopted an idea, there was little likelihood of modifying his views, unless it was framed by his staffers as a
change in tactics and not a change in strategy (Waldstein, 1990: 55). It was
also understood that once a decision was made, Reagan did not care for dissenters. In the case of Grenada, this decisionmaking mode took the form of a
risky shift, especially among the JCS and the Defense Department personnel,
who were initially concerned about the time pressure, insufficient intelligence, and the absence of rehearsal by the troops but who did not press for
changes after the operation was decided on (Edel, 1992; Hill and Williams,
1990; Waldstein, 1990).

In summary, a rigid value system, a strongly held set of beliefs about the
world of nations and the U.S. role in it, the freshness of the Iranian hostage
situation in Reagan's mind, and simplistic reliance on analogies combined to
trigger a simple but vivid scenario (the simulation heuristic) of what would
happen. This scenario defined the high stakes and high probability of the
emergence of a second Cuba in the Caribbean if the United States failed to
take forceful action to overthrow the Grenadian regime. In Reagan's opinion, the Grenada case put the commitment to protect the U.S. backyard to
the test (Gilmore, 1984: 28), yet the intervention did not present a high risk
of confrontation with the Soviet Union. Backing away was unthinkable; an
active policy had to be pursued. According to George Shultz: "He [Reagan]
had agonized about intervention in Surinam, and that frustrating experience
probably prepared him to move definitively in Grenada" (Shultz, 1993:
344). Because the president had made up his mind about the military intervention prior to consulting with his advisers and other cabinet members, he
used consultation merely to reinforce and justify his decision. Moreover, because his decision was ideologically driven, new information had little or no
impact on it. This is best demonstrated by Reagan's response to the news
about the bombing in Beirut and the question whether this should affect the
Grenada decision: "If this [Grenada] was right yesterday, it's right today,
and we shouldn't let the act of a couple of terrorists dissuade us from going
ahead" (Adkin, 1989: 121).

His conviction in his beliefs and his stereotypical view of the Soviet Union
and its allies reinforced his assessments of the negative consequences of nonaction. At the same time, a lack of interest in policy details prevented his exposure to information that could undermine his confidence in the ability of
the military to accomplish a quick, successful operation. In other words, he
used the advisory and decisionmaking procedures more for establishing legitimacy for his preferred policies within the administration than for accomplishing decision-relevant purposes, that is, for meeting managerial goals
rather than for establishing policy priorities and preferences (Anderson,
1985). A number of factors thus converged synergistically to produce a risk-
taking process quite different from the one in the Panama case, which is to be
discussed next in this chapter. These factors included a leader who was low in
cognitive complexity and predisposed toward risk acceptance, who had a

strong foreign-policy-related belief system based on ideological convictions, who recognized that Grenada was a low-risk operation, and who was supported by a small-group and organizational structure that tended to reinforce his preferences. Not only did these factors produce the decision to intervene, but, more important, the decision was not preceded or accompanied by agonizing indecision, as was the case with Bush's Panama decision, and once Reagan made his decision, he pursued its implementation with great determination.

The Panama Case

AN OVERVIEW

Since the beginning of the twentieth century, relations with Panama have been a salient concern of U.S. foreign policy. The strategic value of the Panama Canal has led to several U.S. interventions. From the Panamanian perspective, the overriding issue since the Hay-Bunau-Varilla Treaty of 1903 has been sovereignty over the canal. In the Panama Canal treaties of 1978, concluded with General Omar Torrijos, Panama's nationalistic military dictator, the United States agreed to return the canal to Panamanian sovereignty by the year 2000, and this removed the principal irritant in U.S.-Panamanian relations. The planned course of events was derailed by General Manuel Noriega's policies and actions.

General Noriega emerged as strongman in 1983 after the death of General Torrijos in an airplane crash in 1981. Noriega's years in power brought a gradual change in the U.S. attitude toward the military regime and toward him personally. Though beginning in a spirit of cooperation, by mid-1987 the United States was determined to remove Noriega from power. It first attempted persuasion, then economic and diplomatic sanctions, then support for the civilian and military opposition to the dictator. When these tactics failed, the United States decided on military intervention. Panamanian militarization was itself a development the United States could live with and even support as long as several important U.S. security interests were served by reaching an accommodation with successive military regimes, first Torrijos's and then Noriega's. The U.S. security interests included the conclusion and implementation of the Panama Canal treaties, continued access to U.S. bases and intelligence collection facilities in Panama, and support for the Contras in Nicaragua. The change in the American attitude toward Noriega and his regime occurred during 1985 and 1986 and eventually led, in 1987, to a shift in American priorities from security-dominated concerns to an emphasis on democratization and on combating the drug traffic. The new priorities led to a reevaluation of the balance of perceived benefits and liabilities of working with Noriega.

As U.S. relations with the Noriega regime worsened, both the United States and Panama focused their disagreements on the canal issue, even when additional interests gained prominence. The treaties that were supposed to remove a major irritant in the dealings between the two nations became an issue of contention. Noriega used the issue of access to bases to project the self-serving populist image of a nationalist leader who faced a colonial superpower intent on maintaining Panama's colonial status. For the United States, the question of Panama's political stability and democratization became closely associated with the security of the canal after U.S. withdrawal. By June 1986, American perceptions of Noriega had changed: he was not considered a useful ally but an unbearable burden. In 1986 the prevailing view was still that he could be removed without resort to force. The so-called ouster crisis, which lasted from June 1986 to October 1989, was characterized by a fundamental disagreement within the Reagan administration with regard to what means would be most effective in influencing Noriega. The Departments of State and Defense advocated different policies. Secretary of State George Shultz and Assistant Secretary for Inter-American Affairs Elliot Abrams urged military and paramilitary operations. Admiral William Crowe, chairman of the JCS, along with National Security Adviser Frank Carlucci, vigorously objected, tying their concerns to the potentially adverse consequences of continued access to military bases and the serious risks to some 50,000 Americans in Panama (Donnelly et al., 1991: 24–25; Scranton, 1991: 46–47).

These conflicting views resulted in an emphasis on nonmilitary measures, which included trying to convince Noriega to resign, supporting local opposition movements and covert paramilitary operations, and imposing economic sanctions. None of these measures brought the hoped-for removal of Noriega. Finally, with the failure of the October 1989 coup, serious consideration began to be given to the military option. Until then the prevailing view was that eventually the Panamanians would oust Noriega in a people's revolution similar to the one in the Philippines. The military option developed from the perceptions of four key decisionmakers—President Bush, National Security Adviser Brent Scowcroft, Secretary of Defense Richard Cheney, and General Colin Powell, chairman of the JCS—who viewed the risk of uncontrolled escalation as acute and shared the assessment that the likelihood of the Panamanians removing Noriega was extremely low, especially in light of the abortive coup and the postcoup purge.

How did the change in U.S. perceptions of Noriega come about? A number of events during 1985 made the policy of supporting Noriega difficult to sustain. In September 1985, Dr. Hugo Spadafora was murdered and beheaded by officers of the Defense Forces of the Republic of Panama (PDF), sparking demonstrations in Panama. Spadafora, a longtime opponent of Noriega's, publicly and pointedly criticized Noriega and compiled extensive evidence of his involvement in drug trafficking. The second event was the forced resignation of President Nicolas Ardito Barletta, who, under public pressure, had called for investigation of the Spadafora murder. These events were in themselves not reason enough to cause a policy review by Washington, but they triggered two separate investigations in the United States, by Congress and by the Pulitzer Prize–winning investigative journalist Seymour Hersh. The latter focused attention on corruption, drug trafficking, and the serious human rights violations by Noriega; he depicted the activities of the general and the PDF as conspicuously incompatible with U.S. security interests and in defiance of the basic norms of democracy—thereby weakening any argument for a reasonable trade-off between U.S. security interests and moral imperatives.

When conflict between Noriega and his chief of staff, Colonel Roberto Diaz Herrera, who was denied succession to the position of commandant, led the colonel to denounce Noriega, there were large-scale riots and demonstrations in Panama. The U.S. Senate consequently took a strong rhetorical stance against Noriega, but the PDF remained loyal to him, making the Reagan administration reluctant to take more active steps to oust Noriega. Late in 1987, although Washington officials became increasingly concerned that the PDF might attempt to harm the 50,000 U.S. citizens in Panama, the preference was still for quiet diplomacy directed toward Noriega's ouster.

Several emergent issues affected Washington's attitude toward Noriega. For one, the Iran-Contra affair undermined the policy of covert support to the Contras, thus removing a major U.S. incentive for cooperation with Noriega. Another was the culpability of the Panamanian government in drug trafficking, which had been a subject of contention between Panama and the United States since the 1970s. By 1988 the U.S. government perceived Noriega as a major target in the national campaign against drugs. "Noriega was not just on the wrong side of this issue, he was in the enemy camp" (Scranton, 1991: 105; see also Hathaway, 1991). The same year the last remaining Noriega supporters in interagency deliberations, the Drug Enforcement Agency and the Department of Defense, withdrew their support. By late 1988 the dominant position in the Reagan administration—as it was later in the Bush administration—was that Noriega had to go, but "that U.S. military force should not be used to achieve this objective" (Scranton, 1991: 106). Until October 1989 various diplomatic, economic, and legal measures were applied, to no avail.

As pressure on the general to resign grew and as the Panamanian government lost even moderate and right-wing supporters, Noriega signaled that he might shift to the left, both domestically and internationally. By early 1988 it had become clear that he had rejected a negotiated retirement deal. The government power base continued to erode, leaving him more and more isolated, and in February 1988, he was indicted in U.S. courts. The two federal indictments on drug-related charges carried a maximum sentence of 165 years in prison and $1.65 million in fines if he was convicted (Scranton, 1991: 128). These indictments had an immense effect on U.S. public opinion and narrowed the options for a deal between Noriega and the Reagan administration. But the U.S. government was still unwilling to seriously consider the use of force. Following the failed attempt of Panama's president to dismiss Noriega in February, the U.S. government tried using financial pressure to force Noriega out of office. The measures, implemented in March, were designed to create a cash shortage so that Noriega would be unable to meet the payrolls of the PDF and government employees, the two main sources of support for his regime. Then, it was argued, public discontent would force Noriega out. But again the Reagan administration refused to take more extreme measures, such as a trade embargo, and the Panamanian opposition complained that "the sanctions are too few and far between" (Scranton, 1991: 138). Economic sanctions were finally imposed in April but were not implemented until June. The sanctions did not achieve much. Regional mediation initiatives taken by three former presidents of Venezuela, Costa Rica, and Colombia, later joined by the president of Costa Rica and the prime minister of Spain, also failed. Yet the United States continued to rule out military intervention except to protect the Panama Canal from attack or to safeguard American lives; there was, however, a slow buildup of U.S. forces in the region.

The United States next commenced covert operations against Noriega. Five such operations failed for lack of consensus among the various branches of the Reagan administration and lack of support in Congress. The Bush administration mounted more covert operations, focusing efforts on helping the opposition to win the 1989 election. In the May 1989 election the opposition won in spite of fraud attempts by the Noriega regime. Noriega's unwillingness to accept a loss at the polls led to bloody clashes between the opposition and Noriega's military forces. Eventually the head of the electoral tribunal, acting on behalf of the regime, declared the May 7 election null and void. But the enduring damage to the image of the regime was severe. The sight of "bloodied candidates and dead protesters defined the opposition and Noriega in new terms, especially to audiences in the United States" (Scranton, 1991: 163). U.S. military forces in Panama were reinforced, and preventive measures were established to minimize the number of American targets. The beginning war of nerves pitted U.S. forces against Noriega's forces in a series of incidents and provocations. At the same time, the United States became aware of an impending coup planned by Major Moises Vega Giroldi, chief of security for PDF headquarters, aimed at forcing Noriega to retire. The coup attempt on October 3 failed when U.S. forces in Panama did not intervene to prevent forces loyal to Noriega from suppressing the coup on the same day. But even though the United States did not intervene on that occasion, it was clear that U.S. objectives had changed (Cottam, 1994: 157–60).

Criticism from the American public increased pressure on the Bush administration to play a more active role. The ban on U.S.-initiated or U.S.-supported assassinations, initiated years earlier by President Gerald Ford, was reinterpreted in a more flexible manner to allow U.S. support in the elimination of Noriega (Buckley, 1991: 220–21). New rulings also expanded the scope of legal military operations against terrorists, drug lords, and fugitives abroad. It was now agreed that the United States must have readily available military options, even though the decision for foreign military intervention had not yet been made. Military planners envisioned an invasion that would be fast and cause minimal death, destruction, and collateral damage. The plans were rehearsed during November and December 1989, during which time the National Assembly appointed Noriega chief of government, giving him broad powers. Both the Assembly and Noriega asseverated a state of war between Panama and the United States. Then, on December 16, a U.S. Marine lieutenant, Robert Paz, was killed by Panamanian soldiers at a PDF checkpoint. The United States considered the murder a grave provocation and part of a threatening pattern of escalation. On December 17, President Bush decided on intervention. Operation Just Cause began on December 20. It had two phases: first, a special forces operation whose objective was to arrest Noriega, and, second, an invasion by 26,000 U.S. troops whose objective was to neutralize the PDF. The invasion was successful, and by Christmas Eve the PDF had ceased to exist. Noriega remained in hiding until January 3, 1990, when he turned himself in to the U.S. Army and was flown to a jail in Florida to await trial.

U.S. INTERESTS

The United States perceived three types of interests to be at stake in Panama: tangible, reputational, and ideological. The balancing of these interests was one reason that the decision to act militarily was preceded by a lengthy search for alternatives. It was not a change in the perception of U.S. interests in Panama nor necessarily a change of priorities among these interests that brought about the change of attitude toward the Noriega regime. Instead, a reevaluation of the effectiveness of U.S. policy and its personal and national implications redirected strategy and promoted a willingness to accept risks that both the Reagan and the Bush administrations had refused to consider taking before December 1989. In U.S. perceptions, the balance of risks from direct military intervention compared with the risks associated with alternative policies shifted in favor of the first, and the level of acceptable risk shifted in the same direction, leading to a preference reversal in favor of foreign military intervention.

The immediate threat was to the life and security of American military and civilian personnel in Panama. Their welfare became more salient with the Panamanian declaration that a state of war existed with the United States and with the increasing harassment of U.S. soldiers and their families, which culminated in the killing of a marine officer and the detention of a naval officer, his wife, and four other Americans. The threat to American lives was closely associated in the minds of U.S. decisionmakers with the threat to the American presence in Panama and therefore to the Panama Canal—that is, to keeping it in operation—and to the integrity of the Panama Canal treaties, which are vital to U.S. economic and strategic interests (Quigley, 1992: 251; *Washington Post*, December 19 and 21, 1989). Less important but still significant was the damage to American reputational interests. Noriega's blatant defiance of the United States over many years had made his presence a reminder of U.S. weakness (*Washington Post*, December 21, 1989).

After the war on drugs had become a key issue on the American domestic agenda, it served the purposes of the administration to eliminate Noriega's drug-trafficking activities, which would also end Panama's role as a transshipment point and center for laundering drug profits (Hathaway, 1991; *Washington Post*, December 23, 1989). Though an important consideration, the traffic in drugs would not have triggered foreign military intervention in and of itself. The commitment to democracy in the hemisphere was likewise of secondary interest and was more a justification than an actual motive. It is a fact that the United States has, over the years, been able to cooperate with and willingly support military regimes, including Noriega's. But convergence of multiple interests and the perceived linkage between the tangible, reputational, and ideological issues led to consideration of the military option.

The decisions to intervene militarily, then, represented a complete turnaround in U.S. policy. Until December, Bush had rejected advice to "unleash the full military and go in and 'get Noriega.'" He stated: "That's not prudent and that's not the way I plan to conduct the military or foreign affairs of this country" (*Washington Post*, December 21, 1989). Given a choice, he would have preferred a collective hemispherewide approach, but the Organization of American States resented U.S. military intervention and eventually condemned the invasion by a twenty to one vote on December 22 (*Washington Post*, January 5, 1990). The incentive to use force to resolve a persistent and embarrassing problem came to dominate the thinking of the Bush administration only when Washington policymakers came to compare the costs of a military operation with the costs of the status quo in Panama. An increasingly defiant Noriega had shown a lack of inhibition in dealing with both his own people and the United States. The state of affairs threatened important substantial and reputational interests of the United States, including the reputations of the president and his administration (Quayle, 1994: 141–42).

The U.S. intervention in Panama was on an even larger scale than the intervention in Grenada. Operation Just Cause was a coercive, state-organized use of military force directed at the Panamanian political power structure with the intention of ousting and bringing to justice Panama's head of state and replacing the military regime with a democratic one. Following the removal of Noriega, U.S. military forces were to replace him with Guillermo Endara, a Panamanian lawyer associated for many years with the legendary politician Arnulfo Arias. Endara had won the May 1989 election by a landslide margin of nearly three to one, but Noriega had declared the election invalid. The active U.S. role in reinstating Endara sent a clear message of American commitment to democratization in the region.

NATIONAL CAPABILITIES

Once the United States had decided to engage in a large-scale military operation, the die was cast. The mismatch between the military capabilities of the two nations was huge, although a number of risks were involved. From the very beginning the U.S. military planned to have a decisive edge over the PDF; in the event, 26,000 troops were arrayed against Panama's 19,600, of which 6,000 were in the standing army and the rest in police and paramilitary units, including the Dignity Battalions. Panamanian forces were neither trained nor armed to combat an invasion force. Their largest guns were 60-millimeter mortars, and they had only 29 armored cars and no tanks (Crowell, 1991: 70). The U.S. forces also had uncontested air superiority over Panama.[27] The Military Airlift Command (MAC) could drop troops wherever required, and provide uninterrupted close support to ground troops.[28] Given the heavily tilted balance of military forces, the outcome of the invasion was never in doubt. But two matters were: the number of casualties on both sides and the amount of time required to achieve the goals of the operation, that is, to neutralize the PDF and to capture Noriega. Both could seriously affect the domestic response to the operation and worldwide opinion of it. A quick, decisive defeat of the PDF was essential not only to achieve these goals but also to ensure that the soldiers loyal to Noriega remained incapable of organizing a guerrilla campaign.

The disparity in size between the U.S. and Panamanian armies made the question of the U.S. force structure a minor issue. Within a military establishment as large as that of the United States, in a confrontation with a very small military establishment like Panama's, it is highly likely that enough specialized units will be available to deal with most reasonable contingencies, be the warfare in cities or jungle. This was the case. The Seventh Army, which had been specially trained for urban combat, was sent into Panama City. Army Ranger units and Green Berets were assigned to deal with PDF units in the countryside. Tank and armored units were deployed to deal with the few armored units of the PDF if they became a factor in the operation (*Washington Post*, December 21, 1989). In short, the availability of a broad range of specialized military units combined with the modest numbers required for each specialized task made the structure of the armed forces less than crucial. In fact, the command and control dimension of the operation could have been made simpler by reducing the number of participating services. Considering the limited number of Panamanian troops, there was no need for both the army and the marines to participate. As one military officer observed: "The Army can do the Marines' job, and the Marines can do the Army's job. We've just gotten so everybody has got to have his face in the picture" (*Washington Post*, December 21, 1989).

Because Panama was an area of long-standing strategic concern to the United States, the U.S. Southern Command was based in Panama, U.S. troops had often trained in the region, and military planners had become exten-

sively involved and acquainted with the Panamanian combat environment as
the Panama issue became acute in the months preceding the intervention.
Since early 1988 the U.S. military had been planning an operation against
Noriega, and the troops had had a chance to rehearse their roles. By November 1989 a military "concept plan" and detailed contingency planning had
been submitted by Lieutenant General Carl Stiner, commander of the 18th
Airborne, and General Maxwell Thurman, commander in chief of the Southern Command, and approved by General Colin Powell, chairman of the
Joint Chiefs of Staff, and by the Joint Chiefs. Stiner, Thurman, and Powell
were Vietnam-generation officers who loathed piecemeal operations (Donnelly et al., 1991: 66). All three were active, aggressive, and ambitious —and
very different from Thurman's predecessor, General Frederick Woerner, an
experienced Latin American hand who was much more concerned with the
long-term implications of military actions for Panama's domestic politics
and with Latin American international response. Woerner opposed the use
of force in Panama and was convinced that the objective was not worth the
cost. Thurman, in fact, made no secret of his belief that Woerner suffered
from too much knowledge and therefore had too much empathy for the region and its political leaders (Buckley, 1991: 189, 222). Stiner, Thurman,
and Powell, like many in their generation, felt that if civilian-imposed constraints had been removed in Vietnam and if the military had been given a
free hand, the United States could have won the war. They wanted no half
measures in Panama. They wanted the opportunity to prove that supplied
with the necessary resources and allowed to exert their professional skills to
the maximum, they could control the whole country in a short period and secure the objectives of U.S. foreign policy at an acceptable cost.

Earlier plans by Thurman's predecessor, General Woerner, were unacceptable to JCS operations staff because of the very complex system of command, control, and communications required. Two other concerns with
Woerner's plan were the lack of combat experience of U.S. Army South
(USARSO), designated to run the operation, and the slow pace of the buildup
of forces (Donnelly et al., 1991: 25–26). The plan was also not satisfactory to
the 18th Airborne, the quick reaction force of the Army. Stiner, the corps
commander, believed that both the responsibility structure and the plan were
not in keeping with what one corps planner called "our war fighting philosophy" (p. 29). Stiner articulated his philosophy of fighting as follows:

> Strike the enemy before he strikes you. Make him fight your kind of battle, not
> his. Attack at night. Force your way into his backyard. Maintain surprise but hit
> him with overwhelming combat power at precise points of decision. Paralyze his
> ability to react. Get it over with quickly. Speed would win you the support of the
> American people, save your soldiers' lives, and end hostilities on favorable terms
> without undue enemy casualties. (p. 56)

Stiner's military background in special operations made him effective in accomplishing unity of command between special operations and conventional

forces. The lack of an integrated command was a key weakness of Woerner's plans for military operations (code-named Prayer Book), as well as the earlier Grenada operation (pp. 59–60, 62).

The new plan called for a massive use of force, with attacks at 27 different locations to destroy the entire military command structure of the PDF. The U.S. forces would then overwhelm the PDF, Noriega's main source of support. Overwhelming force, speed, and precision were assumed to reduce the risks of what was foremost on Thurman's mind—what he called non-war-winners—like the risk that an American soldier's wife or child might be killed on a U.S. military installation or that part of the Panama Canal would be disabled. In his view, any such event would offset any other achievements, and the operation would be judged a failure (Donnelly et al., 1991: 72).[29] General Powell argued that a full-scale invasion was less risky from a military point of view than a small-scale precision operation to extract Noriega, for it would be over swiftly and would be more likely to accomplish the operation's goals (*Washington Post*, December 25, 1989, and January 7, 1990). Yet detailed military planning aside, there was a lack of detailed attention to the turbulent economic, social, and political conditions in the post-Noriega period in the new plan—and that had been a key strength of Woerner's plan.

The lengthy planning and training, which included simulated assaults, and the creation of a simplified and effective command, control and communications system helped to avoid the foul-ups and miscoordination among units that took place in the Grenada intervention (Seitz, 1991; *Washington Post*, December 29, 1989), even though Just Cause was a much more complex military operation and included extensive nighttime activities. Intelligence support for the decisionmaking and planning processes provided the broad strategic and political picture, although there were a number of problems with operational intelligence. The lack of reliable information about the Panamanian armed forces was due to the narrow target of intelligence efforts: American strategic interests in the region. The PDF's order of battle and its Byzantine politics were a low priority (Donnelly et al., 1991: 37). The gap in knowledge was compounded by the inadequate intelligence sharing between the CIA and the Pentagon. The military also failed to tap the resources of the CIA for assessments of PDF's military capability, its armament, and its morale and motivation—all information that should have come from the network of agents in Panama (*Washington Post*, January 7, 1990). Nonetheless, the actual adverse consequences of these deficiencies were limited.

The slow evolution of the decision to intervene allowed enough time to ship tanks, armored personnel carriers, and helicopters before H-hour. Both logistics and training were made significantly easier and more effective by the network of U.S. bases in Panama. The logistic task still posed a number of problems, however, the main being the airlift of troops and supplies to Panama. The airlift required mobilization of Military Airlift Command resources from all over the world. The task of supporting large forces was made more difficult by the need to maintain tactical surprise, which did not allow for a full logistic buildup (Donnelly et al., 1991: 86–87). Another problem was the anticipated heavy air traffic over Panama, which would require care to avoid accidents. Similarly, with many task forces operating simultaneously on the ground and performing dozens of missions at night, an efficient communications network had to be established to prevent the deployment confusion that occurred in Grenada.

Once Just Cause was under way, unforeseen bad weather conditions in the United States disrupted the smooth flow of the operation. Bitter cold and ice caused delays in the departure schedule, and the armada was forced to leave, not en masse, but in groups. This, in turn, complicated the jump plan, so more than half the paratroopers missed the drop zone. Yet these unanticipated mishaps did not seriously disrupt the planned combat operations (*Washington Post*, December 29, 1989). The U.S. military had access to Howard Air Base in Panama, and Army rangers and the 82nd Airborne were to seize Torrijos International Airport and adjacent Tochumen Airfield. Because these vital airstrips could be under mortar fire by the PDF and would have to be secured quickly and efficiently, the 82nd had to bring in fire-finding radar equipment designed to pick out mortar locations, and was prepared to employ countermortar measures (Donnelly et al., 1991: 92–93).

As it turned out, then, context and conditions allowed the buildup of the necessary number of troops and the requisite hardware, the appropriate force structure, and detailed operational planning and logistic support systems to take advantage of the immense U.S. military superiority in the most effective way to assure success at a minimal cost in casualties. The direct military risks were therefore not high.

One dilemma that faced the planners of Operation Just Cause stemmed from the risks involved in an operation conducted largely in a heavily populated urban area. Serious constraints had to be imposed on the intensity of firepower that could be brought to bear in support of combat action against the PDF and the Dignity Battalions. On the other hand, fighting in a heavily populated area without close artillery and air support to prevent indiscriminate killing of civilians increased the chances of heavy American casualties— for example, from snipers' fire. Even if American military casualties were minimal, a large number of Panamanian civilian casualties could antagonize the population of Panama—whose support for the American-backed Endara government was sought—as well as public opinion in the United States, South and Central America, Third World countries around the world, and Europe (Donnelly et al., 1991: 99). One way Stiner attempted to resolve these dilemmas was to use advanced weapon systems that could target firepower precisely.

> [The] rationale for using the F-117s was that the planes' night-attack role and precise delivery system were ideally suited to a surprise attack that would minimize both friendly and hostile casualties, by stunning instead of killing hundreds of Panamanian troops—and ones who might not even be loyal to Noriega. The Air

Force officer said that Stiner was proven right when paratroopers found the Panamanian troops running around in their underwear, completely disoriented and
asking, in effect, "What the hell was that?" . . .

The precision offered by the Apaches and the F-117s as well as other attacking
elements of the U.S. force was a key factor behind the operation's success. Because many of the assaults were being conducted in heavily populated areas
within Panama City, heavy emphasis was put on limiting the impact on the civilian population and containing the collateral damage resulting from attacks on
PDF- and Noriega-related facilities. There was surprisingly little damage that occurred in the slums adjacent to Comandancia (Ropelewski, 1990: 32).

What made the operation more difficult was the goal of destroying the PDF
command and control structure while keeping the Panamanian army as in-
tact as possible, so that it could be reformed to serve the needs of the new
government in the preservation of law and order. Consequently, General
Thurman of the Southern Command instructed Lieutenant General Stiner,
the overall military commander of Operation Just Cause, to apply minimum
force, and he forbade the use of mortars, artillery fire, aerial bombing, or
strafing unless an officer with the rank of lieutenant colonel or above approved it. Infantry forces attacking PDF barracks were accompanied by a
psychological operations team. Once a PDF unit was surrounded, the U.S.
forces played a recording on a loudspeaker explaining that the purpose of
the operation was to liberate Panama and that no harm would come to
Panamanians if they surrendered within fifteen minutes. Those orders were
carried out in most cases, but not all (Crowell, 1991: 81; *Washington Post*,
December 21, 1989, and January 7, 1990).

Carrying out operations in urban areas required soldiers to assume tasks
for which they were not prepared. According to 1st Lieutenant Clarence
Briggs's eyewitness account:

Problems arose when we suddenly had to change roles. For the most part we were
infantrymen, trained primarily "to close with and destroy the enemy." Then suddenly we were expected to act as diplomats and policemen. Behavior deemed
meritorious under one set of rules could be construed as unacceptable under another set. It's not difficult to understand how a soldier can become confused when
he is praised for an act in one instance but is then reprimanded for a similar act in
another. This is especially true in an environment where hesitation or a lapse in
judgment could very well kill you or your fellow soldiers. The result was often
frustration, tension, and ambivalence that further complicated an already confusing state of affairs. (Briggs, 1990: 4)

The soldiers' confusion, frustration, and tension could have inadvertently
caused human rights violations, jeopardizing the legitimacy of the operation.
Problems of this type are common to military operations that take place in a
civilian environment, and pose one of the major risks confronting decisionmakers.

As in Grenada, the economic costs of the Panamanian operation did not
play a role in decisionmakers' considerations. The immediate costs were lim-

ited. The overall incremental cost—the cost over and above the cost for normal operations for each service participating in the invasion—was $163.6
million: $155.0 million for the Army, $5.7 million for the air force, and $2.9
million for the navy. The Department of Defense did not request additional
funds to cover these costs but funded the operation from existing resources.[30]

Domestic politics did, however, play a central part in changing the policy
preference from covert operations to military intervention. In that context,
the most important change was in President Bush's position. According to
Elliot Abrams, assistant secretary of state for inter-American affairs in the
Reagan administration, in May 1988 there was an opportunity to negotiate
the ouster of Noriega. A proposal called for the United States to drop its indictments of Noriega on drug charges in exchange for his departure from
power. But 1988 was a campaign year for Vice President Bush, who was already suffering criticism for his past contacts with Noriega, and he opposed
the deal. President Reagan rejected his advice, but Noriega refused to accept
the offer (Baker with DeFrank, 1995: 170–80; Kempe, 1990: 338–40). An
alternative get-tough option—to use any means, including U.S. bases in Panama, to get rid of Noriega—was opposed by then Vice President Bush, Secretary of the Treasury James Baker, Secretary of Defense Frank Carlucci, and
JCS Chairman Admiral William Crowe. The position of the Defense Department was summarized in a single statement: "Noriega is in fact not worth
one U.S. soldier's life" (*Washington Post*, December 21, 1989; Donnelly et
al., 1991: 16). The Pentagon argued that Noriega did not pose a strategic
threat regionally or locally and that the ongoing negotiations over bases in
Spain, Greece, and the Philippines would be jeopardized if the governments
of these countries suspected that foreign bases could be used for interference
in domestic affairs (*Washington Post*, December 21 and 31, 1989). Once
Bush was elected president, the anti-intervention coalition faltered because
of growing domestic criticism of his soft line on China following the Tianan-
men massacre and his policies toward El Salvador. The president was particularly criticized for not taking the necessary steps, during the failed Giroldi
coup attempt in Panama in October 1989, to act more aggressively and de-
pose Noriega (Baker with DeFrank, 1995: 186–87; *Washington Post*, December 21, 22, and 25, 1989). Even though change in the president's position on the use of force was the most important factor in the shift of U.S. policy preferences, especially in light of Bush's imperial style of presidency, a
change in the military's stance might have been instrumental in reinforcing
the trend. During the Reagan administration and early in the Bush administration, the U.S. military considered Noriega no more than a diplomatic nuisance; with the new role assigned to the military by Bush in the war on drugs,
however, this perception changed. "Southern Command authorities concluded they could not wage their drug mission without confronting and dealing with the Noriega problem" (*Washington Post*, December 29, 1989).

Panama occupies a special place in U.S. calculations. Carved from neighboring Colombia in 1903 with the encouragement of President Theodore

Roosevelt, it soon became the focus of U.S. strategic and economic interests, with attention centered on the canal. To most Americans, the importance of maintaining stability in Panama is apparent. Noriega had also come into public disfavor because of his connection with drug trafficking and his brutality and corruption. Add to this the U.S. court indictment, and U.S. decisionmakers could easily validate the legitimacy of Operation Just Cause. The president recognized that as long as the operation was short and casualties minimal, "rally 'round the flag" patriotism would work in his favor (*Washington Post*, December 22, 1989).

Furthermore, congressional criticism of the administration's failed response to the October coup made it safe for the administration to assume that a short, decisive military operation would likely earn bipartisan support. The administration expected only minor opposition in Congress. Indeed, when Bush, a Republican, contacted House Speaker Thomas Foley and Senate Majority Leader George Mitchell shortly before the assault began, both of them Democrats, they immediately agreed to support the decision. The president also won the support of the chairs of the House and Senate Armed Services Committees (*Washington Post*, December 21, 1989). Although President Bush's policy did not fully conform to the requirements of the War Powers Resolution, Congress supported intervention because it was a popular and short-term policy. Public opinion surveys indicated that the intervention was even more popular than the Grenada operation. According to the *New York Times* / CBS post-invasion polls, 55 percent of respondents supported the Grenada operation, whereas 74 percent approved of the Panama intervention (Burgin, 1992). In spite of this anticipated support, the lessons of the past regarding media scrutiny of military operations persuaded the Pentagon to limit journalists' access to live combat reporting (*Washington Post*, January 7, 1989; Watson, 1991). The purpose was to avoid the potential risk of losing domestic normative legitimacy through on-the-scene reports of gruesome details.

The government's claim for the normative legitimacy of the intervention was also aided by the emphasis on the restoration of democracy, which appealed to the messianic impulse in the American national character. Americans believe that their national identity is bound up with acceptance of responsibility and commitment to the spread of American-style democracy. Thus the "president put himself squarely in tradition with a distinguished pedigree" (*Washington Post*, December 21, 1989). The warm welcome that the U.S. military received from the Panamanian public enhanced the positive image of a benevolent hegemon and added credibility to its attempts to establish normative legitimacy. In fact, the Panamanians welcomed intervention for two reasons: relief from the despotism of Noreiga and their highly Americanized culture (*Washington Post*, January 2, 1990).

To justify the intervention at home and abroad, Secretary of State James Baker also distinguished between two types of intervention: one is launched to provide choice, and the other to deny it. The first was justified; the latter was not. The U.S. intervention was launched, according to this line of reasoning, to allow the people of Panama the right to decide who would govern them. Yet this distinction between types of intervention was not well received in many Third World countries. Those in the nonaligned bloc denounced the U.S. incursion. Their views were expressed by Yugoslavia's ambassador to the United Nation, chair of the nonaligned group of nations, who described the intervention as a "gross violation of sovereignty." The Soviets expressed skepticism about the necessity and justifiability of the U.S. attack; some western European and Scandinavian countries also spoke against it (*Washington Post*, December 22, 1989). Particularly critical were the members of OAS, who voted twenty to one, with the United States casting the only dissenting vote, to censure the United States for the intervention. The Spanish text of the resolution was framed in stronger language than the English text. In the English text the members "deeply regret the military intervention in Panama"; in the Spanish text they "deeply deplore" it. Even countries friendly to the United States made clear their belief that the principle of nonintervention takes precedence over all other principles, including the defense of democracy (*Washington Post*, December 23, 1989). In the U.N. Security Council a resolution condemning the United States was supported by a majority of the members, including the Soviet Union and China, but was vetoed by the United States, France, and Britain. France, quick to denounce U.S. policy toward the Third World in the past, stood in support because it deeply disliked Noriega, was distracted by the revolution in Romania, and needed U.S. support in facing the dramatic transformation occurring in Europe (*Washington Post*, January 4, 1989).

In short, domestic legitimacy for the intervention was easy to establish and sustain, but international legitimacy was more troublesome to secure. Third World countries as a group and all Latin American countries in particular, including U.S. allies, vehemently condemned the United States. Milder criticism was heard from the Soviet Union, the Scandinavian countries, and several western European countries. The main American allies in Europe supported the intervention. Because the operation was short and successful, however, the lack of broad international legitimacy had a minimal impact— and no effect on domestic support for the intervention.

THE MARGINAL ROLE OF INTERNATIONAL FACTORS

The Panama intervention took place within a global system that was undergoing transformation. The key country in this process was the Soviet Union, whose domestic and international priorities were dramatically changing. A new emphasis on cooperation with the United States and its allies for achieving Soviet goals was emerging. The Soviet power base at home and abroad was showing obvious signs of the impending collapse of what had once seemed an invincible empire. In the United States the structural weakness of the USSR generated apprehension about the long-term implications for international stability, uncertainty about the applicability of the familiar rules and modes of interaction between the two superpowers, and, at the same time, a cautious sense of relief in the diminished threat that it presented. The United States therefore attempted to reformulate its global and regional roles and maintain its predominant position in the emerging centro-polyarchic system. Events in Panama were a source of continual embarrassment, reflecting as they did a loss of national self-confidence, and they posed questions about whether the United States was capable of seizing new opportunities and asserting a leadership position worldwide. Indeed, being able to respond effectively to the changes in the international environment was construed as a personal test of the leadership qualities of the new President Bush, who was already acquiring a "wimp" image.

Panama presented a challenge and an opportunity for the president to show his mettle. Confrontations between American and Panamanian soldiers convinced the risk-averse Bush to be more risk acceptant. A military intervention, which had been considered unacceptably risky, became acceptably risky—even if the perception of risk itself did not change substantially. Panama, which was within the U.S. sphere of influence, could be isolated from external interference by third parties like the Soviet Union, Cuba, and Nicaragua because all three were vulnerable and preoccupied with domestic problems. The United States had easy access to Panamanian territory, so the likelihood of an effective response by a third party in support of Noriega was very low. Noriega's treatment of the opposition and his illegal drug activities made him an unsavory associate, making it safe to assume that European allies of the United States would support the policy as long as the operation was brief and as long as Panamanian civilian casualties were minimal. The United States anticipated heavy criticism from the OAS and individual Latin American countries, judging from the emergency May 1989 meeting of the OAS (Buckley, 1991: 185–88; Woodward, 1991: 170), and from most Third World countries for violating the principle of nonintervention. In other words, the United States could not expect to make an undisputed case for external normative legitimacy by using a self-defense rationale or the argument that the legally elected government of Panama had invited it to intervene and restore democracy (*Washington Post*, December 21, 1989). U.S. veto power in the Security Council would nonetheless block any attempt at an unfavorable U.N. resolution.

Given the new priorities set by Mikhail Gorbachev, Soviet response could not be anything but mild. The Soviets were compelled to condemn the U.S. intervention, but U.S. officials correctly assumed that the USSR was not going to jeopardize the improvement of relations between the two superpowers. The United States could expect Soviet criticism and was therefore willing to recognize the reciprocity principle. In fact, the United States indicated that it would understand a Soviet decision to intervene in Romania (*Washington Post*, December 28, 1989). Whether the U.S. use of military means to achieve

foreign policy goals would affect the domestic debate in the Soviet Union between Gorbachev's supporters and the hard-liners who contested the large
reduction in defense expenditures was an open question. Indeed, although
the Soviets were forthright in criticizing the United States, they stopped with
that (*Washington Post*, December 21, 1989). The United States could reasonably expect that if the operation was short and successful and led to the establishment of a broadly supported democratic government in Panama allied
to the United States, both Soviet and Third World criticism would dissipate,
especially if U.S. troops pulled out quickly. Yet the possibility of a bloody
fight with Noriega's Dignity Battalions in urban areas could not be ruled out.
Heavy civilian casualties would incite an international backlash, which
might even affect the supportive position of European allies. However, once
the president revised his preferences, international political consequences
would practically be ignored. Informed administration sources are quoted as
saying: "It was a military decision to do it, not a political one. . . . George
Bush made the decision regardless of the risks on the political side" (*Washington Post*, December 25, 1989). The neglect of international considerations could have been explained as resulting perhaps from the low risk attributed to them by the Bush administration, if Bush had not in the past consistently attributed great importance to the anticipated negative responses of
the Latin American countries to a military operation. The explanation for
the change in policy preferences must therefore be more complex.

EXPLAINING THE CHANGE IN PREFERENCES

Why did President Bush change his mind after vehemently opposing military intervention during his years as vice president and during most of his
first year as president? There are two related questions. What caused him to
become risk acceptant, or, in other words, why did he opt in December for
an American solution to the Noriega problem instead of staying with the
search for a Panamanian solution? And, why did the once risk-averse president choose, not a surgical operation to capture Noriega and bring him to
trial, which he and his advisers seemed inclined to accept, but a full-scale
military intervention, which General Powell persuaded him to accept? His
choice is particularly puzzling because a complex military operation with "a
lot of moving parts," with many things that could go wrong, seemed a much
riskier option.

First, it must be recognized that Bush was consistently risk averse; left,
however, with no other viable choice, he very reluctantly became risk acceptant. Being risk averse, he opposed the use of military force during the
Reagan administration, earning a reputation as the most outspoken among
Reagan's advisers who argued against it. As president, he claimed expertise
in foreign policy. On one occasion, when he considered sending troops to
Colombia to combat illegal drug trafficking, he said: "I know enough about
this hemisphere and have had enough experience in dealing with countries in

Latin America and South America, Central America, all through the Caribbean, to know the constraints that exist in terms or should exist in terms of dispatching troops" (*Washington Post*, December 24, 1989). In the case of Panama, he suggested a number of reasons why force should not be used. Chief among them was the serious implications for U.S. relations with all other OAS members, specifically, revival of the image of Yankee imperialism (Donnelly et al., 1991: 48). He also argued that U.S. goals could be achieved without resorting to overt military intervention, which identified him with a policy that was later to prove transparently ineffective.[31] For this and other reasons, discussed earlier, his presidential performance was increasingly criticized. By late 1989 he found himself in the domain of losses in both foreign and domestic affairs. As "prospect theory" predicts (Levy, 1992a, 1992b; Stein, 1992), the once extremely risk-averse president who was hostage to the status quo began to heed suggestions for a drastic change that would allow him to recoup his losses (Baker with DeFrank, 1995: 187; David, 1996). His change in policy preferences was reinforced by his strong dislike for Noriega and driven by an emotional reaction to the killing of a marine lieutenant and the brutal harassment of a navy officer and abuse of his wife by reportedly drunken PDF soldiers. He said: "This guy [Noriega] is not going to lay off. It will only get worse" (*Washington Post*, December 21, 1989; Powell with Persico, 1995: 417, 425). The problem had become vividly real to Bush, as a comment by one of his advisers indicated: "When you're talking about the abstract, he's abstract. When it gets into real crimes against real people, something switches and it's a different Bush" (*Washington Post*, December 24, 1989). The emotional effect probably also contributed significantly to Bush's perception of now being irreversibly in the loss domain.[32] It was now clear to him that action would have to be taken.

To understand how Bush was convinced by Powell to endorse the intervention option requires a look at the decisionmaking process on December 17, 1989, the day the decision to intervene was made.[33] Prior to the meeting with Bush, Powell met Secretary of Defense Richard Cheney and his deputies, together with representatives from the State Department and the National Security Council. Cheney was initially undecided, and the others present tended toward nonaction or a surgical strike. Powell convinced them that other alternatives were worse and that a large force reduced the probability of high casualties (Powell with Persico, 1995: 422).

Having gained the support of Cheney, Powell had to convince the president to adopt the intervention option. At the meeting where the decision to authorize Operation Just Cause would be made were: President Bush; Vice President Dan Quayle; Secretary of State James Baker; National Security Adviser Brent Scowcroft; Robert Gates, Scowcroft's deputy, who was also acting as the representative of the intelligence community; Secretary of Defense Richard Cheney; General Colin Powell; Lieutenant General Thomas Kelly, director of operations for the JCS; and White House Press Secretary

Marlin Fitzwater. Conspicuously absent was CIA Director, William Webster. Bush was not happy that an earlier plan for a covert operation against Noriega, called Panama 5—formulated and approved by Bush in the aftermath of the failed Giroldi coup attempt in October—had been leaked to the press from the CIA (Donnelly et al., 1991: 98).

Bush's performance at the meeting was consistent with his modus operandi. One observer noted that the president would "exhaust every policy option almost to the point of humiliation, then strike decisively when he felt his back against the wall" (Donnelly et al., 1991: 98). There was a strong feeling among the participants that something must be done. The surgical strike option was explored by some of those present, especially Scowcroft, who played devil's advocate. But Powell argued that eliminating Noriega alone would merely open the door to another strongman. Besides, Noriega might be an elusive target, and what the troops knew best how to carry out was a full-scale military operation. Bush was reassured that in Just Cause the number of casualties among U.S. soldiers and Panamanian civilians would be low. Even inaction, he was told, would result in continued casualties in Panama, but in that case the United States would still have to deal with a Noriega who had become desperate and paranoid. His claim to the title of Maximum Leader and the declaration of war on the United States seemed to confirm this (Baker with DeFrank, 1995: 189; Donnelly et al., 1991: 98–99).

To convince the president and his advisers to accept the intervention option, Powell framed the comparative risks of the two options in terms that would make the intervention choice more attractive, with more acceptable risks. He first reframed the main goals of U.S. policy in Panama. All through the Reagan administration, even while Powell was national security adviser, and during most of the first year of the Bush administration, even after the May 1989 election in Panama, the dominant approach has been to leave Panamanian affairs in the hands of the Panamanians (Baker with DeFrank, 1995: 185). Although a coup against Noriega might have brought another military strongman to power, Washington found the approach acceptable because the removal of Noriega was the main objective, even though direct American military intervention was eschewed. Nobody in the U.S. government explicitly admitted that restoration of democracy was a secondary goal. Powell, who understood Washington politics, took advantage of this in shaping his argument. In his presentations on December 17, 1989, to the secretary of defense and the president, he compared the two options and emphasized that even if Noriega could be captured in a surgical operation, the tradition of modern Panamanian politics would cause another military strongman to take his place. To restore democracy it was therefore necessary to decapitate the PDF, which could be done only with a larger operation.

But Powell still had to convince the president that the probability of success was higher, and the long-run costs were lower, for the intervention option than for the surgical operation. He detailed the cunning skills of Noriega, who, as a former intelligence officer, had evaded all attempts to dispose of him. Powell hinted that a failed operation to capture Noriega would prove a major embarrassment to the Bush administration, which was already under heavy fire in connection with other foreign policy issues. At the same time, Powell stressed that the careful planning and preparation for Operation Just Cause would make it a quick, resounding success, although he would not estimate how long the operation would last or how many casualties there might be (Powell with Persico, 1995: 425). He made highly salient the risks of the surgical operation and underplayed those of an invasion while expressing his confidence in the leadership capabilities of Thurman and Stiner and the readiness of the troops. Invasion was the prudent course, he told the president. And the president, assured of a quick military victory with minimal loss of life, both to U.S. troops and Panamanian civilians, and aware that all four military chiefs, as well as his key advisers, supported the invasion, knew, too, how critically the American public viewed his overall performance.

All these arguments convinced Bush. His confident belief that every possible measure would be taken to minimize adverse outcomes and reduce the probability of their occurrence resulted in reframed outcomes and reassessed probabilities. His confidence in Powell and Thurman and the forceful logic of Powell's arguments contributed to the credibility of the military intervention option (Donnelly et al., 1991: 59; Woodward, 1991: 167–71). The costs of the operation were now framed as unavoidable investments in the promotion of hemispheric stability and the democratization of Panama, as well as the defense of American lives. The reframing allowed Bush to disengage himself from the damaging passivity with which he had begun to be identified and to accept risks that he had hitherto refused to take, deciding: "Okay, let's do it. The hell with it" (Powell with Persico, 1995: 423).

As a rule, risk-averse decisionmakers are slow to react to new opportunities to advance their goals at low costs. In policymaking, windows of opportunity often close before they are able to make a decision. The failed October coup attempt was one such missed opportunity.[34] As a problem grows, Bush-type decisionmakers find themselves forced to adopt policies that are more risky than those that could have been effective if a bolder approach had been taken at the beginning.

One question remains: Why did Powell, who, as national security adviser initially opposed using military force in Panama, change his views when he became chairman of the Joint Chiefs of Staff? The most obvious answer is the influence of role playing. Civilians who become national security advisers tend to be more hawkish than military officers appointed to the position. Civilians want to prove that they can be tough in matters of national security, whereas officers want to prove that they can be cautious despite their military background. So Powell was different as national security adviser than as chairman of JCS. One observer said: "In his eighty days as JCS chairman, Powell had changed, and the world had changed. . . . When he was National Security Adviser, he had agreed with his predecessor at JCS, Admiral Crowe, that Panama was of secondary importance. But now, he felt that America should hang out a shingle saying Superpower lives here" (Donnelly et al., 1991: 96). This changed view of the stakes was emotionally reinforced by the harassment of American troops in Panama and a deep dislike for Noreiga. In Powell's own words: "I found Noriega an unappealing man, with his pock-marked face, beady darting eyes, and arrogant swagger. I immediately had the crawling sense that I was in the presence of evil" (Powell with Persico, 1995: 415).

THE MODUS OPERANDI OF RISK-AVERSE LEADERS

The nature of risk-aversion stands out in low-risk situations. In high-risk situations, in contrast, it is more difficult to separate the impact of situational effects on risk aversion from inherent preferences patterns.

The essential difference between risk-averse and risk-acceptant leaders lies in their approach to risky problems. Risk-averse decisionmakers are concerned with process and with the path from the current state of affairs to an improved one. If the process is too costly, they show no interest in the end result no matter how great the reward. In contrast, risk-acceptant decisionmakers are much less concerned with process than with goals. The costs of reaching the goals are of less concern. Risk-averse leaders follow a distinctive modus operandi when they respond to challenges that present risky decisions. Initially such leaders attempt to portray themselves as astutely cautious rather than merely risk averse. This may work temporarily, especially when they conceal their inability to make timely risky decisions behind a facade of aggressive rhetoric. Over the long term, however, decision-support organizations (e.g., intelligence agencies, departments of defense and external affairs), as well as advisers to the leaders in question, recognize the reluctance of the leaders to accept risk for what it is and, for self-serving reasons, provide the leaders with justifications for risk avoidance, interpreting and framing information accordingly. This, in turn, is effective as long as the situation or problem is ambiguous and not very threatening. When the situation changes and the problem seems unavoidable, the risk-averse leader is forced to respond more forcefully.[35]

Reluctant to take action, this type of leader takes substitute action by escalating verbal aggressiveness, eventually committing herself to take action at some unspecified time. If the problem does not subside, words-with-no-action threatens the leader's credibility. The risk-averse leader then engages in active procrastination, represented by lengthy, sometimes unnecessary preparations for a decisive risky action (e.g., slow force buildup, long consultations with other nations, new diplomatic initiatives that are not likely to produce substantial results). When all further options for credible procrastination have been exhausted, the leader, needing to avoid serious reputational damage ("the wimp factor"), will authorize the implementation of a risky option. Time becomes a major factor. The risk-averse leader, uncomfortable in the risky circumstances, impatiently tries to quit them and, consequently, declares success or victory as soon as it is feasible, then returns to business as usual.

Such an approach is flawed by ad-hocism; it does not take into account long-term interests and developments and usually leaves important issues unresolved. If the outcome of the policy allows the general public to lose interest, the leader can ignore the residual negative consequences and move on to other issues. But if the results pose a serious new problem, the pattern repeats itself: understatement of the problem, aggressive rhetoric, active procrastination, and, eventually, reluctant activism.

If low-cost options appear before risk-averse decisionmakers have reached the phase of reluctant activism, they avoid those options. In other words, when opportunities appear, risk-averse leaders pass them up and eventually find themselves forced to take higher risks, represented by higher-cost actions, at a later date. Then they feel that they cannot avoid a risky decision any longer. Sometimes the inefficiency of their prior risk-averse policies becomes so transparent that it threatens their credibility and position. In this case, as leaders come to feel personally threatened, they become furious with the adversary that has placed them in that untenable situation, and seek emotional compensation for past failures by engaging in extreme action against the adversary. Thus, for example, the Panama invasion was intended to compensate for Bush's failure to take advantage of the October coup attempt, and the Gulf War grew out of the failure to anticipate the Iraqi invasion of Kuwait (Campbell, 1991: 217; Mullins and Wildavsky, 1992: 44–45).

Risk-averse leaders are often highly competent tacticians and managers of foreign policy. They are, however, much less effective in shaping and executing broad strategic policy initiatives, because they are more likely to be consolidators than visionaries. The reason is obvious. Because avoiding risk is uppermost in their minds, they do not consider the difficult-to-control challenges that a broader vision often entails. Their need for control also shapes the advisory process. Within the decisionmaking group debates over policy are tame and timid. Risk-averse leaders do not approve of heated debates because they fear losing control. Dissension and multiple advocacy are acceptable only within well-understood limits.

The appointment of close friends to key positions is one way risk-averse decisionmakers make sure that they will not be surprised or embarrassed by policy initiatives that run counter to their risk-avoidance dispositions, a way for them to assure control over the policy process. If they do not want to micromanage decisionmaking, they can delegate responsibility to these trusted friends. In many cases, debates over issues will lack vitality and depth. When the group leader plays an active role in decisionmaking and when the leader's preferences for risk aversion ripple through government agencies, the emphasis on collegiality and loyalty is likely to produce polarization in the direction of risk aversion. In more extreme cases, it will result in groupthink.

The need for control is such a crucial motive that risk-averse decisionmakers tend to get involved and acquire as many details about military operations as possible.[36] In the crucial briefing of December 17, President Bush demanded that General Powell go into great detail. According to one account:

> Throughout the briefing, President Bush asked him detailed questions about the plan: What kind of troops are involved? What kind of equipment do the paratroopers carry with them when they jump? Why are you taking out this target? How are you going to do it? How many helicopters are already there? What intelligence do you have on Noriega? Do you know where he is? Are you sending in enough troops? The president continued to ply General Powell with innumerable questions and to pry into all aspects of the plan. (Flanagan, 1993: 52–53)

George Bush clearly fits the profile of the risk-averse leader. He was cautious and reactive, and he rejected visionary goals and ideological programs. "George Bush is always active, but it is activity without a sense of long-term direction" (Rose, 1991: 309; see also Deibel, 1991). He was gregarious and talkative but inarticulate about goals, preferring to leave his own views obscure as long as possible to make sure that he was not identifying himself with the losing side of an issue. Bush, in short, "would prefer to be a no-lose President. The prudent maxim of a *no-lose* president is this: If you don't play, you cannot lose" (Rose, 1991: 310). Bush believed in the limits of knowledge and prediction, especially long-term prediction. He did not appear to be convinced that knowledge is cumulative. Therefore, "his references to history are few, strained and narrow, generally centering on perceived mistakes" (Duffy and Goodgame, 1992: 72); he did not attempt to gain insight into opportunities for action. He believed that the future was too uncertain to gamble on. Long-term programs seemed inordinately risky because he saw the events as only partly controllable and the validity of outcome predictions as very low (Mullins and Wildavsky, 1992: 55–57; Schafer et al., 1996; Winter et al., 1991).

The ad hoc manner in which the Bush administration operated was apparent in the handling of the October coup attempt. President Bush failed to convene a meeting of his national security advisers and worked "through a crazy quilt of phone calls and informal gatherings" (Campbell, 1991: 207). The emphasis on loyalty, the smallness and homogeneity of the advisory circle, and Bush's penchant for secrecy shut out from the decisionmaking process necessary expertise and information (Berman and Jentleson, 1991: 103; Campbell, 1991: 208; David, 1996; Swansbrough, 1994). This typical risk-averse style is captured in the following description.

> As a decision maker, particularly in foreign policy, where he makes his heaviest investments, Bush likes to work around a small set of decision makers with whom he is comfortable. If Jimmy Carter's problem was to comprehend and resolve the diversity he built around him, George Bush's problem is to create a diversity of perspectives around him. In foreign policy making, Bush is surrounded by people much like himself—pragmatic, skeptical, experienced, and political. They, like Bush, have difficulty articulating objectives publicly. They, like Bush, may not

avail themselves as widely as necessary of specialized experts from within the bureaucracy or from the outside academic and research communities. (Rockman, 1991: 17–18)

Many of the symptoms of risk aversion can also be observed in the behavior of Leonid Brezhnev in connection with the Czechoslovakia intervention. The symptoms are, however, most salient where the institutional context is one of individual preponderance, as in the American case, rather than where it is an oligarchy with one member of the decisionmaking group being first among equals, as with the Soviet Politburo.

The Czechoslovakia Case

AN OVERVIEW

The independent state of Czechoslovakia was created at the Paris Peace Conference after the First World War.[37] Two Czech nationalists, Thomàš Masaryk and Eduard Beneš, united ethnically and culturally diverse peoples under a provisional government. With the support of Slovak leaders, notably General Milan Štefánik, they proclaimed the Republic of Czechoslovakia on October 28, 1918. Between the world wars, from 1918 to 1938, Czechoslovakia was the only eastern European nation to remain fully democratic. During the years of parliamentary rule, under President Masaryk and then his successor, Beneš, strong democratic foundations were laid. The parliamentary system was perceived to be the logical culmination of the national experience and the answer to decades of desire for self-government. Although there was still some stress on a separate Slovak identity, Slovak nationalism was less pronounced than Czech nationalism.

In 1938–39, Nazi Germany occupied the Sudetenland and dissolved the Czechoslovak union. Hitler made protectorates of Bohemia and Moravia and supported an independent Slovakia, which existed from 1939 to 1945. In May 1945, Czechoslovakia was liberated, mainly by the Soviet army, and its borders were restored to essentially the same as those established at the Paris Peace Conference in 1919. In the May 1946 election the Communist Party emerged as the dominant party and the major partner in the National Front, having won 38 percent of the votes. Other non-Communist parties in the National Front Coalition included the Czechoslovak National Socialist Party, the People's Party, the Labor Party, and the Social Democratic Party. President Beneš accepted Klement Gottwald, a Communist, as prime minister. In February 1948 the Communists took advantage of a crisis within the coalition government; they seized power and eliminated all other political parties. A new constitution was approved in May 1948. Beneš refused to sign the constitution, and when the Communists gained complete control in a single-slate election at the end of May, Beneš resigned and Gottwald became president.

The new constitution adopted in 1948 established a Soviet-style single-party system. More important, the scene was set for a complete break with the political, economic, and intellectual traditions of Czechoslovakia—and a move toward Stalinization. In the 1950s there were severe political repression and a series of staged political trials, which resulted in the execution of hundreds of dissidents, including prominent Slovak leaders. The pre-Communist legal traditions, as well as any "socialist legality" incorporated in the constitution, were, for all practical purposes, nonexistent. Economically, the Communists' program represented a radical departure from the socioeconomic system that prevailed between the wars. The industrial sector that produced textiles, glass, and other consumer goods—products that made western Europe a natural trading partner—was transformed. Management was centralized, and factories were retooled for heavy industrial production. Agriculture was collectivized, too. For some years Czechoslovakia was able to maintain a standard of living as good as, if not better than, that of any of its Soviet-bloc neighbors. Eventually, however, the economy deteriorated; by 1961 the crisis was severe enough to call for a fundamental adjustment.

Culturally, the people of Czechoslovakia had been strongly influenced by their geographic position in the center of Europe. The country had a diverse intellectual heritage that included contributions from its own ethnic groups as well as a profound western European influence. The Communist takeover and the Stalinism of the 1950s meant a crackdown on liberalism, social democracy, and nationalism. There was also an intense atheistic campaign to combat religion. One result of this rupture with the past was widespread dissatisfaction and fear; popular support for the regime disintegrated. When the systematic political excesses became known during the brief de-Stalinization period, acute disillusionment set in.

In 1953, Gottwald died, and the leadership passed to a partnership between Antonín Zápotocký (president) and Antonín Novotný (Communist Party first secretary). The new partnership came to power at the same time as the Malenkov-Khrushchev duumvirate declared a New Course in the USSR. The relaxation of political control that occurred in the Soviet Union, Poland and Hungary was only briefly permitted in Czechoslovakia, and this short experiment with liberalization was completely reversed in 1955. When Hungary was invaded by Soviet forces in 1956, the Czechoslovak leadership soundly approved the military action. In November 1957, Zápotocký died, and Novotný succeeded to the top state position of president while retaining his supreme party position of first secretary. During the next several years de-Stalinization was pursued only perfunctorily; instead, there was a sustained effort to stamp out dissident ideas.

The years 1962–64 saw a deepening of the economic crisis and the spread of public disaffection. The slow pace of de-Stalinization, the ineffective response to economic problems, Slovak discontent, intellectuals' protestation of repressive controls, student unrest, political apathy, especially among the youth, and strains in Czechoslovak-Soviet relations—all created widespread resentment against the existing system. Between 1964 and 1967, in response, the government introduced a regulated market system. As the procedures for implementing the program of economic reforms became known, however, it became increasingly clear that the old system required structural changes. Hard-line conservatives set themselves in opposition and impeded reform.

At the Central Committee plenum in Prague in October 1967 the practice of duplicating top party and state posts was subjected to criticism and the resignation of First Secretary Novotný was demanded. Lacking Soviet support and unable to generate confidence in his ability to produce change, Novotný was reduced to his single remaining position of president in January 1968. He was replaced as first secretary of the party by a relatively unknown member of the Politburo, Alexander Dubček. Dubček, a Slovak by birth who was brought up and educated in the Soviet Union—indeed, was a graduate of the Higher Party School in Moscow—was considered a reliable Communist, and his election received Moscow's concurrence. Dubček visited Moscow in January, and Secretary General Leonid Brezhnev visited Prague in February, thereby cementing Dubček's position. During the February ceremonies of the twentieth anniversary of the Communist takeover in Czechoslovakia, however, Dubček's keynote public address emphasized the desire of Czechoslovakia to normalize relations with all the countries of Europe. He also spoke of the "specific path" to socialism in Czechoslovakia, which would require "a new type of democracy" (Skilling, 1976: 188). Prague's general foreign policy orientation did not overtly challenge Soviet interests, but there was a desire to improve relations and extend economic and cultural contacts with western European neighbors. These developments were considered with growing apprehension and suspicion by the Soviets and other eastern European leaders, but opinions within the Soviet Politburo and in intrabloc consultations were divided over the extent of the threat posed by the Czechoslovak reforms and the best way to respond to them.

In the months following Dubček's rise to power, Soviet apprehensions grew. A crisis seemed inevitable, precipitated mainly by the sweeping personnel changes initiated by Dubček without prior Soviet approval (Jones, 1975). The changes involved critical positions in the military, the police, the party apparatus, the government, and the labor unions, all of them touching upon the informal political structure and networks that the Soviet Union used to sustain control of Czechoslovakia. The new appointments were liberals, and some, like Čestmír Císař, new head of the Central Committee's department of education, science, and culture, were reputed to be reformists. Party and government officials, the newspapers, and Dubček himself began to speak out in favor of more progressive policies. Political reform culminated in the introduction of broad measures of democratization and liberalization by the Action Program in April 1968.[38]

Initially, the Soviet response to these events was muted, largely because the leadership was preoccupied with domestic and foreign policy problems elsewhere. In the Ukraine there were political trials and unrest. The government held a series of trials in January 1968 to condemn a small group of intellectuals—Alexander Ginzburg was a prominent defendant—for criticizing Soviet censorship, among other things. Abroad, large-scale riots in Poland in March generated concern, not only in the Soviet Union but also in other bloc countries, that the unrest might spread.

In March 1968, Prime Minister Oldřich Černík traveled to Moscow for talks with Soviet Prime Minister Aleksei Kosygin and Gosplan Chief Nikolai Baibakov. Ostensibly, the main purpose of the talks was to coordinate their economic plans for 1971–75 and to discuss a possible loan to Czechoslovakia for economic revitalization. This was the first time that Czechoslovak and Soviet leaders discussed the Czechoslovak plans for economic reform. Public announcements after the meeting did nothing to reveal any dissatisfaction on the part of the Soviet Union. Upon Černík's return to Prague, Novotný resigned from his remaining post as president. The resignation seemed logical, for Novotný epitomized the abuses of power and centralized control developed under the Stalinist structures of Gottwald. Novotný's demise, though not unexpected, was sudden. His removal without Soviet consent or even advisement brought the domestic situation in Czechoslovakia to center stage in the minds of the Soviet leaders and spotlighted the sharp decline of Soviet influence in Prague.

The resignation of Novotný also brought into clear view the popular support for reforms associated with personnel changes in Prague. In an unprecedented outburst of energy, journalists filled the papers with articles about the abuses of the past two decades: the trials of the 1950s; discrimination against Slovaks; restrictions on the press, on religion, on Czechoslovak traditions. They openly discussed the meaning of democracy, open competitive elections, the treatment of minorities, and other taboo subjects, and the debates continued in hundreds of public meetings. One such meeting, organized by the Prague Union of Youth on March 20, was attended by 15,000 people and lasted more than six hours. The coming months saw more mass gatherings and the organization of non-Communist groups. The Presidium of the party, heady from the public excitement, endorsed the entire "movement of revival" as "healthy" and "creative" (Skilling, 1976: 207).

The alarmed Soviet leaders called an emergency conference of bloc leaders. The Dresden conference lasted one day, March 23, with the participation of representatives from the USSR, the German Democratic Republic (GDR), Poland, Hungary, Bulgaria, and Czechoslovakia. On the agenda were joint military operations, economic relations, the international situation, with an emphasis on central Europe, and, finally, events in Czechoslovakia. In Dresden the Soviet Union and other bloc countries revealed to Dubček for the first time their deep misgivings about the course of events in Czechoslovakia. Public awareness came five days later when Kurt Hager, the leading ideologist in East Germany's ruling Socialist Unity Party, delivered a blistering speech critical of developments in Czechoslovakia. By then, all attention was focused on one issue: Soviet interference. The Czechoslovak leaders were warned that they must reverse course to prevent a repetition of the Hungarian counterrevolution of 1956.

Dresden was a watershed in more ways than one. It forced the Czechoslovaks to realize that reforms must slow down or the country could face Soviet interference, and it created a sense of crisis within the Soviet bloc. The Soviet Union and other bloc members made very clear the seriousness with which they viewed events in Czechoslovakia. The bloc leaders believed that Dubček and the other Czechoslovak reform leaders had understood their concerns about the comprehensive personnel changes, the successor to President Novotný, and the elimination of censorship, which had brought trenchant liberal criticism into the media and party plenums. After Dresden the world began to pay attention to events in Czechoslovakia and to recognize that an international crisis might be at hand.

On April 8, in an atmosphere of tension, the new government of Prime Minister Černík was sworn in. Some of the chief advocates of reform were included in the government, and additional pro-reform administrators were elected regional party secretaries by secret ballot at party conferences in April. Militant articles in the newspapers did little to assure Soviet leaders that the press would be reined in. At a meeting in Moscow on May 4, the Czechoslovak leadership—Alexander Dubček, Josef Smrkovský, and Oldřich Černík, among others—requested a hard currency loan in the amount of $300–500 million to speed up the restructuring and modernizing of Czechoslovak industry (Skilling, 1976: 646). There was little prospect that the Soviets would extend the loan, so Prague authorities made contact with Western sources, including the World Bank, as an alternate source of financing. The Soviets were opposed to the restructuring of Czechoslovak industry and were reluctant to allow the manufacture of consumer goods for export to the West. According to Vasil Bil'ak, a conservative member of the Presidium of the Communist Party of Czechoslovakia (CPCz), the Soviets "begged" the Czechoslovak leaders to forgo economic reforms until the political situation was in hand (Dawisha, 1984: 75). They used the May meeting to demonstrate the vulnerability of the Czechoslovak leadership to them; the military option was articulated. The situation became even clearer when, in July, Dubček defeated anti-reformist forces at a CPCz Central Committee session. With the defeat of their Czechoslovak supporters, the Soviet leaders realized that they would have to either live with Dubček or intervene. An internal coup d'état was no longer deemed possible (Valenta, 1991: 66). A series of Warsaw Treaty Organization (WTO) troop maneuvers and exercises, which had begun in June and continued into July and August, resulted in the massive presence of WTO forces on Czechoslovak territory and along its borders. This would allow the implementation of a military solution on short notice.

A last-chance attempt to resolve the untenable situation took the form of two conferences—the first in Čierna nad Tisou and the second in Bratislava (both on Czechoslovak territory)—at which attempts at a negotiated settlement were made. The Čierna conference, held from July 29 to August 1, brought together the top Soviet and Czechoslovak leaders: almost the entire Soviet Politburo and the full Czechoslovak Presidium attended. The resulting communiqué declared anew Czechoslovak commitments to the Warsaw Pact, to the Council for Mutual Economic Assistance (CMEA), and to the restraint of antisocialist forces. The Bratislava conference began just forty-eight hours later, on August 3, and ended the same day. Participants from all the Warsaw Pact countries except Romania attended. The resulting communiqué, the Bratislava Declaration, reflected the agreement resulting from the Čierna conference; it restated the common principles of economic and military cooperation among WTO countries. There was no public display of discord, but an insertion to the declaration stressed that every fraternal party would decide its path based on its "specific national features and conditions" (Valenta, 1991: 87). In the days that followed, some in the Czechoslovak leadership, including Dubček, indicated that their understanding of what was said at the conferences was that a national interpretation of socialism and an individual path to it was acceptable (Lowenthal, 1968; Skilling, 1976: 306, 311). The meetings suggested Soviet ambiguity and indecision, especially on the part of Brezhnev, and convinced the Czechoslovak leadership that there was little risk of invasion. The Czechoslovak leadership did not, therefore, take steps to undo some of the reforms, reinstating censorship, for instance, and the Soviet interpreted inaction as a matter of reneging on the commitments made at Čierna and Bratislava.

Less than three weeks later, on the night of August 20, the combined forces of the Soviet Union, Poland, Hungary, Bulgaria, and the German Democratic Republic executed a lightning-fast invasion code-named Danube-68. Meeting at midnight, the Czechoslovak Presidium called upon the people not to resist the invading forces and ordered the Czechoslovak army to stay in its barracks. Prime Minister Černík, First Secretary Dubček, and other Czechoslovak leaders were arrested the morning after the invasion and flown to Moscow. The Fourteenth Party Congress, scheduled to meet in September, met in extraordinary session on the second day of the Soviet bloc occupation and denounced the invading forces. President Ludwík Svoboda led a delegation of Presidium members to Moscow to negotiate for the withdrawal of foreign forces on Czechoslovak territory, the restoration of the leadership, and a return to normal conditions. Apparently surprised that a new pro-Soviet leader did not step forward, the Soviets allowed Dubček and the other Czechoslovak leaders to return to Prague just seven days later, after signing a document prepared by the Soviets. The document detailed a pledge to put into practice the "principles and obligations" of the Bratislava agreement and had the Czechoslovak leaders "temporarily" accept the stationing of Soviet troops on home soil. The invasion, then, was a military success but a political failure (Eidlin, 1981; Skilling, 1976: 800–809). It did not resolve the political and social problems that had occasioned the Prague Spring in the first place, and deep resentment and discontent remained.

Soviet intervention in Czechoslovakia was a large-scale ground operation that involved the massive use of combat forces, regular and reserves, consisting mostly of Soviet troops but supplemented by Warsaw Pact troops. It was, by the definition used in this book, a hostile intervention primarily intended to bring about an overthrow of the reformist Czechoslovak leadership and a transfer of power to the hard-line group in the Presidium of the CPCz. The intervention was intended to be a brief and decisive military operation, as indeed it was.[39]

SOVIET INTERESTS

Soviet policy toward the reform movement in Czechoslovakia was driven first and foremost by the domestic interests and politics of the Soviet Union and its east European allies. The domestic concerns sprang from the collective leadership style (*Kollectivnost*) of the Politburo, produced organizational parochialism, and required aggregation of incongruent situation assessments and interests of participating decisionmakers to set operational policy guidelines (Valenta, 1991: 8–9). The differences within the group of key Soviet decisionmakers concerned less the range and essence of the threatened interests than the importance of those interests compared with other interests that would be put at risk by military intervention.

Despite the high salience of external threats in Soviet thinking, the more important threats that emerged as the crisis evolved were of a domestic (Soviet and intrabloc) nature. Although the leadership of the reform movement in Czechoslovakia emphasized that the Communist Party would not lose overall control, some of the most powerful members of the Soviet collective leadership, including Brezhnev, secretary general of the Communist Party of the Soviet Union (CPSU), perceived the reforms as an ideological challenge to the foundation of the socialist system. Developments in Czechoslovakia between January and April 1968 raised the possibility of a spillover effect in other east European satellites, which would threaten Soviet hegemony in eastern Europe, indeed, the legitimacy of all Soviet-style systems, as well as the unity and integrity of the Socialist camp. This perception of threat was reinforced by the March riots in Poland, when students and intellectuals protested the government's repressive policies and demanded a "Polish Dubček." Both Polish and GDR leaders attributed the protests to the reforms in Czechoslovakia and exhorted the Soviet leaders to halt the reform movement.

The Czechoslovak slogan, "Socialism with a human face," challenged socialism not only in other Communist countries but in the Soviet Union itself. The Soviet leaders also had closely linked reputational concerns that became more prominent after the Čierna and Bratislava agreements. They feared that an accommodation with the Czechoslovak leaders would be interpreted as a sign of government weakness (Dawisha, 1984: 26–31, 86–87, 300; Mlynář;

1980; Paul, 1971; Valenta, 1991: 134–36).[40] By July 1968 influential members of the CPSU Central Committee had come to regard the problem as primarily a domestic issue, not merely a foreign policy issue. Of major concern were the effects of the Prague Spring on the Ukraine. Apparently, in Soviet leaders' minds economic and other ties between Czechoslovakia and the Ukraine and the commonality between Czechoslovak reforms, including recognition of Slovak agitation for genuine federalism, and national self-assertion in the Soviet Union made the Czechoslovak problem part of a similar domestic problem. That is, they saw a threat of Prague-induced contamination effects on the Ukraine (Hodnett and Potichnyj, 1970: 115–25).

At the extraordinar, session of the CPSU Central Committee in July, almost half of the speakers were party bureaucrats in charge of internal affairs, political stability, and ideological supervision of the intellectual community. The speakers (e.g., party secretaries from Latvia and Lithuania) expressed concern about the revival of nationalism, especially in the western part of the Soviet Union, and the encouragement that the Czechoslovak example might give to nationalistic sentiments. There was a similar concern over the effects on the intellectual community, which saw a precedent for similar reforms in Soviet society; Academician Andrei Sakharov had issued a manifesto urging the Soviet leadership to adopt some parts of Dubček's reform program. Realizing that controlling Soviet dissidents and intellectuals would be very difficult if Dubček proceeded successfully, the bureaucrats responsible for supervising the local intellectual community considered the elimination of Dubčekism imperative (Valenta, 1991: 58–62, 99–104, 185–86).

To Soviet leaders, national military-strategic interests and, to a much lesser degree, economic interests seemed to be threatened, too. Czechoslovakia was considered one of the most important Soviet satellites in central-eastern Europe, because of its geostrategic position as a buffer between the Soviet Union and West Germany. It also linked the northern and southern Warsaw Pact countries. The loss of Czechoslovakia would have isolated Soviet forces in Hungary from those in the GDR and would have exposed western Russia, the Baltic republics, and the Ukraine in the event of an attack by the NATO countries, whose invasion force would be funneled from West Germany into the Soviet Union (Aspaturian, 1970: 36–37). The potential threat was exacerbated by the absence of Soviet troops from Czechoslovak territory. Soviet requests to station troops there had been turned down repeatedly since 1945. Nor did the Soviets have any nuclear warheads permanently deployed in Czechoslovakia, which left a gap in WTO defenses. Agreements signed by Czechoslovakia and the Soviet Union in 1961 and 1962 entitled the latter to deploy nuclear warheads in Czechoslovakia in the event of a crisis. More important, a treaty signed in 1965 provided for the storage of nuclear warheads in western Czechoslovakia; they would remain under exclusive Soviet control. With the reformist trends in full swing the Soviet military was concerned that the construction of the facilities for the warheads and a Soviet contingent to guard them would not be completed in

1969 as scheduled. There was also concern for the security of the nuclear warheads in light of what the military leaders saw as laxity in Czechoslovakia's border security system. To them, the only viable solution was a direct, permanent large-scale Soviet military presence in Czechoslovakia—which could be achieved only by invasion (Kramer, 1993).

Soviet concerns about the strategic importance of developments in Czechoslovakia were compounded by major changes in command personnel in the Czechoslovak military; the criticism within the Czechoslovak military establishment of the hegemonic Soviet role in the Warsaw Pact; widespread support for far-reaching changes in Czechoslovakia's defense policy that would provide for a more independent national military doctrine; and the voices for reduction of defense expenditures. The man behind these developing trends was, according to the Soviets, the popular Lieutenant General Václav Prchlík, head of the Military-Administrative Department of the Central Committee of the CPCz. In July 1968 the draft form of Czechoslovakia's Plan for Future Changes in Military and National Security Policies, prepared by his office, was leaked to the Soviet embassy in Prague. The report was particularly critical of the nuclear dimension of the military policy of the Warsaw Pact:

> The notion of a general war in Europe that involves the massive use of nuclear weapons is, from Czechoslovakia's point of view, purely senseless. This form of war would bring about the total physical destruction of the CSSR [Czechoslovakia], irrespective of the scale of losses to its armed forces and also, ultimately, irrespective of the final results of the war.
>
> For this reason the aim should be to achieve a pragmatic stability of the state's defense system and the army's structure, a stability that flows from political demands and a dual goal: to prevent excessive danger on the part of the potential adversary, and to preserve the existence and sovereignty of the Czechoslovak Socialist Republic.

This argument challenged Brezhnev's emphasis on close coordination by Warsaw Treaty members (Anderson, 1993: 101–5) and the cohesion of the alliance. It could only raise doubts, too, about whether the Czechoslovak government would comply with the agreement that allowed the Soviets to deploy nuclear weapons on Czechoslovak territory under Soviet control, a high-priority goal for the Soviet Union.

The document called for major reforms in Czechoslovak military and security services. It argued that the external threat was exaggerated for manipulative purposes.

> But in fact this supposed threat was always a superfluous external factor, which served as a basis, on the one hand, for strengthening the unity of the socialist camp and, on the other hand, for justifying the extraordinary human and material resources that were being demanded by the armed forces. The military factor in many respects compensated for the inadequate development of economic cooperation and other ties among the socialist countries.

The State Security Services, which were among the most loyal to the Soviet Union, were criticized:

> These organs were not genuinely controlled by the Party leadership or the social-ist society, and they committed violations of the law with impunity, even after 1953. In their work the State Security organs often did not take account of basic principles of professionalism, science, law and legality, and socialist ethics and morality, and did not take account of the political consequences of the results of its work. (National Security Archives [NSA], Document no. 73, 1994)

Under heavy Soviet pressure for Prchlík's demotion, the Czechoslovak leadership abolished his department but reassigned him to another military position (Dubček, 1993: 167). But Soviet confidence in the credibility of Czechoslovakia as an alliance member was undermined; specifically, the long-term combat readiness of its army was seriously questioned. The Soviet high command—in particular, Marshal Andrei Grechko, the minister for defense, and Marshal Ivan Yakubovsky, commander in chief of the WTO forces—was less concerned with the domestic implications of the reform movement and more concerned with its effects on existing security arrange-ments. These Soviet military leaders questioned the capability and resolve of the Czechoslovak army to hold a defensive line against a massive NATO at-tack for 72 hours, as dictated by Warsaw Pact doctrine. Other influential military officers, such as General Aleksei Yepishev, chief of the Main Politi-cal Administration in the Ministry of Defense, and Marshal Ivan Konev, general inspector of the Soviet army, also expressed the view that imperialist subversion in Czechoslovakia justified Soviet military intervention. The mil-itary, therefore, was in agreement that the crisis should be resolved quickly through the use of force.

This position was reinforced by the evaluation of the situation by the mil-itary command: that the modus vivendi would become in the long run a heavy economic and logistic burden. Preparing for a possible invasion en-tailed mobilization of a large number of reservists, transportation equipment, and other civilian resources, which would cause severe disruption of the na-tional economy (Dawisha, 1980b: 116–18; Dawisha, 1984: 305–8; Whetten, 1969). Other potential economic costs stemmed from the key position of Czechoslovakia as one of the two most industrialized countries in the Com-munist bloc: it was the manufacturer of industrial equipment for bloc coun-tries, as well as an important supplier of military aid to Soviet clients in the Third World.

The Soviets emphasized that they were not going to allow a "second Yu-goslavia" or a "second Romania." They also feared the potential emergence of a "Communist Little Entente" of Yugoslavia, Romania, and Czechoslo-vakia that would destabilize the Warsaw Pact (Remington, 1971: 97). The Czechoslovak leadership was keenly aware of Soviet fears, and Dubček re-peatedly said that the reforms did not challenge immediate and direct Soviet security interests, nor did they involve changes in foreign policy orientation.

He repeatedly stated that Czechoslovakia would retain its membership in the Warsaw Pact and the Council for Mutual Economic Assistance, but to no avail.

Reputational interests were of minor importance in the early phases of the crisis, but this changed as the crisis progressed. Seemingly concerned but undeterred by the show of force by the Soviets and their allies, the Czechoslovak leadership continued with the reform program. Dubček and his colleagues, apparently unrepentant, seemed to be thumbing their noses at the powerful Soviet Union. Brezhnev himself, who had shaped the understandings reached at the Čierna nad Tisou and Bratislava conferences, probably felt personally challenged when the Czechoslovak leadership did not conform with these agreements by undoing some of the reforms. From the point of view of some Soviet and eastern European leaders, the stakes in the Czechoslovak crisis were substantial and the risks considerable; others disagreed. The key members of the Soviet decisionmaking group shared most of the concerns discussed above, but they viewed them with different degrees of gravity; and although some advocated intervention, others urged a more cautious and moderate approach.

NATIONAL CAPABILITIES

In April 1968 appeared the first public indications that the Soviet leadership was seriously considering the use of force. The decision to begin invasion preparations was taken in early May. Marshal Ivan Konev, who was visiting Czechoslovakia, emphasized that Soviet armed forces were ready to protect not only the security of the bloc but also its socialist character. To carry out these threats the Soviet military had at its disposal a massive number of Soviet and allied troops in countries bordering Czechoslovakia: Poland, East Germany, and the Ukraine.[41] In East Germany alone, twelve tank and mechanized divisions of the Soviet army and two East German divisions were available for military intervention on very short notice (Dawisha, 1984: 103–5). The small size and elongated shape of Czechoslovakia put vital regions within easy striking distance of an invading army. What later became a major advantage, which reduced substantially the military risk to the Soviets, was the presence of extensive WTO military forces in Czechoslovakia prior to the invasion. Under the pretext of holding joint staff exercises, code-named Sumava, WTO troops entered Czechoslovakia in June. The presence of Soviet and other WTO troops was intended to make intervention an easier and less costly task and to assist the conservatives in the Presidium of the CPCz to regain control. The military exercises allowed Soviet commanders to tour the country and assess the political mood of the people and the morale of the Czechoslovak army, as well as draw up detailed maps of all military facilities and air bases. Marshal Yakubovsky's headquarters and communications center for the exercises, set up at Ruzyně Airport outside Prague, was not dismantled upon completion of Sumava and were instrumental in landing Soviet airborne troops during the invasion.

Although a decision to withdraw troops gradually was made on July 10, the withdrawal was effectively halted on July 14, a day after it started.[42] In fact, Soviet military procrastination in setting the date for withdrawal was largely due to the anticipation that troops might have to be sent in again if the situation was not resolved peacefully. The leaders were aware that sending a large number of troops across Czechoslovak frontiers, possibly uninvited, would pose enormous military and political problems.

Starting in late July, WTO members exercised their logistic capabilities for sustaining large-scale military operations, including transport of troops as well as technical, medical, and fuel supplies. "Nemen," the largest rear-services exercise ever executed in Soviet history, was intended to test communications between military units. Nemen, and an extensive air-defense exercise in the Baltic–Moscow–Black Sea regions code-named Skyshield, was followed on August 11 by an expanded joint exercise of communication units involving Soviet, Polish, and GDR troops. These exercises camouflaged the logistical and air-power buildup for the August intervention, the positioning of more troops on the Soviet-Czechoslovakia border, and horizontal C^{3}I coordination by the the armies of participating WTO members, which included Soviet, Polish, East German, Bulgarian, and Hungarian troops (Dawisha, 1984: 162, 186–87, 250–51, 274, 280–82). The invasion plan did not impose strenuous demands on the eastern European armies, an indication of Soviet lack of trust in their ideological and operational reliability. Even so, the invasion resulted in much disaffection and confusion once the Hungarian and Polish troops realized that they were not actually defending Czechoslovakia against American "imperialism" and West German "revanchists" (Kramer, 1992b: 9).

The Soviet high command had extensive first-hand knowledge of Czechoslovak military capabilities (weapon systems, doctrine, command structure, and training). Soviet officers were present at all levels of command in the Czechoslovak People's Army. What was unclear to Soviet decisionmakers, however, was the level of support for the reform movement within the military and, consequently, the probability of armed resistance to a WTO invasion. This risk dimension was important, because the Czechoslovak army was considered the most capable in eastern Europe and should have been able to establish a defensive posture that would substantially raise the costs of an invasion (Kramer, 1993: 9; Grigorenko, 1982: 359).[43] Lieutenant General Prchlík, head of the powerful Eighth Department of the Central Committee, the Military-Administrative Department, had called in July for an independent military doctrine, a move implying that a Soviet attempt to station troops permanently in Czechoslovakia would meet with armed resistance, and submitted a plan to that effect. The Soviets were informed of this development by Vasil Bil'ak, conservative head of the Slovak Communist Party, and his supporters.[44] But the Soviets learned that General Martin Dzúr, minister for national defense, refused to even mention the possibility of armed defense, implying that neither he nor President Ludvík Svoboda would issue such an order, and that Dubček had immediately vetoed Prchlík's plan. The Ministry of National Defense publicly denied the rumors that military measures would be taken in the event of Soviet military intervention. These moves reinforced the view of Kremlin hard-liners that the risk of serious armed, organized resistance was low and, in particular, confirmed the view of Soviet Minister of Defense Marshal Grechko that the Czechoslovaks would not resist militarily but would behave like the "good soldier Švejk" (Dawisha, 1984: 233–36, 246–47; Kramer, 1993: 2–3; Rice, 1984: 139–56).

Using overwhelming force in blitzkrieg fashion and ascertaining that political conditions in Czechoslovakia were not conducive to resistance served the Soviet purpose of quickly stabilizing the situation. The Soviet political leaders foresaw problems of control over the military operation and were skeptical of the military's ability to manage the political aspects of the invasion, especially to prevent civil war and find a successor to Dubček. Therefore, they sent a Soviet Politburo member and retired lieutenant colonel, Kirill Mazurov, to Prague to oversee and control operations. To make sure that instructions were followed, the invading forces were extensively indoctrinated about the objectives of the invasion and instructed to treat any persons not manifesting violent resistance as friends (Eidlin, 1980: 82, 127–28; Valenta, 1991: 185).

By early August the Soviets had a massive force in place in Czechoslovakia and on its borders. An extensive logistic infrastructure and communications network was ready to support an invasion.[45] The command structure was ready to coordinate combat operations by troops from the Soviet Union, Poland, Bulgaria, Hungary, and the GDR. An overwhelming force was arrayed against a country that was unprepared for armed resistance and only ill prepared for civil resistance. The blitzkrieg strategy depended on the newly increased and upgraded air mobility of the military; Prague and other vital centers were to be captured by landing Soviet airborne forces while Warsaw Treaty units were moving along a number of strategic highways and railways toward the targeted centers of population and communication. Aware of the political and military costs of Czechoslovak military resistance and of potential external intervention, the Soviets made deception and surprise key components of their strategy. They succeeded by using the large-scale military maneuvers under way to disguise their intentions and, under the military exercise pretext, removed stocks of fuel and ammunition from the control of the Czechoslovak military and transferred them to East Germany. They convinced the Czechoslovak Ministry of Defense that a military exercise would take place on August 22 with the participation of Czechoslovak units, thus the distracting attention of the Czechoslovak military command from their preparations for invasion. The Soviet command also withdrew forces from Czechoslovakia in early August to mislead the Czechoslovak leadership. Up to the last minute Dubček believed that the Soviets would be deterred from military intervention for international reasons—for instance, the risk of jeopardizing the World Communist Conference scheduled for November 1968 (Valenta, 1980: 134; Valenta, 1991: 189–91).

As planned, the invasion caught the United States, as well as Czechoslovakia, surprised and unprepared. In a memo to the president on a trip to Europe, written just a day before the Soviet invasion, the Senate majority leader Senator Mike Mansfield predicted confidently that "quite apart from words it is not likely that the Soviet government will move directly in a military sense to upset the Czechoslovak revision unless Dubček is forced to an extreme anti-Russian position—a position which he had heretofore avoided."[46] From the report of Secretary of Defense Clark Clifford to the Cabinet on August 22 it is clear that the Joint Chiefs also believed that an invasion was unlikely. Clifford admitted: "Militarily, it has been a very clever and sophisticated strategy. This is the time of the year when they usually hold their maneuvers on the Czech border. The fact that the Red Army had divisions there did not disturb the Joint Chiefs."[47] In the emergency meeting of the National Security Council on August 20, when response options were considered, General Earle Wheeler, chairman of the JCS, stated: "There is no military action we can take. We do not have the forces to do it."[48] Thus, for the Soviets, surprise was important, but even more important was reduction of the potential risk of a protracted military action. By invoking the sheer number of troops and the extensive logistic preparations, the military was able to reassure those in the Politburo and others who had not made up their minds about the use of force that the operation would be quick and successful. Military preparations, together with assurances by military leaders, were instrumental in shifting the balance of preferences in the Politburo toward intervention.

In spite of, or perhaps because of, the demonstration of massive force by the Warsaw Pact countries, casualties on both sides were low. Eighty-two Czechoslovak citizens were killed, 300 were severely wounded, and 500 suffered minor wounds between August 21 and September 28, 1968. Around 20 Warsaw Pact soldiers were killed, only one by Czechoslovaks, the rest in accidents. There were morale problems among the invading forces, who were unprepared for nonviolent resistance and found themselves doubting the legitimacy of the invasion, their indoctrination notwithstanding.[49] A handful of Soviet soldiers were executed for refusing to participate in the invasion, and a few others committed suicide (Eidlin, 1980: 39–40; Kramer, 1993: 10; Mlynář, 1980: 186).

Maintaining a state of high military preparedness over a protracted period and involving a huge number of troops resulted in heavy direct financial costs and substantial indirect costs to the Soviet economy. The rear-services exercise Nemen that started on July 23 required calling up thousands of reservists and requisitioning thousands of civilian vehicles to bring Class II and III divisions to full alert. Farm trucks were commandeered, as were all freight trucks, "and both the summer harvest and civil transport practically came to a standstill" (Dawisha, 1984: 250, 275). The grain harvest in western Russia and the Ukraine was totally disrupted. Because the costs of maintaining 500,000 troops deployed in and around Czechoslovakia were too high logistically, strategically, and politically to be sustained indefinitely, interventionists had a powerful argument against adopting a wait-and-see policy (Valenta, 1991: 111–12). In addition, they could compare the economic costs of intervention with those of nonintervention. For both the Soviet Union and the GDR a reorientation of the Czechoslovak economy toward the West would seriously affect vital trade links, deprive them of needed capital goods, and, in general, deal a blow to the Council for Mutual Economic Assistance (Dawisha, 1984: 30, 32, 83–84).[50]

On the political side, a decision for military intervention posed a number of problems that were embedded in the nature of the Soviet political system. The Soviet decisionmaking process had political and institutional constraints.

> As centers of partial power in the system, the various bureaucracies have claim to be heard; the way they marshal their arguments and the skill of their advocacy can help to structure the issues as they are presented to the top leadership, so that in a sense the policy options open to it are already somewhat circumscribed before they become a matter of decision.
>
> Although the Soviet government is not one of formal checks and balances, when viewed in terms of this model, the proliferation of power within a large and complex bureaucratic system like that in the Soviet Union may in some sense serve as a haphazard substitute for constitutional checks upon central authority. It (the bureaucratic proliferation of power) tends to beget potential vetoes upon policy and may lead to immobilism in action, especially innovative action that breaks with established ways of doing things. In effect, this model places the top leadership at the center of a bureaucratic process which may encumber response to new problems and situations as often as it facilitates their "solution," and it suggests that the policies which emerge from the process may represent something less than the product of optimum choice among a full array of alternatives. (Wolfe, 1969: 4)

Specifically, under Secretary General Leonid Brezhnev the Politburo, the key decisionmaking body in the Soviet Union, was basically an oligarchy, with Brezhnev as first among equals. Within the Politburo a smaller inner group made important decisions, like the one to intervene militarily in Czechoslovakia. "Normally when this group found it possible to agree among themselves on policy issues, they could determine policy outcomes" (Stewart et al., 1989: 40). In the absence of agreement, a decision was not likely to be made quickly. In the case of the Prague Spring, time was essential for the emergence of an operational consensus through arguing, persuading, building coalitions, imposing situational constraints, and bringing a variety of pressures to bear on opponents. Five members of the Politburo were crucial to the decision on Czechoslovakia: Leonid Brezhnev, Aleksei Kosygin, Nikolai Podgorny, Mikhail Suslov, and Petr Shelest.[51] In March 1968 the core decisionmaking group was divided among hawks and doves, with Brezhnev himself cautiously undecided and still reluctant to use military force. A decision on intervention was consequently delayed until late August, when an operational consensus could be formed within the Politburo. Other actors—the Central Committee of the CPSU, other bloc leaders, leaders of various organizations (e.g., Minister of Defense Andrei Grechko; the head of the KGB, Yuri Andropov; Minister for Foreign Affairs Andrei Gromyko), and some ranking military officers— supported or influenced one or more of the key actors, but these non-Politburo participants could not decide policy direction by themselves.

Between March and August advocates of both intervention and nonintervention accumulated and disseminated information supporting their views and attempted to establish their authority as experts through visits to Prague, discussions with Czechoslovak and other bloc leaders, and intragroup discussions. Those leaning toward intervention, the riskier alternative, were by far more active than those against intervention, who played the role of cue takers; the latter weighed new evidence and tried to decide whether to persist in objecting to intervention, to abstain from deciding, or to let the hawks go ahead. Brezhnev, though unenthusiastic about accepting the risks of an invasion, "realized that it was critical for him to be seen as the decisive leader of a Politburo consensus when it finally emerged from the fragmented Soviet leadership" (Gelman, 1984: 100).[52] Until a consensus was arrived at, he did not commit himself to any position. He allowed the military to prepare for an invasion while at the same time he negotiated with the Czechoslovak leadership, built intrabloc support, and tried not to alarm the United States and its western European allies. Once the decision to intervene was reached, it was implemented in the usual Soviet style, by using surprise and overwhelming military superiority to advantage, which was a way of achieving rapid success while preventing other risks from emerging.

From the beginning of the crisis to the invasion a considerable effort was made to establish domestic and international legitimacy for military intervention. Those who needed to be convinced of its domestic legitimacy were the CPSU members and the public at large, especially the intelligentsia. On July 17 a plenum of the CPSU Central Committee decided to undertake a countrywide campaign to prepare the public for an invasion of Czechoslovakia. Those that needed to be convinced of external legitimacy were other bloc members, particularly Yugoslavia, Romania, and Hungary. The first two supported Czechoslovakia, and the last was highly reluctant to apply force, both because bitter memories of Soviet intervention in Hungary in 1956 were still vivid and because the Hungarian leaders were themselves introducing market reforms and a degree of liberalization. Others that needed to be convinced were Western bloc countries, non-aligned countries, and nonruling Communist parties. The different audiences required different arguments for legitimacy, although the arguments overlapped.

On the domestic front, cognitive legitimacy was based on the argument that all other means of reaching a peaceful solution had failed owing to the intransigence of the Czechoslovak leaders; intervention was the last resort, and it would be quickly and successfully accomplished. Furthermore, a group of Czechoslovak conservatives had written letters inviting the Soviets to intervene. The five pro-Moscow hard-liners in the CPCz, led by Vasil Bil'ak, had addressed two letters to Brezhnev in August 1968. The more important one, secretly passed on by an intermediary at the Bratislava conference on August 3, stated: "The very existence of socialism in our country is under threat. . . . The right-wing forces have created conditions suitable for a counterrevolutionary coup." It went on to say that "we will struggle with all our power and all our means. But if our strength and capabilities are depleted or fail to bring positive results, then our statement should be regarded as an urgent request and plea for your intervention and all around assistance."[53] The second letter urged the CPSU to respond positively to the first letter. It was signed by only one of the original five (Antonín Kapek) on behalf of all and reached Brezhnev on August 19. On August 14–15, Alois Indra and Oldřich Pavlovský, two senior hard-liners, met the Soviet ambassador in Prague and assured him that as soon as Soviet troops moved into action a majority of Presidium members, the Central Committee, the National Assembly, and members of the government would align themselves with the conservatives in the Presidium. This majority, it was promised, would oust Dubček, take control, and form a revolutionary workers' and peasants' government. This government would appeal for fraternal assistance and thus provide Soviet intervention with a legitimate pretext (Kramer, 1993: 3–4).[54] Ranking party officials were also assured that the extensive military preparations guaranteed that the invasion could be executed quickly and efficiently and with little risk of Western response.

The effort to establish normative legitimacy had two components. One was the use of Marxist-Leninist ideology both to emphasize the extent of the ideological deviation by Prague and to justify the employment of force to deal with the deviation.[55] The underlying ideological argument came to be known as the Brezhnev Doctrine. Stated briefly, it holds that the concept of sovereignty among socialist states must not be understood separately from the laws of class struggle and is subordinate to them. Thus, the Soviet Union has the right and obligation to intervene in any fashion necessary if developments in any socialist country pose a threat to socialism in that country or to the basic interests of other socialist countries (Remington, 1971: 109). In the case of Czechoslovakia, the justification for intervention was the threat of "revisionism." The initial stage of establishing the ideological foundation for intervention took place in the April 9–10 plenum of the CPSU Central Committee, which was devoted to a discussion of foreign policy. Using ideological arguments, the participants framed the main danger from the imperialist bloc not as resulting from direct East-West military confrontation but as embedded in imperialist subversion, "sneaking counterrevolution" (Valdez, 1993: 56), which weakens the unity of the socialist countries and undermines socialist societies from within. The plenum resolutions linked domestic dissent, bloc disunity, and the security interests of the Soviet Union and other bloc members. This link provided legitimacy for Soviet interference in Czechoslovakia if the reforms proved to have adverse effects on the Soviet Union or other bloc members. After the appearance of the "2,000-word" manifesto on June 27, it was not difficult for the hard-liners in the CPSU to label the reform movement as counterrevolutionary. The manifesto called for continuing reform and elimination of the remaining conservatives in the Czechoslovak government, by force if necessary, and promised grass-roots support.

By defining developments as counterrevolutionary the hard-liners prepared the ground for justifying military action (Dawisha, 1984: 56–60, 166–68). The accusations were lent credibility by the letters from the Czechoslovak hard-liners. Their invitation to the Soviets to intervene enhanced the claim that the Soviets had the right to intervene. The ideological polemics were supported by non-ideological arguments revolving around national security and the right of self-defense. The moral-historical basis of this non-ideological normative argument was the right of the Soviet Union to intervene when its western borders were liable to be breached. This right was invoked when the Soviet Union reminded Czechoslovakia of the "blood debt" that it owed for the sacrifice of Soviet 145,000 soldiers in the liberation of Czechoslovakia from German fascism (Dawisha, 1984: 222, 347) and recalled, too, the sanctity of the secure border that resulted from the sacrifice.

The elaborate attempts to justify and explain the military intervention highlight the importance and perceived risk of the lack of adequate legitimacy for intervention even at home. For example, 643 leading officials of the Central Committee apparat and regional branches of the CPSU received a message from the Politburo, to be disseminated among party members, informing them of the invasion and the reasons for it. This message shows how concerned the Soviet leadership was with the potential domestic consequences of the invasion for the "cohesion of the Party and people" and for "the moral and political unity of Soviet society" (NSA, Document no. 86, 1994). It also attests to the power and salience of the norm of sovereignty among nation-states of all political and ideological persuasions. In the final account, the arguments for legitimacy were sufficient as far as the party apparat was concerned but were less convincing to the intelligentsia.

The policy debate and risk calculations of both hard-liners and soft-liners in the Soviet leadership were shaped by time constraints. Because the key decisionmakers were divided over how to respond to the emerging crisis, Brezhnev needed time to build a consensus and make himself part of it. The military needed time to mobilize, coordinate, and organize for a quick and efficient operation. Time was also required to forge agreement within the Warsaw Pact countries—so that the risk of a split within the bloc and the risk of the unilateral Soviet action could be avoided—and to assess more accurately the direction of the Prague Spring, so that the risk of premature overreaction to a situation that might resolve itself could be avoided.

As the crisis progressed, there was a growing sense that time was running out for the Soviet bloc. Risk perceptions focused increasingly on the possibility of reacting too late and having to absorb irreversible damage to Soviet vital interests. Some individuals, most notably Brezhnev, were probably concerned that failure would result in their being ousted from power, or worse.[56] As time went by, it also became clear that the reform movement was enjoying an upsurge in support among the rank and file in the CPCz, whereas the conservatives were being systematically purged from party and state organizations and positions; it became highly likely that key conservative politicians were not going to be reelected to the Central Committee at the forthcoming Fourteenth Party Congress (Dawisha, 1984: 168–69). If intervention was to prevent this outcome, the decision was subject to strict time constraints, for September 9, 1968, the scheduled date of the Congress, was seen as the absolutely final deadline for dealing with the Prague Spring. Prior to that date there were two important intermediate deadlines: August 20, when the Czechoslovak Presidium was to meet, and August 26, when the Slovak Party Congress was scheduled to convene. The latter event was considered the probable first step toward a counterrevolutionary coup led by Dubček. August 20 was therefore perceived as the final logical opportunity for the anti-reformist group to stage a coup and legitimize a simultaneous Soviet intervention. Technically, it was also an occasion when all top Czechoslovak leaders would be assembled in one place and could easily be rounded up (Kramer, 1993; Mlynář, 1980; Valenta, 1991: 179). An additional time factor concerned the large-scale mobilization of military forces and the costs of keeping them in readiness, which entailed high direct and indirect costs to the Soviet economy and required a decision on whether to allow the invasion to take place or to order the military to return to normal deployment.

The time constraints were a source of stress and, to a certain extent, prevented procrastination in and avoidance of decisionmaking. Apprehensions and uncertainties regarding the intentions of the reform-minded Czechoslovak government were not resolved with time, so the demands for action led to a growing dependence on the vivid but inappropriate analogy of Hungary in 1956 (Dawisha, 1984: 350–51). This analogy was used first in interpreting the situation in Prague as potentially counterrevolutionary and later in assessing the probability that a Hungary-like solution would succeed—that is, that following an invitation for intervention by a pro-Soviet "revolutionary government" that government would manage, with initial Soviet assistance, to establish its authority and gain respectability over the long term.

To summarize, the Soviet leadership had to face and consider a number of capability-related risks in connection with the military intervention option. At the early stages of the crisis, the assessments focused on the possibility of armed resistance by the Czechoslovak People's Army, a competent, well-equipped, Soviet-trained military force. The possibility raised the specter of a costly and bloody intervention. The extensive military preparations, the campaign of deception, and the Czechoslovak decisionmakers' resolve to avoid bloodshed substantially reduced this risk, but it remained a matter of serious concern.[57] At the same time, the extensive mobilization of military and civilian resources, to keep the troops of the USSR and its allies in a state of readiness, raised the direct and indirect financial costs of maintaining preparedness and required a quick resolution of the crisis.

The division within the inner group in the Politburo raised the political risks of imposing a decision and the costs of potential failure. Brezhnev, who had not yet fully consolidated his position at the top of the Soviet hierarchy, was especially vulnerable.[58] Two of his rivals, Podgorny and Kosygin, were members of the inner group. A faulty decision would likely result in demotion for the scapegoat, probably Brezhnev (Musatov, 1992). Thus, forging a consensus was necessary both to break the deadlock and to share responsibility in case of failure.

A decision to intervene could have had serious repercussions at home and abroad. Because the use of military force against another socialist country had to be explained, legitimacy arguments were important. The moral and political right of the Soviet Union and its allies to intervene had to be established. With the escalation of the crisis, time became part of the risk calculus. The Soviet leaders had to reach a consensus within a restricted time frame. They realized that procrastination would not be a workable strategy, for beyond a certain point there was an acute risk of irreversible adverse consequences from developments in Prague if a policy decision was not reached.

THE MANAGEMENT OF CONFLICTING INTERNATIONAL INCENTIVES AND CONSTRAINTS

In assessing the risks posed by the international environment, the debate among Soviet decisionmakers was affected by three sets of considerations: their perception of systemic constraints, their assessment of the likely American and NATO responses, and the conflicting positions taken by Soviet bloc members and by nonruling Communist parties in western Europe.

The loose bipolarity of the international system during the Czechoslovakia crisis was viewed with mixed feelings by Brezhnev. The cohesion of the Western bloc was affected by the emergence of Europe and Japan as increasingly independent power centers with considerable autonomy in foreign policy, dramatically expressed by France's withdrawal from NATO's military structure in 1967. Moscow sought to improve relations with the western European countries, including West Germany, to encourage the trend toward polycentrism in the Western bloc. The Soviet bloc, too, loosened up; salient expressions of that were Yugoslavia's revisionism, Romania's independent positions on certain foreign policy issues (e.g., refusal to sever diplomatic relations with Israel after the 1967 war), and issues concerning management of the Warsaw Pact. Yet the Soviet bloc was still much more hierarchically structured than the Western bloc, and Soviet hegemony in the East surpassed that of the United States in the West. Despite the relative decline of the dominant position of the United States, Brezhnev and other key leaders still thought that the essential class solidarity of the imperialists would prevail as the centripetal, unifying tendency in the Western bloc.[59]

Thus, Brezhnev's position on defense since his rise to power emphasized the need to upgrade Soviet defense capabilities, conventional and nuclear, in order to tilt the balance of power in favor of the socialist camp. He did not believe that limited wars would necessarily escalate into nuclear exchanges, so he supported and authorized an upgrade and increase in Soviet ground forces. By 1968 he claimed that the power balance tipped toward the socialist camp (Murphy, 1981: 263–70). The possibility that Czechoslovakia might drift away from the Warsaw Pact and thus adversely affect the Soviet geostrategic situation was a matter of grave concern because it could adversely affect Brezhnev's commitment to achieving military parity with the West. Not taking advantage of the increased and upgraded military forces to meet the threat would, in addition, play into the hands of his critics, such as Kosygin, who questioned the need for the military buildup.

Influential members of CPSU feared that an emergent western European power center would prove an attractive model and source of support to eastern Europe, as the Prague Spring had seemed to indicate (Wohlforth, 1993: 194–95). In the Soviet-Czechoslovak leadership meeting in Moscow in May 1968, Brezhnev stated this concern: "We believe that the events of the present are being organized and directed by forces linked to the West. The thread that controls them clearly leads to France, to West Germany, in a word, you yourselves know where" (NSA, Document no. 27, 1994). Events in Czechoslovakia could cause a domino effect if not dealt with effectively; they could encourage the West to take advantage of its economic strength to woo other eastern European countries, like Poland, and undermine the Soviet power position in the international system. On the other hand, Soviet military intervention would likely put an end to Soviet diplomatic efforts to encourage Europeans to distance themselves from the United States and would result instead in their greater dependence on the United States due to an increase in the perceived Soviet threat. These possibilities had to be weighed even if the risk of direct U.S. military counterintervention was not high. The implicit understandings between the superpowers allowed each to protect the integrity of its bloc against desertion by key members without having to worry too much about the other power's counterintervention.

Still, the question of U.S. response was an important reason for procrastination. In contrast to Soviet foreign policy makers since 1945, Brezhnev did not tend to employ international crises as a means of focusing attention on, and creating an incentive for, resolving outstanding international problems. When the crisis escalated in the spring of 1968, the Soviet Union was trying to reach a strategic arms limitation agreement with the United States. For the United States, the temptation of counterintervention was less salient in the late 1960s than it had been in Hungary in 1956 in light of the parity in nuclear capabilities, a fact mutually conceded by the two superpowers. U.S. involvement in the Vietnam War, and especially the Tet Offensive of early 1968, supposedly further reduced the risk of a Western military response to a Soviet intervention. But the costly Vietnam intervention had also demonstrated American resolve and willingness to take risks even when the costs were high, and because President Johnson had already decided not to run for a second term, he had much more room to take risks, especially if the Czechoslovaks reacted to a Warsaw Pact invasion by taking up arms.[60] If the Czechoslovak army made an effective defense effort, there was at least the remote possibility of U.S. military intervention with West Germany–based forces. "Thus from a military point of view the risk for the Soviet leadership seemed to be greater than it had been during the Hungarian crisis" (Valenta, 1991: 129). However, once it became clear in July that the Czechoslovak army would not fight, the risk of U.S. military intervention dropped substantially. By then, the war in Vietnam and the adverse domestic situation— race riots and the failure of the Great Society programs—made the deterrent power of the United States to prevent an intrabloc intervention seem less worrisome to the Soviet leadership.

Even if the United States did not counterintervene, the Czechoslovak crisis still jeopardized a number of important foreign policy initiatives with the United States. In June 1968 the Soviets ratified a consular treaty with the United States, resumed negotiations on a two-year cultural exchange agreement, reached an accord with the United States that made it possible to pass a U.N. resolution endorsing the Nuclear Nonproliferation Treaty, and announced its historic decision to enter in Strategic Arms Limitation Talks (SALT) with the United States. At the same time, they signaled continued interest in pursuing their *Westpolitik* diplomacy with West Germany and the dialogue on the mutual renunciation of force. These initiatives, inherent impetus aside, were prompted by fear of China and apprehension of a U.S.-China rapprochement.

Although these policy initiatives had the support of a majority of the Politburo members, they were not uncontested, as was evidenced from Brezhnev's own blend of support and skepticism regarding détente with the United States (Valenta, 1991: 42–48). Regardless, the Soviet Union would have risked important bilateral issues had the United States decided to respond strongly to a Soviet intervention. President Johnson, however, adopted a low-profile approach during this period; he was very cautious and clearly reluctant to reverse the process of Soviet-American détente. These developments—plus the U.S. perception of Czechoslovakia as a troublemaker, the provider of economic and military assistance to national liberation movements—caused the United States to refrain from statements or actions that might have had a deterrent effect on Soviet debate about intervention.[61] Speaking after the intervention, Brezhnev recalled that "I asked President Johnson if the American government still fully recognizes the results of the

Yalta and Potsdam conferences. And on August 18 I received the reply: as far as Czechoslovakia and Romania are concerned, it recognizes them without reservation; in the case of Yugoslavia, it would have to be discussed" (Mlynář, 1980: 241).

The combined effect of the American hands-off approach and the renewed confirmation of American support for the status quo in Europe probably enhanced the position of the Soviet hard-liners by reducing, if not eliminating, an important risk dimension. The trend encouraged Brezhnev to move closer to the conservative line. The Soviet Union consequently adopted the two-track strategy of deterrence and reassurance. The mobilization of a massive military force and the extensive military maneuvers not only pressured the Czechoslovak leadership but also signaled an unwillingness to tolerate outside interference. At the same time, Soviet diplomacy attempted to reassure the United States that events in Czechoslovakia notwithstanding, the USSR was interested in pursuing détente and in preserving the post–Second World War status quo. Thus, although by August the Soviet leadership was highly confident that an invasion would not trigger a military response by the United States and its NATO allies, some influential Soviet decisionmakers (e.g., Kosygin, Gromyko) still believed that an invasion would have other serious repercussions for East-West relations. By late August their arguments were not considered serious enough to warrant a dismissal of the military intervention option.

Establishing the legitimacy of the intervention with extrabloc states, especially the United States and western Europe, was practically impossible. The national security argument that "external and internal forces of aggression" conspired against Czechoslovakia was used to establish cognitive legitimacy (Johnson, 1971: 488). The false argument that the Soviet Union and its allies had been invited by legitimate Czechoslovak authorities to send in troops was used to establish normative legitimacy. But realizing how weak these arguments were, the Soviet leadership sought to contain the diplomatic damage of the coming invasion by having Kosygin inform President Johnson one day prior to the invasion that the Soviet Union was prepared to host a summit between Johnson and Brezhnev to discuss limitations on strategic arms (Johnson, 1971: 487–88). In this way, the Soviets sought to maintain their freedom of action within their sphere of influence without jeopardizing détente with the United States.

The role played by Soviet eastern European allies in the Czechoslovak crisis demonstrates that the Soviet bloc was not rigidly hierarchical. Eastern European leaders took the initiative in an attempt to shape Soviet policy and influence the debate within the Politburo in a manner that resembled in many ways the behavior of America's European allies in policy debates in the Western bloc. In trying to shape a consensus on the Czechoslovak reforms the Soviet leaders faced the complex issue of how to reconcile conflicting positions within the bloc and the larger Communist camp, that is, responses by allies in eastern and western Europe. If they decided to intervene, they would probably risk antagonizing the problematic bloc members, Romania and Yugoslavia, as well as nonruling Communist parties in western Europe. If they did not, they might destabilize the positions of ruling Communist parties in other bloc states, as Wladyslaw Gomulka (Poland) and Walter Ulbricht (East Germany) warned.

From the beginning of the crisis the Soviet leadership was aware that the Czechoslovak reforms could have a destabilizing effect on the Soviet bloc, which already had a history of challenges to Soviet authority. Some challenges had been suppressed (Hungary and Poland in 1956, East Germany in 1953), and others continued to pose a problem (Romania, Yugoslavia). The riots in Poland in March 1968 reminded the Soviet leadership how fragile the facade of Communist control was. The quick and broad support gained by the Prague reformists within Communist Party ranks, even among those whose loyalty to communist ideology seemed unquestioned, only underlined this point. The assessment of the situation and positions of other bloc leaders became, therefore, an important input into the debate within the Soviet Politburo on the risks of the Prague Spring for intrabloc cohesion and stability.

The Soviet leadership also had to consider the possibility that within the context of the looser bipolar international system and emerging détente—as well as the growing number of newly independent Third World countries, which were deeply wedded to the concept of sovereignty—a unilateral Soviet military intervention would have much more adverse implications for the Soviet global position and interests than a multilateral act of intervention involving other eastern European allies. A multilateral intervention could more credibly be explained in terms of collective security, as an act intended to preempt Czechoslovak–West German cooperation and forestall a potential threat to the security of eastern Europe.

From the early stages of the crisis, between January and April 1968, the GDR and Poland pressed for blocwide consultations; they wanted to try to persuade the Soviet leadership that Moscow's view of the Prague reforms was too cautious and optimistic. The main concerns of East Germany, the most ideologically rigid member of the bloc, were, first, ideological contamination and political spillover, with resultant destabilization in East Germany, and, second, recognition of West Germany by Czechoslovakia, which the East German leadership feared would undermine the united bloc front against West Germany and eventually result in isolating East Germany.

The large-scale riots that broke out in Poland on March 8 left a lasting impression on Gomulka. Students and intellectuals had protested repressive censorship policies, demanding reforms and calling for a Polish Dubček. At the same time, Gomulka's leadership had been challenged from within the party. Both events had shaken Gomulka. He was not sure that he could survive if the reforms in Czechoslovakia had a spillover effect in Poland. Gomulka's views were summarized in a cable from the Soviet ambassador in Warsaw: "Comrade Gomulka once again expressed the need for us to intervene immediately, arguing that one cannot be an indifferent observer when counterrevolutionary plans are beginning to be implemented in Czechoslovakia" (NSA, Document no. 23, 1994). Thus, as early as March and consistently thereafter, Gomulka of Poland and Ulbricht of the GDR, and later the Bulgarian leader, Todor Zivkov, advocated tough measures, including military intervention. On August 13, shortly before intervention, Ulbricht met with Dubček in Karlovy Vary. The results were disagreement, growing suspicions, and a highly alarmist report to Soviet officials on Prague's relations with Bonn and the intentions of Czechoslovakia with regard to Warsaw Pact membership (Dawisha, 1984: 24–31, 278–79; Kramer, 1993).

Both Ulbricht and Gomulka believed that a counterrevolution was emerging in Czechoslovakia as in Hungary in 1956. János Kádár, the leader of Hungary, took a different position. Although concerned by the events in Czechoslovakia, he did not believe that the situation was similar to the earlier one in Hungary, so he supported a political solution. Until the Soviet decision to intervene, Kádár had the support of Brezhnev and the soft-liners in the Soviet Politburo in his repeated attempts to convince Dubček and his colleagues to be reasonable—to slow the pace of reforms and undo some of their consequences. Kádár supported military maneuvers to pressure the Prague leadership; force was to be used as a last resort. But he was not convinced, even in August 1968, that force was yet required. Kádár eventually agreed to participate in the intervention because of heavy Soviet pressure and a threat of economic sanctions (Gomori, 1972; Musatov, 1992: 39; Valenta, 1991: 175–78).

This debate among eastern European allies of the Soviet Union was tied to the two competing approaches to the crisis within the Politburo, and it also posed the risk of a split within the group of bloc countries still loyal to the Soviet Union. Romania and Yugoslavia posed another problem. Romania, though a WTO member, was excluded from the Dresden Warsaw Pact meeting at which Czechoslovakia was discussed. The Romanians supported reforms in Warsaw Pact management that would reduce Soviet control and had refused to sign the Nonproliferation Treaty (NPT), which the Soviets feared would provide justification for West Germany to refuse to sign the treaty, too. Now the Soviets had an added concern: that Czechoslovakia would follow the lead of Romania in taking a more independent foreign policy posture. Romania had recognized the Federal Republic of Germany (FRG) in 1967, and now Czechoslovakia was trying to improve relations with Bonn. Dubček's insistence that Romania and Yugoslavia be invited to participate in the July meeting of the Warsaw Pact countries as a condition for Czechoslovakia's participation intensified Soviet concerns. In a visit to Prague between August 15 and 17, the president of Romania, Nikolae Ceausescu, renewed the bilateral treaty binding the two countries. The new treaty did not mention the FRG as a potential aggressor and declared that in the event of an armed attack on Czechoslovakia or Romania by a third party, all necessary assistance, including military aid, would be provided (Dawisha, 1984:

276–77). The implied military cooperation could not but enhance the perception that a coalition of Yugoslavia, Romania, and Czechoslovakia would emerge to defy Soviet hegemony and cooperate with West Germany. It also raised the spectre of a military confrontation within the bloc. Romanian support thus inadvertently became a disadvantage for Czechoslovakia in the policy debate in the Kremlin, rather than adding to Prague's bargaining power.

Yugoslavia, like Romania, adopted a line of firm support for Czechoslovakia's right to decide on its own domestic policies without having to worry about external interference. Marshal Josip Brož Tito of Yugoslavia, who symbolized defiance of Soviet hegemony and who had built a unique brand of socialism, visited Prague in August, and the popular enthusiasms and outpouring of affection for him emphasized how fast Czechoslovakia was drifting away from the Soviet Union.

The western European Communist parties joined Yugoslavia and Romania in opposing military intervention, which would adversely affect their electoral prospects. Their influence was related to the World Communist Conference scheduled for November 1968. Some in the Soviet leadership, such as Suslov and Arvid Pelśhe, members of the Politburo, and Boris Ponomarev, member of the Secretariat of the Central Committee, envisioned the conference as a cornerstone of Soviet diplomacy. It was supposed to bring about a closing of ranks with revisionist Yugoslavia, Romania, and the western European Communist parties to confront dogmatist China. The most active among the Eurocommunist parties were those of Italy, Spain, and France. In July the French party leader, Waldeck Rochet, flew to Moscow accompanied by two senior Italian Communist party representatives to express their opposition to interference in Czechoslovakia (Dawisha, 1984: 198–99, 226–28, 259; Valenta, 1991: 26–27). Representatives of eighteen European Communist parties also signed a letter in which they threatened to call a conference to condemn Soviet behavior in case of invasion. As it turned out, the Eurocommunist parties failed to play an effective mediating role. They were marginal in Soviet cost assessments, although their position carried weight with individual Soviet decisionmakers, notably Suslov.

Yugoslavia, Romania, and the Eurocommunist parties, all opposing military intervention, had much less influence over Soviet policy than the GDR, Poland, and Bulgaria, though at the long-term cost of further weakening the cohesion of the Communist camp. The loss of cohesion became obvious at the post-invasion conference in Moscow on August 24–27, when Brezhnev addressed the Czechoslovak leadership. "Comrade Tito and Comrade Ceausescu will say their piece, and so will Comrade Berlinguer. Well, and what of it? You are counting on the Communist movement in Western Europe, but that won't amount to anything for fifty years" (Mlynář, 1980: 241). When the decision for intervention came to be made, the possible effects of the Prague Spring on other bloc countries outweighed the less immediate goal of building a large, united Soviet-guided socialist bloc. The concern with short-term losses over long-term losses reflects national policymaking that was controlled by politicians rather than by statesmen.

With the socialist camp divided, the Soviets felt a need to establish the legal right to intervene, an action at odds with the prevailing norms of sovereignty and noninterference in the affairs of another socialist state. They had to forge an intrabloc consensus on a new principle of conditional sovereignty and the right to intervene under specific conditions, known as the Brezhnev Doctrine. Not surprisingly, they applied the same argument used by the United States to justify intervention in the Dominican Republic in 1965 (Franck and Weisband, 1971: 129). The Bratislava Declaration of August 3 invokes the duty of all socialist countries to defend the gains of socialism; it mentions "respect for sovereignty" but does not refer at all to the principle of nonintervention. Signed by all the players in the invasion drama, both the interveners and the eventual victim—the Soviet Union, Poland, East Germany, Hungary, Bulgaria, and Czechoslovakia—it contained the theoretical and ideological justification for an invasion. Indeed, it was used for this purpose in Soviet propaganda after the invasion.[62] Combined, the Bratislava Declaration and the prior Warsaw Letter of July 1968—which stated that national interests were subordinate to the interests of the international communist movement and that, irrespective of whether the source of a threat was domestic or external, socialist countries had a duty to come to the defense of socialism by applying any measures necessary (Dawisha, 1984: 209–13, 265–70)—established a formal, explicit normative consensus for intrabloc intervention and collective responsibility for military action if the CPCz did not take appropriate measures on its own. With the Brezhnev Doctrine, intrabloc legitimacy for the use of force was created.

To summarize, the international environment posed a number of risks. The loosening of tight bipolarity and the increasing autonomy in setting policy demonstrated by some of the main allies of the United States posed an opportunity to weaken the U.S. position in Europe. At the same time, it also posed a new threat by inspiring similar trends in the Soviet bloc and by encouraging bloc members to explore new, extrabloc networks of assistance and cooperation. The main initial and more immediate concern, however, was the likely U.S. response to intervention. The perceived risk decreased in intensity as the crisis in Czechoslovakia escalated. By August it had become obvious that the risk of American counterintervention was minimal. What remained uncertain was how an invasion would affect the progress toward détente between the two superpowers.

An important external risk was the potential effect of the Prague Spring on the domestic stability of two key Soviet allies, Poland and East Germany. An invasion to deal with that threat might deepen the rift within the socialist camp and deal a heavy blow to the chances of convening a successful World Communist Conference, which some Soviet leaders considered very important. Given a divided eastern European bloc and the risk of criticism and delegitimation, a multilateral intervention became a high-priority goal, even if the actual military contributions of some of the allies would be negligible.

THE FORGING OF A CONSENSUS: THE RISKY SHIFT

Unlike the two earlier cases in this chapter, Grenada and Panama, in which the decision for military intervention eventually depended on the preferences of a single person, Reagan and Bush respectively, the Soviet decision to intervene in Czechoslovakia was a collective decision that resulted from consensus building. Brezhnev played an important role and bears much, but not exclusive, responsibility for the decision. For a leader as cautious and inclined to avoid difficult decisions as Brezhnev, who operated within the confines of a oligarchic collective decisionmaking system, the Czechoslovak crisis could have had costly political and personal repercussions. His personality attributes and the oligarchic nature of the Soviet political system together generated caution and incrementalism (Daniels, 1982; Roeder, 1984). Uncertainties and ambiguities in the situation resulted in intensive policy disagreements among Soviet leaders as well as among bloc members—disagreements over where the main risks lay, with military intervention or its avoidance. Choosing one interpretation and risk perception over the other could have, in case of error, serious policy consequences at the national and bloc levels, as well as personal consequences for those involved. Time was required to sort out which line of argument was the more valid, to build some sort of domestic consensus, and, at the same time, to establish bloc support for, and active participation in, implementing the agreed-on decision. In the six months since the Dresden meeting, Brezhnev vacillated between a hard and soft line but eventually came down on the side of the hard-liners for reasons of both political expedience and growing frustration with the Czechoslovak leadership and disappointment in Dubček. Brezhnev avoided taking a position as long as possible, almost until the last minute. Finally, with no room left for further procrastination, he took a position, forged a consensus of the Soviet leadership in support of it, and brought on board the eastern European leaders, including Kádár. Only then did he authorize an invasion.

The decision to intervene was made on August 17, and the operation was put under the command of General Ivan Pavlovsky, commander in chief of the Soviet Ground Forces. The decision was confirmed, and the principle of collective responsibility was reaffirmed, on August 18 in a meeting of leaders from five bloc countries. The final details of the invasion were probably worked out after this meeting, and the go-ahead signal was not given until the night of August 20. The majority of the evidence indicates that there were no dissidents in the final Politburo vote on intervention. There were, however, expressions of skepticism regarding the wisdom of intervention. The decision emerged as an "unhappy necessity" for the moderates, who remained uncertain about the utility and efficacy of invasion (Dawisha, 1984: 286–90, 293; Kramer, 1993; Valenta, 1991: 173–74, 181). It reflected a painful and probably uncomfortable consensus among the collective leadership, with Brezhnev in the center. Even after a wait-and-see period of approximately six months, during which time the major decisionmakers had plenty of time to influence each other, the Politburo required three days of deliberations (August 15–17) before making a decision, which indicates how difficult it was to forge a consensus. How the policy pendulum of preferences swung between March and August 1968 is the subject of the following discussion. I attempt to capture motivations and causes for the decisionmaking process and the choice of the intervention option and its timing, as well as the interactions between these factors, in order to explain the risky shift in the Soviet decisionmaking group.

The difficulties that the Soviet leadership encountered in reaching agreement on how to react to the developing situation in Czechoslovakia were rooted in the ill-defined nature of the problem. Compared with the challenge of Hungary in 1956, the challenge of the Prague Spring was more gradual and less violent, and the reform movement was led by the party rather than directed against it. This made the challenge to Soviet interests more ambiguous and the intensity and immediacy of the risks posed by it much more difficult to delineate or agree on (Larrabee, 1984: 120–21). Responding harshly was perceived to have serious implications for relations with the West; possibly, too, it might cause a backlash in Czechoslovakia of the kind that the Soviet leaders were eager to avoid. On the other hand, taking too benign an approach could lead to political destabilization in Poland and East Germany and possibly a spillover effect in the Soviet Union itself.

Each view had powerful supporters within the Politburo, but the available information did not allow for conclusive support for either—in part because the Czechoslovak leaders themselves, especially Dubček, probably did not have a clear idea how far they intended to deviate from Communist orthodoxy. They attempted to convey an impression of control but in many ways did not determine the course of events. As the crisis progressed, the transgressions of the reforms—that is how the Soviet leadership perceived them—strengthened the position of the hard-liners in the Politburo, who detected a counterrevolutionary conspiracy. The Soviet leadership was divided on whether the Action Program introduced in the CPCz Central Committee plenum in April 1968 reflected necessary reforms to pacify a restless population or whether the reforms were intended to challenge the leading role of the Communist Party and by implication the Communist system in eastern Europe. Nor could the Soviet leadership determine whether support for the reform movement was based on the intelligentsia or the working class, that is, whether it had narrow or broad appeal. The Soviet leadership was not sure whether conservatives could return to power or were more likely to be swept away by the tide of reforms, nor could they determine what the personnel changes meant. In the first phases of the crisis, Dubček himself was still considered a loyal Communist who could rein in the excesses of his radical colleagues. In a frantic attempt to resolve these puzzles Soviet leaders looked for historical analogies, but found it difficult to settle on the situation in Hungary in 1956 or the situation in Poland in 1956 as comparable. (The uprising in Poland in 1956 was resolved through internal measures and had not required Soviet intervention.) The inability to agree on a definition of the situation led to divergent views of what the risks were and how best to respond.

Within the Politburo, which was the formal focus of all important policy decisions, the driving force was a small group of leaders who drew their strength from various constituencies and power bases inside and outside the Politburo. Some members of this core group were almost permanent owing to their formal positions. Others were brought into the group to help deal with particular issues. Disagreements in the Politburo resulted in slow and cautious policy formulation until consensus was reached (Paul, 1971). Although key decisions were made in the Politburo, outsiders such as government ministries, other party organs, pressure groups, and influential individuals could affect policy formulation by employing suasion, applying direct or indirect pressure, building coalitions, disseminating information in support of their views, or combining these measures.

The process was no different in the Czechoslovakia crisis. From the beginning, the members of the inner group were divided in their analysis of the reform-minded regime in Prague. In general, the hard-liners believed as early as March 1968 that the stakes were high and the options were limited to coercing the Czechoslovak leadership—with military force if necessary—to back away from the reforms. Their risk-taking preferences were a direct result of their construction of the situation. The soft-liners did not see the immediacy of the threats embedded in the situation and therefore considered a broader range of options, from wait and see, to an attempt to influence the pace and nature of reforms, to a range of coercive measures short of the direct use of force (e.g., verbal threats and troop maneuvers), in support of the conservatives in the CPCz Presidium. The soft-liners considered the use of force too risky and thought it would have substantial adverse implications at home and abroad. What becomes clear from analyzing the Czechoslovak crisis is that those members of the decisionmaking group who were most concerned with short-term risks (e.g., immediate developments in Czechoslovakia, their personal position) were liable to prefer a riskier policy option than those who focused on long-term risks. The former were particularly risk acceptant once they perceived strict time constraints, which amplified the immediate risk and gave rise to a perception that the possible negative consequences of not taking action would be irreversible. The temptation under these circumstances was to employ military intervention as a quick once-and-for-all solution, because the balance of capabilities was so heavily tilted in favor of the Soviet Union.

Nikolai Podgorny, chair of the Presidium of the Supreme Soviet, took a strong stand on Czechoslovak revisionism and the threat to the leading role of the CPCz and advocated military intervention early on. Mikhail Suslov, secretary of the Central Committee, took a middle-of-the-road position. He was ready to accommodate the less radical elements of the Prague Spring because he viewed China as the most dangerous threat to the Soviet Union and thought coping with that threat required unity of focus.[63] Petr Shelest, the Communist Party leader in the Ukraine, was highly sensitive to the spillover effect of the Prague Spring on the Ukraine and was therefore a consistent advocate of intervention. Unlike Brezhnev, Prime Minister Aleksei Kosygin viewed the reform movement in a broad perspective and took a pragmatic approach to the Prague Spring.[64] He was less concerned with the threat to socialist bloc unity and more concerned with the consequences of an invasion for two issue areas which he felt personally committed to: reform of the Soviet economy and East-West détente (Anderson, 1993: 127–31, 184; Dawisha, 1984: 300–305). Both reform and détente would, in his mind, be jeopardized in the event of a military intervention. Brezhnev's position was initially closer to that of Kosygin and Suslov, although he failed to commit himself, reserved judgment, and vacillated almost to the last minute.

The risky shift that took place among these decisionmakers' preferences, especially in the case of Brezhnev, entailed a reassessment of national and personal stakes. In the end, this led to a reformulation of preferences toward more risk acceptance, which allowed them to support an option with highly adverse outcomes if it did not work out. In other words, Brezhnev and others became convinced that the risks of military intervention were at an acceptable level because of the long and detailed political and military preparations, because of the time constraints posed by approaching critical dates, and because of the apparently very high personal costs of not supporting the military option.

The reassessment was in part the result of attempts at suasion by various organizations and individuals. Among participants outside the Politiburo, probably the most important was the military. The Ministry of Defense exerted great influence and consistently supported a military solution and the deployment of Soviet troops on the Czechoslovak–West German border. Why was the military so important? A decision *not* to intervene would have had to gain the support of a military that, for strategic reasons, pressed for immediate intervention. Alternatively, a decision to intervene would have had to be implemented by the military with swift precision and within the political constraints imposed by the political leadership. Marshal Grechko, the minister for defense, was also a close ally of Brezhnev's, and Brezhnev had decided as early as 1965 to base his political power on an intimate alliance with the military leadership. Brezhnev was chair of a joint Politburo-military leadership committee, the Defense Council. Membership gave the defense minister, who was not a member of the Politburo, increased leverage and, in many cases, encouraged a mutual accommodation of views on defense matters before they were brought to the Politburo (Gelman, 1984: 63–65, 92–94). This state of affairs guaranteed the military viewpoint a close hearing by the secretary general of the Communist Party.

Other important consultation groups in which a majority advocated a tough line were the Central Committee of the CPSU, a group of eastern European leaders, and the KGB. A similar position was taken by the Soviet embassy in Prague. In the Ministry of Foreign Affairs the minister, Andrei Gromyko, took a hard line, whereas some middle-level officials, including Anatoly Dobrynin, ambassador to Washington, doubted the wisdom of an invasion.[65] Officials in the International Department of the CPSU, notably the deputy head of the department, Vadim Zagladin, took a moderate line because they were concerned with the effect on the international communist movement (Dawisha, 1984: 88–89, 357–58; Valenta, 1991: 58–63, 182). The views of both the inner circle and other players were influenced by their personal beliefs, life experiences, functional responsibilities, organizational interests, and personality attributes.[66]

In analyzing the shift toward a common position it should be understood that both hard-liners and soft-liners perceived the Prague Spring as a grave threat to the cohesion and stability of the Soviet bloc and to Moscow's hegemony (Brecher, 1993: 187). In framing the situation specifically, however, the difference between the two approaches was substantial and led to different policy preferences.[67] What allowed for a common ground in discourse and interpretation, in spite of personal, functional, and organizational divergencies, was the shared political culture provided by Marxism-Leninism. Ideological argumentation was not merely a facade for cynical or self-serving political maneuvering. The key decisionmakers in the Soviet Union were educated and socialized within the intellectual and conceptual framework of Marxism-Leninism, which formed a decisive cognitive and affective influence on the way issues and expectations about bloc members' behavior were socially constructed and framed. The philosophy and practice of Marxism-Leninism contained powerful shared verbal and behavioral symbols and myths that shaped perceptions of risk (Eidlin, 1984).

This shared ideological perspective had a triple effect. It brought particular risks to the attention of all participants, it defined what was normatively and politically correct, and it made deviation from the situational diagnosis that flowed from this framework difficult, if not impossible. In the debate and discourse among Soviet decisionmakers the use of such terms as *revisionism*, *dogmatism*, *international proletarian solidarity* and the recurring warnings by Politburo hard-liners and others of the threat of "counterrevolution" provided a powerful verbal weapon. Labeling the Czechoslovak reforms revisionist and the Prague Spring a counterrevolution associating events in Czechoslovakia with other threats to bloc unity, such as dogmatism, triggered associations of threat perceptions, assigned to them a higher probability than they actually had, reduced doubt in the validity of these assessments, and set the normative grounds for the argument that international proletarian solidarity overrules the norm of sovereignty. Framing the policy debate in these terms made attention to information selective and the message of hardline consultation groups more salient, powerful, and politically influential.

The risk-framing discourse could not have had such influence had it not been a main source of the interpretation in a gradually evolving situation that contained puzzling ambiguities and uncertainties. Particular events were amplified because of this framing—the riots in Poland in March, the Action Program in April, the sweeping personnel changes, the 2,000-word manifesto in June, Prchlík's statements in July—and the framing then became the cognitive anchor against which was assessed the validity of the Czechoslovak government's insistence that it would not withdraw from the bloc or abandon the security and economic arrangements of the bloc and in terms of which that insistence gradually came to be perceived as not credible. The riots in Poland made the Hungary 1956 analogy highly vivid and reinforced the worst-case perception of a similar counterrevolution taking place in Czechoslovakia that had the potential of spreading to other eastern European countries and the Soviet Union. The sweeping personnel changes at the most influential levels in the government, party, army, and state security organizations were not only highly visible but also highly symbolic. That the changes were implemented without consulting the Soviet leaders was taken almost as a personal insult, especially by Brezhnev, and increased the salience of the perceived threat.

In reporting on the post-invasion Moscow meeting of August 1968, a Czechoslovak participant, Zdeněk Mlynář, who was secretary of the Central Committee of the CPCz, wrote:

> According to Brezhnev, all the other sins followed from this main transgression— that Prague did not seek the approval of the Kremlin for everything it did. "Antisocialist tendencies" were rife, the press wrote what it liked, "counter-revolutionary organizations" were formed, and the party leadership was under pressure from these forces, to which it continually gave ground. If Dubček had only acted with Brezhnev's approval or on his advice, if he had cut from his speeches the words and passages Brezhnev had suggested, and if he had appointed the ministers and secretaries Brezhnev had agreed to, none of those horrors could have happened. That, in a nutshell, was Brezhnev's view of the Prague Spring.
>
> Then Brezhnev explained to Dubček that the end result of all this was Moscow's realization that the Dubček leadership could not be depended upon. Even he himself, who had long defended "our Sasha," had to admit that this was so. Because, at this stage, matters of the utmost importance were involved: the results of the Second World War. (Mlynář, 1980: 239)

The shared ideological framework thus provided more than an anchor for the interpretation of ambiguous events; in combination with the growing salience of the Hungary 1956 analogy—which became highly available as the crisis escalated—it also contributed to the projection of dire consequences through a simulation-like process. By July the direction of the Prague Spring, if decisive measures were not taken to stop the reforms, was projected with growing confidence: it would lead to a Hungarian-type counterrevolution that threatened vital security gains of the Soviet Union, which were the result of sacrifices made in the Second World War.

Although the Soviet leadership realized that there were differences between Hungary in 1956 and Czechoslovakia in 1968, the comparison focused on the definition of the threat—that is, counterrevolution—and the goals of the enemy within. A *Pravda* article entitled "Attack of the Socialist Foundations of Czechoslovakia" made the similarities explicit: "In actual fact they [the reformers] are seeking to undermine the very foundations of the new state. . . . [Similar tactics] were resorted to by the counterrevolutionary elements in Hungary that in 1956 sought to undermine the socialist achievements of the Hungarian people. . . . Now, twelve years later, the tactics of those who would like to undermine the foundations of socialism in Czechoslovakia are *even more subtle and insidious*" (Dawisha, 1984: 350). Focusing on these similarities allowed the analogy to be extended: a Hungarian solution to the crisis in Czechoslovakia was implied and made salient.[68]

The Soviet Politburo and the core group of leaders within it formed a collective leadership that operated on the principle of consensus. Consensus building involved extensive interaction and attempts at persuasion when important issues were contested. The resources that the members of the Politburo brought to bear in the process of persuasion included their organizational power base, their functional responsibilities and roles, and the allies that supported, or could provide information that would support, their position. In the Czechoslovak crisis the length of the policy formulation process allowed for extensive interaction among the members of the decisionmaking group. Information about the direction and implication of events in Czechoslovakia was of key importance in the persuasion process. Members of the Soviet decisionmaking elite traveled to Czechoslovakia, met the Czechoslovak leaders at conferences or in bilateral contexts, or, like Brezhnev, exchanged letters and had long telephone conversations with Dubček and other Czechoslovak leaders and then shared their impressions with their colleagues.

Dubček and the other Czechoslovak leaders failed to take advantage of these opportunities to convince key Soviet decisionmakers of their limited reformist goals concerning bloc membership and the future of the leading role of the CPCz. Rather than use the interactions to influence the debate in favor of the Soviet moderates, the Czechoslovak leaders made their position vulnerable. Three miscalculations were significant. First, by actively denying the possibility that the Czechoslovak army would resist invasion, rather than allowing uncertainty and ambiguity on this point to prevail, they made the interveners' decisions less difficult by reducing the perceived risk that an invading army might be confronted with resistance by the formidable Czechoslovak army. Second, they miscalculated the growing risk of intervention, betting on the international repercussions to Soviet interests to deter intervention. They ignored warning signals from Brezhnev, Kádár, and other Soviet and Soviet-bloc leaders. Third, by taking positions that looked like active defiance of the warnings given them and by not even pretending to take steps—even symbolic ones—to slow down the reforms, Dubček and his colleagues weakened the influence of those who looked for signs of Czechoslovak compliance as an excuse to avoid intervention. Eventually they left the soft-liners no credible argument that the Czechoslovak reformers would stop before crossing the line into counterrevolution.

More important, the main sources for information on foreign involvement and the intentions of Czechoslovak leadership—the GRU (military intelligence), the KGB, the Soviet embassy in Prague, and the Department of Liaison with Communist and Workers' Parties in the CPSU—were controlled by hard-liners. Consequently, much of the information that flowed to the top levels of the Soviet decisionmaking system was highly skewed. In cases like this, the sharing of information within the group does not necessarily improve the quality of the decisionmaking because the information is often selective and distorted. Although KGB officials like Major General Oleg Kalugin, station chief in Washington, reported that neither the CIA nor any other Western agency was involved in the events in Czechoslovakia—an important point for proving or disproving the imperialist conspiracy theory—such reports were suppressed. The KGB leadership engaged in intentional distortion and manipulation of intelligence and deliberately offered alarming assessments. These assessments were reinforced by biased reports from the Soviet ambassador to Prague, Stepan Chervonenko; his deputy; and the head of the Department of Liaison with Communist and Workers' Parties in the CPSU Central Committee, Sergei Kolesnikov.

The KGB viewed the removal of key personnel in the Czechoslovak State Security (StB) who were loyal to Moscow as a threat to its influence and control. The personnel changes also undermined its intelligence network. Furthermore, a spillover from the Czechoslovak reforms into the Soviet Union itself would have had serious effects on the KGB's task of maintaining domestic order. Thus, the perception within the KGB was that the Prague Spring compromised both the external and internal security of the Soviet Union. As the reforms continued, even KGB officials who were less enthusiastic about the military option gradually came to support it. Once the KGB leaders, Yuri Andropov in particular, decided to support the invasion option, they used the means at their disposal to bolster their case, emphasizing the risks of continued reform and underestimating the difficulties of regaining popular support for the post-invasion regime and of reestablishing political control even if the invasion was highly successful (Kramer, 1993: 6–8).

The long preparations for the invasion, combined with distorted information about the domestic situation in Czechoslovakia, led to confidence that an alternative government under the control of pro-Soviet conservatives would be established as soon as the Czechoslovak Communists were faced with overwhelming military force and isolated from external help. This script was based on assurances from the hard-line conservatives in the CPCz

Presidium but did not take place. The conservatives in the Presidium lost the vote to endorse the Soviet intervention seven to four, rather than winning it as anticipated, six to five. The incorrect assessment of local support was "a key factor in the Soviet decision to invade and in timing the invasion for August 20, 1968" (Valenta, 1991: 181). Instead, the CPCz Presidium and all party and state bodies condemned the invasion. No new Czechoslovak government or leadership emerged that was willing to cooperate with the occupation authorities, leaving the Soviets no choice but to negotiate with Dubček and his colleagues (Eidlin, 1980, 1981). Having attended to all possibilities of armed resistance, the Soviet military command had not accounted for the "unsqueaky-wheel trap." That is, they did not pay enough attention to the degrading effect on morale that the Czechoslovak unarmed resistance would have on their and their allies' troops. In spite of earlier indoctrination the soldiers did not encounter an armed enemy or see signs of American or German troops attempting to occupy Czechoslovakia. Instead they encountered a hostile but nonviolent population that argued with the Warsaw Pact soldiers over the legality and morality of invading a friendly ally.

The diplomatic and intelligence reports from Prague enhanced Brezhnev's own impressions from his frequent interactions with Dubček that Dubček was a weak leader who often acted under the control of extremist reformers, such as Josef Smerkovský and František Kriegel (Musatov, 1992: 37; Williams, 1996). As time went by, he had growing doubts about whether Dubček could sustain the status of the CPCz and undo the excesses of the reforms. It became clear to him by July that removing Dubček from office would require military intervention. Being an extremely cautious politician, Brezhnev desired full concurrence by the Politburo before acting, so he needed to involve as many members as possible in the negotiations and decisions. Brezhnev arranged for those wavering or opposing intervention to be engaged in direct interaction with the Czechoslovak leadership and share in the negotiations— and thus share in the responsibility for their outcome. Thus, nine of the eleven members of the Politburo participated in the critical Čierna meeting.

The hard-liners in the Politburo were not averse to making sure that Kosygin was excluded from important discussions by taking advantage of his traveling schedule. On at least three occasions Kosygin was excluded from the inner group: at the beginning of April, when he was in Iran; at the end of May, when he was in Karlovy Vary; and in mid-July, when he was in Sweden. While he was on his trip to Sweden the decision was made to convene the Warsaw meeting in July to formulate a bloc policy that would allow the use of military measures, as was the decision to convene an extraordinary plenum of the CPSU Central Committee to discuss the results of the Warsaw meeting.

At the extraordinary plenum the threat to ideological control in the Soviet bloc was emphasized, and none of those known for their moderate views on Czechoslovakia were permitted to participate or speak in the discussions (Dawisha, 1984: 199–201, 217–18). Kosygin did not have a chance to express his views at either the Warsaw meeting or the plenum. In an authoritarian system such manipulation has an added political signaling effect: it tells which way the wind is blowing and what is politically correct. This increases the incentives for taking the "correct" side in the debate.

Although Brezhnev was convinced by July 1968 that military intervention might be unavoidable, his personality affected the manner in which he went about building a consensus for this option while allowing himself room to maneuver. Thus he was both the architect of the hard-line Warsaw letter and of the Čierna compromise. His personal style had always been accommodative and nonconfrontational. He had difficulties making tough decisions and preferred to keep his options open and avoid commitment (Burlatsky, 1989; Dornberg, 1974: 19; Goldgeier, 1990: 78–84). The accommodative style prevented him from imposing a decision to intervene when the core decisionmaking group was divided. This approach was also congruent with his disposition to avoid early commitment and his highly sensitive fear of failure. Brezhnev was prepared to authorize military intervention only after the risk of failure was substantially reduced through extensive military, diplomatic, and political preparations. The tempered risk of the intervention option induced the soft-liners to become timid in their opposition at a time when the risks of nonaction loomed large and as critical dates neared. In the meeting of Warsaw Pact leaders in Moscow on August 18, Brezhnev was ready to demonstrate a much more directive leadership. He read the Czechoslovak conservatives' letter inviting intervention that has been given to him in Bratislava. The letter was kept secret until that meeting for security reasons. He concluded by saying, "Is it worth arguing? Here is a letter and everything is clear in it" (Musatov, 1992: 39). The die was cast.

Over six months or so Brezhnev had evidently moved from risk aversion to risk acceptance. This reversal was apparently associated with a reframing of the situation that involved a growing concern with prospective Soviet losses and an increased threat to his personal political position. Distorted information about the situation in Czechoslovakia nurtured his perceptions. As the information on adverse developments in Czechoslovakia became ever more ominous, Brezhnev tended to overweigh the certainty of a counterrevolutionary development in Czechoslovakia, with all its attendant policy and political risks. In that context, the Hungary 1956 analogy enhanced these political and personal concerns and persuaded him that he was in the domain of losses. He was now ready to take risks.

The change in Brezhnev's preferences also substantially affected the balance in the core decisionmaking group in favor of the hard-liners. This contributed to three types of risky shifts. The hard-liners became more convinced that there was absolutely no effective solution but a military one. Those who had wavered for months adopted a hard-line position, having reluctantly come to believe that there was no other viable choice. The soft-liners became less vocal. They were not convinced that it was necessary to invade Czechoslovakia, but they believed that they had to comply with the military solution. They had no credible alternative, and they faced time pressure and peer pressure to reach an operational consensus.

In brief, the decisionmaking process was driven by multiple forces. The Soviet Union practiced collective decisionmaking, but the process was shaped by bureau-organizational pressures for and against invasion, the hopes and fears of political leaders in eastern Europe, and the informational and normative influences within the small group of key decisionmakers.

Conclusions

A comparison between the three low-risk to moderate-risk cases, Grenada, Panama, and Czechoslovakia reveals telling similarities and dissimilarities. The observations that follow from the comparisons have important implications for foreign military intervention and for risk taking in general. These observations will be tested further in the following chapter.

• The cases suggest that reputational interests at the national and individual level play an important role in triggering decisions that involve risk taking in foreign military intervention. The influence of reputational interests is at least as important as the impact of tangible strategic considerations and is even more important in some cases. This conclusion does not conform to the expectations of the neorealist school in an issue-area for which neorealist assumptions and inferences would seem to be most relevant. This study, therefore, adds to a growing body of evidence questioning the validity of the neorealist premises and their accuracy in reflecting foreign policy motivations.

• In all the cases the risk of hostile responses by the international community as a whole, allies, and hostile third parties inside and outside the region were considered marginal or secondary. Considerations of domestic politics played a much more important role than considerations of third-party responses in terms of the risk assessments affecting intervention decisions and their implementation. The United States and the Soviet Union were both superpowers with veto power in the U.N. Security Council and, as such, could afford to ignore international reactions. This might not be the case with smaller powers. In the case of Grenada, however, domestic concerns were much less important than in the Panama and Czechoslovakia cases. In the first case, the main motivation was probably embedded in the Reagan administration's grand design for foreign policy, and in the latter cases, the domestic political conditions were predominant in shaping policy preferences. The prominence of domestic concerns is another important deviation from common neorealist premises.

• Decisionmaking for the Czechoslovak intervention shows that third parties can shape the intervener's decisions when the decisionmaking group is divided and at least one faction is anxious for external support. Under these circumstances, the policy advocacy of external actors gains in effectiveness. If foreign actors are unable to manipulate competing factions, their opportunities to affect the intervener's policy are very limited, as is indicated by the U.S. interventions in Grenada and Panama.

• The cases indicate the importance of having the top military command support the intervention. Support from the military, regardless of the type of regime, is a necessary condition for intervention when the political leaders have doubts about the wisdom of intervening and prefer to share the risks involved in committing troops. The testimony of military commanders will affect the credibility of claims that the operation will be quick and successful. Thus, the position of the military commanders becomes a critical input in deciding the level of acceptable risk. But when the top political leaders are driven by beliefs in the necessity and rightfulness of intervention, as was the case in the Grenada intervention, even reluctance on the part of the military leadership is not likely to prevent intervention.

• A comparison among the three cases illuminates an interesting but somewhat puzzling phenomenon. In the two U.S. cases, the military did not show great enthusiasm for a policy of military intervention. The Department of State was much more aggressive in advocating the use of force. In the Soviet case, the roles were reversed and more in line with conventional expectations. The military played a key role in lobbying for intervention, whereas the Soviet Ministry of Foreign Affairs and the International Department of the CPSU took a more cautious and reluctant position. The reasons can be traced to the different historical experiences of the two superpowers. The U.S. military was deeply affected by the Vietnam War, which dealt a heavy blow to its prestige and for which it took much of the blame. The State Department was the clear winner because of the emerging predominance of the role of diplomacy over military power. The policymakers were concerned with the Vietnam War in a different way. In their view, the main objectives of U.S. diplomacy in the post-Vietnam era were to rebuild U.S. credibility, to contain the erosion of U.S. power and influence that resulted from the failure in Vietnam, and to reverse the attendant reluctance to use its still considerable military power in support of U.S. interests. For the State Department, then, a credible willingness to demonstrate military force would reestablish the reputation of the United States and preserve its vital interests.

The situation was different in the Soviet case. In 1968 the Soviet military and its high command still enjoyed their legacy as heroes of the Second World War. The Hungarian intervention in 1956 had done little to diminish their glory. The armed forces considered themselves the guarantors of Soviet security through the preservation of the territorial spoils of the Second World War, and they were ready and anxious to do all that was necessary to make sure that none of the countries constituting the Soviet security belt slipped away. It would take a "Soviet Vietnam," the bloody intervention in Afghanistan, for them to become cognizant of the limits of military power. The Ministry of Foreign Affairs and the International Department of the CPSU, on the other hand, considered their mission one of integrating the Soviet Union into the international community and of gaining legitimacy and respectability for their country, so that it could enjoy the benefits of global influence and amicable economic and political relations beyond its immediate sphere of influence. They realized that it would be difficult to achieve these objectives if the Soviet Union persisted in projecting the image of an international bully.

• Unlike the United States in the cases of Grenada and Panama, the Soviet Union badly miscalculated the domestic response to intervention in the target country. In the first two cases, the enthusiastic welcome of U.S. military intervention was a powerful source of domestic and international legitimacy. For the Soviets and their allies, encountering intense hostility from the Czechoslovak public as well as from Czechoslovak state and party institutions became a source of embarrassment and undermined the credibility of the argument that the invasion came in response to a legitimate invitation from Czechoslovakia. This left the Soviets no choice but to deal with Dubček and his colleagues.

• Comparison between the two U.S. cases and the Soviet case also shows that institutions can have a great deal of influence. The institutional framework, formal and informal, is important in deciding what drives risk-taking preferences. In the U.S. cases, a single individual, the president, eventually decided if and how much risk would be taken. His beliefs and values were key to the decision. There are ways by which the president can be influenced by others, but influence and persuasion are mediated by individual cognitions and motivations. In the Soviet decision to intervene in Czechoslovakia, individual cognitions and motivations were the starting point, but they were not the critical determinants of the end results. What was most important in shaping risk-taking preferences were the social interactions by which the members of the collective leadership engaged in consensus building. The chosen policy option did not necessarily reflect any single individual's cognition and motivation but did reflect what was achievable within the constraints imposed by the distribution of power, the roles, and the manipulative competence of the key group members.

• In all the cases, but particularly in the Grenada decision, historical experience and analogies contributed significantly to the political motivations and operational preferences of the decisionmakers.[69] It is highly probable that cognitively simple leaders (e.g., Reagan) rely more heavily on analogies than more cognitively complex leaders (e.g., Bush) do.

• As a rule, after short successful military operations both domestic and international criticisms tend to fade quickly. Because decisionmakers are aware of that and plan interventions to be swift successes, at the early stages they tend to give only moderate weight to the risk of such criticisms.

• In short-term interventions domestic public support and legitimacy have little to do with the quality and finesse with which the legal correctness of the decision is endowed. The public, with few exceptions, is not interested in, or at least is inattentive to, this aspect of legitimacy. What is crucial for establishing domestic legitimacy is presenting the issue in terms of the national interest in a way the public can understand and identify with. Taking a legalistic approach is of little value.

• In short-term interventions a risk-acceptant decisionmaker is not likely to be interested in and interfere with the military implementation of an intervention decision and tends to trust the military to carry out the operation successfully. A risk-averse decisionmaker (e.g., Bush, Brezhnev), fearing escalation and needing control, is likely to interfere with the military details.

• Although the Panama invasion was larger than the Grenada invasion, it was less risky in some important ways. The Grenada invasion was the first of its kind since Vietnam, so President Reagan created a precedent. President Bush relied on this successful precedent in calculating the riskiness of international and domestic responses. At the same time, the military drew on lessons from the Grenada experience to avoid costly errors in Panama. Both intelligence and logistics posed less of a problem in Panama, yet the decision process in the Panama case was slower and more agonizing. The reason is obvious: Bush was a risk-averse president; Reagan, a risk-acceptant one. The implication for the analysis of risk-taking preferences is that leaders' levels of risk aversion can be judged by their decisionmaking style. The less risky the decision and the more agonizing the decision process, the more risk averse leaders can be adjudged. In many cases, risk-taking preferences are better revealed by the process of making the decision than by the substance of the decision reached, because the substance often reflects contextual incentives and constraints rather than the actual risk-taking preferences of the decisionmaker.

• All else being equal, cognitively complex leaders (e.g., Bush) are more likely to be risk averse than are leaders characterized by low cognitive complexity (e.g., Reagan). The logic behind this hypothesis is that cognitively complex individuals are more aware of complexities and uncertainties and are therefore more likely to be pessimistic about having control over risky situations.

• A comparison between Reagan and Bush provides support for the proposition that personality characteristics can enhance or attenuate risk acceptance dispositions. As argued in Chapter 3, Reagan, the actor hungry for attention and an extrovert always in need for feedback from others, turned to higher risk policies that satisfied these needs. Bush, surer of his own value and place in society because of his social background, an introvert, capable of sustained waiting and willing to settle for incremental gratification (Winter et al., 1991), attempted to avoid risky decisions unless forced to take them. In that sense Brezhnev seems to be very similar to Bush.

• The conventional wisdom is that hard-line beliefs are correlated with risk acceptance, and soft-line beliefs with risk-aversion. This, however, is not necessarily the case. Hard-liners may be risk averse when it comes to actual policy preferences, and soft-liners may find themselves eventually more risk acceptant than their initial beliefs would indicate. The lack of congruence between beliefs and actions can be explained as sheer opportunism on the part of the believer or attributed to the objective difficulty that individuals have in predicting accurately how they will react to a real-life test of their beliefs. They therefore assume that when the time comes, their actual behavior will conform with their beliefs. This is often not so. In the cases of Grenada and Panama, two hard-line presidents demonstrated very different risk-taking preferences: Reagan was risk acceptant, and Bush was risk averse. Observers should therefore be very cautious in deducing actual risk-taking behavior from beliefs alone, no matter how explicit these beliefs are.

• Last, but not least, risk-averse decisionmakers seem to have a distinctive cognitive and operational style. They are focused on process, serve as competent tacticians and managers of foreign policy, encourage only tame and timid criticism, emphasize collegiality and loyalty, and usually have a hands-on managing style.

CHAPTER 7

High-Risk Foreign Military Interventions

VIETNAM AND LEBANON

The empirical analysis of risk and intervention can be taken one step further by comparing U.S. intervention in Vietnam with Israeli intervention in Lebanon. The two cases have much in common. Unlike the interventions in Grenada, Panama, and Czechoslovakia, which were brief and low-cost operations, those in Vietnam and Lebanon were protracted and very costly. In fact, both started as cases of intervention and ended as wars. In each case, decisionmakers had in mind a short, limited military action but ultimately committed national resources and reputations to a degree they had never anticipated or intended. In each case, intervention and escalation had multiple objectives. Both the United States in Vietnam and Israel in Lebanon were the superior military power, yet they failed to save their proxies from defeat and failed to establish and maintain the regional order that they believed would best serve their national security interests. Eventually, the American and Israeli governments succumbed to domestic pressures and withdrew from Vietnam and Lebanon. And long before the final recognition of failure resulted in military withdrawal, these interventions ended the political careers of the leaders who led their nations into the quagmire—President Lyndon Baines Johnson and Prime Minister Menachem Begin. The actual results of withdrawal were not as far-reaching as the worst-case analysts had warned over the years, but the consequences for the national psyche were still significant. The humiliating withdrawals left behind a great reluctance to use military force even when it could have been beneficial to do so. This phenomenon came to be popularly known as the Vietnam syndrome in the United States and the Lebanon syndrome in Israel. Thus, the decisions made during the Vietnam and Lebanon interventions had long-term effects on the foreign policies of both countries.

In this chapter I map the military, economic, and political risk dimensions and their assessment by American and Israeli decisionmakers. I also

identify the reasons for these assessments and for the risk-taking preferences
that shaped policies. What makes the comparison interesting are not only the
many similarities but also the substantial dissimilarities in political systems—
presidential versus cabinet—and in political institutions and personal leadership styles. Analysis of the individual cases and comparison between them
allow us to examine, test, and extend the theoretical analysis to cases where
the stakes are high, the costs are substantial, and the time frame is years
rather than days. Owing to the broad theoretical objectives of this study, the
discussion on Vietnam and Lebanon does not cover these conflicts in their
entirety. The focus is on the decisions to take the risk of intervention and
then on the entrapment in failing policies, which took place in spite of the
high risks already recognized and being experienced by key American and Israeli decisionmakers. For the Vietnam intervention, I address decisions taken
between Lyndon Johnson's election as president of the United States in November 1964 and the Tet offensive in January 1968. For the Lebanon case, I
focus on decisions taken between the launching of Operation "Peace for
Galilee" in June 1982 and Prime Minister Menachem Begin's resignation in
September 1983.

The Vietnam Case

AN OVERVIEW

On June 4, 1954, South Vietnam gained independence from France after
the latter's defeat at Dienbienphu made its hold on the Indochinese colonies
impossible to maintain. Within the month, the titular head of the Republic of
South Vietnam, Emperor Bao Dai, asked Ngo Dinh Diem to head the new
government as prime minister. Diem was an Annamese Catholic who had
been an early member of the nationalist movement for the independence of
Vietnam. Diem soon discovered that he had little or no power. Three political-
religious sects, the Cao Dai, the Hoa Hao, and the Binh Xuyen, controlled
much of the countryside, as well as the police in Saigon and Hué. To compound problems, Diem had to deal with the massive influx of refugees from
North Vietnam and the small Vietminh contingent that remained in South
Vietnam to incite trouble for the new government. In all of these matters,
Diem received immediate and crucial aid from the United States. By August
1954 the Eisenhower administration had authorized $300 million to be used
to equip, train, and pay for an army for the new republic. Additional help
was furnished by the U.S. Office of Strategic Services, which was ordered to
assist Diem with intelligence, the conduct of clandestine operations, and re-
location programs involving the transportation of nearly a million refugees
from North Vietnam to South Vietnam. The authorization for assistance
was given against the advice of the U.S. Joint Chiefs of Staff, who wanted instead to wait until Diem had proved his leadership abilities (Eisenhower,
1963: 332–44; Karnow, 1983: 219–20).

In the spring of 1955, when Diem decisively put down a coup attempt by the Binh Xuyen, American policy changed from vacillation to unequivocal support. With U.S. concurrence, Diem held a referendum in October 1955. He won 98.2 percent of the vote and promoted himself to chief of state. In May 1956 the United States augmented the Military Assistance and Advisory Group (MAAG) presence by 350 military men under a pretext, calling the new effort the Temporary Equipment Recovery Mission. The July 1956 date for elections and reunification of the two Vietnams passed, and the seventeenth parallel, which divided North and South Vietnam, became an internationally recognized boundary, at least de facto. U.S. aid was running between $300 million and $500 million a year. Assured of American support, Diem ignored American demands to spend more aid money on social and economic programs and carry out serious land reforms (Young, 1991: 49–59). Between 1956 and 1963, Diem stemmed opposition to his corrupt and nepotistic regime with repressive measures, took full advantage of U.S. aid, including the training of paramilitary police forces, and imposed the extremely unpopular reform program, the Strategic Hamlet program, on the countryside. The purpose of the program was to secure government control over the population by relocating peasants to centers that were to provide security and distance from the Vietminh. The centers were unpopular for many reasons, foremost among them the peasants' aversion to being moved away from traditional villages, the forced labor that accompanied the program, and the political control exerted through local governing councils.

In December 1960 the establishment of the National Front for the Liberation of South Vietnam (NLF) was announced at a secret base near Saigon. The NLF set forth a program of reforms designed to appeal to those in the South, especially those hurt by Diem: higher wages for civil servants, land reform, jobs, equality for national minorities, a neutral foreign policy, and reunification. By October 1961 the NLF had been successful in spreading its influence, and cadres, throughout most of South Vietnam (Young, 1991: 70–73). The recently elected U.S. president, John Kennedy, responded by radically building up programs aiding South Vietnam, particular those addressing counterinsurgency. Kennedy instructed the military to study guerrilla warfare. He authorized a significant increase in the Army Special Forces Group, which became known as the Green Berets. Most important, he formed the Special Group Counterinsurgency, whose members included the secretary of defense, the chairman of the JCS, the heads of the U.S. Information Service and the Agency for International Development, the special assistant for national security affairs, and the attorney general. The Special Group Counterinsurgency was headed by a retired JCS chairman, General Maxwell Taylor, whose advocacy for a policy of flexible response—in opposition to Eisenhower's policy of massive retaliation—was close to Kennedy's own thinking. In October 1961 a fact-finding mission consisting of General Taylor and Deputy Special Assistant for National Security Affairs Walt Rostow was dispatched to learn more about the situation in Vietnam. They recommended that the U.S. presence in Vietnam be increased by 8,000 combat troops in the guise of a flood control team. Kennedy rejected the recommendation that combat troops be introduced but accepted some of Taylor's recommendations to increase the number of advisers and expand military training. An additional 100 advisers were sent to assist the South Vietnamese government.

By December 1961 there were 3,000 U.S. military personnel serving in South Vietnam, by 1962 there were 11,000, and by the time of Kennedy's assassination there were 16,000. Although domestic opposition to Diem increased, his reluctance to institute reforms was steadfast; yet his refusal to meet U.S. demands never stemmed the flow of U.S. aid to the Army of Vietnam (ARVN). American pilots were flying U.S. helicopters to ferry South Vietnamese troops into action. These flights were disguised as training exercises for the Vietnamese army. From 1955 to 1962 the United States spent $2 billion on military, economic, and social reform programs in Vietnam. By 1963 military and economic aid was pouring into Vietnam at a rate of more than $500 million per year (Karnow, 1983: 268, 294). Although Kennedy never authorized combat missions for U.S. troops in Vietnam, his words and actions bolstered American commitment to the region. U.S. assistance seemed to make a significant difference at first, especially the use of armored personnel carriers and helicopters, but the Vietcong soon learned how to counter the technological advantages of their adversary effectively; and the Army of Vietnam was never able to recapture the strategic initiative.

At the time of the Taylor-Rostow mission, Kennedy's foreign policy was suffering from several setbacks, including the failure of the Bay of Pigs invasion, the construction of the Berlin Wall, and the neutralization of Laos. Vietnam became fixed in the minds of American policymakers as a place where the United States could succeed in promoting its foreign policy aims with a limited risk of escalation. Kennedy espoused a strategy of flexible response, designed to respond to wars of national liberation without resort to nuclear weapons or confrontation with the Soviet Union. This was a major conceptual change in American strategic thinking. It was deemed important that the United States prove to communist insurgents around the world that wars of national liberation were costly and exhausting, both economically and politically.

The U.S. policy of containment and the role of the United States as guarantor of national sovereignty to countries threatened by communist-directed or communist-inspired armed insurrections was put to the test in Vietnam. In 1962 there was great optimism that U.S. assistance could tip the balance of power in favor of the South Vietnamese government. Many U.S. military leaders thought that simply helping the South Vietnamese army outstrip the Vietcong forces in firepower would cause such attrition that the communists would sue for peace. As it turned out, however, there were several major obstacles to the successful prosecution of a war in South Vietnam, including corruption, political factionalism, civilian-military mistrust, absence of national spirit or motivation, lack of social cohesion, and inexperience in government. There was also a sharp difference between American aspirations for a stable South Vietnamese government and the reality of the corrupt, myopic Diem regime.

A May 1963 uprising by Buddhists monks revealed widespread enmity against the Diem regime, which took both U.S. officials and Diem by surprise. When Diem ordered counterattacks against the main Buddhist temples in Hué in August, the United States approved a plan by Vietnamese generals to have him removed from power. On November 2, 1963, Diem and his brother Nhu were shot dead after being picked up by the military at a Catholic church in Saigon. Between November 1963, when Diem was overthrown, and June 1964, South Vietnam was rocked by a series of coup attempts led by military officers, which left two men in charge: Air Vice Marshal Nguyen Cao Ky, commander of the South Vietnamese air force, who was appointed prime minister and General Nguyen Van Thieu, the armed forces chief of staff, who was made chief of state.

Diem's assassination came just three weeks prior to Kennedy's, leading both Vietnamese and Americans briefly to consider a negotiated withdrawal of U.S. forces, but President Johnson's top cabinet members and aides, upon whom he was initially highly dependent, almost unanimously advocated continued escalation. Those in favor of intervention included Secretary of Defense Robert McNamara, Secretary of State Dean Rusk, Director of the Central Intelligence Agency John McCone, National Security Adviser McGeorge Bundy, and the entire JCS. Those in favor of withdrawal included the Democratic majority leader of the Senate, Mike Mansfield, and Undersecretary of State George Ball. Johnson viewed the Vietnamese conflict as a test case of U.S. commitment to contain and defeat communist wars of liberation. When reports coming through military, intelligence, and diplomatic channels increasingly disagreed in their assessments of the situation in South Vietnam, Johnson sent Secretary of Defense McNamara to Vietnam to assess the situation. McNamara, who went there in February 1964, reported a government ripe for communist takeover if the United States did not act quickly. More important, he thought that the national security and worldwide reputation of the United States were tied to the survival of the Saigon regime (Karnow, 1983: 342). Johnson agreed to an increased aid package and the provision of more modern aircraft and other equipment.

In response to two purported unprovoked attacks by North Vietnamese torpedo boats on U.S. destroyers that were on patrol in the Tonkin Gulf off the northeast coast of Vietnam in August 1964 Johnson introduced a bill in Congress, later called the Gulf of Tonkin Resolution, to gain broad bipartisan support for his actions in Vietnam. The allegations that the North Vietnamese were responsible for two successive attacks and that they had attacked without being provoked first were based on incomplete and misleading information, but Johnson took the opportunity to push through the resolution. It had two purposes: to establish a legal foundation for continued military operations and to demonstrate legislative support for active American participation in the Vietnamese conflict. The resolution was rushed through in two days by an overwhelming majority in both the House (416–0) and the Senate (88–2). Nonetheless, because of Johnson's sensitivity to the upcoming presidential election, it was not immediately put to a full test.

U.S. military action in Vietnam escalated with the passage of the Tonkin Gulf Resolution, at first very slowly. In response to the incidents that had instigated the resolution, Johnson ordered limited air strikes. In October five American officers were killed in an attack against an airbase in Bien Hoa. In December 1964 seven Americans were killed and fifty-eight wounded in two incidents. Johnson's advisers proposed various retaliatory measures. One option was a program of massive escalation in which the full force of American military might would be deployed against North Vietnam. The second was a program of graduated military moves in which the United States would increase current force strength but would negotiate should an opportunity arise. The third option was to continue the current policy of providing arms and advisers but not escalate direct American involvement (U.S. Department of State, 1992: 916–29).

The third option was the least popular. Most U.S. officials believed that the South Vietnamese could not withstand the constant infiltration from the North. Persisting in the present policy was widely perceived to be equivalent to conceding a communist victory, an alternative most advisers rejected for fear that Vietnam's fall would lead to the gradual consolidation of communist influence throughout Southeast Asia—the domino theory. Although Johnson approved a secret bombing campaign against southern Laos in December 1964, he remained undecided about direct intervention in Vietnam. Just one month later, however, reflecting increased concern from both military and civilian advisers, Johnson approved the concept of ad hoc retaliatory strikes in response to "spectacular enemy action" (Karnow, 1983: 411). The first such retaliatory strike (Flaming Darts) occurred in February 1965, after a strike against a U.S. base near Pleiku in which eight American soldiers were killed and over a hundred wounded. Later that month Johnson authorized and then personally supervised a systematic bombing campaign (Rolling Thunder) against North Vietnam to demonstrate American might and resolve. The use of military force against North Vietnam was also intended to express commitment to, and provide political support to, the government in the South (Cable, 1991: 90; VanDeMark, 1991: 58–66).

The decision to initiate a systematic bombing campaign eventually forced an American decision to increase the presence and activities of ground combat forces. To protect the U.S. air base at Danang, where many of the bombing sorties originated, General William Westmoreland, the U.S. Commander in South Vietnam, requested 3,500 marines for base security duty. The U.S. ambassador, General Maxwell Taylor, opposed the request, anticipating that the troops would become involved in more than base protection. He argued that U.S. ground forces should not be committed to Vietnam because they were logistically and strategically unfit for guerrilla warfare. Despite his warnings, the first marine battalion splashed ashore at Danang on March 8, 1965, with presidential promises to Westmoreland that more would be at his disposal. Within a week, requests for additional forces were on Johnson's desk, and by April Westmoreland had convinced Johnson that American base security would be better served by having the troops perform forward patrols in the countryside. Taylor again objected, but Johnson agreed with Westmoreland. The administration also began to follow a "policy of minimum candor" (Karnow, 1983: 414) with journalists, who clamored to know more about the long-range plans in Vietnam. Public apprehension grew, but in May overwhelming majorities in both houses of Congress approved Johnson's request for $700 million in appropriations to conduct the war.

The summer of 1965 was pivotal in deciding the future of American involvement in South Vietnam. Johnson sought the advice of officials in every branch of government and received contradictory recommendations. Liberals in Congress pleaded for restraint. The CIA issued a report indicating that the use of U.S. forces would be of negligible consequence for the South Vietnamese, at best. Undersecretary of State Ball wrote a long memorandum to Johnson expressing his view that U.S. forces would never be able to turn the conflict around. Westmoreland and McNamara appealed for authority to prosecute the war with greater latitude. In June 1965, Westmoreland requested an additional 180,000 troops, without which, he said, the country would be jeopardized. In a television broadcast on July 28, 1965, Johnson announced that additional U.S. troops would be sent to fight in South Vietnam. He ordered 44 combat battalions to reinforce the U.S. contingents already there.

Initially approved for a period of eight weeks, the bombing campaign continued, with brief pauses, from March 1965 until November 1968. In its course, thousands of sorties were made against targets in the North, and a million tons of bombs, rockets, and missiles were dropped on North Vietnam. Yet, bound by the need to maintain a modicum of legitimacy for this policy, Johnson would not permit bombing of certain key sites, such as the irrigation systems, whose destruction would almost certainly have caused widespread famine. As U.S. forces in Vietnam grew, the requirement for ever greater logistical support increased. By 1967 a million tons of supplies a month were being sent to Vietnam to sustain U.S. troops.

Through all this escalation, little consideration was given to sociopolitical factors underlying the support for the Vietcong guerrillas. The nation-building and pacification programs, incorporated into the painstaking 41-Point Nonmilitary Program, received the lowest priority of all American missions (Cable, 1991: 125–27). After only a few months, reports from diverse sources—the CIA, the National Intelligence Board, and the RAND Corporation—indicated that the air campaign was having little effect on the Vietcong capacity to continue the struggle and would continue to have little effect. At the same time, Vietcong snipers, night probes, booby traps, land mines, and punji stakes were not only maiming or killing significant numbers of American soldiers but also having a devastating effect on American morale in Vietnam. Just as important, as early as 1965, reports by U.S. military commanders showed some basic flaws in American military doctrine. The operational doctrine focused almost exclusively on the destruction of matériel and the disruption of enemy organizational integrity, yet neither of these targets was susceptible to permanent damage in a guerrilla warfare environment. U.S. military reports emphasized statistical data, such as body counts, matériel captured, supplies intercepted, and targets destroyed; they did not acknowledge that favorable statistics alone did not constitute victory (Cable, 1991: 54–76). On the contrary. Not once during the heavy Rolling Thunder bombing did either threats of its continuation or promises of its cessation induce the North to seek peace negotiations on American terms—an unmistakable sign of ineffectiveness and failure.[1]

Negative reporting on the effectiveness of continued air operations increased. Some reports went as far as to state that the bombing was counterproductive in that it served to consolidate the political will of the North Vietnamese. Yet contradictory reviews, usually released by military organizations like the Institute for Defense Analysis or the JCS, demonstrated that the bombing was working or that in some way the United States was winning the war. In January 1967 the discrepancies between the various agencies reporting on Vietnam became so blatant that the chairman of the JCS ordered an interagency conference to resolve the differences and to settle disputes on enemy strength. Secretary of Defense McNamara, an early proponent of air strikes, admitted that neither the bombing nor the ground war was working, and advised that the United States seek a way for a negotiated disengagement from the conflict without humiliation or long-term reputational damage. Although McNamara, Johnson, and others supported the nation-building and pacification programs, which had been identified as improving the morale and fighting spirit of the South Vietnamese, they were unable to overcome the "institutional and intellectual inertia of the U.S. military or offset the political strength of the Joint Chiefs of Staff" that relegated the programs to secondary status (Cable, 1991: 134). Thus, despite adequate intelligence and analysis of alternative courses of action, the United States was entrapped in its offensive military policy in dealing with a communist insurgency in an underdeveloped country.

By the end of 1967, almost 500,000 American soldiers were in Vietnam. More than 15,000 had already been killed. McNamara had become increasingly disillusioned with the entire war effort. In November 1967 he wrote a detailed memorandum to President Johnson in which he advocated a unilateral halt to the bombing and a cap to troop levels. Johnson was deeply concerned by signs of failure in Vietnam and circulated the memo to his advisers for comment. General Taylor, who was appointed special adviser to the president after completing his diplomatic assignment in Saigon; General Westmoreland, U.S. commander in South Vietnam; Ellsworth Bunker, the new U.S. ambassador to South Vietnam; Secretary of State Rusk; Abe Fortas, a

Supreme Court justice and a close confidant of Johnson's; and Clark Clifford, an adviser to the president, were all opposed to even a pause in the bombing campaign. By then, Johnson was determined to replace McNamara. The U.S. presidential election was scheduled for November 1968, and Robert Kennedy, a vocal opponent of the war in Vietnam, had declared his intention to run for the Democratic nomination. Opinion polls placed Kennedy far ahead of the president.

The watershed in American opinion came with the Vietcong's Tet offensive in January 1968. When the assault began, the Americans welcomed the opportunity to bring superior firepower to bear on the massed Vietcong and North Vietnamese forces. It soon became apparent, however, that the North Vietnamese were willing to suffer enormous casualties. Polls of Johnson's performance indicated a dramatic drop in approval (Karnow, 1983: 546). The mounting American casualties and the lack of any reliable forecast of when the war would end had been eroding public support for more than two years. The media became universally negative, and public protests against the war reached new heights. Although the Tet offensive was viewed as a military disaster for the Vietcong, it was a political victory: it caused a major shift in opinion among Johnson's top cabinet members and advisers, along with Johnson's announcement that he would not run for reelection.

The offensive was followed in February 1968 by a request from General Westmoreland for an additional 206,000 troops. To meet it would necessitate the mobilization of U.S. military reserves, a step that Johnson had consistently avoided. On February 27, 1968, a meeting was held to discuss the request. Secretary of State Rusk and Special Assistant for National Security Affairs Walt Rostow favored granting the request, but opinions were divided, and McNamara, sitting in during his last days as secretary of defense, made an eloquent appeal for a pullback (Barrett, 1993: 125–26; Cable, 1991: 227). Nicholas Katzenbach, the deputy attorney general, William Bundy, the assistant secretary of state for Far Eastern affairs, and Clark Clifford, who were also at the meeting, were receptive to McNamara's arguments. Clifford was sworn in as secretary of defense on March 1, 1968, and within weeks he was pushing for a cessation of the bombing and for a negotiated withdrawal. On March 25, Clifford called a meeting of some of Johnson's most respected advisers, the so-called Wise Men, to discuss options with the president. In attendance were Arthur Dean (who had helped negotiate the Korean War peace), Douglas Dillon (former secretary of the treasury), Cyrus Vance (former undersecretary of defense), Dean Acheson (former secretary of state), Generals Omar Bradley, Maxwell Taylor, and Matthew Ridgway (all former chairmen of the JCS), Robert Murphy (a State Department officer), Henry Cabot Lodge (former ambassador to Vietnam), George Ball (former undersecretary of state), McGeorge Bundy (now head of the Ford Foundation), and Rostow. The briefing by the CIA and the military was followed by a frank discussion leading to pessimistic conclusions. All except for Taylor and Murphy thought that the United States should seek a negotiated withdrawal from Vietnam. Johnson was profoundly affected by the group's pessimism (Immerman, 1994; Young, 1991: 227–29).

On March 31, in a televised address to the nation, Johnson announced that he was imposing a unilateral partial bombing halt by restricting bombing to the area south of the twentieth parallel. For the first time, he also placed a ceiling on the number of troops that the U.S. government would allow in Vietnam: 543,000. And he announced that he would not run for reelection. The speech signaled the beginning of the American search for a negotiated disengagement from Vietnam. The actual negotiated withdrawal of troops would take another five years. By the time President Richard Nixon took office in January 1969, over 30,000 American had been killed in Vietnam.

U.S. INTERESTS

The American commitment to Vietnam dated back to the Truman administration.[2] In essence, it was intended to prevent a communist takeover of the country and was associated with the view that the loss of Vietnam would trigger a sequence of losses throughout Southeast Asia (Leffler, 1992). This view found expression in Eisenhower's "falling dominoes" metaphor, through which American interests were perceived and analyzed by consecutive presidents.[3] As the Vietnam issue became acute during the Kennedy administration, the broad definition of interests was translated into a set of specific interests and stakes, which resulted in a reaffirmation of the commitment and an expanded advisory role.

It was President Kennedy who in his inaugural address on January 20, 1961, confidently and proudly announced: "Let every nation know, whether it wishes us well or ill, that we shall pay any price, bear any burden, meet any hardship, support any friend, oppose any foe to assure the survival and the success of liberty" (U.S. Government, 1962: 1). But this sweeping promise, when put to the test, was carried out by his vice president and successor, Lyndon Johnson, who upon unexpectedly becoming president was thrust into the midst of a foreign policy crisis. Under the circumstances of Kennedy's assassination and with little foreign policy experience, Johnson had neither the opportunity nor the experience to define U.S. interests but inherited the definition of his predecessor's administration, especially because most of the Kennedy advisers continued to serve in his administration. Johnson accepted the outlines of Kennedy's Southeast Asia policy and committed himself to "seeing things through in Vietnam" (Johnson, 1971: 42). And even as he acquired stature and experience, when faced with a rapidly escalating crisis, he preferred to stick with the original definition of U.S. interests rather than reconceptualize it.

As early as 1956, Kennedy explained his views on Vietnam's importance for the security and reputational interests of the United States. He embraced the domino theory: if Vietnam fell, so would Burma, Thailand, India, Japan, the Philippines, Laos, and Cambodia. Kennedy and Johnson both saw Vietnam as a test case of the resilience of the democratic experiment in Asia, one whose success or failure would set the tone and direction for other Asian nations. Also being tested in Indochina were American responsibility and determination. If Vietnam fell, American prestige in Asia would "sink to a new low" (Podhoretz, 1982: 19–20; U.S. Government, 1966a: 394–96). Subversive revolutionary movements in Africa and Latin America would be encouraged by that failure, so the U.S. reputation as a guarantor of countersubversion had to be protected.[4] It was feared that even in Europe an American withdrawal from Vietnam would bring dire consequences, by, for example, providing President Charles de Gaulle of France with ammunition in his campaign to undermine U.S. leadership in Europe (Chen, 1992: 34; *Pentagon Papers*, 1971, 3: 216–17, 598, 622–24, 658–59). The threat to U.S. reputational interests was closely associated in Johnson's mind with the threat to his own reputation and place in history if Vietnam was lost during his presidency. He simply refused to be the president who lost Vietnam, as Truman had been the president who lost China. He made that concern clear 48 hours after taking the oath of the presidency, saying, "I am not going to be the president who saw Southeast Asia go the way China went" (Ball, 1992: 116). If that happened, he would be branded a coward and "an unmanly man. A man without a spine" (Kearns, 1976: 252).

Since 1961 the Americans had been seeking to achieve their goals by building up the military capabilities of the government of South Vietnam, intending to train and arm a force of 500,000 regular and paramilitary troops by mid-1965. This military force was expected to make the Saigon government strong, effective, and stable enough to prevent victory for the communist NLF. By 1964 it was clear that the strategy was going to fail. But, having stated the case for supporting South Vietnam, American decisionmakers felt committed to military intervention even at a high cost in American lives and material resources. Intervention to buy the Saigon government time would avoid adverse political, strategic, and financial costs of great magnitude. It was also hoped that American intervention would boost morale in the Saigon regime and its military forces and galvanize a new effort to snatch victory from imminent defeat or would at least put the Saigon government into a position where it could negotiate from strength (Kolko, 1972).

Intervention was ultimately rationalized by a view of the United States as the leader of the free world and by the acceptance of the responsibilities flowing from this role, even if it meant taking high risks and bearing exorbitant costs. President Johnson summarized the American position by saying, "We did not choose to be the guardians at the gate, but there is no one else" (U.S. Government, 1966b: 794). Secretary Rusk encouraged the president to view the issues as global in impact. Johnson, who accepted the premise, saw the defense of Vietnam as part of the U.S. commitment to the Southeast Asia Treaty Organization (SEATO)—as no different from the commitment to

NATO. The purpose of both treaties was to contain communism. In Europe the immediate threat was the Soviet Union; in Asia it was the People's Republic of China (PRC). The North Vietnamese were perceived as surrogates acting on behalf of Chinese aggressive ambitions for hegemony (Podhoretz, 1982: 66, 108–9; Simons, 1994; VanDeMark, 1991: 121–22).

Johnson conformed with the definition of interests inherited from the Kennedy administration not merely because of circumstances and a lack of foreign policy experience. The underlying logic of the Vietnam policy was familiar to him and easy to accept. He himself was a product of the generation that witnessed the 1930s, including the rise of Fascism and Nazism in Europe and Asia, the Munich appeasement, the rise of Germany and Japan to the status of world powers with aspirations for world dominance, and the weakness of Western democracies in not standing up to the aggressors at an early stage, leading to a devastating world war that the United States had no choice but to participate in. Almost as soon as the Germans and the Japanese were defeated, a new and no less potent threat was posed—this time by an authoritarian regime driven by communist ideology that contested the core values of the United States and its allies. The challenge of the communist bloc to the determination of the Western bloc, led by the United States, to contain its expansion resulted in a bloody war in Korea. There the United States proved that the post-1945 status quo would not be undone through the use of force.

In a sense, then, the Vietnam policy represented both continuity and change. Continuity was represented by the claim that the policy served such traditional American interests and values as stability, freedom, and democracy. Change was represented by the purposeful attempt to grab the initiative from the adversary in order to maintain American credibility and the status quo rather than losing both and then attempting to restore them. Accepting the validity of this approach in assessing the nature and definition of U.S. interests in the Asia-Pacific region and its most salient flash point required being prepared to respond aggressively to setbacks. To resist challengers of the status quo by raising the stakes meant being ready to intervene militarily, to escalate if necessary, and to pay the economic, political, and military costs involved.[5]

Strategic, reputational, and ideological interests were tightly interwoven. Strategic motivations, which had to do with the threat to the balance of power in Asia and its global ramifications, were reinforced by strong ideological beliefs. These beliefs gave the security issues an intensity and level of commitment they otherwise would not have had. The contested issues and the prospect of communist victory thus became a test of national reputation.[6] All these resulted, at the early stages of American intervention, in a highly favorable public response that resonated on Capitol Hill.

What made U.S. interests even more salient was that since 1955, even without the presence of combat troops, the U.S. investment in the Vietnam policy had accumulated to a substantial size. The repercussions of a decision to desert South Vietnam in 1964 in the face of almost hopeless domestic instability and growing Vietcong attacks were conceived by most of Johnson's advisers as unacceptable. Typical of that attitude was McGeorge Bundy's cable from Vietnam to the president immediately after the bombing of U.S. Army barracks at Pleiku: "The situation in Vietnam is deteriorating, and without a new U.S. action defeat appears inevitable. . . . The stakes in Vietnam are extremely high. The American investment is very large. . . . The international prestige of the United States, and a substantial part of our influence, are directly at risk in Vietnam" (*Pentagon Papers*, 1971, 3: 309).

Indeed, the Pleiku attack initiated a sequence of events that soon led to a strategic bombing campaign and the deployment of American combat troops. On June 7, with 75,000 U.S. troops already deployed, General Westmoreland requested that the United States commit 44 battalions to Vietnam and that the troops take on a general combat role. These requests were approved in July. Within a few months the ground war in Vietnam was Americanized. The debate between a military approach and a political approach to the Vietnam problem was decided in favor of the riskier of the two.[7] For the next decade U.S. military intervention in Vietnam was a major component of U.S. policy in Asia, with comprehensive ramifications on U.S. global and domestic policies.

In sum, the interests perceived to be at stake were a potent mix of the strategic, ideological, and reputational. Most of Johnson's senior advisers defined the interests threatened in the same way. Their positions were not merely easily comprehensible by the inexperienced president but also addressed his own intuitive understandings and biases concerning world politics. Johnson, who worried about being compared to his charismatic predecessor John Kennedy, viewed the Vietnam crisis as a personal test that could make or break his reputation and his presidency. A show of weakness in the face of acute threats to what were perceived to be vital national interests, with both regional and global ramifications and important symbolic value, was out of the question. The demonstrable determination to take high risks sent a message to foreign actors that this American president could not be pushed around and another message to the American public—that their president was assertive.

NATIONAL CAPABILITIES

After winning the 1964 presidential election, Johnson faced the prospect of the imminent collapse of the Saigon regime. He had two basic choices: withdraw the U.S. troops and allow a communist takeover, or authorize a qualitative change in the role of the United States in Vietnam—active participation in combat. In analyzing the risks confronting Johnson and his administration as they initiated and then expanded intervention in Vietnam, I focus on the strategic bombing of North Vietnam, Operation Rolling Thunder, and the decision to deploy combat troops to South Vietnam, starting with the authorization to employ two marine battalions to defend air bases. That decision was quickly followed by the expansion of the combat duties of American troops to the point that in July 1965 the president authorized the extensive utilization of American troops for offensive operations through the decision to send 44 battalions to Vietnam, which marked the Americanization of the ground war. Although the successful employment of strategic bombings was expected to eliminate the need for large numbers of American ground troops, it became obvious a few months after the strategic bombing began that the expectation was unfounded. Thus, the employment of combat troops—which in early 1965 was considered necessary only to supplement the central policy of strategic bombing—became the primary policy by the second half of 1965. Strategic bombing came to play a supplementary role, that of increasing the effectiveness of the ground war and decreasing the risks involved in implementing it. The direct target of Rolling Thunder was North Vietnam, whereas the direct target of the ground war was the Vietcong. The message being sent to North Vietnam was that the NLF could not win. In April 1965, McNamara summarized the prevailing view in a memo to the president: "A settlement will come as much or more from VC [Vietcong] failure in the South as from DRV [Democratic Republic of Vietnam] pain in the North" (Porter, 1979: 370). My analysis of the two policies concludes with McNamara's resignation and the Tet offensive. Within this three-year period, the air and ground wars were begun and then accelerated—apparently independently of increasingly negative assessments of their efficacy.

The air and ground operations began almost simultaneously: strategic bombing started in March 1965, only shortly before the two marine battalions arrived in Danang. These actions were followed by National Security Action Memorandum (NSAM) 328 on April 6, 1965, wherein the president authorized the deployment of two additional marine battalions and an increase in U.S. logistical and support forces by 18,000–20,000 soldiers. NSAM 328 also sanctioned the use of U.S. troops in offensive ground operations, although the change of policy was announced in a very low tone to avoid alarming the American public (Schandler, 1977: 21). Shortly thereafter, Westmoreland's request for 44 battalions was approved, making large-scale military intervention a reality. With troop numbers increasing so rapidly and the conduct of the war in Vietnam taken over by the Americans and with the anticipation that even more troops would be required, the force structure became a major bottleneck and threatened to become a main source of risk to the effective conduct of the war.

Although the number of available troops necessary to achieve American goals was a key to calculating the risks of failure, at no time did there seem to be a broadly shared agreement over what the number was likely to be. In a meeting between the president and the JCS on March 15, 1965, General Harold Johnson, army chief of staff, told the president that winning the war would take 500,000 U.S. troops and five years (Gibbons, 1989: 166). Even though Johnson was considered an officer of superb competence, his pessimistic estimate—after sending quite a shock wave through the administration—was basically ignored. Because the administration did not initially deal with the question of how many troops would be required, Westmoreland could demand incremental increases in their number by implying that without another presidential authorization for a troop increase, American policy was doomed to failure. In mid-1965 there were about 75,000 U.S. troops in Vietnam. By the end of the year there were 184,000 troops. The number rose to 389,000 in 1966, then to 463,000 in 1967 and to 495,000 in 1968 (Berman, 1992: 33; Gibson, 1986: 95). By 1967 the number of American troops in Vietnam was more than twice as great as the number—208,000—that McNamara believed could be deployed without weakening American capability to carry out other worldwide commitments effectively, especially NATO commitments.

The uncertainty about the troop requirement had several causes. Information on the size of enemy forces was unreliable. The size of the Vietcong force, its ability to mobilize and train reinforcements in South Vietnam and to infiltrate reinforcements from North Vietnam into the South in spite of interdiction bombing, and the number of North Vietnam Army (NVA) troops were all unknown. The method used to assess enemy casualties both in aerial bombing and ground engagements complicated the problem. The body count indicator of success created an incentive for American field commanders to exaggerate the number of enemy casualties substantially (Krepinevich, 1986: 202–3). Another source of bias in assessing the troop requirement was the evaluation of the enemy's strategy. Common wisdom held that 10–20 government troops were necessary to defeat one guerrilla insurgent. A joint memorandum of the CIA and the Defense Intelligence Agency (DIA) in 1965 that used this criterion had a sobering effect. Its conclusion, based on the assumption that the Vietcong strength consisted of 50,000–60,000 main force regulars and roughly 100,000 irregulars, was that a force level of more than 1,000,000 men would be required to defeat the insurgency (Krepinevich, 1986: 143–44). The number was staggering, even taking into account the 567,000 South Vietnamese army troops.

Westmoreland's view of the South Vietnamese army as hopelessly inept made it unavoidable that the military containment of the Vietcong become almost exclusively the task of U.S. troops. In that context, Westmoreland's request for 44 battalions was justified because of the mounting crisis in Saigon, the escalating activity level of the Vietcong, and the threat of military intervention by the NVA. Johnson's commitment to give the military commanders the men they needed to win led to ever-increasing demands for more troops. Thus, between July 1965 and February 1966, Westmoreland upped the number of maneuver battalions required from 58 to 79. By June 1966 he was requesting a total of 90 maneuver battalions for 1967. By March 1967 he was claiming that his attrition strategy was working, and requested that U.S. forces be increased to 108 maneuver battalions by mid-1968 (Krepinevich, 1986: 179, 183, 187).

The constant demands for more troops were also justified by the estimates of growth in enemy strength. In 1966 it was estimated that the Vietcong force amounted to 220,000, not including 38,000 NVA troops—a significant increase over the 160,000 the year before. It had been established
that the enemy could increase its military strength by the equivalent of fifteen
battalions a month through recruitment and infiltration (Krepinevich, 1986:
180). Westmoreland's strategy required a buildup of forces to the point—the
so-called crossover point—where enemy losses in battle would exceed the
enemy rate of replacement. For this numbers game to be credible, the JCS
and Westmoreland had to make several assumptions about force ratios.
They assumed that the Vietcong were moving into the third stage in the cycle
of revolutionary war—the conventional stage. Therefore, they inferred that
a 3:1 ratio in favor of the United States and its allies would suffice to win the
war. The force ratios required to win a guerrilla war, 10:1 or even 15:1, were
out of reach. They argued further that "force multipliers," such as firepower
and the mobility provided by extensive use of helicopters, more than compensated for troop number constraints (Krepinevich, 1986: 157–63; Gibbons, 1989: 364).

These premises provided a sense of confidence that the United States
would prevail, but they were false.[8] The Vietcong were not about to enter the
conventional war stage, and force multipliers had limited effects in a guerrilla war. George Ball, Clark Clifford, the CIA, and the DIA challenged the
idea that the war was moving into the conventional phase, but to no avail.
Even supporters of escalation expressed skepticism about the likely outcome
of intervention. McNamara's principal adviser during the war, Assistant
Secretary of Defense for International Security Affairs John McNaughton,
was less than optimistic about the outcome as early as 1965. According to
his analysis, with troop levels between 200,000 and 400,000 the probability
of success was only 20 percent, the probability of an inconclusive outcome
70 percent, and the probability of defeat 10 percent. For 1967 the respective
probabilities were 40, 45, and 15 percent, and for 1968 they were 50, 30,
and 20 percent. Yet he recommended escalation. The chairman of the JCS,
General Earle Wheeler, told the president in 1965 that "pacifying" South
Vietnam would require 700,000 to 1,000,000 men and seven years; if winning meant just staving off defeat, smaller numbers of men would suffice
(Gelb with Betts, 1979: 126–27). In the absence of any valid reference point
for estimating the troop requirement, the decisionmakers lost touch with reality. Once into the hundred thousands, the numbers cited were more wild
guesses than thoughtful estimates that could be validated by analysis.[9]

As the war became protracted, with no end in sight, and demands for
troops mounted rather than abated, the pressure on the manpower resources
of the United States became a problem. The selective draft system was unfair
to the less advantaged, particularly black males, who were more likely to be
drafted, sent to Vietnam, and placed in high-risk infantry combat units. Consequently, black casualties represented almost one-fourth of losses of en-

listed men in the army.[10] In spite of an increasingly problematic draft system and growing manpower requirements, however, the JCS could not mobilize reserves—primarily because of the president's wish to maintain a facade of business as usual. This was necessary because, declaring a state of national emergency could jeopardize the prospects for the Great Society program, which was of great importance to Johnson, who wished to be compared to Franklin Roosevelt, the author of the New Deal. In reality, the U.S. military quickly found itself at a disadvantage because of North Vietnamese and Vietcong determination and their ability to match any increase in American troops. In addition, maintaining the readiness of U.S. forces worldwide, especially in Europe, became progressively more difficult for all services over time.[11] This raised questions about whether the United States could meet military challenges in other parts of the world if required to do so.

The insufficiency of the manpower base was made even more salient by the limitation of a tour of duty in Vietnam to one year. Denied access to experienced and trained National Guard troops and general reserves and given the limited tour of duty, the army lacked the expertise and unit cohesion that come with time and experience. The consequences for military leadership at all levels were even more severe, with the reservoir of available experienced officers and NCOs from among career personnel stretched very thin in the army and other services worldwide (B. Palmer, 1984: 83, 169–70, 175). That officers in command positions served only six months before being transferred was counterproductive as far as military efficiency was concerned. Explanations provided by the army command notwithstanding, most Vietnam veterans confirmed that frequent changes in command were detrimental to morale and discipline. Because the army believed that the war would be short, it apparently tried to give command and staff experience to as many officers as possible (Krepinevich, 1986: 206–7). Whatever the reason, this policy seriously affected the quality of military performance.[12]

As the rationale and purpose of U.S. intervention in Vietnam came to be domestically contested, the draft perceived as unfair, and military service in Vietnam seen as promising futile death or injury, the troops became demoralized. The effects of demoralization on the army were not anticipated, but they became more evident with the passage of time: extensive drug and alcohol abuse, a high rate of desertion (67,000 deserted in 1967 alone), and violent eruptions of racial antagonism between blacks and whites (Levy, 1991: 61). All this had a very negative influence on the quality of military performance. Taken together, the factors that undermined military effectiveness counteracted the effects of force multipliers, but the military command continued to highlight the factors that enhanced force effectiveness and practically to ignore factors that reduced it.

Although the American commitment to Vietnam had been on Washington's agenda as a political-strategic issue since the very beginning of the Kennedy administration, and although the possibility of deploying American combat troops had been raised as an option even back then, surprisingly little thought was given to the doctrine and strategy that should guide deployment of combat troops.[13] Counterinsurgency was given much publicity, but the military actually saw it as a sideshow.[14] It was reluctant to allocate too many resources to prepare for small wars and preferred to concentrate instead on what it regarded as the main task, winning a major war in Europe. Most officers were not interested in having their careers diverted to a sidetrack, and because of the standard focus of military training on conventional wars, they had little training in counterinsurgency wars. The instinctive reaction to the acute crisis in Vietnam was, therefore, to project the strategic concepts and capabilities developed for fighting a conventional war in Europe to the Southeast Asian environment. The inadequacy of the strategic conceptualization soon revealed itself in the air war over North Vietnam and then in the ground war.

The strategic concepts and expectations guiding the bombing of North Vietnam were formulated by Johnson's civilian advisers, who drew on their acquaintance with the theory of limited war, which had evolved in the mid-1960s. Their recommendations, which represented the "gradual escalation" school of thought and which were opposed by the JCS, were for a carefully tailored bombing program against North Vietnamese targets as opposed to the tit-for-tat bombing strategy that had prevailed prior to the implementation of Rolling Thunder. The bombing was supposed to achieve several mutually reinforcing goals, mostly affecting the capabilities of North Vietnam and the policy preferences of its decisionmakers. The first goal was to signal U.S. determination and ability to inflict heavy damages on North Vietnam, to demoralize its population, and thereby to convince its leadership that supporting the war in the South was both useless and very costly. A demonstration of American determination was also supposed to reinforce the morale and will to fight in the South, and to provide the United States with increased leverage to influence the performance of the Saigon government toward higher effectiveness. The second goal was to reduce the logistic capability of the North Vietnamese to continue supplying and infiltrating men and matériel to the South Vietnamese communists, mostly by cutting off the supply and infiltration routes. A third goal was to bring the North Vietnamese to the negotiation table on U.S. terms (Kattenburg, 1980: 122–25).

The policy designers assessed the bombing policy to include three risks: significant U.S. air losses, which would adversely affect American feelings about and attitudes toward the administration's policy; an upsurge of Vietcong terrorism; and a military confrontation on the ground with North Vietnam and China. These risks were considered acceptable when compared with the risk of a South Vietnamese collapse. The chances of success were initially estimated to range between 25 and 75 percent. But it was believed that the policy would be worth carrying out even if failure was the result, because it would prove U.S. credibility and help deter future insurgencies (*Pentagon Papers*, 1971, 3: 687–91). "No one—not the civilians in the Defense Department or the State Department, not the president, and certainly not the generals—believed North Vietnam could endure the bombing for more than six months" (Tilford, 1993: 69; also Thompson, 1980: 29). No one foresaw that the operation would last for years, until November 1968.

The emphasis on the bombing objectives shifted over time, mostly in response to pressure in the United States for results.[15] During the spring and summer of 1965 the main objective was forceful persuasion, that is, convincing the North Vietnamese that their nascent industrial base was at risk. They were in essence requested to make a choice between retaining their industry and continuing their support for the Vietcong. During the summer of 1965 through the winter of 1966–67, when this approach failed, the emphasis shifted to air interdiction, intended to disrupt Hanoi's ability to infiltrate men and supplies into the South. But as early as the summer of 1965, McNamara realized that the bombing had no measurable effect on North Vietnam's support capability, a view shared by the chairman of the JCS and General John McConnell, air force chief of staff, although the generals interpreted the reasons for failure differently from McNamara.

The only positive effect of the bombing campaign seemed to be improvement of morale in South Vietnam. Yet in spite of the limited efficacy of the bombings the policy response was an incremental expansion of the geographic scope and intensity of the bombing campaign and the nature and number of targets attacked.[16] Between June 29 and the end of July 1966, North Vietnam's petroleum storage facilities were intensely attacked and about 76 percent destroyed (Thompson, 1980: 44–46). The JCS argued that destroying these targets, in light of North Vietnam's dependence on the import of petroleum, could affect the North's support of the insurgency in South Vietnam. Yet these expectations were not fulfilled. The North Vietnamese dispersed their oil storage facilities and relied increasingly on imports of petroleum products from the Soviet Union. After the failure to reduce the North's support for the insurgency and faced with pressure from hawks in Congress, notably the powerful Senate Preparedness Subcommittee, chaired by Senator John Stennis, Johnson removed many of the remaining constraints on bombing and authorized bombing industrial and transportation targets in and around Hanoi, Haiphong, and along the Chinese border in order to take advantage of civilian vulnerabilities (e.g., by bombing the single power plant in Hanoi) to coerce the North Vietnamese government to negotiate (Clodfelter, 1989; Pape, 1990).

All these objectives were unachievable given the nature of North Vietnam's regime and society. Threatening the destruction of the industrial sector was embedded in the strategic bombing doctrine of the air force, which assumed the existence of a highly developed industrial base on which the target country was dependent for military matériel. Its destruction, then, should disrupt the war-making capabilities of the enemy. But an underdeveloped country like North Vietnam was not very susceptible to such coercion. Its industry produced only 12 percent of the gross national product. There were few industrial targets worth attacking, and these were not essential to the economy or defense of North Vietnam or the prosecution of the war in the South. Thus, the strategic bombing doctrine was not very effective because it targeted an asset of limited value (Clodfelter, 1989: 125–26; Pape, 1990; Thompson, 1980: 72–73). In contrast, the destruction of the irrigation dikes in the North, which would have inflicted havoc on the agricultural economy, was never systematically attempted. Nor did Rolling Thunder ever pose a serious risk to population centers because of the North Vietnamese evacuation program and because of the U.S. choice not to target the population itself for normative reasons.[17]

Attempts to influence the North Vietnamese government by imposing punishment and suffering on the civilian population were not likely to work anyway, because civilians had already for years been tolerating heavy sacrifices that resulted from being a mobilized society at war. As for depriving them of necessary consumer products by destroying the country's industrial base, North Vietnam was compensated by economic assistance from the Soviet Union and China.[18] In fact, the bombing campaign served to rally the population around the regime. Interdiction, too, was likely to fail because of the small units used by the Vietcong. By 1967 over 96 percent of all engagements took place at company strength or less, and the communists initiated 88 percent of all engagements. The more than 200,000 communist troops in the field required only 380 tons of supplies a day, only 15–34 tons of which came from North Vietnam. In any case, North Vietnam's industry was not a critical source of war matériel. Most supplies came from China and the Soviet Union; North Vietnam was only the funnel.

Bombing was not likely to be effective against a transportation system whose railways and roads were not used to full capacity and therefore had an inbuilt redundancy that reduced the effects of short-term disruptions. In any case, the North Vietnamese effectively repaired rail lines, roads, and bridges by mobilizing approximately 500,000 laborers for these tasks (Clodfelter, 1989: 131–32; Pape, 1990: 127–28). In fact, the transportation system was even expanded during the war. More important, the established routes were supplemented by waterways and trails; on the latter, loads were carried by humans. All the bombing could do, according to one CIA assessment, was to "put a lower 'cap' on the force levels which North Vietnam can support in the South—but the 'cap' is well above present logistic supply levels" (*Pentagon Papers*, 1971, 4: 137); the Tet offensive in 1968 proved the truth of this statement. Members of the Johnson administration doubted the efficiency of strategic bombing as early as 1964, even before it was initiated (*Pentagon Papers*, 1971, 3: 211–15). The risks of failure were, therefore, known when the president authorized Rolling Thunder. But the anticipated risks of failure did not detract from the support for bombing in 1965.

Moreover, the administration had to balance the goal of using strategic bombing to preserve a free and noncommunist South Vietnam against other objectives, such as avoiding Soviet or Chinese intervention, preserving the Great Society program, securing a favorable American image abroad, and maintaining the support of Western allies. All these imposed restrictions on the air campaign. There were limitations on weaponry, including a ban on nuclear weapons and only restricted employment of B-52 bombers. There were restrictions on attacks within a 30-mile radius of Hanoi, a 10-mile radius of Haiphong, and within 30 miles of the Chinese border. There were constraints on the rate of sorties. President Johnson based the tempo of attacks on North Vietnam on the schedule of attempts to arrange negotiations. Preoccupation with other foreign policy crises reduced attention to Rolling Thunder and interfered with target selection, which required direct presidential approval. A similar effect was caused by Johnson's frequent absences from Washington. Finally, there were restraints on the choice of targets, imposed by the desire to minimize the loss of civilian lives.

Although continual intensive strategic bombing of 94 fixed targets for 28 days, suggested by the JCS in 1965, may not have brought North Vietnam to its knees and intimidated it into ceasing to supply assistance to the NLF, the alternative of slow escalation had some disadvantages. "On the psychological level, the people of North Vietnam were given a chance to become gradually accustomed to the air attack; on the practical plane, Hanoi had time to institute effective countermeasures" (Littauer and Uphoff, 1972: 35). In fact, gradualism violated the principles of mass and surprise that were traditionally associated with the effective use of air power. Furthermore, "air technology applied to a guerrilla conflict leads to an interaction in which the tasks that pilots are formally asked to perform are frequently beyond the limits of their actual capabilities, both technical and emotional" (Littauer and Uphoff, 1972: 30).

Military and civilian decisionmakers and advisers had reverse roles in the air and ground wars. Rolling Thunder was designed by civilians in the administration and carried out by a military that rejected gradualism and consistently pressured for an increase in the intensity and scope of the bombings. The strategy of the ground war, on the other hand, was designed and controlled by the military. The civilians in the administration, including the secretary of defense, limited their role to making sure that the number of troops requested was not excessive and could be supported without a reserve call-up (Schandler, 1977: 41). It almost seems as if a tacit understanding emerged between civilian and military advisers. The military agreed to accept a limited air war of strategic bombing even if they occasionally protested. In return, they were allowed to run the ground war with little interference from the civilians, as long as the military accepted the constraint of no mobilization of reserves and offered some sensible strategy for winning the war.[19] Noninterference by civilians in the conduct of the war prevailed even when victory remained elusive and out of reach.

In fact, the administration failed to define a clear military mission or to establish a clear limit to the resources to be allocated, so the military defined its own mission and imposed a micromanagement approach with cumulative effects but only a vaguely defined purpose.[20] "The JCS seemed to be unable to articulate an effective military strategy that they could persuade the commander-in-chief and secretary of defense to adopt. In the end the theater commander—in effect, General Westmoreland—made successive requests for larger and larger force levels without benefit of an overall concept and plan" (B. Palmer, 1984: 46). The number of troops increased in response to Westmoreland's requests, which had no well-defined set of strategic goals behind them: "The figures came from nowhere because the administration abdicated its responsibility to set priorities" (Rosen, 1982: 97).[21]

This state of confusion was to a large extent the result of tunnel vision on the part of the military. "Simply stated, the United States Army was neither trained nor organized to fight effectively in an insurgency conflict environment" (Krepinevich, 1986: 4). The army's concept of war, which called for reliance on a high volume of firepower to minimize casualties, was based on the successful experiences in the two world wars and the Korean War and on the anticipation of a similar war in Europe against the Soviet Union, for which it was considered the army's main task to be prepared. In the absence of guidance from Washington and under pressure to design a strategic plan, Westmoreland followed the path of least resistance. He designed a strategy that followed from the established doctrine and force structure—a strategy of attrition that relied on American abundant resources and technology and required building up a force designed for protracted combat to grind away at the enemy on a sustained basis. The U.S. Army was to seek and constantly engage the enemy's large units, taking advantage of its superior firepower and mobility (Krepinevich, 1993; *Pentagon Papers*, 1971, 4: 299–302). This was considered a quicker and more efficient way to ensure victory in South Vietnam than counterinsurgency.

Westmoreland himself reluctantly recognized the risk that the large enemy units would avoid battle and draw American forces into remote areas, thus facilitating control of the population-dense areas of Vietnam by the Vietcong. In fact, as it turned out, most engagements with the enemy occurred at company strength or less. Moreover, the Vietnam War environment required emphasis on light infantry formations, not heavy divisions; on restrained use of firepower in order not to alienate the local population; and on the solution of social, economic, and political problems rather than mere destruction of the enemy's military force. The extensive use of tanks, armored personnel carriers, fighters for tactical ground support, and helicopters was congruent with the army's concept of war but was mostly ineffective against guerrilla units; it was maintenance intensive and required an enormous logistic tail and many support troops, as well as highly indiscriminate use of firepower (Hatcher, 1990: 105–6; Krepinevich, 1986: 166–70, 192; D. R. Palmer, 1984: 179–85). In spite of these strategic deficiencies, the army stayed preoccupied with the attrition strategy. Efforts at pacification, intelligence gathering, and destruction of the insurgent infrastructure remained peripheral to the main war effort.

The preparedness, training, and hardware of the army and air force tailored for a major war in Europe biased the army's choice of doctrine, strategy, and tactics in Vietnam. The capabilities for strategic bombing, large-unit war, and high mobility provided by the helicopter induced the application of the army's concept of war, the attrition strategy, and search-and-destroy tactics. The marines, on the other hand, lacking much of the army's equipment, most notably helicopters, stressed small-unit action and generally showed a greater degree of creativity and flexibility in adjusting to the conditions of Vietnam (Gallucci, 1975: 118–20; Gibbons, 1995: 197–201).

Although there were plenty of opportunities to learn, there was little incentive to do so. The main criterion for success, body count, continually validated the strategy of attrition. An alternative proven strategy was not readily available. Westmoreland and his staff therefore felt comfortable with a concept of war they knew and understood. When McNamara, disenchanted with the army's performance, decided to ask for a critical evaluation by the civilians in the systems analysis unit of the Office of the Secretary of Defense, their conclusions were received by the JCS with open suspicion and hostility.[22] Besides, the military commanders were not about to admit that they were wrong and give up the military prerogative of conducting combat operations as they saw fit.

The effects of these factors were enhanced by the reluctance of McNamara to interfere with the tradition of field command autonomy. When, in late 1967, he was finally ready to challenge the JCS, his ability to do so was severely limited. The president had already began to distance himself from and undercut the influence of his progressively more dovish defense secretary. From the beginning, the president himself was reluctant to challenge the military, fearing the political consequences in Congress (Gallucci, 1975: 125–31; U.S. Department of State, 1992: 916).

The problems with and fallacies in the military's views on committing combat troops to Vietnam and the failings of the strategy of attrition were not revealed only in hindsight. Numerous reports and assessments dating back to April 1965 warned of the rocky road ahead and the risks involved. For example, a CIA special memorandum on Vietnam of April 1965 warned that the Saigon government would let the United States assume most of the burden of fighting. Similar views were expressed and conveyed to Johnson by Ambassador Taylor. These warnings received no attention. In July 1965, Thomas Hughes, director of intelligence and research at the Department of State, prepared a detailed analysis for Rusk which rejected the army's assessment that the war was about to move from guerrilla to large-unit warfare and that the enemy's vulnerability to U.S. firepower would increase.[23] In other words, according to Hughes, the primary assumption underlying the army's concept of war was incorrect. But his argument was to no avail. In March 1966 a study from the army's staff commissioned by the army's chief of staff criticized the attrition strategy and recommended giving top priority to pacification, but it was suppressed. Similar assessments in late 1966 by McNamara's whiz kids at the Office of Systems Analysis and by McNamara himself had only very marginal effects in encouraging the JCS to rethink its position on the appropriateness of the war strategy, force structure, and number of troops required (Krepinevich, 1986: 150–51, 160–61, 180–85; VanDeMark, 1991: 91–92).

Westmoreland and the JCS persisted in arguing that the attrition strategy was bearing fruit, that the United States was about to reach (or had already reached) the crossover point, and that with more men and more matériel they could turn the situation around in South Vietnam.[24] Similarly, the Air Force generals in charge of Rolling Thunder had not lost faith in the doctrine of strategic bombing, even though "the Air Force was not equipped either with the aircraft or doctrine suitable for fighting a limited and unconventional war against a pre-industrial and agrarian foe" (Tilford, 1993: 103). They refused to recognize that strategic bombing was not affecting the ferocity of the war in the South or crushing Hanoi's morale and determination.

There was a clear asymmetry between the U.S. doctrine of limited war—a war confined to South Vietnam—and North Vietnam's commitment to all-out war. North Vietnam could set the pace of the war and take advantage of sanctuaries in Laos and Cambodia (B. Palmer, 1984: 193). The United States found itself in a competition in attrition with the Vietcong and North Vietnamese; it was increasingly on the defensive—exactly the opposite of the initial intentions of the JCS and the commander in the field.[25] Although the North Vietnamese perceived the limited-war approach of the Americans to be an indication of weakness and did not believe expansion very likely, they prepared for both an American invasion and an air assault on their major cities. The American limited-war approach confirmed the view of Le Duan, party secretary of the Democratic Republic of Vietnam, that the Johnson administration lacked the capability and will to introduce enough troops into South Vietnam to thwart a Northern victory. Thus, what was supposed to be a show of resolve was not understood as such by the enemy and, indeed, encouraged the North Vietnamese government to persevere (Thies, 1980: 280–82).

The problems related to the strategic inadequacy of both the air and ground wars were enhanced by a command and control structure that was not flexible or responsive to the requirements posed by the complexity of managing very large forces. This limitation was due to a lack of vertical and horizontal integration, which increased the inefficient use of the resources available. Air activities over South Vietnam were controlled by the Military Assistance Command Vietnam in Saigon. But Rolling Thunder missions, a much more complicated matter, were coordinated from Commander in Chief, Pacific (CINCPAC), headquarters in Honolulu. Ground commanders and the Seventh Air Force nominated targets and handed their recommendations to MACV; MACV approved them and sent a list to CINCPAC; CINCPAC sent the list to JCS; JCS sent it to the secretary of defense; and the secretary of defense coordinated with the secretary of state before sending it to the president.[26] Delay was inevitable, especially because the president insisted on approving targets on a weekly or at most biweekly basis.[27]

The existence of one command for air operations in South Vietnam and another command for Rolling Thunder did not contribute to the effective management of the war effort. Exacerbating the inefficiency was the lack of a central command even for Rolling Thunder. Interservice rivalry between the Air Force, Navy, and Strategic Air Command precluded it, and the related operations were controlled independently by the participating services. This rivalry, reflected in an intensive contest over which service had the better war machine, caused not only distortion in mission assessments but the flying of unnecessary missions. Lives and aircraft were lost that should not have been, but the respective services undertook the missions to establish a record of action (Mrozek, 1989: 38–41; Thompson, 1980: 76–81).

The command and control structure lacked not only vertical unity of command between and within the U.S. armed services but also horizontal unity of command between the U.S. and South Vietnamese armies—unlike the case in the Korean War, when the South Korean army fought under the American commander-in-chief. Although Westmoreland initially championed the idea, he changed his mind and seemed not to want to bother with the South Vietnamese army, which was consequently allocated responsibility for what became known diminutively as "the other war," that is, pacification.[28] Having two separate command structures violated the basic maxim of effective counterinsurgency operations and allowed the enemy to exploit "seams" between administrative areas. Westmoreland believed that the Americans could do the main job on their own; he considered pacification of secondary importance, although a growing number of assessments indicated its centrality to the outcome of the war and the future of South Vietnam. A consequence of his belief was that all tasks had to be performed by Americans, which increased enormously the administrative and logistic tail behind American combat troops. Excluding the South Vietnamese army from the main war effort also affected how the legitimacy of the war was assessed in the United States and abroad; the war was now an American war rather than an indigenous war with American assistance. It also produced the "takeover effect"; the South Vietnamese stepped aside and let the Americans do the fighting. In the final account, the cumbersome and fractionated allied command structure led to wasteful duplication of effort, sluggish response to enemy initiatives, and lack of overall supervision of the war effort (Herring, 1990; Krepinevich, 1986: 194–96; 1993; B. Palmer, 1984: 179; Wirtz, 1991: 95–98).

In considering the practicality of such a comprehensive war effort, a prominent factor was logistics. In 1965 a decision to Americanize the war was made easier because important parts of the logistic infrastructure were readily available. The network of air and naval bases in the Pacific stretched from the U.S. Pacific Coast to the Indian Ocean, and the technology required for intensive strategic bombing was in place. Unlike in Korea, the United States was not unprepared for the magnitude of the task in Vietnam. Unlike in other intervention cases, in Vietnam speed was not essential. Decision-makers did not anticipate victory within days or even weeks. They anticipated an intervention that would take at least six months to work and probably two years or even longer. Given no urgent need for speed and taking into account the immense resource base of the United States and the preparedness of the American leaders to commit as many resources as necessary, logistic bottlenecks were not likely and were not perceived as a threat to success.

If the magnitude of the logistic task was unanticipated, immense material and technological resources, combined with ingenuity in solving complex management and engineering problems and an awareness of the successful historical record made the task more of a challenge than a threat to the American military. Still, the costs quickly mounted owing to Westmoreland's choice of strategy. The geography of Vietnam and the decision to defend all of the South and confront all enemy units no matter how remote posed substantial logistic problems. South Vietnam is elongated and narrow, with more than 1,400 miles of coastline and about 900 miles of land border. The enemy could infiltrate the country in many places and cut the main roads and single railroad in numerous spots. Consequently, to support large operations in every region, as well as the many small bases of U.S. special forces and South Vietnamese civilian irregulars, it was necessary to establish a half-dozen major ports or base areas plus several smaller ones. This required a great number of logistic troops, as well as security forces to defend the base areas. About half the U.S. military personnel in Vietnam were logistic troops who had the job of supporting almost 1,000,000 personnel. The open-ended commitment of the president and secretary of defense to send anything required to win the war was reflected on the logistic side in the shipment of supplies: large quantities of every conceivable item were shipped automatically or on demand without anyone's necessarily having a firm grasp of what was actually needed for the war effort (B. Palmer, 1984: 43–44, 68–69, 70–71). Unmonitored shipping wasted resources and increased unnecessarily the economic costs of war. Indirectly, the logistic costs contributed to the economic risks of the war in the United States and South Vietnam.

The decision to Americanize the war in July 1965 did not cause undue alarm in Washington about direct economic risks. It was assumed that the United States could simultaneously wage a war in Vietnam and a war on poverty.[29] In July, Gardner Ackley, the chair of the Council of Economic Advisers, reported that the United States could bear the burdens of both. He assumed that the cost of war would not exceed $10 billion (although within a year the cost approached $20 billion). In fact, the Council of Economic Advisers believed that the overall effects were likely to be favorable to the American economy (Gibbons, 1989: 439). In 1966 the president himself expressed similar confidence: "We are a rich nation and can afford to make progress at home while meeting obligations abroad—in fact, we can afford no other course if we are to remain strong" (Schandler, 1977: 225). Thus, economic risks were not in themselves a cause of major concern. Intuitively, perhaps Johnson felt that wars tend to affect social welfare programs and thus could undermine his Great Society program. He said, "Oh, I could see it coming all right. History provides too many cases where the sound of the bugle put an immediate end to the hopes and dreams of the best reformers" (Kearns, 1976: 251). Still, he supported the deployment of combat troops in Vietnam, and McNamara made it clear that economic costs were no object: "Under no circumstances is lack of money to stand in the way of aid to that nation" (Krepinevich, 1986: 139).

Abundant resources made possible the American way of war—technowar. "Much more sophisticated equipment was performing the foot soldiers' role, but not effectively enough to warrant the immense differences in costs" (Kolko, 1985: 190). Rolling Thunder proves the point. By December 1967 the tonnage of bombs dropped on North Vietnam (864,000 tons) surpassed the quantities dropped during both the Korean War (635,000 tons) and in the Pacific theater during the Second World War (503,000 tons; see Lewy, 1978: 385). The costs of bombing in North Vietnam had reached an exceedingly high level. In the heavily defended regions of the Red River Delta the United States was losing one plane every 40 sorties, meaning that each sortie effectively cost $75,000 for replacement of lost aircraft, over and above the normal operating costs of about $8,500. This cost "appears to have exceeded the implicit bounds of acceptability in cost terms" (Littauer and Uphoff, 1972: 41). The direct operational costs reached $460 million in 1965 and $1,247 million in 1966. In comparative terms, it cost the United States $6.60 to inflict $1.00 damage on North Vietnam in 1965 and $9.60 to inflict the same amount of damage in 1966 (*Pentagon Papers*, 1971, 3: 136).

This kind of rich man's war soon had unanticipated policy and political consequences in both the United States and South Vietnam. By late 1966 the introduction of a large number of American troops into Vietnam and the resulting local expenditures had caused major inflation. The cost-of-living index for the working class in South Vietnam had risen by 92 percent. Continued infusion of troops and money could further destabilize the fragile economy and have seriously adverse sociopolitical consequences (Schandler, 1977: 41). A destabilized South Vietnamese economy looked particularly dangerous, for a quick military victory was not in sight.

In the United States the consequences were substantial in their impact on American decisionmakers. The costs of direct American intervention climbed from $0.1 billion in 1965 to $5.8 billion in 1966 and to $20.1 billion in 1967. The opportunity costs to society due to the draft in 1966–72 reached $65 billion, according to one estimate (Riddell, 1975: 98, 170). After 1966 the Pentagon returned to Congress annually for special supplemental appropriations ($11 billion in 1966 and $12 billion in 1967), which became a political liability to the administration. The budget deficit rose from $3.8 billion in

1966 to $8.7 billion in 1967 and to $25.2 billion in 1968. Inflation was inevitable. In 1966 consumer prices rose by 3.4 percent, in 1967 by 3.0 percent, and in 1968 by 4.7 percent. But the administration failed to convince Congress to raise personal and corporate taxes. In addition, between 1964 and 1969 the U.S. balance-of-payment deficit doubled, shaking confidence in the stability of the dollar (Kolko, 1985: 288–89; Riddell, 1975: 298; Schandler, 1977: 226–27). Yet throughout 1964–67 "Johnson had consistently ignored or de-emphasized the economic ramifications of the military pursuit of the war. . . . The domestic economic effects of those escalations were not a policy input in Johnson's decision-making process" (Riddell, 1975: 404). In late 1967 and early 1968, with Congress critical and a presidential election upcoming, these factors could no longer be avoided. When General Westmoreland requested 206,000 more troops in 1967, the request was sent to the Council of Economic Advisers for an evaluation of economic implications; they advised against it (Halberstam, 1972: 610; Gibbons, 1989: 307).

The president viewed many military and economic aspects of the bombing and of combat troop deployment from a domestic political perspective. After all, he had spent all of his adult life in politics and was constantly alert to the implications of domestic constraints and incentives for political behavior. The end of 1964 found Johnson president in his own right after a landslide victory in which he won 61 percent of the votes cast and most of the states. But he was a president whose strength and interests lay with domestic issues and social reforms, and he found himself almost immediately having to address a foreign policy crisis, an issue-area where he had little experience and even less time to learn. But at that point he had room to choose among alternative options concerning Vietnam.

In late 1964 the public, the press, and Congress were unsure what should be done about Vietnam. Although an American responsibility and role were recognized, there was also considerable interest in a negotiated solution. The domestic environment thus allowed the president to make a choice, especially in light of his dramatic election victory. At this turning point he could probably have made the case for any option he preferred (Burke et al., 1989: 147–49). From a formal constitutional perspective as well, there were few constraints on his freedom of decision, especially after the Gulf of Tonkin Resolution was passed in Congress in August 1964.

But this picture of presidential omnipotence is partly misleading. In reality Johnson had to consider and retain the support of Congress, the public, his own cabinet and advisers, and the rest of the government bureaucracy, including the military leadership, especially in light of his inexperience in defense and foreign policy. The need for caution was implied, as was the need to avoid antagonizing any of these constituencies and hence the need to sound them out and monitor their responses before and after any major decision. As I argue below, between 1965 and the end of 1967 Johnson found himself progressively losing, to one degree or another, the support of important parts of these constituencies and consequently facing a mounting challenge to the legitimacy of his Vietnam policy.

A master of the game of congressional politics with a keen understanding of the relation between the presidency and Congress, Johnson anticipated the requirement of congressional authorization as it became evident to him that the United States would have to expand its military involvement substantially in Vietnam after the presidential election. The Tonkin Gulf incident provided the opportunity that he needed to get Congress to approve such authorization.[30] Thereafter, the legal authority for expanding U.S. military intervention in the air and on the ground was embedded in the Gulf of Tonkin Resolution, according to which the United States was "prepared, as the President determines, to take all necessary steps, including the use of armed force, to assist any member or protocol state of the Southeast Asia Collective Defense Treaty requesting assistance in defense of its freedom."[31] It is clear from the proceedings of the debates in the Senate, which lasted for eight hours, that the participants realized the far-reaching ramifications of this authorization. Senator William Fulbright, chair of the Senate Foreign Relations Committee, emphasized that the resolution authorized the president in advance to land as many ground troops in Southeast Asia as he deemed necessary and even to engage in a direct assault on North Vietnam (Lewy, 1978: 34). The resolution gave broad latitude of decision to the president with the bipartisan support of Congress, endowing it with a moral authority that went beyond explicit legal authorization and allowing him to commit U.S. resources to a Vietnam policy of his choice.[32]

The 89th Congress was the most decidedly Democratic Congress since the Roosevelt era. The House had 295 Democrats and 140 Republicans; the Senate had 68 Democrats and 32 Republicans. The test for Johnson's Vietnam policy came in mid-May 1965 over the authorization of a $700 million expenditure on the U.S. mission in Vietnam. The final vote in the Senate was 88 in favor, 3 against (all Democrats). In the House 418 were in favor and only 7 were against. These overwhelming votes were interpreted to mean that Congress was not likely to challenge the policy and that the policy would not threaten the Great Society program.

The 90th Congress, like its predecessor, was dominated by Democrats. In the House the Democrats held 248 seats, the Republicans 187. In the Senate the Democrats held 64 seats to the Republicans' 36. The evident increase in dovish sentiments was not necessarily reflected in votes. Even the dovish legislators who wanted a halt to strategic bombing and a greater emphasis on negotiations did not support withdrawal. The test came in 1967 in a vote on a supplemental appropriations bill for more than $12 billion. In the House 385 members voted in support of the bill and only 11 voted against it. The general sentiment was that a vote against the bill would only give aid and comfort to Hanoi and encourage it to hold out against efforts to bring peace, which would be tantamount to abandoning the American boys in Vietnam.

The views in the Senate were more ambiguous. The debate was dominated by three Democrats: William Fulbright, Mike Mansfield, and Richard Russell. The last was considered extremely important to the outcome. He stuck to his support for quick and decisive escalation to bring the war to an end. Fulbright and Mansfield were critical of the Vietnam policy but felt that they could not vote against the allocations, as advised by Senator Wayne Morse, that would provide the matériel necessary to support the troops. The bill, with an amendment by Mansfield that expressed support for the Geneva Accords of 1954 and 1962 and urged the convening of a Geneva peace conference, passed 77 to 3. Johnson's policy of keeping his door open for any member of Congress w. ɔ wanted to see him for an off-the-record meeting had paid off.[33] The debate itself indicated signs of uncertainty in Congress, but it also demonstrated that the administration had managed to maintain congressional support for its main objectives (Barrett, 1993: 48–51, 63–69).

Public opinion considerations ranked high among the constraints considered by the Johnson administration, which saw "the national interest as best served by avoiding a polarization of public opinion, either in favor of massively punishing North Vietnam or of withdrawing from the struggle. In particular the president wished to preclude political polarization that might divert support away from the Great Society legislation he would propose to the new Congress" (Simons, 1994: 136). This was one of the reasons the president did not acknowledge publicly that U.S. policy in Southeast Asia had changed dramatically with Rolling Thunder. Similarly, the president insisted that the decision to send Marine battalions to secure air bases be kept low key "to minimize any appearance of sudden change in policy" (Rusk, 1990: 450).

A memorandum by McGeorge Bundy on June 30, 1965, assured the president that "the public as a whole seems to realize that the role [of the United States in Vietnam] must be played. Furthermore, open skepticism as to our tactics subsides at times of sharp crisis in the situation" (Gibbons, 1989: 350). The same memorandum also affirmed that the Vietnam policy had support in the press as well as Congress, although there was some criticism in Congress. When Johnson asked Defense Department officials and the JCS in a meeting on July 22 whether the public would go along with having 600,000 men fighting 10,000 miles away at a cost of billions of dollars, he was reassured of public support.[34] The response by the public and Congress to Johnson's announcement on July 28 that more ground troops would be deployed in Vietnam, as well as to the escalatory measures since February, probably reinforced Johnson in his belief that he could handle both public opinion and Congress.[35] The risk posed by domestic disapproval therefore seemed acceptable and manageable until 1967.[36]

The incremental commitment of combat troops was partly a consequence of fears that fast escalation would play into the hands of the right by causing a war frenzy that could lead to pressures to go all out. Chester Cooper, a senior staff member of the National Security Council, observed:

"We were scared not because of the Vietcong, they did not scare us. The American people scared me and the Administration" (Lorell et al., 1985: 45). At the other end of the political map, the antiwar movement was gaining the attention of the White House as it gained momentum and as its incessant hounding of Johnson and his colleagues became personally irritating and emotionally debilitating. But the administration believed in 1966 that the main threat of the movement was not at home, where it could be politically contained, but in Hanoi. The administration feared that Hanoi would get the wrong idea about American resolve to see the war through (Small, 1987; 1988: 62, 86–89). Marches, sit-ins, ad campaigns in the newspapers, and desertion by a growing number of important newspaper and magazines from the supporters' camp—all were picking up in 1966 and were widespread by 1967.[37] The president and his advisers, realizing that they might lose the war at home, turned to overselling the war's progress. General Westmoreland played an important role in building up expectations for imminent success, which were soon thereafter dashed by the Tet offensive (Gibbons, 1995: 843–52; Small, 1988: 93–96, 123–24).

During the 1965 policy debates, concerns that increased casualties would diminish public support or entrap the administration into escalation beyond prudent levels were raised but were not very significant to the decisionmakers. In fact, President Johnson and his key advisers were encouraged by the strong public support for their policy in the early months of the war, interpreting the rally-'round-the-flag fervor as a lasting commitment. They assumed that losing Vietnam might be much more damaging in terms of public support than losing lives to prevent imminent defeat. They trusted that the American public would bear the burden of heavy casualties in order to protect the vital interests at stake (Lorell et al., 1985; Schwarz, 1994). But by 1966 the accumulated casualties were a reality and not an unpredictable speculation, and protraction of the war became a certainty. At that point the risk of further casualties became an important consideration. As the number rose, public support for the war and the overall approval rating of the president declined (Gibbons, 1995: 174–76; Lorell et al., 1985; Lunch and Sperlich, 1979; Mueller, 1973). But in 1967 the public, like Congress, still upheld U.S. goals in Southeast Asia and rejected unilateral withdrawal, even if the *means* used to achieve these goals now came under question. The approval rating for the president's handling of Vietnam policy declined, and an increasing number of those in public opinion surveys stated that sending troops to Vietnam was a mistake.[38] Johnson was well aware of these trends and, although he was not thinking in terms of overall policy revisions, he refused Westmoreland's 1967 request for another major troop increase for Vietnam.

Johnson depended on his inner circle of defense and foreign policy advisers to give his administration the image of competence in an acute crisis. In the policy debates between civilians and military officers, he leaned toward compromises. But several of his civilian advisers felt growing discomfort with his policies and the emerging tilt toward the preferences of the military and decided to resign (e.g., McGeorge Bundy, George Ball, Jack Valenti, and Robert McNamara). Johnson's support for his generals was evidently reinforced by public opinion surveys, which indicated that the public believed, by a margin of 53 to 36 percent, that in wartime the military should be allowed to run the war, and, by a margin of 73 to 10 percent, that in Vietnam the military was handicapped by civilians who would not let them go all out. By the time McNamara resigned, 45 percent of those surveyed in a Harris poll gave him a negative rating for his job performance, compared to 42 percent who gave him a positive rating. General Westmoreland, on the other hand, received a 68 percent positive rating and a 16 percent negative rating on his performance as commanding general in Vietnam (Berman, 1989: 121).

Public and congressional support for the war was determined by the crucial issue of policy legitimacy and the difficulties of reconciling contradictory trends that the issue raised. The debate over the legitimacy of the Vietnam policy touched on both cognitive and normative dimensions. Those who advocated a deeper American intervention in Vietnam justified it first and foremost on national security grounds, emphasizing Vietnam's critical geostrategic position and the domino theory. A second line of justification presented North Vietnam as an immoral enemy subverting a legitimate peace-loving country in violation of the 1954 Geneva Accords, to which the United States had been party. In this context it was important for the administration to deny that the situation was the result of indigenous South Vietnamese insurgency. These two general arguments were brought together by the specter of expansionist demonic communism threatening everything that Americans valued. Thus the war in Vietnam, seen in the broader context of the Cold War, was a just and necessary one. At the same time, the notion of gradual escalation with pauses to allow the enemy to reconsider its policy was attractive to the American public because it seemed to indicate a rational, well-considered, and restrained policy intended to settle the conflict through negotiations. Similarly, consultations with President Eisenhower and the Wise Men, the group of prominent American public figures established in September 1964 in conjunction with Johnson's presidential election campaign, projected the image of a flexible president willing to listen to advice.

Those who opposed the war claimed that American intervention was illegal because the United States was violating its pledge in the Geneva Accords to refrain from the threat and use of force. It was also violating Articles 2 and 33 of the U.N. Charter, which require refraining from the use of force and seeking solutions to disputes through peaceful means. Opponents argued further that the president should have asked Congress for a declaration of war;[39] in its absence, the moral and legal justification of the war were in doubt.[40] They also perceived the regime in South Vietnam as corrupt and nondemocratic and internal instability as a spontaneous popular revolutionary response, with which the United States had no right to interfere. Violent riots, anti-American demonstrations, self-immolation, hunger strikes, and civil disobedience in South Vietnam supplied ammunition for these arguments.[41]

One of the most powerful challenges to the legitimacy of the Vietnam policy was directed at the means of achieving it, the consequences of which were brought into the living rooms of every American family in stark brutality through live media coverage. The reliance on heavy firepower to reduce casualties among U.S. troops was intended to preserve cognitive legitimacy, but it undermined normative legitimacy. Indiscriminate use of firepower caused numerous civilian casualties and massive physical destruction, making it a subject of condemnation both at home and abroad. Thus, the more the issue of legitimacy became central to the debate about the war, the more the army's concept of war became problematic. McNamara expressed great concern about civilian casualties, saying, "There may be a limit beyond which many Americans and much of the world will not permit the United States to go. The picture of the world's greatest superpower killing or seriously injuring 1,000 non-combatants a week, while trying to pound a tiny backward nation into submission on an issue whose merits are hotly debated, is not a pretty one" (*Pentagon Papers*, 1971, 4: 172). Yet those who believed in the communicative value of enemy civilian losses, in the implicit power dialogue with Hanoi, tended to weigh domestic and foreign criticism against the pressure they believed the losses put on the regime in Hanoi, and they decided the policy was worth the risk (Gibson, 1986: 369).

These problems were acutely expressed in Rolling Thunder. Although American bombing of North Vietnam was much more discriminating than Allied bombing of Germany and Japan in the Second World War, and although there was no bombing directed at the civilian population as such, the operation left the impression that civilians were being indiscriminately bombed. This impression was due largely to the massive propaganda campaign of the communists and their supporters in the West, which negatively affected the external legitimacy of the intervention and caused the United States difficulties in relations with its allies but caused the most damage to the normative legitimacy of the war in the United States. In fact, great efforts and sacrifices by the Air Force and Navy went unattended. Pilots regularly risked their lives and were shot down following rules of engagement intended to reduce, if not completely prevent, collateral damage (Lewy, 1978: 394–406). A good example is provided by the preparations that preceded the start of the bombing of North Vietnam's oil storage facilities in June 1966.

> *The list of precautions taken to reduce the non-combatant casualties is a catalog of the factors affecting air strike and anti-aircraft suppression accuracy.* The JCS included the following requirements in their order to begin the bombing: use of the most experienced pilots; detailed briefing of pilots; execution of strikes only when visual identification of targets permitted; selection of the best axis of attack in order to avoid populated areas; maximum use of electronic equipment to hamper SAM [missile] and AAA [anti-aircraft artillery] fire, thereby limiting possible distractions for pilots; use of weapons of high precision delivery; and limitation

of SAM and AAA suppression to sites outside populated areas. (Littauer and Uphoff, 1972: 39–40)

Although decisionmakers and the planners of Rolling Thunder were aware of the serious damage it could do to the U.S. image and the sense of the war's legitimacy, attempts to limit collateral destruction were largely unsuccessful. Most harmful were reports by American journalists like Harrison Salisbury, assistant managing editor of the *New York Times*, which left the impression that the United States was willfully bombing civilian targets of no military value. The effects of these reports were enhanced ironically by a misleading impression created by Department of Defense releases that bombing missions were always perfectly accurate. Television viewers were exposed daily to the impact of artillery fire, aerial bombing, and chemical agents, especially the horrifying effects of napalm, and the environmental degradation caused by the war, all these consequences were perceived as alien to American values of decency, fair play, and the sanctity of life. Opponents of the war also claimed that the war would eventually permeate and destroy the fabric of American society by exposing hundreds of thousands of American soldiers who served in Vietnam to the debilitating moral and psychological impacts of the war (Levy, 1991).

The president and his advisers were sensitive from the very beginning to the political risks of failing to establish the legitimacy of their policy. They were quite successful in convincing the American public that for practical and moral reasons the United States had to carry out its commitment to Vietnam and that this goal could be achieved at a high but acceptable cost and within a reasonable time frame. The means were more suspect but initially raised no undue alarm. But sometime in 1966 this changed. The growing number of American casualties, the inability of the decisionmakers to set a time for American withdrawal, and the continual exposure of the public to wartime sufferings of the Vietnamese, as well as the American soldiers, raised tough questions about the efficacy and morality of the means.

What added to the growing doubts expressed in the media and passed on to the public was the administration's policy style. After 1965 Johnson and his advisers deliberately obscured the magnitude of the escalation to avoid alarming the public and to allow social reform legislation in Congress to proceed without interruption. By 1966 there were alarming signs of a credibility gap between the media and public, on one side, and the administration, on the other. Doubts about the efficacy of the war's management contributed to the undermining of the cognitive legitimacy of the means applied to achieve the objectives, the costs involved, and predictions of imminent success. But a majority still supported the ends, although they preferred a negotiated solution to a coercive one. Criticism of the morality and practicality of the means, suspicion of the optimism emanating from Saigon and reflected by Washington, and worries about an endless war with mounting casualties eventually led to questions about the importance of the strategic and political objectives themselves, apart from the costs of attaining them. The whole structure of policy legitimacy became irretrievably undercut.

THE INTERNATIONAL MILIEU: THE CHALLENGE OF UNCERTAINTY

In 1964, when Johnson was elected president, the United States had reached a zenith in its power. Its policies of containment in Europe, Asia, and the Western Hemisphere were an overall success. Despite occasional setbacks, it had not backed away from superpower crises, engaging the Chinese in Korea and the Soviets in Berlin and Cuba and involving itself in a variety of worldwide efforts to retain pro-Western governments in power as well as prevent anti-Western governments from taking power. The show of resolve during the Cuban missile crisis in 1962 and its favorable resolution gave credence to the view that the United States was the most militarily powerful and most technologically advanced superpower. The resultant sense of potency permeated the perceptions of American decisionmakers. Worldwide commitments combined with a sense of confidence in U.S. conventional and nuclear military capabilities provided decisionmakers with the motivation to use U.S. military power to remove acute threats to U.S. interests. Vietnam was where incentives, motivations, and opportunity converged to allow the United States to demonstrate its power to shape events—by applying direct military force if necessary.

The Vietnam crisis escalated in the context of a loose bipolar international system characterized by mutual deterrence. In Asia, the United States, with its network of alliances and bases and with its ground, air, and naval forces had a decisive advantage in conventional capabilities vis-à-vis the Soviet Union, but it could not match China's human resources. Nuclear capabilities aside, both the United States and the Soviet Union had lost the absolute hegemony within their blocs that they had enjoyed in the 1950s, and their leadership was challenged by France and China, respectively. The overall confrontational posture brought with it the danger of escalation, for both superpowers, egged on by the intrabloc challengers to their hegemonic positions, could be drawn into a spiraling head-on competition. At the same time, the confrontational posture offered the superpowers an incentive for cooperation nested in the common interest of avoiding escalation but also related to opportunities for taking advantage of the intrabloc rift in the rival bloc.

The constraints on superpower competition made it easier and less risky for the United States to get involved in Vietnam, because relations between it and the Soviet Union had reached a level of maturity that was assumed likely to keep conflicts in check and from escalating into nuclear confrontation as long as U.S. policy avoided blatant provocation. The superpowers had learned from the Cuban missile crisis that coercive diplomacy might work and that close encounters need not get out of hand, especially if both sides understood where to draw the line and effectively communicated their stakes to the adversary. Cyrus Vance, a high-ranking official in the Pentagon in 1965, summarized the prevailing mood: "We had seen the gradual application of force applied in the Cuban Missile Crisis, and had seen a very successful result" (VanDeMark, 1991: 115). Introduction of the flexible response doctrine by the Kennedy administration paid off. The United States had the conventional military capability to meet a challenge to its interests without having to resort to nuclear weapons. A policy with the right mix of resolve, daring, and caution represented an acceptable risk.[42] Backing away from the challenge could therefore undermine the U.S. position in the superpower rivalry. The rift within the communist bloc between China and the Soviet Union presented an opportunity that could be exploited for U.S. advantage even if it also held the risk of exploitation by North Vietnam to its own advantage—as was the case. Competition between China and the Soviet Union turned U.S. intervention in Vietnam and the bombing of North Vietnam into an opportunity for the two communist rivals to vie for the reputation of being loyal, helpful, and resolute allies of North Vietnam and use this reputation to their advantage in intrabloc rivalry. The rivalry made it more difficult to reach a negotiated agreement in Vietnam because it reduced the Soviet incentive to play intermediary and help pressure North Vietnam to compromise.

Ideologically, what looked like an American attempt to suppress a movement of national liberation required an uncompromising Soviet stance that operated on three levels: global, intrabloc, and bilateral. Owing to its rigid ideological views of the U.S. escalation in Vietnam and its need to compete with China, the Soviet Union felt obliged to take a hard line on Indochina by significantly escalating the tone of warnings to the United States and by promising to extend assistance to North Vietnam. At the same time, the Soviet Union attempted to forge a more cohesive communist bloc under its leadership by emphasizing "socialist unity" in facing up to American attempts to expand its influence in Asia (Chen, 1992: 35–40). The Soviets did not want a direct confrontation with the United States, but Soviet policy nonetheless exacerbated U.S.-Soviet tensions, violating the spirit of détente that had emerged in the aftermath of the Cuban missile crisis.[43]

China saw the Americanization of the war in Vietnam as a shift of emphasis in U.S. policy from Europe to Asia, a change that it believed was encouraged by the Soviet Union to increase the threat to China. It feared collusion between the two superpowers and aggressively opposed any increase in U.S. military activity. It also resolutely rejected Soviet proposals for socialist unity. The eruption of the Cultural Revolution made Sino-Soviet cooperation practically impossible. The Chinese were unwilling, however, to risk an all-out war with the United States unless directly attacked (Jian, 1995). For its part, the United States attempted to reassure China that it had no intention of provoking war with China, even while the Rolling Thunder bombings were moving closer to the Chinese border.

In the looser communist bloc, made less cohesive by the Sino-Soviet split, North Vietnam found itself well positioned to preserve its freedom of policy formulation in spite of mounting dependence on the Soviet Union and China for crucial military and economic assistance. Thousands of North Vietnamese cadres received Soviet training (Gaiduk, 1995–96). About 3,000 Soviet military technicians were sent to train North Vietnamese to use and maintain Soviet weapons. Other North Vietnamese were trained in Soviet institutions. In addition to training and supplying weapon systems, the Soviet Union sent considerable amounts of food and other agricultural products to North Vietnam. Chinese aid consisted mostly of light infantry equipment, artillery, and ammunition, as well as large quantities of food. More than 100,000 Chinese engineering troops operated in North Vietnam to help repair railway lines, build roads and bridges, construct air fields and other defense works, and maintain and repair communication systems. China also sent antiaircraft artillery units with a total strength of more than 150,000 troops (Chen, 1992: 49–51; Jian, 1995). By taking advantage of both the interbloc and intrabloc conflicts, Hanoi was able to maneuver itself into a strong bargaining position with the United States, which increased the costs of the American intervention in Vietnam and reduced its effectiveness.

In spite of a favorable balance of power, American principals considered the possibility of a Soviet or Chinese counterintervention an acute risk, even though a number of Special National Intelligence Estimates (SNIE) presented in 1965 refuted the likelihood of its happening (U.S. Department of State, 1996a: 763–73; U.S. Department of State, 1996b: 228–30, 404–6). To Dean Rusk, the secretary of state, the risk was of particular concern. Even Rusk did not consider it serious enough to deter escalation of U.S. intervention but saw it as a reason for self-imposed restraints in the use of the superior U.S. military capabilities. Only George Ball considered the risk of military confrontation with China and the Soviet Union an important reason for a negotiated withdrawal.[44]

To prevent counterintervention, the United States relied on a combination of deterrence and reassurance. Deterrence was established by an extensive, powerful military presence in Asia. Reassurance took several forms. The gradual pace of the increase in U.S. ground forces and ground combat operations in Vietnam in 1965 was driven to a substantial extent by concern over the reaction of China and to a lesser extent, the reaction of the Soviet Union. It was believed important not to present either with a sudden policy change that might trigger a Chinese or Soviet military counterintervention and potentially a nuclear war (Lorell et al., 1985: 42–44; Rusk, 1990: 456). One reason the administration never fully and clearly specified its demands and conditions in the form of an ultimatum for halting Rolling Thunder was the fear of putting the Chinese and Soviets on the spot if the ultimatum was not met. The United States constantly emphasized its self-imposed restraints and limited objectives rather than demands for specific North Vietnamese behaviors (Simons, 1994). Both the air war and the ground war had inbuilt limits set by the fear that the conflict might escalate into confrontation with the Soviet Union and China and thence to a nuclear exchange driven by the logic of the situation rather than by choice.[45]

The fear was particularly vivid because of the shared memories among top decisionmakers of a recent political-military nightmare—the Cuban missile crisis—in which many of Johnson's cabinet members and advisers had participated. The Cuban crisis left a mixed imprint on the participants' minds. The outcome convinced many that a policy of applying gradual pressure could work. Yet the crisis that brought the two superpowers to the brink of nuclear war reminded the participants that such crises should be avoided if possible, before they deteriorated into a competition in resolve and reputation. The fear of escalation was probably enhanced by China's membership in the nuclear club, which it had joined in 1964.

The events that might trigger Soviet or Chinese intervention were perceived to be of two types. One was expansion of the war beyond certain geographic or substantive limits, which China or the Soviet Union would perceive as a crucial challenge to their tangible or reputational interests. The second was human error or accident, such as a bomb falling on a Russian ship in Haiphong Harbor (Tilford, 1993: 74). Mining Haiphong Harbor was considered too risky because the Soviets would be forced to choose between sweeping the mines themselves or abdicating to the Chinese the role of primary supplier to North Vietnam if just the supply route through China remained open. The Americans believed that the Sino-Soviet conflict would, in that case, commit the Soviets to deeper involvement; they might be compelled to send new forms of military assistance, or they might initiate a crisis in Berlin or elsewhere to offset U.S. initiatives in Vietnam. Likewise, bombing the northeast railroad, which brought Chinese aid to North Vietnam, and killing Chinese volunteers could expand the war by means of the intervention of the Chinese air force from bases in China and North Vietnam or even by China's sending hundreds of thousands of volunteers, as in Korea (Lewy, 1978: 383; Tilford, 1993: 91).

The risk most feared never materialized, even when President Richard Nixon authorized the heavy bombing of sensitive targets, whose destruction the Johnson administration believed likely to result in Soviet or Chinese intervention. But Soviet and Chinese paramilitary and noncombatant assistance to North Vietnam caused enough trouble. It contributed substantially to the relatively mild impact of the strategic bombing, more than making up for the damage. The Soviets and Chinese replaced destroyed trucks, railroad stock, and construction machinery. They prevented serious food shortages and provided petroleum products, arms, and ammunition. By the spring of 1966 there were already some 50,000 Chinese engaged in maintenance and air defense related to the railroads. Yet the risks of attacking the port of Haiphong and the rail line to China prevented interference with this flow of aid (Lewy, 1978: 392).

The Soviet and Chinese presence did not merely constrain particular military operations and increase the resilience of the Vietcong and North Vietnamese; the Soviets and the Chinese, especially the former, were also perceived as a key to a negotiated settlement. By using the carrot of possible negotiations, China and the Soviet Union affected the tempo and intensity of American military operations. But they could not or would not fulfill American expectations because they lacked leverage with Hanoi. The Sino-Soviet conflict prevented the Soviets from applying effective pressure on North Vietnam to reach a negotiated settlement.

U.S. allies in Asia and Europe posed another problem. To preempt domestic criticism it was desirable to have troops from the Asian and European allies of the United States participate in the Vietnam intervention (Gelb and Betts, 1979: 261). Pressure on these allies, as well as generous commitments of aid, convinced the Republic of Korea, Thailand, the Philippines, New Zealand, and Australia to send troops to Vietnam.[46] Japan did not send troops but consented to the use of air bases on its soil for bombing missions.[47] Ultimately, no European ally participated in the war on the U.S. side, and there was only limited Asian participation. Their absence undermined the external legitimacy of the war (B. Palmer, 1984: 191) and gave credence to the view that the war was a neocolonial one.

Even the diplomatic support of Western allies could not be taken for granted. The allies occasionally threatened to take diplomatic initiatives that would embarrass the Johnson administration at home and abroad, weaken the legitimacy of its policy, and undermine American policy objectives. Specifically, Johnson was concerned that he might be pressured into premature negotiations that would force the United States to make concessions that would amount to a Vietcong victory. In February 1965, for example, the president of France, Charles de Gaulle, called for a conference to end the war, and in March, France and the Soviet Union jointly called for a reconvening of the Geneva Conference of 1954 to achieve this objective. In April the nonaligned heads of state called for negotiations without preconditions. About the same time, the Canadian prime minister called for talks in Geneva. Shortly after the launching of Rolling Thunder the British prime minister, Harold Wilson, also advised the United States to seek a negotiated settlement while proceeding with the bombing.[48]

The attempts by allies and others, such as the U.N. secretary general, to pressure the Johnson administration into accepting formulas for negotiation that seemed to the president and his advisers to be unfavorable to American interests posed the domestic threat of mobilizing support for the formulas in Congress. This threat complicated the administration's tasks—an outcome the president sought to avoid.[49] Thus, although both domestic and international legitimization for the use of force implied and imposed an emphasis on U.S. preparedness to negotiate (U.S. Department of State, 1996b: 594–95), the actual pursuit of negotiations as well as the expressed acceptance of the terms for negotiation had to be cautiously phrased and executed in a manner that would not commit the United States to negotiations prematurely. In general, this state of affairs posed a serious communication dilemma: How was the United States to communicate both its resolve to stay the course and its wish to negotiate and, at the same time, convince allies like South Vietnam, Thailand, and Laos that negotiations would not result in their abandonment?[50] These conflicting demands made it difficult to communicate a consistent and credible policy. Faced with this dilemma and with disagreement over priorities among these goals within the administration, Johnson preferred vague compromises and fuzzy policy statements that in the end did not fully convince either enemies or partners.

All in all, although international factors were discussed extensively in connection with the decision to intervene and at each phase of escalation, they were never considered critical inputs. The possibility of Chinese or Soviet intervention, the refusal of the European allies to send troops, and the initial uncertainty about whether the Asian allies would send troops and how many—these were never considered important arguments against intervention. American decisionmakers considered it desirable to minimize the risk of counterintervention by third parties and to have the participation of troops and at least diplomatic support from European and Asian allies (U.S. Department of State, 1996b: 50), but these international factors were considered of secondary import and thus affected mostly the way decisions were presented and implemented, not the essence of the decision to intervene and escalate.

In summary, between 1964 and 1968 the main concern of U.S. decisionmakers was the possibility of a sudden collapse of the regime in Saigon and a takeover by the communists. This, it was believed, would initiate the collapse of noncommunist regimes in Southeast Asia and then in the rest of Asia and the Third World. But the chief effect would be a collapse of the American reputation and the U.S. position in the global arena. The collapse of the national reputation would cause the collapse of the personal reputation of President Johnson.

The second risk that was a source of deep concern was that getting involved in a war in Asia could severely affect the Great Society program.

The third area of apprehension had to do with public opinion. In 1965 the risk of losing support for intervention was considered, but it was perceived as limited and manageable. At the other extreme, a war frenzy whipped up by the right would do untold damage to the progress of social reforms and would cause escalation into a uncontrollable confrontation of the major powers.

The fourth risk concern was the escalation of the limited war in Vietnam due to Soviet and Chinese intervention. This risk perception affected the management of the war but was never considered a deterrent to the policy of military intervention per se.

What is evident is that, although there was no lack of assessments about the substantial probability of policy failure or the deficiencies of force structure, strategy, and C^{3}I, the risks of failure were dismissed by the president and most of his key advisers. That these risks were ignored is notable considering the obsessive concern of the Johnson administration with reputational interests. The failure of direct American intervention should have been considered likely to have even more devastating reputational consequences than a South Vietnamese collapse without American intervention. Even more puzzling, from spring 1965 through summer 1967 Johnson practically ignored accumulating feedback information that quite convincingly indicated that the early warnings about failure were materializing. Why was the risk of failure ignored to start with? Why was the risk of the collapse of the Saigon regime so salient? Why did Johnson and some of his advisers disregard or miss signals of an impending costly policy failure and persist in the same policy? In broad terms, two issues have to be addressed: the persistent orientation toward escalation, which eventually resulted in entrapment in a failing policy, and the specific policy forms that escalation took—strategic bombing and the ground war. These are the subjects of the following discussion.

LYNDON JOHNSON: A RELUCTANT RISK TAKER

Immediately after the 1964 presidential election, Vietnam policy became Johnson's most urgent and pressing foreign policy issue. Most decisionmakers agreed that South Vietnam was not just another country but a test case, that the collapse of the regime was imminent without immediate American action, and that the collapse would have implications not only for Southeast Asia but for most other regions as well. Although these points of agreement and seeming certainty marked the Vietnam policy as one of the most urgent policy areas to deal with, highly pertinent dimensions of the problem remained hazy and were often contested among the principal players in Washington and Saigon: How acute was the political crisis in Saigon, and could the Saigon government reform itself and thus blunt the appeal of the National Liberation Front? What were the force size and capabilities of the enemy? Did the Vietcong or North Vietnam pose the main threat, and how committed was the latter to the Vietcong cause? What type of war was the enemy conducting, guerrilla or conventional? What would an American winning strategy entail? How committed were China and the Soviet Union to North Vietnam, and to what degree did they have shared or conflicting interests and goals in Southeast Asia? If the United States stepped in, what would be considered victory? These questions were not fully resolved and clarified with the passage of time, but they were important in shaping the policy debate, and they underlay the struggle within the administration for control over the policy formulation process. Formulation of policy options, choice of preferred policies and implementation tactics, evaluations of risks associated with policy options, and risk acceptability—all these were contested.

The uncertainty surrounding these questions deepened the opaqueness of such issues as the duration of the intervention, the number of troops required, and the economic costs of an effective policy, making it difficult to determine the implications of intervention for the president's ability to carry out his domestic agenda. The ill-defined nature of the Vietnam policy problem was exacerbated by its being a wicked problem: there seemed to be no solution or policy that did not promise to generate other similarly or even more adverse outcomes. The choice was among different risks. Attempts to break the issue into subcomponents—for example, dealing separately with the main forces of the Vietcong and with pacification—did not work.

The attempt to neatly divide the problem into parts was the result of incorrectly identifying the motives of the enemy—incorrectly because the Vietnam War combined aspects of conventional and guerrilla war. Chameleon-like, it changed character several times between 1964 and 1968, making it difficult to formulate objectives, define victory, and develop a means-ends theory of victory (Cable, 1993: 109; Thomson, 1968: 50; Vickers, 1993). By 1965 the prevailing view among Johnson's key advisers was that the United States was fighting a guerrilla war, but one about to enter a conventional phase, and that the Vietcong were driven not by a genuine sense of domestic disaffection but were an instrument for achieving the territorial ambitions of an external power—North Vietnam. American decisionmakers refused to recognize that the war in the South was an insurgency. Not being able to define the enemy's strategy and objectives correctly led to miscalculations of its willingness to suffer; Washington did not realize that "Hanoi viewed its losses as not being disproportionate to the goals it sought" (Cable, 1993: 142). Breaking the will of the enemy became increasingly problematic. The difficulties of problem definition account in part for characteristics of the policy that emerged: creeping incrementalism, a deliberate blurring of policy transitions, and a failure to specify objectives and acceptable outcomes.

In the absence of a well-defined problem, the framing of the problem gained special importance in focusing attention on and highlighting the advantages of an escalatory policy. Because the preferred frame would provide a guide for choosing between escalation and withdrawal by suggesting which was the more costly, framing was critical to the debate between hawks and doves in 1965. Responding to George Ball's cost-as-loss perspective and his alarming predictions of the disastrous consequences of escalating the war in Vietnam, Robert McNamara presented a cost-as-investment perspective. He assured the president that bombing was controllable and its risks manageable and far less serious than those of withdrawal (U.S. Department of State, 1996b: 106–13, 514–28).[51] Consistently using numbers to support arguments provided a sense of certainty and, in the debate within the administration, gave their user an edge over the opposition. McNamara was particularly effective in the precommitment period when debating with doves in the administration:

> McNamara was a ferocious infighter; statistics and force ratios came pouring out of him like a great uncapped faucet. He had total control of his facts and he was quick and nimble with them; there was never a man better with numbers, he could characterize enemy strength and movement and do it statistically.

Poor George [Ball] had no counterfigures; he would talk in vague doubts, lacking these figures, and leave the meetings occasionally depressed and annoyed. (Halberstam, 1972: 581)

Various aspects of the costs of escalation were sterilized by language that "immunized" decisionmakers from thinking in terms of real costs. The decisionmaking group often used innocent Orwellian language like *reconnaissance flights* and *targets of opportunity* to describe sorties in which pilots could shoot at anything that moved. They also used obfuscatory qualifiers; on one occasion, for example, McGeorge Bundy said that the bombing in North Vietnam had caused "substantial but limited damage" (Ball, 1992: 146–47; Thomson, 1968: 51). Everyday metaphors contributed to the perception of the situation as a zero-sum game and made withdrawal an option that would not be considered seriously. For Johnson the war was sometimes a boxing match where he was one of the contenders, which meant that there were two basic outcomes—win and lose. In addition, with the boxing metaphor in mind, the situation became much more personalized, and therefore policy revisions became less likely. Another popular metaphor—a medical mission to save the sick patient, South Vietnam—enhanced the sense of urgency and the either-or view of the situation (Ball, 1992: 127–28, 132–33).

Even more powerful in shaping meaning were the historical analogies of appeasement in the 1930s, which invoked visions of impending collapse and disaster on a global scale (Khong, 1992; May, 1973: 113–14; Rystad, 1982). But the most prevalent image in American decisionmakers' minds was the domino theory, which transformed Vietnam from an ordinary country into a high-value asset, a finger in the dike containing communism (Kahin, 1987: 239, 314). From this perspective, the collapse of South Vietnam would be disastrous, even though the country was politically unstable, corrupt, and poor. At the personal level, Johnson believed that if Vietnam collapsed, he would be crucified politically. Even a cautious, politically astute, and domestically focused president could not walk away from the Vietnam problem. He could not frame the situation as an opportunity, as Ball suggested—as the last clear chance to disengage; rather, it was a high-threat situation. Finding himself in the loss domain, Johnson had to take the risks necessary to avoid the imminent collapse of South Vietnam. And once his thinking about the policy tilted toward recognition of the unavoidability of taking risks, Johnson faced more unresolved questions: How much risk should he take? What type of risky policies were preferable (e.g., strategic bombing, combat troops)? How should the chosen risky policy be operationalized (e.g., by stealth, by declaration of an emergency)? And how long and how far should the chosen risky policy be pursued?

As the overall situation in Vietnam became framed as a major threat to American interests as well as to the Johnson presidency, attention to threat-related risks became more acute. The severity of the political crisis in Saigon became a permanent part of Johnson's and his advisers' risk analysis. The historical memories of Munich, where an expansionist power met with ap-

peasement, which led to a world war, was juxtaposed against the lessons of
Korea, where a forceful military response by the United States stopped another expansionist communist proxy of the Soviet Union and China in its
tracks and saved South Korea.[52] This contradiction made very vivid the risks
of demonstrating weakness. Indeed, the analogies had particular power because the key decisionmakers had experienced these events. Johnson was
taken with the Munich analogy, and all his main advisers—McGeorge
Bundy, Rusk, and McNamara—accepted the lessons of these and other analogies.[53] Those who addressed the issue without referring to these experiences
were bound to be less persuasive than those who did. "I [McNamara] cannot
overstate the impact our generation's experiences had on him [Rusk] (and
more or less, on all of us). . . . George [Ball]'s memo failed to address these
underlying concerns, and President Johnson turned away from it" (McNamara
with VanDeMark, 1995: 195).

South Vietnam and U.S. policy toward it acquired symbolic value and
hence inordinate importance. Vietnam was not just a local problem of political instability in a poor Third World country in which U.S. policy might succeed or fail; it was a test of U.S. leadership, status, resolve, and commitment
to core values. The problem was therefore not a Vietnamese problem but an
American one. As such, there were historical precedents for dealing with the
situation. Three persuasive analogies were invoked by Johnson's advisers.
Rusk used the Greek civil war in the 1940s to demonstrate that when the
United States took a resolute position the communist guerrillas realized that
they had no option but to disengage. There were no negotiations. McGeorge
Bundy referred to the Berlin situation to demonstrate that no American president from Truman to Johnson had agreed to a change in the status of Berlin
as a divided city, implying that the same was true for Vietnam. Former president Dwight Eisenhower relied on the American Civil War to emphasize that
Abraham Lincoln had refused to negotiate from weakness because it could
have been disastrous (Gardner, 1995: 174–75). Defining the problem in
these terms made solutions such as negotiated withdrawal unlikely to be seriously considered even as mere options. This definition of the situation also
implied that the president was in the loss domain and implied, too, the necessity of taking risks in order to avoid the worst developments. And it made
the risk of the collapse of the South Vietnamese government a salient risk dimension, overshadowing the longer-term risks of American intervention,
which Ball presented.

The collapse of the Saigon regime and a communist takeover were seen
as irreversible, as in eastern Europe and China. This enhanced its salience.
Johnson's assumption that he would be held personally accountable for the
failure to prevent a communist takeover made this risk a preoccupation of
his. On the other hand, the incremental and piecemeal fashion in which the
policy evolved made it difficult to detect the full extent of the risk of failure.
This difficulty was reinforced by the focus on hard data—number of troops
and enemy casualties—in policy deliberations; softer data, such as the en-

emy's resolve and determination, were not attended to rigorously and systematically (Burke et al., 1989: 233). Thus, the risk of failure was underestimated, and the probability of breaking the enemy's will was consistently overestimated.

The salience and vividness effects brought attention to particular risks and accentuated their gravity. The disorderly nature of information processing and policy formulation in the Johnson administration magnified the influence of vivid and salient information on the president's views. His tendency to accumulate large amounts of data and immerse himself in detail, combined with the highly unstructured approach to the information, made it practically impossible for him to examine it systematically and vigorously as well as analyze the validity of alternative views. Under these circumstances, what gained attention and weight and therefore had the most effect in shaping the president's preferences was the more vivid and salient information, even though it was incomplete and narrow in perspective.

To prevent the collapse of South Vietnam and a communist takeover, to break the will of the enemy, and to prevent a humiliating and confidence-shattering American defeat, the civilian and military advisers, with the president's authorization, devised two strategies.[54] Both were nested in a broad doctrine of limited war and were instances of a more general strategy of coercive diplomacy.[55] At an early stage of the escalation, Johnson, McNamara, Rusk, and McGeorge Bundy realized that victory in the military sense of defeating or eliminating the enemy was impractical, and accepted a more modest goal of stalemate, which would force North Vietnam to accept a negotiated settlement favorable to the United States. The two chosen strategies—controlled and graduated strategic bombing of North Vietnam and a ground war of attrition in South Vietnam—emerged independently but became intertwined as the war progressed.

By the spring of 1967 the question of progress became a major consideration because the administration needed to convince Congress and the public that the Vietnam policy was successful. There were disagreements in this matter among the CIA, the JCS, the MACV, and the U.S. embassy in Saigon. Projections of net gains or losses for enemy forces, which were key to the assessment of the attrition strategy, were unreliable for several reasons: undetected infiltrators, an undetermined number of People's Army of Vietnam (PAVN) soldiers in Vietcong units, unknown total PAVN losses, uncertain Vietcong personnel replacement rates, dubious body count figures, and unreliable intelligence reports by South Vietnam. In spite of the questionable credibility of statistical measurements, these remained the main criteria by which success was measured. Before the Tet offensive, Westmoreland had the president convinced: "We are grinding down the Communist enemy in South Vietnam and there is evidence that manpower problems are emerging in North Vietnam; our forces are growing stronger and becoming more proficient in the environment" (Cable, 1991: 186–87, also 173–82; Johnson, 1971: 257–58). One critic of the war, Daniel Ellsberg, correctly observed that the flow of information from Saigon, intended to convince the public and Congress of progress in the war, eventually replaced "phony and invalid optimism with genuine invalid optimism" (Ellsberg 1971: 262). Then the Tet offensive in January 1968 shattered the president's confidence in military measures.

Some aspects of American culture reinforced risk perceptions and risk-taking preferences in a manner that encouraged or at least did not discourage the general trend toward escalation. The traditional perception of manifest destiny in American history had taken a global flavor after the Second World War, and the global responsibility that it imposed was not only strategic but also strongly moral-ideological. The idea of manifest destiny negated the argument that the United States had no legal commitment to Vietnam (DiLeo, 1991: 67). The responsibility led presidents from Truman to Johnson to assist South Vietnam in preserving its independence from North Vietnam, whose nationalistic claim for unification was considered illegitimate in terms of the higher values of freedom and democracy. American intervention driven by beliefs in moral infallibility and military invincibility, a combination that justified taking risks and raised the expectations of success.

American beliefs about rationality and the projection of these beliefs onto the enemy led to the expectation that, given the favorable balance of power, the enemy's breaking point was within the range of acceptable costs. American decisionmakers did not fully grasp the extent to which the enemy's perspective was different, that for the intensely nationalistic North Vietnamese leaders the war was closer to a zero-sum game than to a mixed-sum game of chicken, which is how the American leaders perceived it.

In American terms, the war was fought for reasons of state, without passion and in cold blood. Its designers believed it to be a perfect occasion for applying theories of limited war. As McNamara stated: "The greatest contribution Vietnam is making—right or wrong is beside the point—is that it is developing an ability in the United States to fight a limited war, to go to war without the necessity of arousing the public ire" (Baritz, 1985: 323). This view was congruent with the American preference for taking an engineering approach to problem solving. Vietnam was seen as a problem of social engineering that could be resolved by expending resources. A senior member of the Johnson administration observed that Vietnam was "another example of that troublesome American belief that the application of enough resources can solve any problem; if no solution is forthcoming, it means the resources are inadequate and the answer is to apply more" (Hoopes, 1969: 70).

These attitudes had two important consequences. One related to the management of the war. The engineering approach to problem solving meant an emphasis on technological factors and the de-emphasis of sociopolitical factors, which contributed to the marginalization of the pacification strategy (Vought, 1982). The engineering approach requires hard data and as little reliance as possible on soft data. To produce hard data, structure and specificity were imposed on qualitative variables, such as the enemy's morale and resolve and the war's progress. Going to extremes in the attempt to quantify what could not be adequately quantified meant that reports based on these measures provided a false sense of confidence in their assessments. The other consequence of the American attitudes was a sense of optimism (Spector, 1993). Viewing the problem through a rationalistic cultural lens made the prospective outcome look rosy, which encouraged risk taking and policy nonresponsiveness even in the face of setbacks.

The broad definition of the problem and its specific framing had disposed the president and other key decisionmakers to consider the issue a major threat requiring risky decisions within a limited time. Although these factors made a high-risk policy acceptable, they did not yet irrevocably lock the president and his advisers into a pattern of progressively escalating military intervention. Much was said by observers of the Vietnam policy on the policy commitment that Johnson inherited from the Kennedy administration. Johnson himself propagated the myth that he was ensnared in a policy not of his own making. That was true, but only between the time Johnson assumed the presidency in 1963 and the 1964 election.[56] After his landslide victory in the election, he could have chosen any course. But he chose to assist the South Vietnamese government and transformed his initial limited commitment into an open-ended commitment to preserve an independent noncommunist South Vietnam (Herring, 1986: 108). Therefore, what drove Lyndon Johnson to risk taking in Vietnam to the extent and in the particular form that he did can be best explained by the interaction between Johnson's personality and the social environment in which he operated and which he had shaped to suit his decisionmaking style.

Although the definition of the problem and its framing implied Johnson's acceptance that the Vietnam problem could not be avoided or put off, he was not thrilled with the prospect of making a difficult, risky choice. He "confronted Vietnam as an unsure and troubled leader grappling with an unwanted and ominous burden. . . . Johnson moved cautiously and warily, constantly shifting and hesitating in the face of momentous decisions" (VanDeMark, 1991: 213). Johnson realized that risky decisions were inevitable, but because of his cautious nature he preferred and searched for low-risk options that held the promise of control. He was not going to go all the way unless he had no other choice (Rosen, 1982: 91). Most important, he was not going to sacrifice his envisioned social reforms by entering an all-out war, as the JCS preferred, as long as he could avoid it.[57] He was going to try to have both guns and butter, as he explained in a conversation with Dorris Kearns.

> I knew from the start that I was bound to be crucified either way I moved. If I left the woman I really loved—the Great Society—in order to get involved with that bitch of a war on the other side of the world, then I would lose everything at home. All my programs. All my hopes to feed the hungry and shelter the homeless. All my dreams to provide education and medical care to the browns and the blacks and the lame and the poor. But if I left that war and let the Communists take over South Vietnam, then I would be seen as a coward and my nation would

be seen as an appeaser and we would both find it impossible to accomplish anything for anybody anywhere on the entire globe. (Kearns, 1976: 251)

That he was committed to preventing the collapse of South Vietnam and a communist takeover as much as he was committed to the Great Society program, his true love, was not surprising. The commitment to Vietnam was not mainly the result of a commitment to the Kennedy heritage; it touched on his own core beliefs. He chose to try to save South Vietnam, but with as few policy and political risks as possible. In reality, he was incrementally driven into progressively riskier decisions even as he tried to avoid them.

Johnson was a globalist. He saw, not merely a hemispheric role for the United States, but a global one. That role, however, was not well specified in his mind. Though lacking extensive foreign policy experience, he had well-established beliefs about international relations based on his life in American politics and his witnessing of international events that had formative effects on his views. The failure to check fascist aggression in the 1930s had left a lasting impression, as did the Korean War, and from these sources lessons on how to deal with communism were inferred. In 1961, as vice president, he traveled to Southeast Asia at Kennedy's request, met with Diem, and upheld personally an American commitment to assist Vietnam.

Johnson perceived life as a struggle between good and evil. He saw communism a menace to U.S. security. It was a global and monolithic force that used so-called wars of national liberation to cloak naked expansionism. The war in Vietnam was part of the communists' grand design. Consistent with an active role in world politics for the United States was Johnson's belief that the battle against communism must be joined in Southeast Asia with strength and determination (VanDeMark, 1991: 9–10). Toward South Vietnam, the U.S. ally, he had a benevolent, patronizing attitude. The North Vietnamese leaders he did not perceive as nationalists but stereotyped as aggressive and authoritarian stooges of China and the Soviet Union. But based on his experience in politics, he believed that a mix of carrots and sticks was the most effective way to deal with opponents. He therefore expected Ho Chi Minh, the North Vietnamese leader, to realize the extent of American determination and be reasonable. For that purpose he offered in addition to the stick of coercion the carrot of economic development, the Mekong development plan, and was surprised when the North Vietnamese did not bite (Gardner, 1995: 177–200; Kearns, 1976: 265–66; Taylor, 1993).

The belief in American manifest destiny and the consequent obligation to deter bullies and defend peace-loving noncommunist countries was closely associated with confidence in American omnipotence, based on its immense economic and military resources and unbroken record of military successes, many of them witnessed by Johnson himself. He therefore believed that the United States had the power to control the behavior of other actors and shape events in a direction of its own choosing, as in the First and Second World Wars and in Korea. U.S. obligations and capabilities were being tested once again in the Cold War, specifically, in Vietnam, and the United States would once again prevail.

This typical Cold War interpretation of the situation in Vietnam was validated by similar beliefs held by Johnson's top advisers, McGeorge Bundy, Rusk, and McNamara. Their shared globalist geopolitical views included a determination to use American power in the service of American foreign policy and in the cause of maintaining the integrity of the U.S. commitment to freedom and social progress everywhere in the world.[58] In that sense, they saw an abiding coincidence between national self-interest and altruistic American goals for other nations (Cohen, 1980; DiLeo, 1991: 94–104; Loewenheim, 1994; Roberts, 1965: 74, 76; Schoenbaum, 1988; Shapley, 1993). Their willingness to use U.S. power was also linked to beliefs in the controllability and manageability of human affairs; these had intellectual and also, in Rusk's case, religious sources.

When applied to foreign policy, these beliefs generated a faith in the ability to set the course and pace of events, to manage crises and control escalation, and to use power as a scalpel to carve out policy results that would serve the virtuous American interests. An engineering approach to foreign policy implementation was a natural extension of such views. Limited war and gradualism in the application of force, as in the bombing campaign, were believed to be possible, controllable, and effective. They had recently worked in Cuba, which seemed to validate their views (VanDeMark, 1991: 115, 219). For Johnson, a president haunted by self-doubt (Barber, 1977: 93–95; Kearns, 1976: 170), having his beliefs validated by men who represented the established wisdom became a source of confidence. This sense of confidence in his operational code beliefs and their use to guide his interpretation of the Vietnam situation and determine policy responses would make policy reassessment extremely difficult once a policy was initiated, because Johnson viewed the reasons for the policy as highly compelling. These beliefs, then, made an active-assertive policy, including the use of extensive military force, acceptable; they made a policy of withdrawal much less attractive.

Johnson's policy preferences were reinforced by two values that he considered important, loyalty and decisiveness. He demanded complete loyalty from his subordinates and expressed little respect for those who were indecisive and tortured by doubts. Doubts, he thought, were a sign of weakness (Halberstam, 1972: 532). When projected into the foreign policy arena, these values required a policy of loyalty to friends and a respect for commitments to allies, even if they were costly. Commitments had to be carried out with decisiveness and determination, nor should there be any doubt about the U.S. intention to carry them out. In Vietnam taking a tough stand was made easier by the hostile stereotypes of the leadership in Hanoi: aggressive, expansionist, and oppressive and at the same time unworthy to be a world power's opponent. North Vietnam was merely "a raggedy ass little fourth rate country," and its soldiers were—to use the the words of JCS chairman

Wheeler—"raggedy-ass little bastards" (Halberstam, 1972: 512, 541, 593; Shapley, 1993: 339). Racial arrogance and lack of empathy made the enemy less threatening and more deserving to suffer the consequences of an American military force unleashed to defend the weak.

But Lyndon Johnson cannot be dismissed merely as a man driven by a simplistic belief system. He had at his service some of the best brains in the country and a network of military and civilian organizations that produced a wealth of intelligence information and assessments about all aspects of the situation in Vietnam. Johnson himself was a voracious reader. His White House office resembled an information command center; at night he took home piles of documents and stayed up late to read them. He consulted extensively with members of his cabinet, members of Congress, trusted friends, and experts, including former presidents and other experienced statesmen (Berman, 1989: 111–13; Rostow, 1972: 362–64). He could not be unaware of the increasingly obvious risks of failure.

Why did so many skeptical, critical opinions and assessments have so little effect on the president's optimistic views on the progress of the war? The answer can be found in understanding how and why Johnson used the advisory system and the available information. Here we must take into account an important aspect of his narcissistic personality and the related motivational biases. Johnson was "a man of stunning force, drive and intelligence and of equally stunning insecurity" (Halberstam, 1972: 432). He had doubts about his adequacy for the presidential task and suspected that others had similar doubts, especially where foreign policy was concerned. On one occasion he commented bitterly, "Now they say that I am not qualified in foreign affairs like Jack Kennedy and those other experts. I guess I was just born in the wrong part of the country" (Graff, 1970: 56). In that context, with the Vietnam policy as the main issue on the president's agenda, it became the ultimate challenge to and test of his competence. His immersion in and preoccupation, almost obsession, with every available piece of information had a ritualistic role—to provide the reassurance that he had everything under control, was aware of all that was going on, and understood every policy nuance and point of view.[59] Johnson was trying to establish confidence in the quality of his decisionmaking skills. He also needed personally to feel comfortable with decisions that involved considerable risk. Politically and personally it was important for him to have others trust his judgment, recognize his decisionmaking skills, and thereby validate and legitimize his decisions, especially the risky ones. Spreading the word and projecting an image of his confident competence was an important part of these needs.

Role acting and impression formation were the essence of Johnson's behavior. He played the role of president as he understood it should be played and attempted to convince those observing his performance that he was knowledgeable about issues, open-minded and willing to listen, cautious yet decisive; at the same time, he wanted to let everyone know that he was the boss and not a puppet carrying out decisions made by the intellectuals and experts in his cabinet. Observable precautions taken to avoid misjudgment and poor decisions were intended to preempt both his own inner cognitive dissonance and to reduce the opportunity for criticism by others. Johnson therefore had both to be sure for his own sake and to demonstrate to others that he did pay attention to dovish points of view. Going through the social ritual of listening to views he detested served a useful psychological purpose and allowed him to make decisions with a greater sense of self-confidence, which eventually made his decisions resistant to change. An illustration is Johnson's treatment of George Ball. He always listened to him attentively, not responding, and then allowed McNamara, Rusk, or Bundy to point out how wrong and ill informed Ball was (Ball, 1982: 390–92; Shapley, 1993: 328–29). On other occasions he pretended to listen to advice but only tuned in to the discrepancies between the adviser's position and his own to find out whether his advisers agreed with him (Burke et al., 1989: 243–44). This elaborate, time-consuming, and demanding process produced an illusion of the control that he craved.

But it was not all show. The information did have essential value. The problem was that Johnson was trying to cope with a prohibitively large amount of it. The amassed information very often contained inconsistencies and contradictory assessments provided by competing agencies and individuals, each hoping to gain the president's support. The information processing problem was exacerbated by Johnson's lack of the analytic skill required for the task. Eric Goldman, a White House aide, said of Johnson that he had "a clear, swift, penetrating mind, with an abundance of its own type of imagination and subtleties [but] . . . the grown man came to the White House with a grab bag of facts and nonfacts, conceptions and misconceptions, ways of thinking and ways to avoid thought which had been gathered largely from his early crubbed [inconsequential] environment" (Goldman, 1969: 525–26). Naturally enough, he imposed on the confusing array of information and assessments his operational code beliefs and simple schemata drawn from broadly shared metaphors, such as the domino theory, everyday games, such as football, and a limited personal inventory of historical analogies, such as Munich, Korea, and, to a lesser degree, Cuba (Khong, 1992).

These cognitive devices produced his confident insights into the risks of alternative policy options. An American defeat did not seem likely. He could not believe that it would happen; it had never happened before, and the enemy was just another small and weak Third World country. Thinking in linear terms and predicting outcomes by the amount of resources invested, Johnson assumed a causal link between the uncontested military-technological superiority of the United States and the desired outcomes of the war. The nontransparent cultural, social, and political factors that would make breaking the will of North Vietnam an elusive goal and would raise the risks of failure did not come readily to mind. For Johnson, Vietnam was just another country. He found it difficult to understand that Vietnam and the United States might not share the same values and that they were different in fundamental ways. The perception of similarity reduced the uncertainty of the situation and increased his belief that he could deal with it the same way he dealt with opposition in Congress—persuasion and negotiation—because everyone had their price (Kearns, 1976: 265–66). Reflecting on why Hanoi was keeping a particular back channel of communications open, the president projected his own possible motives in a similar situation onto Ho Chi Minh: "I know if they were bombing Washington, hitting my bridges and railroads and highways I would be delighted to trade off discussions through an intermediary for a restriction on the bombing. It hasn't cost him one bit."[60] On another occasion he referred to a staggering number of Vietcong casualties and observed: "Think of how terrible it would be if our losses were proportionately as high as theirs. We don't think Hanoi knows how serious it is. . . . They are looking at things through rose-colored glasses" (Graff, 1970: 98). His comment reflected the supposition that once the enemy realized the actual number of their casualties, they would relent, as he would have done.

The risks that were easily reconstructed and simulated in his mind were those that he observed in everyday life or that had been experienced by other presidents, but in his case he could easily imagine that the consequences would be worse.

> I knew that Harry Truman and Dean Acheson had lost their effectiveness from the day that the Communists took over in China. . . . And I knew that all these problems taken together, were chickenshit compared with what might happen if we lost Vietnam. For this time there would be Robert Kennedy out in front leading the fight against me, telling everyone that I had betrayed John Kennedy's commitment to South Vietnam. That I had let a democracy fall into the hands of the Communists. That I was a coward. An unmanly man, a man without a spine. Oh, I could see it coming all right. (Kearns, 1976: 252–53)

On one occasion he used a rape sequence metaphor: " 'If you let a bully come into your front yard one day,' he explained, 'the next day he'll be up on your porch and the day after that he'll rape your wife in your own bed.' " The conclusion was that the only way to prevent conflict is to stop the aggressor from the very first moment that his intentions are perceived or even suspected (Kearns, 1976: 258). Similarly, his nightmares of political crucifixion, the Chamberlain historical analogy, and his grief over mounting casualties simultaneously made the risks of withdrawing from and losing South Vietnam very vivid and prominent.

At the same time, the benefits of making concessions, halting the bombing, negotiating, and the like were uncertain. What Johnson viewed as American gestures of goodwill, like pauses in the bombing, were not reciprocated by the North Vietnamese (Johnson, 1971: 254–56). The enemy did not seem to understand how difficult it was to impose limitations on the conduct of the war, to halt the bombing unilaterally when the JCS, CINCPAC, and the field commander resented it and when some civilian advisers, like Rostow, influential Senators, like Richard Russell and John Stennis, or his old friend Abe Fortas opposed it. There was little reason to believe that winding down the U.S. war effort would bring the enemy to the negotiating table. For example, discussing a possible bombing pause at a luncheon meeting on September 26, 1967, the president expressed exasperation, saying, "I think they are playing us for suckers. They have no more intention of talking than we have of surrendering. . . . We get nothing in return for giving all we have got."[61]

In the absence of hard information and the presence of ambiguous information, Johnson depended increasingly on his own judgment and preferred the advice of individuals who offered him certainty that was congruent with his beliefs and values, like Rostow, Westmoreland, and the JCS, whose advice sustained his own judgment. If they got most of what they asked for, the situation would be stabilized and Johnson could go back to the woman he loved, the Great Society program. Setbacks were discouraging, but the doves could not yet offer a proven alternative policy that would deliver acceptable outcomes. But if he agreed to send in more troops and matériel, the situation might be turned around, and then the cost of the Vietnam policy could be counted as an investment in securing a noncommunist regime in Vietnam, keeping dominos from falling all over Asia, and strengthening the U.S. reputation and position globally.[62] In fact, Johnson perceived South Vietnam as an American asset and therefore was willing to pay a much higher cost for not losing it than he would have for acquiring it, as predicted by the endowment effect.

All these cognitive and motivational biases in Johnson's information processing and decisionmaking resulted in a selective, self-serving approach to information that supported his predispositions and anchored and perpetuated his interpretations of the situation in Vietnam. Although Johnson sometimes had doubts about the direction he chose, he overcame them by telling himself what he told Jack Valenti in 1965 with regard to recommendations made by experts: "All these recommendations seem to be built on pretty soft bottom. Everything blurs when you get almost to the gate" (Valenti, 1975: 341). In other words, his own choices were at least as good as those offered by others.

He adapted his positions mostly in response to domestic political requirements, to which Johnson paid close attention. Policy adjustment did not result from improved cognitive information processing or new information; it was driven by such motivational incentives as the preservation of Johnson's position in the game of domestic politics. Johnson's lack of the will or ability to respond to policy failure or learn from it stands out against the responses of two of the architects of the Vietnam policy, McGeorge Bundy, who left the administration in 1966, and McNamara, who became increasingly disenchanted with the policy but lacked the courage and will to challenge the president's policies until 1967 (Herring, 1993c). It is not surprising, therefore, that Johnson continued to focus on the same risks that preoccupied him in

1965 with minor modifications that were mostly tactical politics. Policy drifted while Johnson regarded it in this preoccupied, even trancelike state, until the Tet offensive shook him awake.

Johnson's cognitive closure was helped by the fact that he as well as his advisers were operating under heavy psychological and physical stress that came from the immensity of the problems and the fatigue of working long hours for months on end. This state of affairs made him more rigid, irritable, and defensive and thus more wedded to his view of the world (Ball, 1992: 115; Thomson, 1968: 50). The extent of Johnson's open-mindedness was occasionally also affected by his moods. If he felt good about something, he could focus well and hard on the decisions ahead. But if he was depressed, he spent hours in rambling talk, shifting from one extraneous subject to another. In such moods, he required constant encouragement, which undermined any chance of critical discourse (Kearns, 1976: 322).

Being insecure also affected Johnson's postdecision behavior. He seldom explained why he had decided to do something, even less, why not (Simons, 1994: 147). Evidently, once he made a decision, he did not want to be exposed to questions about his rationale that would raise doubts in his mind. He wanted to move ahead with implementation. "By temperament Lyndon Johnson was a deliberate and cautious man to the point of decision, and thereafter totally action-minded and hard-hitting" (LBJ Library, William Bundy, manuscript, 1971: ch. 27, p. 14). Deliberation was lengthy, implementation of a decision immediate, making a behavioral commitment that was resistant to doubts. Policy reconsideration, especially policy reversal, was unlikely.

Why was gradual escalation preferred over other such options as withdrawal or going all out? I have already indicated that cognitive and motivational factors made either withdrawal or continuation of the pre-1965 policy unattractive. Some personality attributes also bolstered this policy preference. Johnson's exhibitionism, aggression, and dominance required an active, highly visible escalatory policy. The Vietnam policy also triggered deep-seated fears of shame and humiliation. The advice that Johnson was given from 1965 on repeatedly emphasized the theme of humiliation to the United States if it withdrew from its commitments or allowed itself to be defeated by a poor, insignificant country. Johnson saw a direct link between the humiliation of the United States, the failure of his policy in a war that had become so personalized, and shame and humiliation for himself. The narcissistic streak in his personality could not allow his honor and credibility to come under attack, as they would if he reneged on his commitments to South Vietnam (Steinberg, 1996). William Bundy's description of Johnson's response to the Wise Men's advice on July 8, 1965, to commit the 44 battalions requested by Westmoreland seems to confirm this point.

The President probably expected that *most* of the Panel would be *generally* in favor of a firm policy. What he found was that *almost all* were *solidly* of this view, and this must have had a distinct impact on his personal and private delib-erations. There can be no doubt that a large strand in the President's make-up [was] that he should not fall short of the standards set by those who had played leading parts in World War II and throughout the period of American successes in the Cold War. Now a fair sample of these men, of American "Genro" if you will, had advised him to see this on through. (LBJ Library, William Bundy manu-script, 1971: ch. 27, p. 21)[63]

Johnson's emotional insecurities thus reinforced cognitive insecurities that stemmed from his unfamiliarity with foreign policy and made him turn to strength, certainty, and hope where he saw it—in what he hoped would be a decisive, even if high-cost military intervention.

We may understand why Johnson was driven toward a risky policy of es-calation. But why didn't he chose all-out war? One reason is that Johnson was cautious. For him, gradual escalation in the air war, which was a show of toughness and determination, also held a promise of control and re-versibility. Along the same lines, when faced with the need to deploy combat troops, he attempted to introduce a policy of incrementalism. The situation in South Vietnam, however, overtook his intentions and imposed a choice between withdrawal and fast escalation. He chose the latter.

SOCIAL INFLUENCES AND POLICY PERSISTENCE

So far I have emphasized the prime importance of Lyndon Johnson's per-ceptions, motivations, dispositions, and role in directing U.S. policy toward escalation, but the picture is somewhat more complex than that would imply. Because the president was both insecure and inexperienced in 1964 and because he craved for and depended so much on others' positive evalua-tion of his behavior, the inner circle of advisers was instrumental in his mak-ing an explicit choice to escalate first in the air war and then on the ground. The interaction between the president and his advisers also sustained the president's insensitivity to signals that warned of impending failure. It might well be the case that, had the advisory system operated differently, some change in policies might have been introduced in early 1967, even before the Tet offensive, because by then Johnson was aware that the war was costing him the social reforms he was so keen on, that his public support was slip-ping, that his critics in Congress were linking up with the protest movement in the streets, and that even within the Departments of Defense and State there was dissent. If the advisory process had been different, it is quite possi-ble that even for a president reluctant to change his policies once they were set in motion, the risks of the Vietnam policy might have been more salient and vivid than the promises that success was around the corner offered by the JCS, the field commander, and the special assistant for national security.

To be sure, Johnson changed the atmosphere and management style in the White House, especially after the 1964 election, when he came into his own. Although he retained the three key players in the arena of foreign and defense policies and continued to look to them for guidance, the texture of his relationship with these men was very different from President Kennedy's. Johnson's feelings toward them were a blend of respect for their skills and resentment for their differences in mentality—detached and reserved where he was outgoing, and confident where he often felt unsure and inadequate. He depended on them and felt uncomfortable with dependence because it only emphasized his own insecurity.[64]

In that mixed atmosphere both sides were driven to engage in impression formation and management, so the relationship between the president and his advisers often assumed a staged quality. In the Kabuki-like theater of power there were rigidly structured roles played with hyperactive grand gestures; the president played at being presidential and his advisers at being meek and deferential. Occasionally, however, the relationship became competitive; the president's position and the advisers' expertise were pitted against each other, and neither side was completely frank with the other. Like sumo wrestlers, the president and his advisers circled, waiting for an opening, both sides wanting to be sure that they took the right step and did not embarrass themselves. William Bundy describes the president as follows:

> His style had become more and more personal throughout 1965; his illness in the fall of 1965 raised this element to new heights, from which it was never thereafter to descend. One can track with some clarity his decisions of February and July. By December, it was almost impossible to tell how he was reacting and how he might decide to act. The war, it is not unfair to say, was beginning to get to him. His innate preference for keeping his subordinates both guessing and familiar only with a part of the picture was at its height in these months of semi-isolation in Texas. My own hunch is that he came to prefer to operate this way, and thenceforth did it even when the machinery of government was ready to his hand in Washington. (LBJ Library, William Bundy manuscript, 1972: ch. 33, p. 36)

The advisers, for their part, were eager not to lose the president's attention and support and apprehensive of earning his displeasure and inviting his intimidating anger.[65] Eventually the heavier weight of the presidential position won. Those who disagreed with the Vietnam policies were shut out of the policy process (e.g., Vice President Humphrey) or found it impossible not to resign (e.g., Ball, McGeorge Bundy, and McNamara) (Gallucci, 1975: 99; Kearns, 1976: 320). The environment was not conducive to a frank exchange of opinions.[66]

In those early days after the election, when the Vietnam policy was being given its military flavor and when the president had not yet decided which direction to take, the core group of advisers favored an increase in the risks taken by the United States, a bias resulting from their earlier involvement in setting Vietnam policy.

> During the fall of 1964 . . . it had become starkly evident to Johnson's advisers that the policies they had been shaping ever since John F. Kennedy assumed the presidency had failed. However, since their own reputations were so closely

bound up with these policies, it was difficult for them to call for a shift from a military to a political track. . . . Such a move would have exposed their own previous counsel and very possibly been costly to their tenure in the administration. . . . Moving to a policy of heavier and geographically expanded military intervention would be perceived [they believed] by the American Congress and public as having consistency and continuity. (Kahin, 1987: 245)

For Johnson, being in charge of Vietnam policy was a new experience; for his most senior advisers, it was a direct extension of what they had been doing before. For them, therefore, the new presidency did not represent a chance to rethink and reconceptualize policy but was linked to their prior commitment to a progressively riskier policy.

No wonder, then, that the National Security Council Work Group that was established before the election suggested only three options: continuation of the present policy, dramatic escalation, and gradual escalation. A political solution was not considered an option. McNamara led the risky shift; he was joined by McGeorge Bundy and eventually Rusk, who was at first a little hesitant. Rusk was, as a rule, unwilling to confront the secretary of defense. This was a result of his experience in the Truman administration as liaison between Secretary of State Dean Acheson and Acheson's foe Defense Secretary Louis Johnson (Burke et al., 1989: 139; DiLeo, 1991: 102). The policy direction that was congruent with Johnson's disposition now had the seal of approval from the three topmost foreign and defense policy experts in his administration, Rusk, Bundy and especially McNamara, whom Johnson described as the best secretary of defense in the history of the United States. Because this group of key advisers had a set of shared beliefs that conformed with Johnson's predispositions, there was an added impetus to reject arguments raised by Ball, Mansfield, and other dissenters who pointed to the domestic political risks, the potential for defeat, the unreliability of the South Vietnamese ally, and the capabilities of North Vietnam and its allies. McNamara and McGeorge Bundy, powerful advocates of the supremacy of American power, the irrelevance of the French experience, and, above all, the manageability of escalation, discredited the dissenters' arguments logically and empirically. The leading dissenter was Ball. McNamara recollects that "because he was recognized as having a strong European bias, Dean, Mac and I treated his views about Vietnam guardedly" (McNamara with VanDeMark, 1995: 156). Ball himself mentions that in front of the president McNamara "would shoot me down in flames with all these new statistics I hadn't heard before. He tried to give the impression I didn't know what I was talking about" (Shapley, 1993: 332).

In theory the larger meetings of the NSC group—being an arena where more points of view, different arguments, and additional information could be shared—should have provided another opportunity to challenge the principals. In reality this was not the case. Johnson disliked large meetings with freewheeling discussions. When the NSC met, the agenda was tightly controlled and the meetings were short. Consequently this body was neither a central decisionmaking forum nor a forum for systematic discussion of policy options (Burke et al., 1989: 141–43; Herring, 1994: 13–14). The atmosphere at the larger NSC meetings is captured by Chester Cooper.

> The NSC meetings I attended had a fairly standard format: the Secretary of State first presented a short summary of the issues, the Secretary of Defense added his comments, and there was some fairly bland and desultory discussion by the others present. Because many around the table had not participated in, nor indeed been told of, the detailed advance discussions, "gut" issues were seldom raised and searching questions were seldom asked. The President, in due course, would announce his decision and then poll everyone in the room—Council members, their assistants, and members of the White House and NSC Staffs. "Mr. Secretary, do you agree with the decision?" "Yes, Mr. President." "Mr. X, do you agree?" "I agree, Mr. President." During the process I would frequently fall into a Walter Mitty–like fantasy: When my turn came I would rise to my feet slowly, look around the room and then directly at the President, and say very quietly and emphatically, "Mr. President, gentlemen, I most definitely do not agree." But I was removed from my trance when I heard the President's voice saying, "Mr. Cooper, do you agree?" And out would come "Yes, Mr. President, I agree." (Cooper, 1970: 223)

McNamara's influence became even more prominent after February 1965, when the NSC shifted to small, informal meetings, such as the Tuesday Lunch Group. The limited number of participants narrowed the range of considerations. Rusk was often reluctant to express himself frankly, and Bundy often sided with McNamara (Burke et al., 1989: 184–87). Thus important issues failed to be raised, and alternatives to escalation, such as negotiation or even communist victory, were not substantially discussed.[67] It was taken for granted that the first was hopeless and the second unacceptable.

With the decision to initiate Rolling Thunder and then deploy combat troops, the shift in the balance of influence in the decisionmaking groups in favor of the secretary of defense and the JCS, mostly at the expense of the secretary of state, was even more pronounced.[68] Johnson respected Rusk and felt comfortable with him because of the similarity in their backgrounds, but he was infatuated with McNamara's talents. More important, now that the security of U.S. forces was involved, the military represented nonpartisanship and expertise; given its responsibility for the situation having more influence was thus appropriate (Cooper, 1970: 275; Gallucci, 1975: 32–33, 88–89; Shapely, 1993: 283).

The balance of influence soon shifted again. As McNamara became progressively disenchanted with the war, he was more often on a collision course with the JCS and General Westmoreland, who wanted faster and more extensive escalation.[69] Johnson, now fully committed to a military solution, distanced himself from McNamara; on the rise now was the influence of the JCS and, in particular, General Wheeler (Clifford with Holbrooke, 1991: 456–59). Several factors account for this second shift in the balance of influence, all primarily related to the intensity of the military engagement. At this point, almost half a million American troops were in Vietnam. With the United States at war, the president became more dependent on the advice and support of the senior military officers, who represented expertise and non-partisanship, to legitimate his decisions with Congress and the public.[70] Moreover, General Wheeler prevailed on the JCS to always present a unified JCS position in policy debates, unlike the civilian advisers, who were often divided (Baral, 1978: 232–33; Herring, 1994: 33; B. Palmer, 1984: 34–35; Shapley, 1993: 325). JCS influence contributed to the persistence of their policies in spite of their lack of successful results.

In 1967 the Tuesday Lunch Group remained the focal point of the advisory system, with a larger number of participants than in 1965–66, including CIA Director Richard Helms and General Wheeler. But the luncheons were not the exclusive forum for consultation. By this time Johnson relied on no single individual or group for crucial advice, as opposed to the situation in 1965, when McNamara played a crucial role in the escalation decisions (Barrett, 1988). In essence, Johnson's advisers represented two main lines of argument. One group wanted to hold to the present course, avoiding escalation and seeking a negotiated end to the war (e.g., Hubert Humphrey, Robert McNamara, Dean Rusk, Richard Helms, Clark Clifford). The other called for all-out escalation leading to victory (e.g., Walt Rostow, Earle Wheeler, Abe Fortas, Richard Russell). Practically no one advised unilateral withdrawal, and almost all assumed that preserving the status quo was possible even without further escalation. A majority believed that the United States had the situation under control and that the military situation was progressively improving (Barret, 1993: 83–108, 160–65; Humphrey, 1984; Johnson, 1971: 257–58). The Tet offensive of January 1968 gave them a rude awakening.

Making the advisory process even less valuable was Johnson's known preference for consensus; essentially, participants had to accept compromise positions. No doubt Johnson's congressional experience biased him to resolve policy debates in this fashion, but he was often unaware of the disagreements underlying the compromises; he heard only the watered-down versions of arguments (Ball, 1992: 151–54). From his point of view, a compromise pleased or displeased all equally. The policy of step-by-step escalation that emerged was supposed to satisfy the hawks, who wanted victory, and avoid alarming the doves, who wanted a negotiated solution.

In the final account, compromises favored the hawks because the hawkish position, even if not accepted in full, scored a few points. General Wheeler actually believed that a "foot in the door" approach would eventually gain for the military the strategic freedom that the JCS wanted, and he prevailed on the JCS to accept the compromises. Each concession had its own momentum, carrying escalation further (Baral, 1978: 305; Herring, 1994: 41). This process of policy formulation did not allow a systematic and comprehensive review of purpose and method because the middle-of-the-way option that emerged was not an alternative discussed on its own merits or even favored by the competing players but was the outcome of averaging other options. It often carried contradictions or inconsistencies. Still, the averaged option was the president's option, and, as such, carried special weight and was difficult to reject even if it was not workable.

Whatever the flaws of the Vietnam policy-formation process, it was not a result of groupthink, as some have suggested (Janis, 1982). The two key decisions of 1965, the strategic bombing of North Vietnam and the deployment of combat troops to South Vietnam, were the result of an intense debate that began in the Kennedy administration and continued after Johnson resumed the presidency. Initially, the policy of strategic bombing was not congruent with the president's own best judgment—he thought that it was not going to be effective. It was more in tune with the views of the civilian advisers. The strategy of gradualism was also a subject of debate between the JCS and the civilians. With the combat troop deployment decisions, both general objectives and such specifics as number of troops and their missions were contested.[71] Group members had access to diverse intelligence sources, which, in many cases, provided alternative and competing definitions of the situation or even disagreed on the most basic facts, such as the size of the enemy force. To the extent that the groupthink phenomenon can be discerned, it relates to the JCS. Even those, like General Johnson, who had doubts about the efficacy of the ground war strategy did not seriously challenge it (Herring, 1994: 44–45, 49–50). The JCS stifled debate among themselves, took a common stand against the divided civilians, and thereby acted as a faction within the administration.

Although there was little groupthink, there were many occasions when individuals preferred to conform rather than challenge decisions. In some of the most evident cases, conformity reflected not the primacy of normative over informational influences but personal values of correct or responsible behavior. I have already mentioned Rusk's reluctance to contradict McNamara due to his experience in the Truman administration. But the most notable example is that of McNamara (Herring, 1993c, 1994: 48). As of late 1965 the secretary of defense became more and more pessimistic about winning the war. He came to believe that the war could not be won in the military sense but that continued military pressure would be conducive to an acceptable negotiated agreement with North Vietnam. The extent of his pessimism was revealed in a long secret memorandum to the president dated May 19, 1967, in which he rejected Westmoreland's demand for an additional increment of 206,000 troops and recommended a negotiated end to the war. McNamara emphasized his commitment and loyalty to the president to explain why he had not pushed earlier for a substantial revision of policy by threatening to resign or actually resigning (McNamara with VanDeMark, 1995: 314). In an interview he said: "You have a tremendous obligation as an officer of the government dealing with an issue of national security, to be careful that you don't press on the president views that carry the risk of substantial error. . . . And there is a question how far an adviser should go in publicizing his personal opinion when it differs from the majority view. . . . I had an obligation to recognize that I was in a minority and that I may be wrong" (Shapley, 1993: 421).

This argument raises the possibility that members of decisionmaking groups conform with majority views not necessarily because they fear the majority's sanction but because they do not feel confident enough to present a dissenting view that could be wrong and could result in high costs. When an ambiguous high-risk decision is to be made, with normative and informational influences pulling in contradictory directions, the effect of the informational influences will depend on which of the decisionmakers' values are invoked and to what effect. In the Johnson decisionmaking group the values of truth and validity were set against the values of patriotism and loyalty, the latter made it easier for members to justify why they preferred to conform rather than to dissent.[72] Hence would be dissenters, who were not completely sure of their position because of the ambiguity of the situation, were likely to be swayed by the invoked values of loyalty and patriotism. Those like George Ball who did not harbor any doubt that the Vietnam policy was wrong were not affected and not swayed.

The policies advocated by the Defense Department and the military services, and consequently their risk-taking preferences, were driven to a great degree by organizational imperatives and interorganizational relations. When, after the active militarization of the American presence in South Vietnam, the importance and role of the State Department in the decisionmaking process declined, the military (army, navy, air force, Strategic Air Command) and the Department of Defense dominated policy debate and implementation. Before 1965 the Department of Defense had demonstrated a healthy caution regarding recommendations for direct military intervention, but this changed when it became clear that the president was tilting toward a military resolution of the crisis. At this point the Department of Defense embraced the new and risky policy and became one of the main sources of optimistic assessments of the use of force. It did have disagreements with the military, which wanted faster escalation, but the two agreed on the desirability of the use of force and supported Westmoreland's troop requests, in spite of the risks involved. Later, the failure of the policies and the tensions and dislike mounting between the secretary and other senior members of the Department of Defense and the military establishment in Washington and Saigon resulted in a progressively skeptical approach to the efficacy of military means by the Department of Defense, particularly McNamara.[73] The consequence was an effort to reduce the near exclusivity of force and to proffer a strategy more focused on negotiation, with force levels capped at 470,000 men or so—but still no support for withdrawal. The JCS consistently and persistently presented a much more sanguine view of the situation than the civilians.[74] The military was more optimistic about Soviet and Chinese thresholds of tolerance and less concerned about civilian bombing casualties (Gallucci,

1975: 93–96). Escalation, they claimed, would present few new risks and would make it possible to break the will of the enemy.

All three services were conducting operations that they perceived to pertain directly to the essence of their organizations in the interlocking ground and air wars. To shrink from the risks involved or even seriously question the chances of success would have been considered a blow to the underlying rationale of their missions. This reluctance to hold back was reinforced by the traditional competition among the services. None of the services would admit the possibility that it might fail to accomplish its task.[75] The risk-taking behavior advocated by the military was thus driven not only by each service's need to prove itse. capable of carrying out its mission but also by the need to avoid the impression that it was less capable than the other services.[76] The combined effect brought organizational cognitive closure within each service, which created the organizational equivalent of interservice groupthink; this was reflected in the JCS and resulted in policy entrapment.

A change in this collective state of mind was particularly difficult to induce because the organizational culture of the American military since before the Second World War emphasized a "can do" attitude, an aggressive offensive doctrine, and the extensive use of firepower. Westmoreland, a typical product of West Point who accepted the prevailing military organizational paradigm, could not have been expected to deviate from the organizational culture and introduce a strategy adjusted to a low-key conflict against a highly motivated enemy (Komer, 1986: 48–60; Perry, 1989: 136–37; 174–75). His attrition strategy, search-and-destroy tactics, and emphasis on large-unit war were almost inevitable. Even Generals Wheeler and Johnson found it difficult to explicitly criticize him, even though Westmoreland was not their preferred choice for field commander in Vietnam, because his strategy represented mainstream thinking in the military.

In spite of being at odds, the services and the Department of Defense both still agreed even in 1967 on the need for extensive American combat troop employment, with its associated risks. Therefore, even though the Department of Defense recognized the futility of armed intervention, it would not recommend a withdrawal from South Vietnam or even substantial disengagement. As the relationship between the services and the Department of Defense deteriorated, not changing the established division of labor or the command and control structure became a matter of practical expedience; to negotiate and reach new arrangements would have been extremely difficult. Thus both sides had a tacit interest in not rocking the policy boat too hard. They were satisfied with merely stating their divergent positions and leaving things at that, thus contributing to a policy stalemate.

In the final account, however, the lack of an overall management structure for the war facilitated the predominance of the American and South Vietnamese military perspectives (Hammond, 1992: 180–90; Komer, 1986). Every agency acted independently, constraining the emergence of a comprehensive view of risks and particularly the interaction effects of separate risks being taken by different agencies. For example, pacification required a discriminating use of firepower to reduce South Vietnamese civilian casualties and wide-ranging damage to property. But the large-unit war and the strategy of attrition conducted by the army called for exactly the opposite, the use of massive firepower. Because no one had a vested interest in coordinating these contradictory policy requirements, the two approaches continued to undermine each other.

Why did Johnson and his advisers pursue military strategies that were obviously not delivering results but were raising critical assessments not only from outsiders but even from important agencies and individuals within the administration? Particularly puzzling was Johnson's continued support for Rolling Thunder when he did not believe in the efficacy of air power. His position on the impact of the bombing was summarized in his cable of December 1964 to Ambassador Taylor.[77]

> Every time I get a military recommendation, it seems to me that it calls for a large scale bombing. I have never felt that this war will be won from the air, and it seems to me that what is much more needed and would be more effective is a larger and stronger use of rangers and special forces and marines, or other appropriate military strength on the ground and on the scene. I am ready to look with great favor on that kind of increased American effort, directed at the guerrillas and aimed to stiffen the aggressiveness of Vietnamese military units up and down the line. Any recommendation that you or General Westmoreland take in this sense will have immediate attention from me, although I know that it may involve the acceptance of larger American sacrifices. We have been building our strength to fight this kind of war ever since 1961, and I myself am ready to substantially increase the number of Americans in Vietnam if it is necessary to provide this kind of fighting force against the Vietcong. (Berman, 1982: 34–35)

The failure of Rolling Thunder to achieve its objective came as no surprise to the president. But he was incapable of calling it off. Instead he found himself authorizing its expansion, even though he occasionally imposed a complete or partial halt as part of the stick-and-carrot bargaining strategy.

Several complementary explanations shed light on the persistence of failed or ineffective strategies. Let me address the air war first. The concepts behind the air war were embedded in academic theories of limited war that were widely discussed and accepted in their time. The civilian intellectuals who dominated the advisory process in 1965 deeply believed in the validity of those concepts and in the efficacy and necessity of gradualism in order to avoid the potentially devastating effects of uncontrolled escalation (Herring, 1993a). Johnson's lack of enthusiasm for the air war apparently stemmed from a lack of trust in intellectuals' prescriptions for managing men's work— war—and a lack of confidence in intellectuals as practitioners. These and his knowledge of the limited effectiveness of strategic bombing in past wars caused him to be skeptical. On the other hand, gradual controlled bombing kept the war contained and allowed Johnson to proceed with his domestic agenda, as well as retain the support of the trusted retired general and ambassador to Saigon, Maxwell Taylor, and his key cabinet members. So he accepted and authorized the air war as an experiment.

Once the air war was on, the president found himself personally involved in planning it and running it and responsible for authorizing weekly or biweekly lists of fixed targets. At this point, he became behaviorally committed to the success of the strategy, or at least to not admitting failure. He could hope that the accumulated effects would justify the costs in lives, planes, and planning efforts.[78] Thus, he was willing to authorize increased geographic scope and more targets for bombing in response to pressure from the JCS.

By the summer of 1965, although there was growing skepticism about the efficacy of strategic bombing, "virtually all [Johnson's civilian advisers] tied the air campaign to the ground effort. . . . As long as the United States maintained troops in the South, Johnson's advisers had difficulty opposing any measures that supported the ground units" (Clodfelter, 1989: 71; see also Gibbons, 1995: 160). The president, who had decided to engage a massive number of American troops in ground combat, found it hard to reduce an effort that, in the minds of the public and many in Congress, was intuitively associated with curtailing North Vietnam's support of the insurgency in the South, even though various reports and studies repeated the claim that the bombing had only a marginal effect. The prevalent view was still the counterfactual one expressed by the military. CINCPAC Admiral U.S. Grant Sharp's 1968 assessment is typical.[79] "Perhaps the most important measure of the effects of the bombing, however, would be the consideration of the situation if there had been no bombing at all. The uninhibited flow of men, weapons, and supplies through North Vietnam to confront our forces in South Vietnam could have had only one result for the United States and its allies—considerably heavier casualties at a smaller cost to the enemy" (Lewy, 1978: 390; also Cable, 1993).

The facts show otherwise. In 1965–68 almost half a million men infiltrated from North Vietnam into the South (about half of them by January 1968). Only 5 percent of the infiltrators were killed in bombings of infiltration routes. During this period the main force strength of the communists in South Vietnam increased about 75 percent, enemy attacks increased fivefold, and overall activity levels increased ninefold. The bombing destroyed only about 10 percent of the supplies moving south and did not put a ceiling on the volume of supplies that North Vietnam was able to move in (Lewy, 1978: 391). Obviously the JCS blatantly neglected to take base-rate information into account and preferred the more vivid reports and photographs of supposedly successful bombing missions.[80] The failure to stop the infiltration of men and supplies was difficult to document vividly and therefore did not undermine the well-established belief in the potency of America's awesome air power. With the Joint Chiefs' belief in bombing unshaken by base-rate information, it became practically impossible to convince the military to agree to a bombing halt, and it was impossible for the president to order such a halt as long as the military argued that it saved American lives.

Entrapment in a failed air strategy, which opened the door to deployment of combat troops, can also be attributed to the incremental manner in which the military strategy evolved. The problem with an incrementally emerging policy that involves the large-scale use of military force and has a high cost in lives is that decisionmakers find it hard to look at the prospect of failure. They cannot admit that the strategy might fail and, at the same time, hope to gain legitimacy and authorization for the strategy. It was in the interest of those who supported bombing (e.g., Taylor, the JCS) to play down its potential for leading to combat troop deployment with all its attendant risks (Burke et al., 1989: 175–76; Halberstam, 1972: 507–8). Therefore, the question of what would happen if Rolling Thunder failed was never fully discussed until it was implemented and did not deliver the anticipated results. By then it was too late. The commitment was made, and the next step up the ladder of escalation had to be taken. Incrementalism was thus conducive to misperception of the full extent of the risks involved.

A somewhat different process took place with regard to the ground war. In the absence of direction from the top, the strategy emerged from below, shaped by the field commander.[81] The attrition strategy devised by Westmoreland became associated in his mind with his reputation as a military commander and strategist. As the war became more costly, Westmoreland's commitment to his strategy, which had critics even within the army (Gibbons, 1995: 201–12), became more difficult to revoke or revise. The strategy had the support of the JCS, and President Johnson, in spite of being the commander in chief, did not feel competent to challenge JCS judgment on purely military matters. He therefore refused to question the military management of the ground war, especially one that was conducted by "a casting director's dream for the role of a general" (McNamara with VanDeMark, 1995: 121).[82] Besides, Johnson, as an astute politician, evidently saw the wisdom of Jack Valenti's advice in a memorandum written in 1964:

> If . . . something should go wrong later and investigations began in Congress, it would be beneficial to have the Chiefs definitely a part of the Presidential decisions so that there can be no recriminations at these hearings. . . . At one of the Bundy-McNamara-Rusk Luncheon Meetings, you might have Wheeler present. . . . That way . . . they will have been heard, they will have been a part of the consensus, and our flank will have been covered in the event of some kind of flap or investigation later. (Ball, 1992: 154)

Ironically, the worse the war went, the more dependent Johnson became on the support of the JCS and his field commander, and the less he was capable of interfering in the conduct of the war, which was his obligation and privilege.

Consequently, there was little incentive to rethink the attrition strategy. Johnson was, however, ultimately accountable for its results, and he understood this. The effect was increased entrapment. As the war progressed, outsiders came to see it as "Lyndon Johnson's war." At the same time, Johnson himself personalized the policy process (Barber, 1977: 51–52). He referred to the American soldiers in Vietnam as "my troops," the NSC and the Department of State as "my Security Council" and "my State Department." This personalization of the war also found expression in his commitment to win it in order to justify the casualties (Herring, 1993b; Reedy, 1982: 147). As casualties mounted, his commitment increased, despite decreasing domestic support for the war and a growing confidence gap between the public and the Johnson administration. Johnson was emotionally trapped. The personalization of the policy and the prohibitive investment of national and personal resources and emotions in it created a strong incentive for "effort justification" (Milburn and Christie, 1990). Failure and withdrawal were unthinkable. The only acceptable response was pouring in more resources by way of gambling for resurrection, at least as far as the domestic constraints would allow it.[83] Escalation was accompanied by optimism about results, and doubts were shoved aside, especially because Westmoreland claimed progress and promised success. The president had invested so much in the Vietnam policy that "failure, therefore, was a challenge to the rightness of beliefs, to some integrity of self, which must be even more fiercely defended when under attack" (Kearns, 1976: 257).

Finally, there was a practical reason for the persistence of air and ground strategies. The strategies, which had emerged from a lengthy debate within the administration, reflected a compromise reached after a heavy intrabureaucratic and interbureaucratic infighting. The compromise was resistant to change. Nobody wanted to go through that process again or to rework a complex policy with several elements that required orchestration (Thies, 1980: 295–99). Given the deepening civil-military rift in the defense establishment, the policy status quo also assuaged Johnson's fears of a military revolt and allayed concern on the part of the JCS that they might be perceived to be challenging civilian authority (Herring, 1994: 49–50). Under these circumstances policy persistence became the shared objective of most key players—they saw it as the lesser evil—and prevented necessary reforms in strategy.

This analysis of U.S. policy and risk-taking behavior in 1964–68 in connection with Vietnam exposes an individual-dominated decisionmaking process. But an understanding of the complex set of decisions that unfolded, and their evolution over time, requires the recognition that, as important as Johnson's personality was, other social, contextual, and, to a lesser degree, cultural variables must be factored in, as must the interaction of all these with the personality variables. Such a broad explanation accounts not only for the general preference for escalation over withdrawal but also for the preference within this policy orientation for limited war over all-out escalation. Each option represented different types and levels of risks. The explanation offered here clarifies why Johnson and his advisers were so insensitive to the alarm signals that warned of the imminent risks of failure until they were overtaken by events in January 1968.

The Lebanon Case

AN OVERVIEW

The modern origins of the state of Lebanon and its communal structure can be traced to a political entity that was created by the Ottoman Empire and recognized by the international community in 1861. Nineteenth-century Lebanon had a clear Christian majority, which made it an anomaly in a Muslim region, and the country required European assistance to assure its continued existence. After the First World War, France was given a League of Nations mandate to govern Lebanon and neighboring Syria. In 1920 France drew a border to separate the two territories, strengthening Lebanon and increasing its territorial space by including land that was historically part of Syria, thus creating what came to be known as Greater Lebanon. The new territories were populated primarily by Muslims and non-Maronite Christians, thereby diluting the Christian Maronite character of Lebanon and destabilizing the demographic balance in that country.

The independent state of Lebanon, established in 1943, was made possible by the formation of the National Pact. Under the agreement signed by leaders of all Lebanon's communities, the Muslims recognized the legitimacy of a sovereign Lebanon under Christian political primacy in return for the Christian communities' willingness to share power and recognize the Arab character of the country. It was agreed that a Maronite Christian must be president and that there must be six Christians to every five Muslim deputies in Parliament. The system resembled a confederation of proto-national communities, each of which commanded the ultimate allegiance of its members. The main religious groups recognized in the system were the Maronite Christians, the Druzes, the Sunni Muslims, and the Shi'ite Muslims. There were several smaller religious groups as well, including Greek Orthodox, Armenians, and Greek Catholics. Yet extremists in each group rejected the compromise and continued to seek a purely Muslim or Christian Lebanon. The foundations for long-term instability and communal conflict were inherent in this status quo–oriented and archaic political system.

A severe civil war erupted in 1958 when a coalition led by the Druze leader Kamal Jumbalatt and the Sunni leader Saeb Salam challenged the authority of the Lebanese government headed by President Camille Chamoun, a Maronite Christian. The challenge and resultant crisis were sparked by the flaring of pan-Arab Nasserite ideology and by Chamoun's successful attempts to exclude some important rival politicians from election to parliament. At Chamoun's request, President Dwight Eisenhower sent U.S. Marines on July 15 to help the government reassert its authority. The commander of the Lebanese army, Fu'ád Shihab, credited with maintaining a neutral position during the crisis, was elected president shortly thereafter.

A popular president, Shihab led Lebanon during the critical years 1958–64 and effectively preserved a moderate balance between Lebanon's Christian identity and an Arab nationalist orientation. Nonetheless, his presidency was marked by a return to the traditional political status quo. Shihab did not attempt to reform the power distribution or relations between community and state. When he reluctantly stepped down in 1964, the inherent flaws in the Lebanese system became apparent, but at the same time attempts at reform became increasingly difficult to push forward and threatened stability. The conservatives in the Christian communities insisted on maintaining Christian primacy and preserving their dominant socioeconomic status, while forces in the Muslim communities sought to improve their political, economic, and social position. Shihab's successor, Charles Helou, was a weak and ineffectual president who did not have a stable majority in parliament and who, on the foreign front, had to deal with growing inter-Arab rivalries and an escalating Arab-Israeli conflict. In 1970 a conservative, Suleiman Faranjiyya, was elected president. Faranjiyya soon faced a series of domestic and external developments with the renewal, extension, and intensification of activities by the Palestinian organizations. A full-scale civil war broke out in 1975; in the next year and a half more than 10,000 Lebanese died and many more were wounded (Rabinovich, 1985: 57).

The civil war was precipitated by several events, most of which were external to Lebanon. The most important was the establishment of the headquarters of the Palestine Liberation Organization (PLO) in Lebanon in 1970 after the organization was exiled from Jordan, following its failed attempt to overthrow King Hussein, in what became known as Black September. Lebanon had already accepted 180,000 Palestinians into its territory after the 1948 war between the newly established state of Israel and its Arab neighbors, but they were considered primarily refugees and, in line with prevailing Arab policy, were not integrated into Lebanese society. The presence of large numbers of Palestinian refugees exacerbated tensions between Lebanese Christians and Muslims. The PLO had organized an independent armed force in Lebanon and staged limited operations from Lebanon for years, that in 1968 instigated a retaliatory Israeli raid on Beirut's international airport. Efforts by Gamal Abdel Nasser of Egypt to regulate relations between the Lebanese government and the PLO led to the Cairo Agreement of 1969, but this did little to ease the problem. The Palestinians made Lebanon's southern region a staging ground for military operations against Israel, which resulted in more Israeli retaliations and often disrupted the daily life of the predominantly Christian and Shi'ite Muslim populations. Many Shi'ites fled north and to Beirut, further unsettling the demographic balance among communities, particularly in Beirut.

Several other external factors also influenced events in Lebanon significantly. The oil embargo of 1973 and the subsequent accumulation of huge financial resources in the Arab world had vastly increased the resources available to causes promoting Islamic solidarity and power. Lebanon's Muslims thus were inspired to be assertive in pursuit of their political goals. During the same period, the growth of the Muslim population brought into relief the unjust distribution of political power, which favored the Maronites and other Christian minorities. There was also a significant shift in the position of the Western powers in the Middle East. Unlike during the 1958 crisis, the Western powers were now cautiously analyzing the newfound power of the Arab world and were not inclined to intervene on behalf of minority interests in an archaic political system that clearly required major reforms. Syria, which was now enjoying a spell of domestic political stability under President Hafiz al-Assad, had evolved as a strong regional military power with an ambitious foreign policy. Syria had never abandoned its implicit claim on Lebanon, especially on those areas annexed to Lebanon in 1920. Speaking in July 1976, President Assad said: "Historically Syria and Lebanon are one country and one people" (Avi-Ran, 1986: 14). One of Assad's main policy objectives was to extend Syria's influence over neighboring countries, including Jordan and Lebanon, and over the Palestinians. The Syrian leadership attempted to capitalize on Egypt's declining influence in the Arab world and to prove, to the United States as well as to the Arab allies of the United States, that it could provide effective, moderate leadership in the region (Rabinovich and Zamir, 1982: 34–39).

By December 1976 the collapse of the Lebanese political system was imminent. Fighting had raged between various factions since April the previous year. Prime Minister Yitzhak Rabin of Israel had declined to intervene on behalf of the Maronite community, which had requested assistance; he decided that the possibility and cost of war with Syria were too high. A Syrian-dominated Arab Deterrent Force, sanctioned by the Arab League, intervened instead and brought the intense fighting to a halt. The move was endorsed by both the United States and Israel, the former having mediated to obtain Israel's tacit agreement to military intervention by Syria. In return, Syria accepted limitations on its presence, specified in the "red lines" understanding of 1976, which established a modus vivendi that lasted until 1981. By the terms of the Syrian-Israeli agreement, Syria would not dispatch forces south of the Litani River, use its air power in Lebanon, or deploy ground-to-air missiles in Lebanon.

Eighteen months of civil war took their toll. The population was exhausted, and political authority existed only in Beirut. The rest of the country was controlled by the Arab Deterrent Force (primarily Syrian, but with other token Arab contingents), local militias, or the PLO. The country lost many of its functions as a financial, cultural, and communications center in the region. Large numbers of Christians, believing their sheltered status to have been lost forever, emigrated and took large sums of capital with them. The Lebanese army collapsed in 1976, and rival factions could not reach an agreement on its restoration.

As political order broke down, four critical problems became salient: the Syrian desire for hegemony in the region; the need to resolve the Palestinian issue; the ongoing conflict between Christians and Muslims in Lebanon; and Israel's quest for security (Rabinovich, 1985: 89). In the immediate aftermath of the civil war, Syria seemed to have gained the most. The United States and the regional Arab powers endorsed de facto its preeminent position in Lebanon, and there was the possibility that Syria could further consolidate its power. The agreement with Israel appeared to allow Syria a free hand while satisfying both countries' desire to minimize friction with each other. There were occasional military skirmishes between Syria and Israel, but largely because of Israeli raids into southern Lebanon in retaliation for PLO raids into Israel. Then, between 1977 and 1980, the political opposition in Syria attempted to overthrow the Syrian regime by instigating a series of crises, culminating in an attempt to assassinate President Assad in 1980. This renewal of domestic instability, combined with a substantial reduction of economic support for Syria from the Arab world, led Syria in 1980 to announce a partial withdrawal of its forces from Lebanon and their redeployment.

During this period several events took place in Israel with important results for Israel's policy toward Lebanon. In a 1977 election the right-wing Likud coalition defeated the Labor alliance and came to power with Menachem Begin as prime minister. The Likud coalition was generally believed to be more nationalistic, conservative, and hard-line in pursuing Israel's foreign policy goals in the region. Since the 1975 civil war, Israel had been building an elaborate system of defenses along its border with Lebanon to protect its northern settlements in the Galilee. Israel was also responsible for helping to establish, and providing economic and material support to, Major Sa'd Haddad, a Greek Catholic and the leader of a local militia recruited from the Christian population of southern Lebanon.

A Palestinian raid into Israel, north of Tel Aviv, in March 1978 caused heavy casualties—37 people killed and 78 wounded—and led Israel to launch Operation Litani. This large-scale retaliatory raid into Lebanon had two objectives: to destroy PLO bases that were a continuing source of harassment to towns and villages in the Galilee and to extend the buffer territory under the control of Major Haddad in order to improve Israel's border security. The Israeli operation was only a partial success, however. Heavy artillery shelling and air strikes resulted in a large number of civilian Lebanese casualties, extensive property damage, and a significant refugee problem (Yaniv, 1987: 72). The operation failed to engage large numbers of PLO forces, who had sufficient time to retreat north. With the Israeli political reputation strained internationally, particularly with the United States, there was effective pressure on the Israeli government to withdraw and accept the presence of a United Nations Interim Force in Lebanon (UNIFIL) south of the Litani River. Major Haddad's forces were allowed, at Israeli insistence, to occupy a five-kilometer security belt along the border. Israel considered the U.N. forces a liability because they proved incapable of halting the PLO's strikes at Israel but restricted Israel's ability to launch retaliatory raids into Lebanon. Thus, during 1977–82 there was a significant hardening of Israel's attitudes vis-à-vis its northern neighbors.

In 1978 the Israeli government had begun to forge a closer working relationship with, and expanded support for, the Lebanese Front, the umbrella coalition of all Christian parties in Lebanon. The leader of the Front, Bashir Jumayyil, appeared to be gaining in both effectiveness and popularity, and among the Lebanese faction leaders he was the most vocal in his determination to bring about departure of Syria and the Palestinians from Lebanon. In 1980 Syria turned over some of the area occupied by its forces in Beirut to the PLO and redeployed its troops along the strategic Beirut-Damascus road and in the Beqa'a Valley. With greater Israeli support the Lebanese Front grew in power and confidence; in March 1981 it extended its military and political presence into an area considered by Syria to be of vital strategic importance, a town just north of the Beqa'a Valley and east of Beirut known as Zahle. An immediate Syrian attack to retake the town ended in the shooting down of two Syrian helicopters by Israel, an intervention in support of Front forces. In retaliation Syria placed four Soviet-made surface-to-air (SAM-6) missile batteries in the area, in direct violation of the red-lines understanding with Israel. As a crisis atmosphere developed, third parties took pains to articulate their positions. The Soviet Union made it clear that it would not intervene in Lebanon on Syria's behalf. Egypt signaled its noninvolvement by proceeding with the implementation of the Camp David peace accords.

In July 1981 the PLO shelled the northern Galilee, inflicting heavy damage on Israeli towns and villages. Israel responded with massive air raids on Palestinian positions in Beirut and southern Lebanon. The July shelling, however, proved to all parties that the PLO was still able to operate effectively against Israel from its bases in Lebanon and that Israel was unable to eliminate this threat through retaliatory strikes. A cease-fire was worked out in July with the active mediation of both the United States and Saudi Arabia, but the Syrian missiles remained in place. Meanwhile, the United States renewed discussions with Egypt and, less directly, with the PLO, which sought to use the Egyptian-Israeli peace initiative to develop a peace process that would include the Palestinians. These discussions alarmed the Israeli leadership, who feared that the PLO might persuade the United States to engage in a direct dialogue. In April 1982 the last Israeli troops completed their withdrawal from the Sinai; with peace on its southern border with Egypt, Israeli attention could now focus on the conflict in Lebanon (Yaniv, 1987: 104–5).

Between December 1981 and June 1982, Prime Minister Begin and Defense Minister Ariel Sharon failed on several occasions to win support from the Israeli cabinet for a large-scale invasion of Lebanon code-named Big Pines (later referred to simply as the Big Plan). Despairing of obtaining authorization for the invasion, they contrived to win approval for a limited operation (Yaniv, 1987: 107–9). The immediate pretext for the Israeli invasion came from the cease-fire negotiated in July 1981. According to Israeli interpretations of the cease-fire, the PLO would not attack any Israeli position or asset, either in Israel or anywhere else. The PLO objected to this interpretation, arguing that the PLO was an amalgam of political factions, not all of which could be fully controlled. The attempt by a Palestinian to assassinate Shlomo Argov, the Israeli ambassador to London, on June 3, 1982, brought heavy retaliatory air strikes against PLO positions in Beirut. The PLO countered with a heavy shelling of the Galilee. Just a week earlier, Alexander Haig, U.S. secretary of state, had made a speech in which he indicated support for an Israeli action in Lebanon. This was one of a number of signals that Begin and others interpreted to indicate an American supportive position (Naor, 1993: 249). Emboldened by the signals of support, affronted and enraged by the attack on the Israeli ambassador, and deeply concerned about continued PLO shelling of and incursions into the Galilee, Prime Minister Begin requested and received Israeli cabinet approval for a limited operation against PLO forces in Lebanon. Operation Peace for Galilee began on June 6, 1982.

The approved plan was designed to achieve four objectives, only the first of which was explicitly stated to either the cabinet or the public. The announced objective was to destroy the PLO military infrastructure in southern Lebanon and create a security zone of 40 kilometers, the effective range of the PLO's artillery and rocket launchers. The remaining objectives became apparent over the course of the next several months and as a result of actions taken by the Israeli Defense Forces (IDF) in Lebanon. One was to destroy the PLO's position in the rest of Lebanon, particularly Beirut, in order to eliminate its hold on the Lebanese political system and to diminish its role in the Arab-Israeli conflict. Another was to defeat the Syrian army in Lebanon and to effect its full or partial withdrawal. The last was to facilitate the reconstruction of the Lebanese state and political system under the hegemony of Israel's allies, Bashir Jumayyil and the Lebanese Front (Rabinovich, 1985: 122). Although most members of the Israeli cabinet were initially opposed to the large-scale invasion, Begin and Sharon argued that limited strikes would have no effective use, just as with Operation Litani, and that as long as the PLO retained its headquarters in Beirut, it could harass Israeli settlements. The cabinet eventually approved the launching of Operation Peace for Galilee.

As with Operation Litani, the PLO forces in southern Lebanon withdrew quickly to the north. Within a few days, Israeli troops advanced to the outskirts of Beirut without engaging PLO forces in any significant battles. The Syrian air defense system north of the Beirut-Damascus road was attacked and destroyed both to back Israeli advances by allowing its air force complete freedom to operate in support of the ground operations and to minimize the risk of heavy casualties. By maneuvering Israeli tank divisions to outflank the Syrian forces, Defense Minister Sharon had placed the Syrian leadership in the difficult position of choosing between a humiliating retreat and standing and fighting. Asad chose to do the latter, and Sharon demanded and received the government's approval for a direct assault on Syrian positions in the Beqa'a Valley. When Israeli forces reached the suburbs of Beirut,

Sharon and Begin, who believed that they had an agreement with the Lebanese Front to support the IDF by attacking PLO positions within the city, found themselves disappointed. It soon became clear, however, that the Front leader Bashir Jumayyil was becoming evasive and was not going to attack the PLO forces. At the same time, the Reagan administration increased its pressure on Begin to accept a cease fire, and the Israeli cabinet members also realized that they had been duped into incrementally approving operations beyond the one stated in the original plan. The cabinet withheld permission for Sharon to continue. By the end of June the city of Beirut was under siege, and the Israeli polity was divided on how to proceed.

The PLO forces under siege in Beirut were still in a position to extract a political or moral victory if the IDF withdrew. During a period of six weeks Sharon vainly tried to gain cabinet approval to tighten the siege. Without approval, he authorized massive air strikes on the Lebanese capital on August 6 and 12. An angry American response and growing criticism within the Israeli cabinet and from the public of his conduct of the war led the Israeli government to agree to cease all military activity in and around Beirut and allowed the completion of the negotiations for the withdrawal of PLO and Syrian forces from the city. By August 21 the PLO was rapidly leaving for safe havens in eight different Arab countries; the withdrawal was completed by September 1.[84]

As the war quickly escalated beyond the initially declared intentions, the United States became concerned that it might adversely affect American interests in the region. Its concern constrained Israel's freedom of action, eventually resulting in the announcement of the Reagan Peace Plan on September 1, 1982. The plan called for self-government, in association with Jordan, for the Palestinians in the West Bank and Gaza. Implementation of the plan required a quick settlement of the situation in Lebanon and an Israeli withdrawal. The government of Israel rejected the plan and its main provisions outright (Ben-Zvi, 1993: 140–44; Medzini, 1990: 180–85). Tensions in U.S.-Israeli relations increased, and the United States took an even more active stance in the crisis.

Next came the assassination of Israel's charismatic and astute ally Bashir Jumayyil and massacres by the Christian Lebanese militia—the Phalangists—in the Palestinian refugee camps of Sabra and Shatilla. These actions resulted in intense domestic and international pressures on Begin's government, which was held accountable for not taking preventive measures to avoid the massacres. These pressures compelled Israel to withdraw from western Beirut. Israel had lost the initiative in Lebanon. Its forces found themselves combining retrenchment and retreat (Feldman, 1992: 150). The IDF was caught in a cross fire between various Lebanese factions and incurred a growing number of casualties. Public support in Israel for the Lebanon intervention eroded, and relations with the United States suffered, but the government remained committed to its policy. By early 1983 Lebanon, the PLO, and Syria recognized the weakened position of Israel and were unwilling to concede to its demands: a complete withdrawal of the PLO and Syria from Lebanon, security arrangements in a 40-kilometer-wide strip along the Is- raeli-Lebanese border, and a de facto peace with Lebanon. It would take nine more months for the Israeli government to recognize that it was unable to pay the political, military, and economic costs of continued occupation. Fi- nally, on September 4, 1983, it undertook a unilateral withdrawal to the Awali River. On September 15, Prime Minister Begin resigned. The Israeli policy toward Lebanon was redefined, and its objectives became much more modest, focusing on the immediate threats in the southern part of Lebanon bordering Israel. Still, it took almost two more years (until June 10, 1985) for the IDF's presence in Lebanon to end.

Israel's intervention in Lebanon was a large-scale military operation in- volving all three main branches of the Israeli Defense Forces—the ground forces, the air force, and the navy. It was, however, principally a conventional ground operation that involved a large number of troops, regulars and re- serves. At least formally, it started as a neutral intervention, that is, one not intended to bring about a change of government in Lebanon. But it turned into a hostile intervention after the first week, when the objectives were broadened to include the establishment of a strong Christian-dominated government friendly to Israel in place of the weak Syrian-controlled govern- ment. Although Israel's intervention in Lebanon lasted for more than three years, it was initially intended to be a short-term operation. In fact, Prime Minister Begin believed that it would be over within a few days, and a similar view was expressed by the Chief of Staff, General Rafael Eitan (Eitan with Goldstein, 1985: 211; Naor, 1986: 16, 48–49).

ISRAEL'S INTERESTS

Israel's military intervention in Lebanon was the most controversial and divisive war in the history of the state. Controversy was itself unusual in a state where the majority of the population was united by a common external threat. Israel's wars, broadly perceived as ultimate acts of self-defense, were characterized by national unity, even though the society otherwise exhibited a lively diversity of opinions among various social sectors and political par- ties and groups. The war in Lebanon was different. It resulted in deep, last- ing chasms within society. The controversy focused on the objectives of Op- eration Peace for Galilee and their relevance to national interests. Supporters and critics contested three issues: the war's objectives, definitions of the na- tional interest, and the instrumentality of the war's objectives in advancing the national interest. The war turned out to be costly and protracted, with- out achieving important stated goals—and it occurred after the even costlier Yom Kippur War of 1973. Had the war been shorter, less costly, and more successful, like the Six Day War of 1967, the controversy might never have reached such an intensity or resulted in such dramatic political fallout.

To explain the contested issues requires defining Israel's interests and objectives in four contexts: perceptions of overall national defense and security; the perceived security threat to northern Israel; the politics of the Arab-Israeli conflict as they developed after the peace process with Egypt began in 1977; and domestic politics, especially the rivalry between the rightist ruling Likud Party and the defeated Labor Party, a rivalry that reflected a broader political-ideological divide between right and left nurtured from deep historical roots.

A Palestinian terrorist's attempt to assassinate Israel's ambassador to London triggered Israel's intervention in Lebanon. But the planning for the operation preceded that event by many months and represented a change in the government's strategic assessment of how to deal with the military and political threat posed by the autonomous Palestinian territorial base in Lebanon. The first stage of the intervention, from its start to the siege of Beirut, was driven by the objective of destroying the PLO command and infrastructure in the area stretching up to 40 kilometers from the Israel-Lebanon border. The most prominent Israeli interests were military-strategic.

Three developments were associated with the emergence of Lebanon as a source of considerable threat. First was the rise in influence of the PLO and its decision after Black September in 1970 and the PLO expulsion from Jordan to relocate its headquarters and most of its fighting forces and training facilities from Jordan to Lebanon. The second development was the inherent weakness of the Lebanese polity, the deepening intercommunal cleavage within Lebanese society, and the gradual weakening of state institutions. These allowed the PLO to operate as an autonomous actor practically without interference from the Lebanese government. For Israel to either pressure or negotiate with the government of Lebanon to resolve the problem was useless.

These two developments had serious implications for a third—the growing insecurity of Israel's northern border (Yaniv and Lieber, 1983). After 1968 attacks on Israeli civilian targets from Lebanon increased, which had serious adverse morale effects on the population in the towns and villages along the border. What followed was a cycle of PLO attacks and limited punitive reprisals by Israel against Palestinian targets in Lebanon. Eventually, the government realized that the policy of reprisals did not achieve the desired goal of deterrence. Even when the costs to Lebanon were considerable, the Lebanese government was incapable of imposing restraint on the PLO. Some key Israeli decisionmakers therefore believed that a large military intervention was necessary to deal with what had become both an acute security threat and a domestic problem. Between March 1978, when the Litani operation was undertaken, and 1981 the PLO had acquired a variety of artillery pieces, including multiple rocket launchers and long-range 180-millimeter guns, which allowed it to shell the Israeli population over the UNIFIL positions in Lebanon and from beyond the security buffer zone carved out of southern Lebanon. The shelling escalated and in July 1981 led to the "mini-attrition," when PLO artillery simultaneously bombarded 33 Israeli communities from Nahariya on the Mediterranean to Kiryat Shmonah at the foot of Mount Hermon. Israel responded with heavy bombings by artillery and from the air. Nonetheless, the civilian population along the border was vulnerable to panic, and the government was apprehensive that the public at large would view it as weak and incapable of dealing with the security challenge. Both concerns required rethinking the strategy of reprisals that seemed to have only limited and short-term effects.

Military confrontation with the PLO became inevitable, in spite of the July 1981 cease-fire agreement. In the long run, Israel could not allow the PLO to hold the civilian population in the north hostage or to get away with sporadic acts of terror abroad. The credibility of Israel's deterrence posture and its reputational interests, among friends and foes alike, were in danger (Naor, 1986: 46; Yaniv and Lieber, 1983: 131). The attack on the Israeli ambassador provided Israel with the excuse to respond militarily. Broader political objectives—concerning the Syrian presence, the PLO headquarters in Beirut, and the political power structure in Lebanon—were suggested by the defense minister, but they were not approved by the government. The cabinet, focusing on strategic-military gains, approved an operation that would push the PLO 40 kilometers back behind the Awali River, putting most Israeli civilian targets beyond PLO artillery and rocket range. By June 11, 1982, six days after the operation had begun, this objective was achieved, and the Syrian missile batteries were destroyed. Success did not, however, make it impossible for the PLO to shell the Galilee. The Syrian forces in the Beqa'a Valley provided sanctuary to the PLO, and the Galilee could be shelled from there, as happened even during the operation (Yaniv, 1987: 112). Hawks in the Israeli government further assumed that only by destroying the PLO's main and only autonomous base in the region would its influence be diminished to such a degree that self-government for Gaza and the West Bank, called for in the Camp David Accords, could be negotiated directly with the populations of these territories, thus assuring that Israel would have long-term control over the territories (Evron, 1987: 107–8).

The emerging logic of the political-strategic situation required another extension of the scope and objectives of the operation. To prevent a Syrian attack on the IDF flank, it was considered strategically imperative to induce the Syrians to retreat, yet this had to be accomplished without a direct military confrontation, which was specifically forbidden by the cabinet decision that authorized the operation.[85] Accordingly, the plan was to outflank the Syrians, forcing them to retreat, while reassuring Assad directly and through Reagan administration channels that the IDF would not attack the Syrian forces unless attacked first. This flanking move, which cut off the Beirut-Damascus highway and Syrian reinforcements, brought a direct military confrontation with the Syrians and carried the IDF all the way to Beirut. At this point, it made sense to augment the goals of the operation to include expelling the PLO headquarters from Beirut, so that the PLO would not salvage a moral and political victory from military defeat and thus retain the bargaining power to become a negotiating partner in any peace settlement in the region. Finally, to make sure that the decision to withdraw the PLO and the Syrians from Beirut would not be overturned as soon as Israeli forces withdrew, it seemed necessary to pursue as an additional objective that of establishing a strong pro-Israeli, anti-PLO, anti-Syrian Lebanese government.[86] In practical terms, that meant a government controlled by the Christian Lebanese Front and led by Bashir Jumayyil. Israel, in extending its interests and objectives, was sliding down a slippery slope. These interests and objectives were part of a predesigned strategy for dealing with Lebanon, Syria, and the PLO that preceded the war, and they were set by the defense minister. But domestic constraints prevented the architects of the policy from openly declaring the nonmilitary interests and objectives from the start, so they emerged only incrementally (Feldman and Rechnitz-Kijner, 1984; Shiffer, 1984: 111, 114–20; Yaniv and Lieber, 1983; Yaniv, 1987: 110–17).

Several risk dimensions associated with these interests were prominent in the minds of Israel's decisionmakers: (1) the risk of getting involved in an all-out war with Syria, including engagement on the Golan front; (2) the risk that, if the vulnerability of Israel's northern cities and villages was not addressed decisively, much of their population would migrate to the already overpopulated central and safer parts of the country; (3) the risk that the peace process with Egypt would be negatively affected by intervention in Lebanon; (4) the risk that nonaction or something less than total victory over the Palestinians and their Syrian allies would be detrimental to Israel's reputational interests and deterrent credibility; and (5) the risk of an unbearable and unacceptable number of casualties in a military operation—always a dominant concern of Israel's decisionmakers. This last factor had been particularly acute since the Yom Kippur War; the large number of casualties then was still vivid in the minds of the public and the politicians.

NATIONAL CAPABILITIES

The change in government from left-wing Labor to right-wing Likud in 1977 represented a structural strategic change in approach to security policy. The implementation of the Likud approach was delayed because the foreign and defense policies of the first Begin government (1977–81) were influenced by Defense Minister Ezer Weizman, Foreign Minister Moshe Dayan, and Chief of Staff Lieutenant General Mordechai Gur, three individuals with worldviews either rooted in, or close to, Labor ideology and conceptions of national security. The Likud security doctrine was not implemented until the second Begin government (1981–83), with Ariel Sharon as defense minister, Yitzhak Shamir as foreign minister, and Lieutenant General Rafael Eitan as chief of staff. The security doctrine of the Labor Party held that war was essentially a defensive and preemptive instrument. In the Likud view, in contrast, war was a legitimate means of achieving broader political objectives, including reshaping the regional political order.[87] This basically Clauzewitzian approach, as expressed in the Lebanon War, amounted to an offensive military doctrine. It represented greater willingness to accept higher risks when the situation offered a window of opportunity for achieving important political objectives and when the balance of military power was decisively in Israel's favor. These two very different doctrines found expression in the public debate on whether only "no choice" wars or whether both "no choice" and "by choice" wars were legitimate.[88] The Likud doctrine posed a much more difficult challenge for establishing domestic legitimacy for war and its costs than the traditional defensive-preemptive doctrine, which legitimized war only when Israel was clearly threatened (Horowitz, 1985; Inbar, 1989; Lanir, 1985). Begin and Sharon underestimated the gravity of this problem. They had not anticipated the extent to which the public would resent the moral dilemmas and the costs and therefore the degree to which policy legitimacy would be challenged.[89]

Israel's policy toward Lebanon went through three distinct phases in the late 1970s and early 1980s, each representing the acceptability of more risk. Begin's first government, which took office in 1977, continued, under the guidance of Defense Minister Weizman, the basically defensive policy of Weizman's predecessor, Rabin. This required a 10-kilometer buffer zone controlled and defended by Major Haddad and his militia, UNIFIL forces that extended north to the Litani River, and selective military operations by the IDF against Palestinian targets north of the Litani in reprisal for Palestinian attacks. With regard to the Christians, the position was that "Israel will help the Christians help themselves" through military and economic assistance but without direct intervention (Rabin, 1983: 10, 58). The second phase started when Begin took over the Defense Ministry himself in 1980, after the resignation of the minister of defense, and Chief of Staff Eitan persuaded him to adopt a proactive offensive policy. This meant continual preventive offensive ground, naval, and air operations against Palestinian targets, which culminated in the summer of 1981 in heavy bombings of Palestinian headquarters and bases in Beirut and throughout Lebanon. The third phase, which emerged when Ariel Sharon was appointed minister of defense in the second Begin government, was much more ambitious. Envisioned were the total elimination of the PLO's infrastructure and military capabilities in Lebanon and the restructuring of the Lebanese political order with a strong pro-Israeli central government. These objectives would require large-scale intervention by Israel to destroy Palestinian military power, deter Syrian interference in Beirut and other parts of Lebanon, and provide a military umbrella for Israel's Lebanese ally, Bashir Jumayyil, to seize power and maintain it.

Barring an Israeli agreement to recognize and negotiate directly with the PLO, a military solution that would take advantage of Israel's great military superiority to defuse the Palestinian threat was almost inevitable. The entrenchment of Palestinian military organizations in southern Lebanon made the area practically an autonomous Palestinian region, from which were staged military operations against Israeli targets and even more often in the buffer zone established after the Litani operation and defended by Major Haddad and his Israeli-trained and Israel-supplied militia. During the national election campaign in early 1981 and the mini-attrition in July 1981, Begin made public commitments to remove the threat of Palestinian artillery to northern Israel. Thus, when the newly appointed minister of defense in the second Begin government came into office, he found that the IDF General Staff had prepared plans for a range of military operations against the Palestinians under the assumption that a large-scale military intervention was unavoidable (Gabriel, 1984: 60–61; Shiffer, 1984: 65).

In and of itself the Palestinian military forces in Lebanon were no match for the powerful, modern, and well-trained IDF. Although the outcome of a direct all-out military conflict could never be in doubt, three risks were pertinent in Israeli decisionmakers' minds. First and foremost was the number of casualties that could be expected. Asked by government ministers about this, the IDF chief of staff refused to give an estimate. The second was the risk that the time allowed to complete the military objectives would be very limited. It was assumed that within days international pressure for a cease-fire would mount. But pushing ahead quickly would increase the number of casualties, whereas proceeding cautiously would run the risk that international pressures would be effectively applied before the military and political goals were achieved, thus giving the Palestinians a claim for victory. Finally, there was the risk of Syrian intervention and expansion of the war to the Golan front (Eitan with Goldstein, 1985: 210–11, 229–32).

To deal with these risks it seemed best to use massive military force capable of attack on a wide front, from the Mediterranean to the Beqa'a Valley; those IDF units could simultaneously engage Palestinian forces, local militias, and the Syrian forces in Lebanon while other IDF units reinforced the Golan front and established a deterrent-defensive posture there. This decision entailed the mobilization of a large number of reservists, which in turn would adversely affect domestic politics. It also made it unlikely that preparations for the operation could be kept secret; forewarned of an impending invasion, the Palestinians could prepare for it and thus raise the costs to Israel.[90] Yet both Palestinians and Syrians were tactically surprised. Israel had been concentrating large military forces in the north since December 1981, claiming that they were needed to deter possible Syrian retaliation for Israel's annexation of the Golan Heights. So the exact timing of the operation could not be inferred from troop concentrations. And in Israel the adverse consequences of large-scale reservist mobilization were initially not deemed too threatening, because Defense Minister Sharon and Chief of Staff Eitan were confident that the combination of firepower and maneuverability, especially in light of the decisive advantage in numbers and technology, was bound to bring a quick (96 hours), low-cost victory that would settle the Lebanon problem once and for all.

On the eve of Operation Peace for Galilee, Palestinian forces in Lebanon numbered 15,000 combatants. They were equipped with about 100 tanks, mostly T-34s (an obsolete Second World War model) and a small number of the more advanced T-54s. They also had 350 artillery pieces, about 80 Katyusha rocket launchers, about 150 armored fighting vehicles and armored personnel carrier vehicles, more than 200 antitank rocket launchers and about 200 antiaircraft guns. About 8,000 of the combatants were positioned in southern Lebanon, and most of the rest were in Beirut and the surrounding refugee camps. The Palestinian strategy was to avoid frontal open-ground engagement, in which the IDF could bring to bear its quantitative and qualitative superiority, and instead to retreat in an orderly fashion to the refugee camps and urban areas and draw the IDF into fighting there, where its advantages would be largely neutralized. This was also the type of fighting with which the Palestinians had acquired extensive experience during the civil war in Lebanon. The strategy had another intended benefit: it would slow down the progress of IDF forces and allow the international community to impose a cease-fire, which would prevent the IDF from achieving its objectives (Avi-Ran, 1987: 197–98).

Syrian forces in Lebanon on the eve of the war included two armored brigades, two mechanized infantry brigades, an armored unit, and two infantry brigades of the Palestinian Liberation Army, commando battalions, artillery units, anti-aircraft missile batteries, and other anti-aircraft and anti-tank units. The total deployed force consisted of 30,000 men, 612 tanks, 30 commando battalions, 150 armored personnel carriers, 300 artillery pieces and anti-tank guns, and 30 SAM batteries. The war caught the Syrian army tactically unprepared. It had not anticipated the timing and scope of the invasion. Their force in Lebanon was widely dispersed, suffered from morale and discipline problems, and was unprepared for a large-scale Israeli attack. The Syrian leadership anticipated an attack on the Palestinians but not on their own forces in Lebanon. Thus the Syrian army in Lebanon found itself in a decidedly inferior position, both quantitatively and qualitatively, compared with the IDF.[91] Only toward the end of the first week of the invasion, when Syria became convinced that Israel did not intend to expand the war and would not attack along the Golan front, was another reinforced armored division transferred from the Golan front to the Beqa'a Valley (Avi-Ran, 1986: 155–56; Avi-Ran, 1987: 239–41; Gabriel, 1984: 233).

As Israel's military presence in Lebanon extended into 1983 and 1984, Syria, with the assistance of the Soviet Union, engaged in a military buildup and reduced, but could not eliminate, its inferior military position vis-à-vis Israel. Although the Syrian army did not reach strategic parity—it could not expect to defeat the IDF and gain an endurable victory—it became a formidable opponent that could not itself be defeated at an acceptable cost and in a short time. This deadlock in the military situation in Lebanon was more detrimental to Israel than to Syria because (1) to maintain substantial forces in Lebanon, Israel had to call up reserves, and (2) because Israel was a democracy, domestic disenchantment and criticism were formidable obstacles to sustaining the Lebanon policy (Heller et al., 1984: 246–48, 266–67).

In comparison with the military forces of Syria and the PLO, the IDF seemed decisively superior on practically every dimension. The Israeli force had about 80,000 troops and 1,240 tanks operating in nine divisional structures and two provisional task forces (Davis, 1987: 78). Its technological advantage, proven on the battlefield, was substantial—especially in the air. Using advanced electronic warfare capabilities, in a single afternoon the Israeli air force destroyed or damaged seventeen of nineteen SAM batteries, cleared the skies, and established total air superiority by shooting down 92 Syrian Migs (25 percent of their first-line fighters) with no loss of aircraft to itself. All Syrian and PLO forces were thereafter exposed to air attacks, whereas Israeli ground forces could advance without the threat of air attacks, and the air force was free to engage in ground support tasks. Similarly, the innovative use of remote-piloted vehicles made real-time intelligence on enemy force structure and positions possible, which helped improve planning and reduce casualties (Carus, 1985; Gabriel, 1984: 97–100, 195). The preponderance of military power gave the government and the IDF high command confidence in their assessment that the war would be short and decisive. Yet, although the technological edge reduced casualties and helped achieve some objectives, it could not assure a quick victory, as soon became clear.

Operation Peace for Galilee was one of the longest and most extensively prepared Israeli military operations.[92] This makes the inadequacies in planning and execution even more significant. The preparations and planning provided a strong sense of confidence that the anticipated results would materialize, regardless of criticism from within the IDF of serious qualitative deficiencies in the IDF's capabilities.[93] The plan was to take advantage of the overwhelming numerical and firepower superiority in a combined operation that involved all branches of the IDF. Attacks were launched on a long front in three main thrusts in the western, central, and eastern sectors almost simultaneously by advancing on land, landing from the sea, and leapfrogging enemy positions by helicopter. These attacks were preceded and supported by heavy air force bombing and intensive shelling by mechanized artillery.[94]

The emphasis was on mobility and firepower through the use of a massive force of tanks and armored personnel carriers. Speed was essential, so pockets of resistance were left to be cleared later. On reaching Beirut the IDF put it under siege, for storming any part of the city could have entailed heavy casualties, which would have raised a storm of protest within both the government and the IDF. Furthermore, occupation of an Arab capital by an Israeli army would have caused unforeseeable complications and possibly an international backlash. It was therefore expected that Christian Phalange forces would carry the brunt of the fighting in the city and surrounding refugee camps.

The strategic planners did not take into account several important pitfalls. They assumed that the Palestinian forces would either be destroyed or flee. This happened in some cases, but in many other cases Palestinian retreat into urban centers and densely populated refugee camps forced the IDF to fight precisely where it could not bring to bear its advantage in numbers and firepower. This type of fighting is slow and involves a large number of casualties, and the IDF had little experience with it. Moreover, "the IDF showed great concern for human life, sometimes conceding surprise and initiative to the enemy as a result of the effort to reduce civilian casualties. . . . The caution and concern shown by the IDF to minimize civilian casualties may well have cost its troops more dead and wounded than normally would have been the case had they used the great concentrations of firepower at their disposal" (Davis, 1987: 113–14; also Gabriel, 1984: 194).

Eastern and central Lebanon are mountainous. The progress of tank divisions along the few available, narrow roads was frequently slow and easy for the adversary to anticipate, giving the advantage to the defender. Because there was little room for maneuvering, the attacking forces were vulnerable to much smaller but determined enemy units and were exposed to ambushes, greatly reducing the effectiveness of the armored divisions. The IDF forces had scant experience in mountain warfare and consequently suffered from the absence of the training, force structure, and tactics required for mountain campaigns. Both mountain and urban fighting require large numbers of well-trained infantry, which the IDF did not have (Gabriel, 1984: 193; Kober, 1995: 413, 415–17). Inexperience, lack of appropriate training, and scarcity of infantry units resulted in improvisations that were both inefficient and costly. As the number of casualties mounted, the objectives of the operation came under fire domestically, and the opposition to the war grew both among the public at large and within the military.[95] The public outcry, at home and abroad, grew when the use of heavy artillery and air support to reduce casualties among the troops resulted in extensive civilian casualties, large-scale damage to property, and a wave of war refugees (Yaniv, 1987: 103). In some cases, the use of these casualty-saving measures was in stark contradiction to instructions given to the IDF to minimize civilian casualties even at the cost of efficiency.

After the PLO was expelled from Beirut and it became evident that a quick Israeli withdrawal from Lebanon was not likely, the Israeli army found itself becoming routinely enmeshed in intercommunal peacemaking and peacekeeping tasks for which it was neither trained nor prepared.[96] The initial bias in favor of the Christian militias antagonized the Druze, Shi'ite, and Sunni communities. In the case of the Lebanese Druze community, the conflict threatened to spread into Israeli society and disrupt the delicate and valued relations between the IDF and the Druze community in Israel, whose members had served in the Israeli military and paramilitary forces. That problem eventually forced the government to take a more bipartisan position in conflicts between Druzes and Christians, which further strained the relations with the Christians.

The Shi'ites emerged as the more serious and enduring risk. Since the early 1970s the power of the traditional leadership in that community had declined, and Imam Musa al-Sadr had emerged as its leader. Influenced by the ideas of Iranian Shi'ite activists, Sadr built a powerful mass movement—the Movement of the Disinherited. The movement articulated a set of demands, including a larger share of power and more positions of influence in the Lebanese political system, protection for the Shi'ites in southern Lebanon, and a much larger share of state funds and development projects for Shi'ite areas, as well as remedies for various specific complaints. Israel's pro-Christian bias was perceived as a threat to the community, so it took an aggressively hostile position toward Israel and determined to make Israel's presence in Lebanon as costly as possible. Long lines of communication made the IDF vulnerable to hit-and-run guerrilla attacks by local militias—tactics for which the troops were unprepared. The attacks also raised daily normative humanitarian dilemmas for Israeli soldiers in pursuit of their perpetrators, who often had the support of the local population. These events and problems increased the unpopularity of Israel's military presence in Lebanon; both IDF soldiers and Israeli citizens had growing doubts, which, for the government, raised the policy and political risks of the Lebanon operation.

It was the emphasis on quantitative superiority in each theater of operation that meant mobilization of a large number of reserves. This requirement had several significant implications. In the first place, it had an economic cost in terms of lost GNP. Second, the reservists were much more politically aware and active than the younger conscripts and were less reluctant to criticize the justifiability and management of the operations. Some reservists contacted members of the Knesset, Israel's parliament, and passed information to government ministers when they perceived a discrepancy between announced government policy and actual implementation by the minister of defense. Thus the pressure for accountability increased, and Sharon found it difficult to deviate from the explicit policy without cabinet awareness. Eventually his authority over the conduct of the war was curbed. As the war became protracted and its goals contested, reservists exhibited a growing reluctance to serve in Lebanon for long periods. Many of those who served in combat units felt they were being called on to risk their lives for goals that were no longer justified. Consequently, the issue of burden sharing became an important component of the public debate over the government's policy. Disaffection in the military and among the public at large weakened Israel's bargaining position and encouraged the Syrians, Palestinians, Shi'ites, Sunnis, and Druzes to oppose Israel's objectives and presence in Lebanon and to raise the costs of its presence by frequently attacking Israeli units.

The conditions on the battlefield required that units often be switched around from one command to another, and field commanders were moved around as well. These shifts not only created confusion about who was in command of a particular unit at a particular time but also violated the IDF's principle of maintaining unit integrity. In addition, centralization of command—one of the lessons of the 1973 war—was in many cases inappropriate for the terrain and battlefield conditions in Lebanon. Greater centralization of decisionmaking and the development of corps-level formations, both the product of the technological and organizational innovations in the IDF before the war, together with the proximity of headquarters to the front, meant that headquarters intervened frequently in the conduct of operations. The tighter control nets afforded the minister of defense opportunities to micromanage the conduct of the campaign. The chief of staff and the corps commander of the eastern sector likewise interfered with the field commanders' decisions. Such vertical interference required twelfth-hour revisions in battle plans, which had serious adverse effects on the performances of affected units (Yaniv, 1987: 134–35; Gabriel, 1984: 155–56).[97] These complications only added to the overall confusion and conflict at all levels of the command structure. Making this state of affairs even worse were inadequate training and unsound professionalism at the high command level, aversion to risk taking on the part of commanders, and the lack of vital intelligence at the field-unit level, not because the intelligence was unavailable but because it was inefficiently distributed (Wald, 1992: 70–74).[98]

An even more critical problem—which was partly responsible for runaway escalation—was the lack of full government knowledge, command, and control of the operation from the planning stage through implementation. In an important war game code-named Roses, for example, played out in March 1982, the Israeli command learned that the probability of achieving the political objectives in Lebanon were minimal even if the military operation was a total success, that fighting the Syrian army in Lebanon would be unavoidable, and that the number of casualties in the first four days of the operation could be as high as 100. These conclusions were not brought to the attention of the government.

There is an even more important example. On May 13, 1982, the chief of military intelligence, Major General Yehoshua Saguy, warned against war in Lebanon.[99] He predicted that war with the Syrians in Lebanon would be unavoidable and that the gains from the operation would be limited because the PLO would reorganize in northern Lebanon or in Syria if expelled from Beirut. He foresaw the division within the Reagan administration, recognized Secretary Haig's vulnerability, and gave only a 60 percent probability to Haig's chances of convincing the administration to support Israel unless the provocation was of a magnitude that clearly justified war. The Soviets, Saguy concluded, were not interested in a war in the region, a view they had expressed to Yasir Arafat and Assad. But if a war broke out and threatened Assad's regime or if the operation became protracted, the Russians would get directly involved. The United States would then blame Israel for inviting Soviet intervention, and both superpowers would demand an end to the fighting and Israel's withdrawal from Lebanon. This assessment was unacceptable to Sharon and was never presented to the government (Schiff and Ya'ari, 1984: 120–23).

During the war itself, the prime minister and his cabinet were unaware of the mobilization of additional reserve units and of continued military operations to improve positions during cease-fires. The government was also unaware when an armored force entered Beirut and took positions near the presidential palace.

These incidents were not isolated but systematic, leading the prime minister to comment cynically that he was informed by the minister of defense of everything that took place, sometimes before and sometimes after the fact (Naor, 1986: 35–36, 86–88, 113–15, 134–37). This loss of control stood in stark contrast to the prime minister's initial confidence in his ability to control the scope of military operations. Responding to early apprehensions to that effect raised by some ministers on June 15, Begin emphatically stated, "The government will keep its hand on the pulse, nothing will be done without the expressed decisions of the government" (Medzini, 1990: 42).

An important component of the plan—the use of a proxy, the Phalangists, to carry the burden of clearing Beirut and its refugee camps of PLO forces—failed to fall into place. It soon became clear that Bashir Jumayyil was reluctant to commit his small military force to this deadly work. In fact, he attempted to keep at arms' length from Israel in order to look like an independent nationalist leader. Even more detrimental to the plan, Phalangists were involved in rape, murder, and theft, all of which increased domestic criticism in Israel of the alliance with Jumayyil. After Jumayyil's murder, Phalangist units retaliated by massacring Palestinians at Sabra and Shatilla— massacres for which Israel was held accountable. This was the final straw; the Israeli government had to withdraw from Beirut. These Phalangist actions did not come as a complete surprise to the Israelis. During the extensive planning that preceded the intervention, decisionmakers in Ministry of Defense and IDF headquarters were warned that the Phalange were completely unreliable and would likely cause trouble "of virtually catastrophic proportions" (Yaniv, 1987: 136; also Naor, 1986: 118–19; Schulze, 1996; Shiffer, 1984: 102–5, 114–15, 124–30). Rather than reduce the intervener's burden, the proxy proved to be unreliable and uncontrollable—quite the reverse of optimistic assessments made by the Mossad, the organization charged with cultivating the Phalange.

The logistics of the operation were complex because of the large force involved but were not anticipated to be as problem ridden as they turned out to be. The underlying premises for a problem-free operation were that it be short and fast moving, and it was taken into account that the distances from the rear bases of the operating forces to the front would be relatively short. In the event, the operation ran behind schedule, the fighting was heavier than anticipated, and combat units had to be resupplied more often than planned. The small number of usable narrow roads were often blocked by the heavy congestion of advancing forces because the large forces were so excessively concentrated in a limited space, creating serious logistic problems for moving vital fuel, ammunition, and water supplies to the front.[100]

As the lines of supply became longer, the pockets of resistance left behind for the sake of rapid advance became sources of trouble, because the enemy often blocked or mined supposedly secure roads. Thus, combat units had to clear roads several times after they were initially taken—including the vital coastal road from northern Israel to the outskirts of Beirut. In 1984 the defense of supply lines became an important argument for withdrawal from the eastern sector; such a withdrawal "would substantially shorten the IDF's internal lines, reduce its overextended logistic spread and improve the ability to move reinforcements from base to front" (Yaniv, 1987: 258).

The economic costs of the operation were way above early estimates, because fighting lasted longer and was much more intensive than expected. This meant greater costs for perishables (e.g., fuel, ammunition, spare parts) and higher attrition of replaceable hardware (e.g., tanks, trucks), plus the indirect cost to the Israeli economy of having to support reservists and, at the same time, divert them from economically productive to nonproductive activities.[101] According to one estimate, the total cost had reached $1.26 billion by March 1983 (Gabriel, 1984: 235). The war was costing almost a million dollars a day at a time when the Israeli economy was already in a deep recession that involved three-digit inflation, large numbers of bankruptcies, falling foreign currency reserves, and growing unemployment.[102] The expenses of the military operation accelerated the economic decline and affected the economic and social welfare of the population. As the war dragged on, the economic implications eroded public support for staying in Lebanon and increased dependence on the United States. Both consequences indirectly undercut the government's bargaining position with the Syrians, for the government needed to convince them of Israel's resolve to remain an active player in Lebanon. Nor, given the dire economic situation, could the IDF expect to be compensated for the outflow of $200 million a year for the operations in Lebanon; the result could only be substantial cuts in training programs (Yaniv, 1987: 176, 256–57). This bleak prospect contributed to growing military resistance to a continued presence in Lebanon.

In terms of domestic political capabilities and constraints, large-scale military intervention in Lebanon could be affected by three main factors: the formal requirements for authorization of the operation; the response of the parliamentary opposition, especially the Labor Party; and the informal societal requirements for legitimacy, which in the Israeli case are of critical importance for decisions involving war and peace.

To start with, the Israeli political system imposes considerable constraints on the key decisionmakers' authority to go to and to manage war. These constraints are embedded in the electoral system, which is based on proportional representation; because of the division within the electorate, no single election victor can realistically hope to win a majority of at least 61 seats in the 120-member Knesset.[103] Thus, aspiring prime ministers must induce other parties to join theirs to establish a ruling coalition. In such a coalition the views of the smaller parties often become crucial to a degree disproportionate to their electoral strength. Prime ministers have to be attentive to the wishes of their coalition partners and careful not to antagonize them in order for their government to survive. They also have to consider the shadow of the future, that is, to take into account future elections and their own need for the support of these parties to build future coalition governments. Small parties that are vital coalition partners thus have a great deal of influence in deciding the outcome of contested issues, even if their ministers are only a minority in the government. In many cases, the prime minister finds it difficult to act decisively and especially to make risky decisions about contested issues. Even if a risky decision gains approval, coalition members are likely to monitor its implementation closely.

The results of the election to the Tenth Knesset in 1981 were no exception to this rule. In spite of the surprisingly strong performance by the Likud, Likud and Labor tied, each winning 48 seats. Prime Minister Begin depended on coalition partners to put together a majority government, a necessity that greatly restricted his and his senior ministers' freedom of decision.[104] Complicating matters was that the Likud itself was an integrated amalgamation of parties with different security perspectives. The two dominant components were Herut, which had a hawkish activist ideology, and the Liberal Party, which had a much more moderate approach to security. Simcha Ehrlich, the leader of the latter and Begin's deputy prime minister, and Energy Minister Yitzhak Berman were among the main opponents of the war in the cabinet.

A large-scale military operation in Lebanon needed the authorization of the government, but the government was divided over this very issue. A large ground operation that might escalate into war with the Syrians could have serious international repercussions, could affect the fragile peace with Egypt, and would probably involve a significant number of casualties was viewed with extreme caution by ministers in the coalition government and, indeed, was not supported by all the Likud ministers. When on December 20, 1981, Begin presented Big Pines to the cabinet for the first time, he quickly learned that it was not likely to be authorized and did not pursue the matter further (Schiff and Ya'ari, 1984: 96–98). The plan and various moderate versions of it were presented in the cabinet several times during the following months, but support was weak. Begin then refused to proceed with it until June 1982, when he gained broad support in the cabinet for the more limited Operation Peace for Galilee.

Political constraints on the operation conflicted with the requirements of an effective military action. If the hidden objective was to force the Syrians out of Lebanon, it would have made sense to attack the Syrian army first and

only then deal with the lesser military threat of the Palestinians.[105] But because the government had refused to authorize the Big Pines plan, which required an assault on Syrian forces, any attack on the Syrian army now had to
be a by-product of the attack on the Palestinian military force. Furthermore,
because the government had not authorized the original Big Plan, the defense
minister resorted to manipulation and misinformation in pursuing the ambitious hidden objectives of the war that were embedded in that plan. Under
these circumstances military planning and implementation were driven more
by political imperatives than by professional military considerations that
would take the fullest advantage of Israel's decisive military superiority with
the lowest cost in casualties.[106] The need to avoid stating the true and full
range of objectives and to implement them only incrementally because of political constraints confused the field commanders, who were not sure what
was expected of them in light of the contradictions between declared government policy and the instructions of the defense minister and chief of staff.
These contradictions, a growing lack of enthusiasm among the troops, and
fuzzy, piecemeal war objectives made the field commanders tend toward
overcaution, avoiding bold initiatives and risky ventures. Their risk aversion
was responsible for missed opportunities on the battlefield, slow progress,
and, in some cases, increased casualties.

The division within the cabinet was particularly significant in light of the
uncertainty about the position that would be taken by the opposition Labor
Party in the public and in the Knesset debates in the event of a large-scale
military operation. In a meeting in April 1982, the prime minister apprised
the top Labor leadership of the change of policy toward Lebanon. Of the
three Labor leaders who attended—two retired generals who had served as
chiefs of staff (Yitzhak Rabin and Chaim Bar-Lev) and a past minister of defense (Shimon Peres)—none expressed explicit support for a war in Lebanon. Qualified support later emerged for the watered-down plan, but in August, after the bombing of Beirut, the Labor position shifted toward outright
criticism.[107] The position of the Labor Party was important for two reasons.
Its 48 seats in the Knesset made a formidable rival to the Likud government
if it attracted member parties of the Likud coalition, a coalition that held
only a slim majority.[108] More important was the broader problem of establishing national consensus, an amorphous but very real requirement dictated
by the political culture of a country that was essentially a nation at arms.[109]
Even the most bitter critics of the Labor leadership recognized that it had
managed Israel's defense policy since independence mostly with competence
and imagination. Among the leadership and members of the Labor Party
were many well-respected retired generals, chiefs of staff, and a former defense minister, as well as a former prime minister and Knesset members who
had served long terms on the Knesset's prestigious defense and foreign policy
committee. Labor Party supporters also held many high command posts at
all levels of the regular and reserve forces, including the special forces. Thus,
Labor's support of the government policy in Lebanon would do more than

simply continue the Israeli tradition of demonstrating solidarity with the government in wartime; it could also affect the positions of those within the cabinet who were opposed or wavering. And it would give the policy domestic bipartisan legitimacy that it would not otherwise have.

The extreme importance of domestic legitimacy in the Lebanon War cannot be overstated. Historically, all Israel's wars had such legitimacy, and so did the Lebanon War at first. But the legitimacy began to erode as early as the first cease-fire, one week into the war. Although the broader goals of the war were a matter of intensive public debate almost a year before the war actually broke out, the limited operation—the war was initially defined as such—was supported by broad consensus, given the unresolved missile crisis, the assassination attempt on Israel's ambassador to Britain, and the heavy shelling of Israeli settlements in the Galilee. Surveys showed public support ranging from 66 to 84 percent. In the Knesset 94 of the 120 members rejected a motion of no confidence in the government on June 8. But as the broader goals of the war became more obvious, when fighting spread beyond the 40-kilometer range and casualties among IDF forces and the Lebanese civilian population increased, and when international criticism grew more strident, domestic legitimacy slipped away (Barzilai, 1992: 190–96).

With the siege and bombing of Beirut in August, the public began to question the political-military rationale for the intervention and the normative justification for its objectives and the means of achieving them. In the process, more fundamental issues were raised. Senior Cabinet members questioned decisions that supposedly were made in the name of the cabinet but in their view were without its authorization.[110] The legitimacy of the intervention was undercut by cabinet disagreement with and distrust of the minister of defense; local and foreign media coverage of the war, which provided the public with both true and false information; opposition in the Knesset; and extraparliamentary opposition groups, including officers and soldiers who were experiencing the war firsthand and communicated their disaffection to the public at large.[111] The erosion of legitimacy greatly diminished the decision flexibility of decisionmakers and increasingly turned war-related decisions, after the first cease-fire, into subjects of acrimonious public debate. As the operation expanded its objectives and scope and extended its duration, the initial legitimizing argument became inadequate, and new or additional arguments were added. The line between cognitive and normative legitimacy was often blurred, and the two were so closely interlinked that there were extensive and frequent spillovers from one dimension to the other.

Supporters of Operation Peace for Galilee intended to establish its cognitive legitimacy by arguing that the PLO was not merely a terrorist nuisance to daily life but an immediate acute threat that made life unbearable to the civilian population in the Galilee and a threat to the existence of Israel because of its attempts to trigger a general war between Israel and the confrontational Arab states within two or three years and its success in deterring moderate Arab states from negotiating with Israel. According to one view, the PLO conspired with and was fully supported by Syria in this plan, which, if successful, would result in heavy casualties even if the Arab coalition was defeated (Medzini, 1990: 117–18; Naor, 1986: 27–28). The severity of this threat, argued the operation's advocates, justified a response that, according to critics, was grossly disproportional to the provocation. Preventive war, said the supporters, would remove the threat. The operation was presented as the final resort; other, milder measures had been tried in vain. The PLO could not be trusted to keep agreements—for example, the U.S.-negotiated cease-fire of July 1981. It was therefore important to establish that terrorist activities abroad against Israeli or Jewish targets or terrorist activities involving penetration from Jordan were violations of the July 1981 agreement between Israel and the PLO and thus justified a decisive response.[112] The operation, it was emphasized, had limited objectives and was to be brief in duration (24–48 hours) and limited in range (40 kilometers from the northern border)—and these limits would guarantee the support of the parliamentary opposition, result in few casualties, and yet remove the threat to civilian population in the Galilee once and for all (Naor, 1986: 47–49).

Normative legitimacy was based on the right and obligation of a government to defend the lives of its citizens. To emphasize this obligation, top officials dehumanized the enemy (e.g., Arafat was described as a murderer and the PLO as not bound by civilized, humane values or normative codes of behavior). Not only was Israel threatened, but the Christians in Lebanon, too, were perceived to face a threat of physical annihilation; one had only to recall Syrian ruthlessness in dealing with its domestic opposition and Syria's past performance in Lebanon. Because of the Jewish history of persecution, Begin's prevailing sense was that Israel had a special responsibility to behave differently from France and Britain, the countries that had betrayed Czechoslovakia in 1938, and to prevent the persecution of an allied minority (Naor, 1986: 27, 119; Naor, 1993: 240). Proponents also argued that preventive wars were just besides being instrumental. Such wars reduced risks that, if not confronted at an early stage, could bring disaster and unbearable casualties later; a preventive intervention was justified by the obligation of a government to defend its citizens (Inbar, 1989). Normative legitimacy was further reinforced by emotive arguments that emphasized the threat posed to children's security and that raised the memory of the Jewish Holocaust. Prime Minister Begin, interviewed on Israel television on June 15, 1982, posed a set of rhetorical questions: "What did they do to our people? What did they do to our children? Why couldn't they go to school like the children of London, or the children of Paris, or the children of Rome, or the children of Moscow? Why did they have to be in small shelters day and night? Why were men, women, and children killed? Why was a man who had been in Auschwitz killed, a man who came to Eretz Israel to live and not be killed?" (Medzini, 1990: 45).

Sensitive to the need for preserving cabinet unity, the prime minister refused to authorize implementation of the revised and diminished Big Plan on

May 10, when the issue was raised in the cabinet. The cabinet was divided, although the supporters of the operation were likely to win a vote by a narrow margin of one. On many other occasions and with many other issue-areas, Begin considered a majority of one sufficient for a decision in favor of that option—but not in this case. Not until June 5, when the vote on Operation Peace for Galilee was fourteen for, one against, and two abstaining (Naor, 1986: 39, 51), did Begin feel that he had the normative legitimacy to commit the country and the military to the costs of the operation, which, at that point, did not include targeting Beirut and the Syrian army in the Beqa'a. This pattern repeated itself when Begin brought the decision to attack and occupy southern Beirut to the cabinet, where there was at first only a simple majority (one additional vote) for the decision.[113] Again Begin did not approve the operation because he saw broad cabinet support as an important source of normative legitimacy (pp. 129–30).

By late June there was a strong sense in the cabinet, Knesset, IDF, and public that the defense minister was misleading the government. IDF spokespersons had informed the public that the Syrians or the PLO provoked violations of cease-fires and other incidents, but ministers received calls from soldiers, including senior officers, that the IDF had initiated them (Naor, 1986: 115–16). The IDF justified incremental forward advances as unavoidable responses to enemy initiatives that threatened Israeli forces, but this argument boomeranged, too. Its obvious disadvantage was it became a self-imposed constraint on military operations, keeping the IDF from exploiting its numerical advantage and deploying large forces for offensive operations or using air support for such operations. The constraints raised the number of casualties, which further undermined the tolerance for the war and its costs in the military and among the public at large (Naor, 1986: 113–16, 132). The result was a further decline in the credibility of the defense minister and, by implication, the government. It looked as if the government was using collective deceit to pursue a war that no longer served the declared war objectives and thus also violated the principle of accountability. Alternatively, if the government was unaware of the violation of its own directives or was incapable of restraining Sharon, then it was grossly incompetent and could not be trusted to run the war. Either way, cognitive and normative legitimacy were doubtful.

The extension of the war to densely populated urban centers and into a major Arab metropolis under the scrutiny of the international media called attention to the growing number of Lebanese civilian casualties and the many dislocated and uprooted civilians. This was most vividly demonstrated by the Phalangist massacre of Palestinians in the refugee camps of Sabra and Shatilla in September 1982. Media stories were often overly dramatic and exaggerated and sometimes outright misinformed, but their effect on domestic and international legitimacy was devastating.[114] For the IDF, urban warfare posed a serious moral dilemma, especially when it became clear that any large offensive operation to take part of Beirut would require almost indiscriminate shelling and bombing of densely populated areas of the city to reduce casualties among the attacking units (Barzilai, 1992: 200–203; Schiff and Ya'ari, 1984: 266–68). Forging ahead with such an operation would, it was feared, irreparably corrupt the moral and humanistic values and contaminate the image of the IDF. Obviously, as the war expanded, the cognitive and normative legitimacy of the war objectives, the means employed, and the anticipated probabilities of success at an acceptable cost were all challenged.

As the protracted deployment of Israeli troops in Lebanon and the growing number of casualties became increasingly unpopular, the war thus became more difficult to justify either morally or practically.[115] It became necessary to exalt the war in Lebanon as a great success and its achievements as worth defending even at great expense (Medzini, 1990: 388; *Yediot Aharonot*, June 17, 1983). Accordingly, the sunk costs of the war became a justification for incurring further costs by staying in Lebanon. Public criticism of government policy became an excuse for Israel's inability to achieve better outcomes in the negotiations. And refusing to withdraw IDF forces was described as a demonstration of resolve and a necessary condition for preserving Israel's reputational assets to deter future aggression. Even Sharon's successor as defense minister, Moshe Arens, stated in May 1983 that "our ability to survive is first of all contingent on one thing, and that is our willingness to go and fight when it is necessary" (Medzini, 1990: 424). Faced with bitter criticism from parliamentary and extraparliamentary opposition, Begin stressed to the Knesset on June 15, 1983, that Israel had been willing to accept a ceasefire as early as June 11, 1982, but the enemy had refused. He also reminded the opposition that this was not the first war in which there were difficulties. Begin emphasized that in the past, when the Likud was in opposition, it had not criticized the government while the war was going on, and he concluded by saying, "We have to stand together" (p. 428).

In spite of these constant attempts to maintain legitimacy, parliamentary opposition, extraparliamentary opposition, and a large segment of the public remained unmoved and sharply critical of Israel's presence in Lebanon and the mounting costs. The pressure for withdrawal was relentless. But the ultimate indication of the failure of domestic legitimacy for the Lebanon policy was provided by the prime minister himself. In September 1983 Begin resigned without explanation, saying only that he could no longer bear the burden. Obviously, his disillusionment with the consequences of the Lebanon policy and, in particular, its high cost in lives was one of the main causes.

Once legitimacy was effectively challenged, suspicion and mistrust of the government became a feature of the relationship between the government and the public. Decisionmakers' room to maneuver became narrower, and divisions within the decisionmaking group widened and were often exposed to public view as some decisionmakers distanced themselves from what seemed more and more an unsuccessful and unpopular policy. The main architects of the operation had anticipated neither the extreme difficulties of attaining nor the eventual disintegration of the support needed to sustain legitimacy.[116]

THE INTERNATIONAL MILIEU: FROM ACQUIESCENCE TO ACUTE CONSTRAINT

The first Reagan term was characterized by a commitment to strengthen U.S. global leadership through a combination of military buildup and resolve to face down Soviet advances. This meant that the United States was willing once again to project power or act forcefully through regional allies. Within the context of a loose bipolar international system, but with a more confrontational relationship between the two superpowers, regional allies of the United States had much more room for autonomous policies, even when dependent on the patron superpower for political support and for military and economic assistance—especially when their policies served to weaken Soviet allies and, by implication, the Soviet position.

The mix of dependence and autonomy was particularly apparent in the Middle East, where Israel was a regional power, close U.S. ally, and a major recipient of U.S. military and economic aid. Yet in Begin's view, Israel had substantial room for independent initiatives. He thought that his predecessors' concern for international constraints on Israel's offensive options was an immobilizing, obsessive fear rather than the perception of a real threat (Schiff and Ya'ari, 1984: 78; Sofer, 1988: 115). He and the other leaders of Israel had thus demonstrated a resolve to act independently of U.S. wishes. The peace process in the Middle East was started by the two most important regional powers, Israel and Egypt, without consultation and even in defiance of their superpower patrons. It resulted in a peace treaty between Israel and Egypt, mediated by the United States, that brought Egypt out from the Soviet camp and placed it firmly in the American camp, enhancing the U.S. position in the region. At the same time, by eliminating the Egyptian threat to Israel's southern border, the Camp David Accords isolated Syria, which now had to face a situation of strategic inferiority vis-à-vis Israel, and placed Israel in its most favorable strategic position since the 1967 war. It could now focus its superior military capabilities on its northern border; Syria could not hope to soon reach strategic parity, which would allow it to initiate and win a war against Israel on its own. Direct Soviet involvement was unlikely. Syria's intervention in Lebanon since 1976 further complicated the Syrian military position with Israel. Yet, based on the experience of the 1973 war, the Israeli government was not fully reassured that Syria would not initiate war, even if its chances of winning were slim, expressly to involve the superpowers and improve its bargaining position.

In spite of being the largest U.S. aid recipient, Israel considered itself a key strategic partner of the United States rather than a client state. In practical policy terms, this enhanced its tendency to act independently of U.S. wishes and interests and to expect post facto U.S. support in the United Nations and its Security Council. Israel's confidence in its ability to act independently of its patron's wishes when it served its national interests was embedded not only in structural factors but also in the backing of a well-established, powerful pro-Israel lobby in Congress, the U.S. administration, and the American public. The sense of omnipotence found clear expression in 1981 in Israel's attack on the nuclear facilities of Iraq, the annexation of the Golan Heights, and the bombing of PLO headquarters in Beirut at a cost of heavy civilian casualties. In all these cases, Israel acted without consulting the United States, using U.S. weapon systems that had been supplied for defensive purposes. In the last two cases Israel acted in defiance of the explicit wishes of the United States, which was apprehensive of the implications of Israel's policy for American interests in the region. The United States responded by suspending aid and the Memorandum of Strategic Understanding of November 1981. But Begin's attitude and belief in Israel's exclusive right to take actions that served its security interests were unshaken. He summarized them sharply in his stormy meeting with U.S. Ambassador to Israel Samuel Lewis on December 20, 1981, after the Golan annexation and the U.S. response. Begin told Lewis that Israel was not some "banana republic" to be treated in such a cavalier manner. More than ever before, Begin and Sharon viewed the traditional patron-client relationship between the United States and Israel that emerged in the mid-1960s as a secondary constraint on Israel's freedom of decision and one that could be overcome relatively easily.[117]

That is not to say that Israel's key decisionmakers were unconcerned by U.S. reaction to a large-scale intervention in Lebanon and the potential costs of U.S. opposition. Their goal was to convince the Reagan administration to acquiesce in, if not actively endorse, a large-scale military operation in Lebanon. In any case, American support would constitute only one incentive for military action in Lebanon and not the most important one. In light of the Reagan administration's emphasis on the need to forge a strategic agreement among like-minded countries about curbing Soviet influence in the Middle East (Glad, 1988; Quandt, 1984), Begin's expectations that Reagan would be receptive to the argument that their two countries shared an interest in defeating their common adversary were not baseless. "As president, Reagan's contribution to shaping American Middle East policy consisted primarily of injecting this theme of the Soviet instigation of regional unrest into the thinking of his subordinates" (Quandt, 1993: 338). Indeed, there was sympathy among some key members of the Reagan administration, including the president and the secretary of state, for a limited, short operation. Sharon and Begin interpreted sympathy as approval and, in 1981–82, doubled their determination to resolve the threat from Lebanon militarily.

The traditional script of past wars and military operations was well known to every Israeli decisionmaker. The sequence was simple. Israel engages in proactive or reactive military operations. The Arab side faces the prospect of defeat and humiliation. The Arabs approach their superpower patron, the Soviet Union, to apply pressure on the United States to impose a cease-fire and withdrawal on Israel.[118] Concerned with a possible confrontation with the Soviet Union and under pressure from the Arabists in the administration, the U.S. president requires that Israel cease all hostilities and negotiate. At the negotiating table Israel has to give up much of its battlefield achievements and is not allowed to enjoy the spoils of military victory to a full extent.

Begin was familiar with this pattern, as was Sharon, and they were concerned with the question of how much time they had before international pressure would mount. Both had experienced such pressure before, and both were determined not to let it cause failure in Lebanon. Begin's approach to the United States, and to superpower impositions in general, was apparently influenced to some extent by his experience as commander of a small resistance organization (Irgun Zevai Leumi) that challenged the British Empire in the struggle for independence and, in his view, forced it to back off. This personal lesson, juxtaposed with the realization of the immense importance of U.S. support and aid and a general benign view of the United States, resulted in a mixed attitude on his part. Sometimes he wanted to defy the United States, sometimes he recognized that defiance could only be applied cautiously, and he accepted that Israel would occasionally have to comply with some U.S. demands. The ascendancy of the preference for defiance, the belief that the Lebanese operation would be over quickly, and confidence in Israel's strategic importance in Reagan's confrontational policy toward the Soviet Union all led Begin and other Israeli policymakers to attribute only secondary importance to U.S. warnings against taking rash action.[119] In this context, the ambiguous wording of U.S. warnings was interpreted as an indication of acquiescence on the part of the Reagan administration, though not necessarily active encouragement to go ahead. The United States could be ignored even if it was concerned.[120]

How did the U.S. position on Operation Peace for Galilee evolve from tacit endorsement to active pressure for termination of hostilities? Because the operation brewed for many months, Begin and Sharon had time to try to convince the president, the secretary of state, and the secretary of defense that the operation was necessary and would serve the global and regional interests of the United States. Their strongest supporter in the Reagan administration was Secretary of State Alexander Haig.[121] In spite of the dangers, Haig viewed the operation as posing a unique opportunity to remove all foreign troops from Lebanon, restore power to the Lebanese government, and allow progress in the peace process in the Middle East, which had been stalled by Syrian and PLO opposition (Haig, 1984: 318). Thus, although the Reagan administration did not explicitly endorse an invasion of Lebanon, it was well aware of the preparations to do so. Israeli decisionmakers knew that its concentration of forces since December 1981 could not have escaped observation by U.S. intelligence. On several occasions Ambassador Lewis expressed concern about a pending operation.[122]

The confidence of Begin and Sharon that the United States would not oppose a limited military operation if Israel was obviously provoked was in part based on the fact that Israel's intentions were public knowledge and were discussed openly in the U.S. media months before the war.[123] Under these circumstances, the failure of the United States to act more firmly was considered a signal of acquiescence. A conversation between Sharon and Haig in May and a meeting later that month between Israel's ambassador to the United States, Moshe Arens, and Haig reinforced the view that Washington was not going to react too forcefully and that Israel's reason for invasion, explained to American representatives on numerous occasions, was well understood and sympathized with by key members of the Reagan administration. Although the opposition of some other prominent members in the administration was known, notably Secretary of Defense Caspar Weinberger, these officials were discounted as isolated individuals known to be hostile to Israel and unrepresentative of the deeper, friendlier sentiments toward Israel felt by the president himself. The American position was interpreted as tacit approval for Operation Peace for Galilee not only by Israelis who supported the intervention but also by those in the parliamentary opposition, the public, and the government who opposed it. The opposition's hand was accordingly weak in the debate that preceded the war. One minister reportedly commented, "I cannot show myself to be less of a patriot than the Americans" (Schiff, 1984: 83). The relatively mild U.S. reaction to the expansion of the war in its first week bolstered the view that the United States endorsed the broader goals of the war and considered the war beneficial to American interests, a point of view that seemed to be supported by President Reagan's remarks at the conclusion of the Begin-Reagan meeting on June 21, 1982, in Washington, which emphasized U.S. support for an independent Lebanon under a strong central government. Thus the risk that the United States would oppose the operation, and the serious ramifications that such a development could have, did not overly concern the Israeli cabinet. The issue "was mentioned but not discussed by the Israeli Cabinet when it voted to go to war" (p. 75).

The effort to convince the United States of the cognitive and normative legitimacy of the war touched on both practical and legal points. The Lebanon intervention was argued to be an important contribution to checking Soviet influence in the Middle East by defeating its two remaining allies, the PLO and Syria. Furthermore, Lebanon had become the main training ground and shelter for international terrorism. By destroying the organized PLO infrastructure and training camps and by establishing a strong central government in Lebanon, international terrorism would suffer a heavy blow. Normatively, Israel claimed the right of self-defense and emphasized that it was morally wrong to deny it a right that was recognized to apply in cases that were less convincing.[124] The attempt to seize the moral high ground was obvious in Begin's letter to Reagan of June 6, 1982:

> The purpose of the enemy is to kill—to kill Jews; men, women and children. Is there a nation in the world that would tolerate such a situation which, after the cessation of hostilities agreement, has repeated itself time and again?

The question is clearly answered in the most recent action of the United Kingdom which is now waging a full-fledged war eight thousand miles from its shores in the name of Article 51 of the United Nations Charter. Mr. President, the bloodthirsty aggressor against us is on our doorstep. Do we not have "the inherent right to self-defense"? Does not Article 51 of the Charter apply to us? Is the Jewish State an exception to all the rules applying to all other nations? The answer to these questions is enshrined in the questions themselves. (Medzini, 1990: 5)

These arguments lost much of their force as the war went beyond 40 kilometers from Israel's northern borders. The combined effects of live media coverage of the misery, destruction, and dislocation of civilians; the declining influence of Israel's main supporter in the administration, Secretary Haig, which led to his forced resignation; the growing pressure from Saudi Arabia and its supporters in the U.S. administration; and the emerging threat of superpower confrontation—all these shifted the American position toward active attempts to bring about a cease-fire and withdrawal of Israel's troops.

The view in Reagan's administration was that Israel's policy could jeopardize the U.S. reputation and its relations with Egypt and Saudi Arabia, making a regional defense alignment against the Soviet Union difficult (Shultz, 1993: 70). To influence Israel toward restraint, the Americans warned it that such actions would put the relation between the two countries at stake.[125] At the same time, the Americans took a surprising unilateral initiative to resolve the Palestinian issue as expressed in the Reagan Peace Plan, which caused much distress to the Israeli government.[126] U.S. Marines were sent to Beirut to secure the safe departure of PLO forces and returned later as peacekeepers. They had a few close encounters with Israeli troops a few times, which strained Israel-U.S. relations further.[127] The United States also made some attempts at coercive diplomacy, such as delaying the formal notification to Congress of the sale of 75 F-16 aircraft to Israel—a delay that lasted from June 1982 until May 1983. Similarly, a scheduled transfer of 4,000 shells of cluster bombs was suspended. These selective sanctions were followed by warnings of additional sanctions. The United States also procrastinated on Israel's request for the transfer of vital technologies for its advanced aircraft project, the Lavi. In the same vein, the administration delayed permission to use U.S. military sales credits for domestic purchases in Israel and prevented third-party countries from using U.S. military credits for purchases from Israel.

All these sanctions were implemented unilaterally by the U.S. executive branch. Yet the president's staff was aware that more painful measures linking economic aid to specific policies could be effectively blocked by Israel's supporters in Congress. That the administration was deterred from linking aid to Israel with its Lebanon policy in spite of harsh criticism in the American media and from the public indicated the strength of Israel's support in the United States. It became painfully clear to the administration that drastic sanctions were not an option. Consequently, Israel had much leverage in resisting American demands for immediate withdrawal and for adoption of the Reagan Peace Plan, but it made tactical concessions, such as withdrawal from parts of Beirut. The demonstration of loyal support for Israel in Congress notwithstanding, Israel's image eroded, jeopardizing its special, long-standing position in U.S. legislative and public opinion. Another unanticipated drawback was the increasingly harsh and vocal criticism of Israel's policy by one of its main sources of support and influence in the United States—influential members of the Jewish community (Ben-Zvi, 1993: 140–58).

The selective coercive diplomacy by the United States had only limited success in influencing Israel's policy. What success they did have could partly be attributed to the reinforcement that they gave to Israel's domestic opposition to the war. But the U.S. policy had two other important effects. It increased the Begin government's sense of stress and isolation at home and abroad, and it reinforced the realization that the cost of the war had exceeded all original estimates. Consequently, the original objectives were eventually reconsidered and modified. At the same time, the obvious cleavages between the Begin government, important segments of the Israeli public, and Israel's main ally, the United States, encouraged first the Syrians and the Palestinians and later the various communal militias—Druze, Sunni, and Shi'ite—to challenge Israel's forces more resolutely and to refuse to endorse the U.S.-sponsored Lebanon-Israel Accord of May 1983.[128] Even a superpower finds it difficult to formulate and fine-tune a policy that is addressed to multiple parties in an intervention situation without causing damage to some of its policy objectives.

Obviously, Israel's key decisionmakers overestimated the degree to which the United States would acquiesce to their actions. This miscalculation was partially due to not fully understanding the inner workings of the Reagan administration. They did not realize that Reagan's modus operandi made him highly dependent on a close group of advisers who were more pro-Arab than pro-Israeli (Glad, 1988); they were not aware how vulnerable Haig's position was; they did not realize that, in spite of Reagan's friendship and strong commitment to Israel, he might take advice that would be damaging to Israeli interests in order to advance broader American interests in the Middle East (Reagan, 1990: 410–11). Although some of this misjudgment can be attributed to a lack of relevant information by the Israeli government, much of it can be traced to the belief that Operation Peace for Galilee would quickly be a fait accompli, making it possible to avoid the type of complications in U.S.-Israel relations that would develop with time. Thus, in spite of Israel's heavy economic, military, and political dependence on the United States, the pattern discerned in our earlier cases regarding the secondary role of external factors holds in the Lebanon case, too. The concern with the U.S. position was substantially lower than one would expect under the circumstances. Israeli decisionmakers focused most of their attention on military and domestic political capabilities. They tried to persuade the United States to endorse the invasion, but even the absence of U.S. support in and of itself probably would not have been a reason for nonintervention.

The relaxed attitude toward U.S. support for the operation was accompanied by even less concern about the possible responses of other international actors. Initially the perception of having been given a green light reduced Israeli decisionmakers' risk perceptions with regard to the reactions of the three other important third parties that posed potential military and political threats to Israel: the Soviet Union, Egypt, and the United Nations.[129] The Israeli government assumed that U.S. acquiescence would affect the positions taken by the other three actors. Egypt was becoming heavily dependent on U.S. military and economic assistance, which gave the United States leverage in influencing the Egyptian response and moderating the implications of intervention for the continuation of the peace process. U.S. support for Israel would also deter active Soviet support for Syria. Finally, the veto power of the United States in the U.N. Security Council and its role as a key player in the United Nations made U.S. support imperative for blocking and restraining punitive U.N. sanctions against Israel.[130] But as the intervention became protracted, the role of international players became more salient.[131]

Immediate Soviet involvement, unlike the escalating intensity of the U.S. response to the Lebanon crisis, was mild, in comparison to its past behavior in major Middle East crises. The Soviets' main concern was preventing an all-out Syrian-Israeli war that would threaten Damascus, jeopardize the shaky Assad regime, and force them to consider direct military intervention and a possible superpower confrontation. The Soviets were disappointed by the lack of Arab unity and the lack of support for Syria and the PLO. They therefore focused their activities in three areas. They attempted to get Washington to restrain Israel by emphasizing their concerns about the proximity of the conflict to their southern border. This exchange did not take the form of an ultimatum, but Brezhnev's message to Reagan shortly before the first cease-fire was that "a most serious situation had been created which contained the possibility of wider hostilities" (quoted in: Spechler, 1988: 174–75). To signal their intentions the Soviets reinforced their Mediterranean fleet and put two airborne divisions on alert. In the United Nations, the Soviets were active in initiating resolutions that would embarrass Israel and force the United States to use its veto power or abstain, thus inconveniencing the United States in its relations with the Arab world. They treated the PLO and Syria differently, but both clients considered the Soviet response disappointing (Golan, 1982–83; Rabinovich, 1985: 148–50; Ross, 1990). Although veiled Soviet threats of intervention were instrumental, if not decisive, in persuading Prime Minister Begin to announce a cease-fire in June, a cease-fire did not satisfy the Soviet allies. From a Soviet point of view, the heavy losses of the Syrians both in the air and on the ground threatened the reputation of Soviet weapon systems among Soviet clients and raised doubts among its adversaries about the effectiveness of Warsaw Pact forces that depended on similar weapon systems. The Soviets agreed to resupply the Syrian army with new and more advanced weapon systems to replace their losses. Later they also provided Soviet crews to man a new generation of SAM-8 and SAM-9 surface-to-air missiles; and in October they decided to supply a state-of-the-art integrated air-defense system, including SAM-5 missiles, which had never before been deployed outside the Warsaw Pact countries. This decision was implemented in January 1983 and was the Soviet response to the U.S. deployment of troops in Beirut to oversee the evacuation of the PLO and to police the cease-fire, to which the Soviets strongly objected. This new air defense system required an additional 1,000 to 1,500 Soviet combat personnel in Syria and signaled an increased commitment by the Soviets to Syria. This commitment later enhanced Syrian resistance to and eventual rejection of the U.S.-brokered accord between Lebanon and Israel.[132]

Brezhnev increased Soviet commitment to Syria but, in contrast, did little for the PLO. Specifically, he rejected their call for direct military intervention, for fear that it would lead to a military confrontation with the United States. Such a measure would also have posed very serious operational and logistic problems, which would have made it an extremely risky venture, especially because the Soviet army would have had to take on the best-trained, most-experienced, and best-organized military force in the region (Ross, 1990; Spechler, 1988).

Neither before nor during the war was the Soviet factor a direct risk consideration among Israeli decisionmakers.[133] Three factors can account for this. First, given Reagan's attitude toward the Soviet Union, a Soviet threat to Israel would have resulted in American countermoves. Second, the prime minister and most cabinet members anticipated a short war that would not involve Syria, the Soviets' main client. Third, even if the Syrians were involved, the view in Israel was that Moscow would not rush to assist Damascus as long as the fighting took place in Lebanon and not within Syrian borders. Thus Israel abstained from attacking SAM batteries on the Syrian side of the border even when they posed a threat, so that there would be little risk of the implementation of Soviet commitments to Syria (Bavly and Salpeter, 1984: 115; Sella, 1983). Even though the Soviets were cautious, their involvement had two effects. The first was to heighten American concerns about the Israeli action. For Washington, the worst-case scenario was that a protracted conflict would bring in the Soviets and lead to a superpower confrontation (Spechler, 1988: 175). The other effect was that the rearming of the Syrian army made Syria increasingly assertive and defiant after the humiliation suffered in the early stages of the war.

The possible responses of other Middle Eastern countries were of little concern. Israeli policymakers judged there to be a minimal risk that local third parties would interfere to aid Syria and the PLO in their quandary. Egypt was bound by the peace process, a process opposed by Syria and the PLO. Jordan, too, viewed Syria and the PLO as antagonists. Iraq and Iran were at war, and Saudi Arabia was focused on events in the Persian Gulf. Thus, an Arab coalition in support of Syria and the PLO seemed unlikely to form, and indeed none did.

But even with the hostility between Egypt and Israel's opponents in Lebanon, the question of how a leading Arab country like Egypt would respond to an Israeli invasion of another Arab country and to the occupation of its capital had to be considered. Invasion could jeopardize the most important achievement of the Begin government, the peace treaty with Egypt. The treaty had strong opposition in Egypt (Greenwald, 1988), as the murder of Anwar Sadat in 1981 suggested, and both the domestic opposition in Egypt and opponents of the peace treaty in the Arab world would probably argue that the treaty made the invasion of Lebanon possible, as indeed was the case. Israel's invasion had put the Egyptian supporters of the Israel-Egypt peace treaty on the defensive. The invasion therefore became an important reason for the cold peace between Egypt and Israel. But these potential risky effects on this important bilateral relation were not taken into serious account in planning the operation (Feldman, 1992: 152).

Begin viewed the concessions that he made to achieve the breakthrough with Egypt as heavy sacrifices in themselves.[134] He refused to allow other foreign and defense policy issues to become hostage to the peace with Egypt; this attitude is demonstrated by the Israeli bombing of Iraqi nuclear installations in 1981 shortly after a summit with Sadat, a matter that caused Sadat great embarrassment. Israeli decisionmakers believed that Egypt recognized the importance of the peace dividends that it had reaped, not the least of which was U.S. magnanimity.[135] If the treaty was abrogated, Israel might attempt to reoccupy the Sinai and thereby jeopardizing the substantial Egyptian revenues from the oil fields. The Egyptian government would have to expend large sums that it could not afford on a military buildup to face such an potential Israeli threat. The Egyptian military therefore supported the peace treaty. In light of these considerations, only a limited Egyptian response could be expected, and Egypt responded to the invasion mainly with diplomatic efforts, working through the United States and the United Nations and practically freezing normalization of the relation between the two states. On the whole, these efforts had little effect (Greenwald, 1988). But as domestic criticism in Israel increased, the opposition used the cold peace with Egypt to demonstrate the counterproductivity of the intervention and Begin's lack of commitment to peace. This claim illustrates the link between a lack of international legitimacy and the undermining of domestic legitimacy.

Another potentially disruptive third party was the United Nations, in which the Arabs and Soviets could count on an anti-Israeli majority. This pro-Arab bias actually made the United Nations ineffectual in influencing Israel's calculations. Israeli decisionmakers assumed that, no matter what the circumstance, Israel would be condemned. The antagonism between the Israeli government and the world organization had engendered an Israeli policy of disregard for the latter's positions and rendered the body largely impotent as far as the Arab-Israeli conflict was concerned. It was obvious that Arab states and their supporters in the General Assembly and the Soviet Union as a permanent member of the Security Council, which had the authority to impose sanctions on Israel, would attempt to pass resolutions condemning Israel, calling for a cease-fire, demanding withdrawal, and threatening sanctions. It could be anticipated that the presence of the UNIFIL in southern Lebanon would add urgency to the reasons for the secretary general and Security Council to get involved.

The main threats posed by U.N. activity stemmed from the possibility of sanctions and the undermining of any hope for international legitimacy. Neither risk received much attention in Israel. Without American support for sanctions, such a resolution could not pass. Support for sanctions was considered unlikely given the U.S. voting record on major issues concerning the Arab-Israeli conflict; the United States had consistently used its veto power in support of Israel. Two Israeli suppositions—the green light given by the United States to the intervention and the hostile attitude of the Reagan administration toward the United Nations together reinforced the view that the United States would prevent any drastic U.N. measures. As for the risk of international delegitimization, experience had shown that defying the United Nations had a very low cost, and, besides, the operation was anticipated to last for only a brief period. Once the invasion had achieved its objectives, the storm of condemnations would die away.

From the very first day of Operation Peace for Galilee, as expected, the U.N. Security Council focused its attention on the invasion at the urgent insistence of the Lebanese government.[136] Starting with Resolution 509 on June 6, 1982, the Security Council called for Israeli withdrawal. This was followed by a Spanish resolution on June 8 calling for a condemnation of Israel and threatening sanctions. The resolution was vetoed by the United States. As the war continued, the Security Council focused on humanitarian measures (Resolution 512, June 19). As the siege of Beirut tightened, a French resolution that called for Israel's withdrawal from Beirut was vetoed by the United States, but a resolution calling on Israel to resume supplies of food and water to western Beirut was approved unanimously (Resolution 513, July 4). As disagreements between the United States and Israel grew over continued operations in Beirut, the United States signaled its displeasure by abstaining on a Security Council resolution that censured Israel, called the intervention an invasion, and hinted at sanctions for noncompliance with a previous resolution (Resolution 517, August 4). But a Soviet attempt to take advantage of Resolution 517 and adopt a related resolution calling on all member states to refrain from supplying Israel with arms and other military aid until it fully withdrew from Lebanon was vetoed by the United States. U.S. disapproval of Israel's invasion and continued presence in Lebanon became more evident after the Sabra and Shatilla massacres (e.g., Resolution 521, September 19), but the United States stopped short of allowing the adoption of resolutions calling for sanctions. The one-sided position taken by the United Nations backfired and made it easier for the United States to avoid support for extreme U.N. resolutions. In the words of the U.S. secretary of state, "We could count on extremists at the United Nations to concoct outlandish language for all resolutions on this topic, so we could abstain or veto at little political cost" (Shultz, 1993: 62). In essence, the Israeli risk assessment regarding U.N. action in the crisis was vindicated. Nonetheless, the constant discussion and criticism of Israel's Lebanon policy and its coverage in the international media contributed to the international delegitimization of the protracted intervention. International criticism provided ammunition to the domestic opposition to the war.

Let me summarize the analysis of the risks posed by the intervention. The designers of Israel's Lebanon policy misperceived the risks associated with their broad objectives and with the instrument applied to achieve these objectives, extensive use of military power. They underestimated their adversaries' resilience and willingness to take risks. They overestimated the IDF's capabilities and the extent to which these capabilities could be applied without restraint, political or situational, in the service of political objectives— hence their overoptimistic assessment of the short time required for the operation to achieve its goals. Because of this optimism, their cost estimates were highly skewed. They never anticipated that the cost in lives would be as high as it was, that domestic legitimacy would be so difficult to sustain, or that the war that was to demonstrate the regeneration of national confidence would instead become another protracted trauma and the cause of much dissent and bitterness. They did not anticipate that the benefits of such a risky policy would be so short-lived, or that the policy would stir up Lebanese communal sentiments against Israel such that a new long-term threat to Israel's northern settlements would emerge—the Shi'ite threat. Because of the inherent religious fanaticism the Shi'ite threat has become in some ways even less containable than the preceding Palestinian threat. Israeli decisionmakers did not anticipate that the two defeated parties, the PLO and Syria, would recuperate from the debacle so fast and even improve their position in the two regional power plays toward which Israeli intervention had been directed, the Palestinian and Lebanon issues, respectively. Their poor assessment of developments and associated costs was not limited to domestic and military issue-areas. The attribution of secondary weight to international consequences, especially the implications for relations with the United States, proved wrong. The major powers became involved in various ways that limited Israel's policy options and its ability to advance its objectives and preserve the gains from the policy.

Inadvertently and without being an explicit part of the planning, the time factor became important to the direction that Israel's Lebanon policy took. With the passage of time, the fallacies of the assumptions underlying the policy were exposed, which resulted in a further decline of support for the policy, both within the government and among the public. Similarly, the mounting costs of staying in Lebanon became a political, military, and economic burden that the country and its government could ill afford to sustain. Even if staying the course had been a politically viable option, such a policy would have been difficult to justify, for over time, trends and developments in Lebanon were made neither more controllable nor less uncertain and unpredictable. The complexity and quasi-chaotic state of local politics, the difficulties of governance, the volatility of communal relations, and the consequences of intervention and meddling by external powers seemed to prove that the only statement one could make with confidence about the future of Lebanon was that uncertainty would prevail. Thus, the Israeli government was unable to offer the public a clear and convincing vision of when and how the Lebanon situation would turn around for the better; its past optimism seemed thinner than ever and quickly evaporated. Burdened with a policy that had already cost a prime minister and a defense minister their jobs, the government had no choice but to unilaterally disengage from the Lebanese quagmire without achieving its primary objective, a stable and peaceful Lebanon.

Unrealistic perceptions of risk are evident not merely in hindsight. Intelligence analysts, military officers, and politicians foresaw and forewarned the government about many of the adverse consequences, as we have seen. Negative feedback effects seemed to influence the underlying assessments, the confidence with which they were held, and the resultant policy too little and too late. These patterns of risk assessment and risk-taking preference are explained next.

THE MAKING OF A QUAGMIRE: THE KEY DECISIONMAKERS

To a large degree, security policy debates and discourse are about divergent risk perceptions and risk-taking preferences. The Lebanon policy debate within the Israeli government was no exception. The estimates of Begin, Sharon, and Eitan diverged from those of some other members of the cabinet regarding the risks that a military intervention entailed, the relative efficacy and risks of alternative policy options, and the acceptability of the risks associated with military intervention. The differences extended from the initial debate on whether to intervene through the stages of escalation and entrapment. The reasons for the differences and for their resolution in one direction or another are explored below.

Three questions have to be addressed. First, how and why was a decision made to take the risks that involved the use of military force on a scale that in the past was used only in major Arab-Israeli wars? Second, once the operation was being carried out, why did a limited and supposedly short-term operation escalate into the first-ever entry of IDF forces into an Arab capital, an act that had always been considered unwise and very risky?[137] Finally, why were the achievements up to and including PLO and Syrian withdrawal from Beirut considered insufficient to terminate the operation? Why did their withdrawal lead instead to an extension of the objectives, so that what was first believed to be an operation of a few days, and then a few weeks, turned into an intervention of three years? The answers to these questions are complex. They require an examination of the sources of risk perceptions and the evolution of risk-taking preferences as they emerged in an open democratic society in which both individual-dominated and collective cabinet-style decisionmaking alternately prevailed and were interwoven. The explanations have also to take into account the contextual circumstances, institutional constraints, and individual manipulations that resulted in escalating policy commitments at resurgent costs. In the first stage of the discussion I explain how policy preferences evolved among the three key decisionmakers, Prime Minister Begin, Defense Minister Sharon, and Chief of Staff Eitan. In the second I show how these individuals attempted, successfully or not, to mold the cabinet's policy to fit their preferences.

Begin's 1981 election campaign promise that the day would come when no Katyusha rockets would fall on the town of Kiryat Shmonah signaled a shift of focus in the security policy to the PLO and the deteriorating security situation on Israel's northern border. This shift became even more pronounced when the threat on Israel's southern border was resolved with the Israeli-Egyptian peace treaty and the completion of the withdrawal of Israeli troops from the Sinai. From the prime minister's vantage point, the problem was well defined. The main objective would be to destroy permanently the state-within-a-state that the PLO had established in Lebanon. A quick, forceful military operation would remove the threat to Israel's northern settlements once and for all and might accomplish other important objectives as well. Sharon and Eitan assured him that it could be done.

Yet the reality was much more complicated. The troubled history of modern Lebanon, with its mosaic of antagonistic ethnic, religious, and political groups often led by cunning, ambitious, and manipulative individuals, the lack of a strong central government, and the extreme difficulties of governability due to constitutional and practical constraints should have been reasons enough for viewing the Lebanon problem as highly ill defined and not given to a quick solution. The enmirement of the Syrian army in Lebanon since 1976 in the quagmire of Lebanese political culture—where political alliances were driven by extreme opportunism, allies were manipulated, and agreements were regularly violated—should have sounded the alarm against commitment to the Lebanese Front and its leader, Bashir Jumayyil. The cost of antagonizing other powerful communities, especially the largest of all, the Shi'ite community, should have been anticipated. Instead, this complex and ambiguous environment was structured and framed as a well-defined, relatively simple opportunity for applying Israel's superior military power—an opportunity that would allow Israel to shape and direct developments in the region to its short-term and long-term advantage at an acceptable risk and that would guarantee its military and political preponderance in the region and its uninterrupted control of the West Bank. Begin, Sharon, and Eitan saw the circumstances as auspicious for defeating the PLO and thus both settling an acute security problem and possibly winning in the long run the battle for Eretz Israel (the Land of Israel, including Judea and Samarea), which was to them a matter of critical ideological, political, and strategic importance. In fact, the same imperatives that drove their foreign and security policies also manifested themselves in a dynamic drive to settle the West Bank and, to a lesser degree, the Gaza area and the Golan Heights, so that giving up the territories, no matter which political party was in power in the future, would have become practically impossible (Sandler, 1993: 203–12).

When Likud came close to losing the 1981 election to the Labor Party, which was not committed to Eretz Israel, it sharpened the sense of urgency and potential loss if an irreversible fait accompli was not established while Begin, Sharon, and Eitan were in a position to shape history. Framing the situation this way meant that they could have found themselves in the loss domain if they did not take advantage of what they considered a unique historical opportunity. The Lebanon intervention was thus intended to secure a benign external environment both for immediate security purposes and for other ideologically driven policy objectives. In Begin's case, an additional source of urgency can be traced to his declining state of health. He had suffered three coronaries and a stroke and apparently realized that time was limited for accomplishing his lifelong ambition of establishing a Jewish state that would survive in security and peace (Post and Robins, 1993: 135). The waning of physical health and the attendant perception that time was running out are known to induce dramatic, risk-seeking political moves, especially when two additional conditions prevail: the decisionmaker has a sense of mission and a personal tendency to take risks. Both conditions existed with Begin.

Broadly speaking, the issue was framed as a two-level problem. At the tactical, short-term level it was framed as a PLO threat to Israel's northern settlements, which had to be addressed in a decisive and permanent manner. At the strategic long-term level it was framed as an opportunity to secure Israel's dominant regional position and its possession of the territorial acquisitions of 1967 and thus to achieve certain ideological goals that had a strong emotional appeal. To miss this opportunity would have been a loss of great magnitude. The combined effects of the tactical and strategic framing of the problem implied that risk avoidance could result in both short-term and long-term losses, a possibility that strengthened the incentive for a risk-acceptant policy. Yet although Begin, Sharon, and Eitan were prepared to take the risks of executing the original Big Pines (the Big Plan) as early as December 20, 1981, the cabinet turned the plan down. Between December 1981 and June 1982 there were at least five abortive attempts by Begin to get cabinet approval for some sort of invasion, either there was no majority support, or the majority was too slim for Begin's comfort (Yaniv, 1987: 107–9).

When the cabinet approved a less ambitious plan on June 5 by an overwhelming majority (fourteen for, one against, and two abstaining), Begin and a majority in his cabinet were encouraged in their tendency to assess the operation from a narrow military perspective, without giving much attention to the broader political picture in Lebanon. The plan was not carefully scrutinized by the cabinet (Naor, 1986: 44). In fact, most of those who had opposed the operation until that point wrongly believed that they had successfully contained the ambitious policy aspirations of Sharon, and so they did not contest the optimistic assessments of what Operation Peace for Galilee would achieve.[138] Later, the incremental extension of objectives and scope resulted in a set of ad hoc decisions that again were reviewed in isolation and not in a broader context. With few exceptions, then, members of the cabinet treated each occasion for decision as involving a relatively simple and well-defined situation. Only after Israel was already deep in the Lebanese quagmire were the complexity and uncertainty inherent in the situation recognized.

It is not that Israel's key decisionmakers were unaware of the religious, ethnic, and political complexity of Lebanon. They simply did not see its immediate relevance for the success of Operation Peace for Galilee. What were vivid and salient to them were the implications of the threat posed by PLO shelling of Israel's northern settlement, the close working relationship with the Lebanese Front and its leadership, the unresolved problem of Syrian missiles in Lebanon, and the stalled Palestinian autonomy talks and their implications for Israel's control of the West Bank. In the background was Begin's, Sharon's, and Eitan's deep-seated dislike for the PLO and its leadership. Begin harbored Holocaust memories that became associated with the PLO threat. All three held the traditional Revisionist vision of an alliance between non-Muslim and non-Arab minorities in the Middle East against the Arab-Muslim threat, and they perceived the Maronite Christians as natural allies. These factors converged to focus attention on the PLO and Syrian threats to Israel, as well as to the Christians in Lebanon, and made them highly salient in information processing. The threats were also the issues that Begin was highly familiar with. He was aware of the many ethnic, political, and religious groups that made Lebanon such a complex and unpredictable arena, but these were not subjects with which he was familiar enough to be affected by their consideration. His selective attention, guided by the availability of historical memories and immediate threats, was reinforced by a cognitive style that centered on generalized patterns and did not concern itself with details. In the absence of military-strategic experience, he was in no position, and had no wish, to question the details of a military strategy outlined by two admired generals, and he was bound to be impressed by the broad sweep of the plan and its appealing emphasis on utilizing Israel's military capabilities to their fullest (Perlmutter, 1987: 379–81; Sofer, 1988: 100–101, 114).[139]

Begin's and Sharon's presentation to the cabinet and the argument for authorizing Operation Peace for Galilee implied a certainty of quick success. The assessment of success allowed for the certainty effect, which led to an overweighing of the outcomes considered certain compared with other outcomes considered merely probable. Expected outcomes were introduced mostly in terms of the benefits of carrying out the operation and in terms of negative counterfactuals that emphasized the adverse consequences of not carrying it out. Although the number of casualties was the most important cost dimension on the cabinet members' minds, Chief of Staff Eitan refused to provide an estimate. Without an estimate that important concern remained abstract and was less significant in decisionmaking than it would have been if the hundreds of casualties had been predicted—as they were for the war games conducted by the IDF, but of which Begin was not informed.[140]

According to the plan the main population centers in Lebanon would be bypassed; the Syrian army would not be engaged; combat operations would be limited in duration; the burden of the intervention would be shared by the Phalange forces; and the massive use of military power would produce a quick, decisive victory and limit the number of Israeli casualties, as well as casualties among Lebanese civilians. As the intervention progressed, the optimistic expectation of few casualties was negated. The cabinet responded with great dismay; it scrutinized Sharon's estimates of the cost of combat operations with increasing suspicion and grew ever more reluctant to blindly approve operations like the occupation of western Beirut.[141] Potential casualties became the major reason for disapproving these requests and a cause for doubting Sharon's and Eitan's, especially the former's, claim to unquestioned expertise on military matters. The decline of confidence caused even Begin to question Sharon's judgments, although Begin could not bring himself to terminate the intervention.

Sharon had worked on his plan from the minute he was appointed minister of defense, and he had great confidence in its viability and probability of success (*Yediot Aharonot*, June 18, 1982). This confidence was embedded in his own overconfidence in himself as a military and political strategist, in what seemed to him the impeccable linear logic inherent in the grand design itself, and in his underestimation of the adversary. Sharon and Eitan despised the PLO and discounted its military capability in frontal combat. They viewed it as cowardly terrorist organization that attacked defenseless civilians, hence assumed that, once it was faced with the awesome power of the IDF, its resistance would be quickly shattered. The same view applied somewhat to Sharon's assessment of the Syrian army.[142]

In his schema the sequence of events was almost inevitable: A forceful, determined military operation in Lebanon would eliminate the PLO as a political actor and cut Syria's influence down to size. This would allow the installation of a strong pro-Western and friendly Christian-led government in Lebanon under Bashir Jumayyil. A victory in Lebanon would serve both American and Israeli strategic interests. The intervention would contribute to the decline of Soviet influence in the region and enhance cooperation between the United States and Israel in forging a long-term strategic alliance. In this new regional context of Israeli preponderance, the Palestinian leadership in Gaza and the West Bank would recognize their own vulnerability and the PLO's impotence and would have no choice but to reach a settlement with Israel on terms that acknowledged Israeli domination. Sharon was alerted by advisers to some risks, but he attributed low probabilities to them.

A belief in the inherent inferiority of the adversary, great personal confidence, and a seemingly faultless plan to deal with despised adversaries gave rise to overconfidence in the success of the plan and a sense of control over the future. There was no room for questions about the validity of the assumptions underlying the Lebanon policy. Thus, rather than taking account of information indicating failings in the plan and adapting it accordingly, Sharon and Eitan, with Begin's support, attempted to engineer the environment to suit their wishes. As criticism of the war in Israel intensified and the end remained out of sight, Begin and Sharon were forced to take stock of the operation's achievements. In public appearances they described the large stocks of weapons and ammunition discovered and captured in southern Lebanon and made a case for what might have been had the operation not been launched. This type of counterfactual argument sustained the commitment to the now much-expanded operation. Negative feedback from the front was attended to selectively or rationalized as indicating only short-term upsets. The original grand design remained the salient anchoring point for decisions. Even the costs, much higher than were anticipated, were framed as necessary investments in Israel's future security and the defense of its northern settlements.

The extensive attention given to the planning of the operation led not only to greater confidence in the high probability of ultimate success but also to underestimation of the time required for the implementation of each stage of the plan. Not taken into account were the adverse implications of incremental implementation, necessitated by the cabinet's nonapproval of Big Pines in its original form. The resulting domestic political constraints were not fully recognized. This failure was compounded by an underestimation of the enemy's resilience and willingness to suffer. Thus, overconfidence and the consequences of incrementalism meant constant underestimation of the time required for carrying out the plan.

Misjudging the time required had several negative consequences. The prime minister and the government repeatedly agreed to scheduled cease-fires based on time estimates provided by the defense minister and chief of staff that turned out to be incorrect in terms of goal accomplishments by the IDF. Consequently, the IDF did not observe the cease-fires, often without the knowledge of the cabinet. Cease-fire violations brought angry responses from Washington, on occasion from President Reagan himself, which embarrassed Begin and his cabinet and adversely affected their credibility in Washington. Cease-fires sometimes had adverse consequences on the battlefield, too, for military objectives that were within reach had to be relinquished. In some cases, attempts to achieve these objectives after the elapse of a cease-fire turned out to be costly in terms of casualties because of enemy reinforcements or improvements in the position of enemy. In other cases, the fighting went on despite the cease-fire, but because the violation of the cease-fire had to be discreet, the IDF could not bring to bear its superior firepower, maneuverability, and numbers, and this circumstance also resulted in higher casualties. The adverse consequences raised resentment among field commanders and soldiers, who saw themselves as paying the price for political games played by politicians who were insensitive and did not care enough for the sacrifices made by the troops (Schiff and Ya'ari, 1984: 245–48). The resultant military attrition undermined the legitimacy of the Lebanon policy.

Because the architects of the operation neglected to take into account the social context within which decisions would be made and public support mobilized, Israel's political culture had unanticipated effects on the severity of the war's domestic political risks. For three decades, since independence, Israel's political system had been dominated by the political left and its allies. Their views on security set the norms that defined the circumstances under which risking the lives of Israeli soldiers was justified. The very high number of casualties suffered by Israel in the war that preceded Operation Peace for Galilee, the Yom Kippur War of 1973, substantially increased sensitivity to this risk dimension. The alternative view on national security, represented by the rightist Likud, was consistently projected by their adversaries on the left as irresponsible, insensitive to human lives, and impractical. This perception of Likud's "irresponsibility" could not be erased in its first four years in power. In fact, the Likud did not rise to power in 1977 because of the attraction or merits of its ideology and political-strategic program, but mostly because of Labor's mismanagement and failures. In its first four years in power (1977–81), the Likud was discredited, and it almost lost the 1981 election. Thus the transfer of power from Labor to Likud did not necessarily mean that the voters also made the transition to Likud political-strategic ideas and values or that the essentials of the popular political culture had changed.

The concept of offensive war for achieving broad political objectives remained anathema to a large and influential section of the Israeli public, who continued to view military power as an instrument to be used only for defensive purposes. It is no accident that Israel's military forces are called Israel Defense Forces. The concept of war by choice was not likely to be a successful quick sell or to replace the deeply embedded concept of no-choice war, and the two concepts imply different types of risk acceptability. The attraction of war by choice became particularly questionable when it was less than successful in Lebanon. As the war drew out, the futility of the new strategic approach became more evident. It undermined public confidence in the government leaders, who, it was thought, had stepped outside the bounds of the historical national consensus on when to take risks, which risks were justified, and what costs were acceptable.

The small geographic size and population of Israel and the tradition of openness have created an intimate society. The network of information exchanges through horizontal and vertical personal and professional contacts and the investigative, well-connected media together make it very difficult for decisionmakers to conceal information for long. Because the war raised highly affective issues that directly involved many Israelis, exposure of deceit and attempts at concealment snowballed. This was the case with the public reaction to the government's attempt to dismiss the gravity of the Sabra and Shatilla massacres. Public pressure forced the government to form a commission of inquiry, which submitted a report that cost the defense minister his job. The report was also critical of and embarrassing to the prime minister, foreign minister, chief of staff, and other generals. Overall, such cultural-societal effects worked to undermine the acceptability and viability of the Begin-Sharon political-strategic approach to the use of force as applied in Lebanon.

What is still puzzling is why Begin and Sharon raised the stakes when they had more than one opportunity to terminate the intervention. I suggest that, where the stakes and costs are high, the policy contested, and the personal reputations of key decisionmakers on the line, the termination of an intervention policy takes place only when its designers can claim victory. As long as they cannot, policy termination will be delayed. Furthermore, as a policy failure becomes salient and costly, it tends to become more exclusively associated with its initiator. Just as Vietnam became Lyndon Johnson's war, Lebanon came to be known as "Sharon and Begin's war." By attributing failure to a particular person, everyone else dissociates themselves from it. Ironically, being associated with the problem makes it more difficult for a decisionmaker to get disentangled from the policy, because this would be construed as an admission of failure. Policy initiators prefer to gamble for resurrection, which entraps them even more. Sharon and Begin fell into this pattern of behavior.

In the Lebanon case, incongruence between formal goal statements (e.g., an operation limited to a 40-kilometer-wide strip) and the much more ambitious private intentions, as well as the vague and indeterminate manner in which the initial statement of goals was given, made a decision to terminate the intervention difficult because of the ensuing confusion about the actual range of policy goals.[143] Military objectives were added as the war rolled on, the product of either hidden designs or circumstances that diverged from the planning scenarios; thus it was unclear at any point what the actual range of policy goals was, whether the objectives had been achieved, or when the operation could be appropriately terminated. Claims that important goals had been achieved were only partly convincing because of this ambiguity. Prime Minister Begin and Defense Minister Sharon claimed that the war permanently removed the threat to Israel's northern settlements, but with the chaotic domestic situation in Lebanon and the continued Syrian presence the validity of the assertion seemed dubious.[144] The assassination of Bashir Jumayyil made intervention outcomes and longer-term implications even more uncertain.

Those who planned and initiated the war needed an event that would transparently confirm that the main goals were achieved and would thus justify the costs. The expulsion of PLO and Syrian forces from Beirut could have been such a confirming event. But PLO and Syrian forces gave a good account of themselves during the siege of Beirut, so their eventual expulsion was less humiliating to them than Begin, Sharon, and Eitan found desirable.[145] Jumayyil's assassination undermined another important success-confirming event, the election of a strong, Christian, pro-Israeli Lebanese president. Thus the search turned to another highly visible, symbolic, and tangible act: a written peace agreement with Lebanon (*New York Times*, October 11, 1982; Yaniv, 1987: 159–60). It could provide an opportunity for termination of the intervention and allow decisionmakers whose reputations had become interlinked with the war to claim victory. But the watered-down Israeli-Lebanese accord of May 1983 had little sting and made termination of the intervention more, not less, difficult.[146] In the end, termination was further delayed, and required the establishment of a national unity government in Israel and a different decisionmaking setting.

The argument so far has emphasized shared contextual incentives and constraints to which most decisionmakers could have reacted similarly. I next argue that personality variables had an important role in determining the impact of contextual effects on preferences—that is, decisionmakers with different personalities reacted differently to the same situational incentives and constraints. Although these contextual factors helped direct and sustain risk-taking preferences, the deeper and more permanent causes of risk taking in the case of Begin and risk seeking, rather than mere risk taking, in the cases of Sharon and Eitan must be sought in the beliefs, values, and personality attributes of these key decisionmakers. The differences between Begin and his Labor Party predecessors in their approaches to foreign and security policies reflect the ideological backgrounds in which those approaches were embedded. Begin and his Likud Party represented the New Zionism (Neo-Revisionism).[147] The Labor Party leaders represented Socialist Zionism. In essence, the latter emphasized incrementalism, whereas the former emphasized the importance of the grandstand and the momentous event. Begin's, Sharon's, and Eitan's beliefs and values were anchored within the framework of the New Zionism. Although the three had some differences in cognitive style and emphasis, they shared certain beliefs, values, and stereotypes.

Begin was the most cognitively complex of the three, Eitan the least so. Begin, an intellectual, spoke several European languages, believed in the importance of cold logic, and emphasized the importance of basing decisions on extensive information. But the combination of a populist oratorical style, intense ideological commitment, and vivid memories of and identification with Jewish suffering through history seemed to neutralize his intellectual bent in certain issue-areas, so he simplified and exaggerated arguments. The Arab-Israeli conflict and, in particular, the Palestinians in the PLO, to whom he often referred as Nazis, were among the most salient subjects of this kind of treatment. Similarly, he distrusted and derogated Syrian President Assad (Peleg, 1987: 68–72; Seliktar, 1988). In the emotionally charged cabinet meeting of June 5, 1982, when the decision on Operation Peace for Galilee was made, Begin tried to convince his cabinet to support his decision, saying: "Our fate in the Land of Israel is fighting and self-sacrifice. Believe me, the alternative to that is Treblinka, and we decided there will be no more Treblinkas. This is the moment when one must choose self-sacrifice. Let the malevolent criminals [the PLO] and the world know that the Jewish people have the right of self-defense like any other people" (Naor, 1986: 47–48).

The three decisionmakers held certain key operational code beliefs in common. They all believed that international politics was basically an arena of perpetual conflict and unavoidable violence. The Arab-Israeli conflict, and in particular the Palestinian issue, was intrinsically a zero-sum game. The main pillars of national security had to be military power and the determination to use it (Benziman, 1985: 267; Eitan with Goldstein, 1985; Peleg, 1987; Seliktar, 1988; Sofer, 1988). Drawing on both similar and dissimilar historical events, they reached identical conclusions. Inspired by world history (e.g., European appeasement in the 1930s) and Jewish history—in particular, vivid powerful personal memories of the Holocaust and other instances of Jewish persecution—as well as the early experience of Arab-Jewish relations in Palestine, all three drew the conclusion that Israel's security could be guaranteed only through self-reliance.[148] The wishes of the international community should be ignored when they did not accord with Israel's vital security-related goals (Peleg, 1987: 86).[149]

Their beliefs in the conflictual nature of international politics were logically complemented by beliefs about the critical importance of accepting, even seeking, risks. All three thought a reactive-defensive policy signaled weakness and invited aggression. Taking the initiative was associated with being in control of events. A reactive-defensive approach was associated with the risk of losing control. For Begin, initiation and risk taking were not only instrumentally correct ways of dealing with threats but also important symbolic steps that broke with the history of Jewish persecution—in the past, Jews had often had to take a passive-defensive low-profile posture. The new breed of Jewish nationalists in the Land of Israel, according to him, had to take the initiative and, if necessary, aggressively seek to make a secure environment.[150] The use of military force for preemptive purposes, and not only for defense, was perfectly legitimate.

Risk taking was thus considered an important leadership-related value. Leaders, then, have to exploit rare historical opportunities when they occur. It is a responsibility of leadership to face up to this task.[151] What reinforced the value of risk taking were the life experiences of all three decisionmakers. Begin had commanded a small, active, and at times very effective resistance organization, Irgun Zevai Leumi (National Military Organization), in the struggle for Israel's independence. The Irgun had faced heavy odds in confronting simultaneously the British colonial power, the Palestinian Arabs, and the much larger rival Jewish resistance group, the Hagana (Defense). The Irgun saw taking the offensive as a key to survival and effectiveness and took the initiative even when risks were high. The lessons learned while with the Irgun (1943–48) left a permanent impression: "Begin always considered the period of the revolt as the summit of his life" (Sofer, 1988: 62). In fact, he applied to Lebanon the same concept of "active resistance" that had guided his strategic thinking in the early years of the struggle for independence (Naor, 1993: 250–51).

In the 1950s, Sharon and Eitan were among the founders of Israel's daring and legendary military Unit 101, and later, of the first paratroop brigade. Through their daring exploits these units became models for the IDF, which at the time was an overcautious and ineffective military force that saw failure too often. The example set by Unit 101 and the paratroops made daring, creativity, and risk acceptance part of the IDF ethos, an ethos that would be instrumental in making the IDF one of the most effective and successful military machines in the world. Sharon and Eitan rose through the ranks to military commands by representing the spirit of risk taking, which they brought to their roles as defense minister and chief of staff and thus to the management and direction of national security policy. The life experiences of all three thus reinforced their view that by not taking the initiative, by avoiding risk, one loses control of a situation. They were not concerned about uncontrolled escalation; they believed that limited, well-planned military operations could be controlled and that escalation and de-escalation could therefore be deliberately manipulated with little risk.

Once committed to a risky policy, the three were not easily dissuaded. They knew from experience that occasional setbacks would occur, but they believed that these should not be allowed to derail policies. Perseverance in adversity was necessary to win. One did not turn away from a policy just because it ran into difficulties, or because the immediate costs were higher than anticipated. The larger stakes had to be kept in mind. Leaders had a responsibility not to be deterred from their set goals by temporary difficulties. Thus, it is not that Begin, Sharon, and Eitan were insensitive to casualties, or to growing parliamentary and extraparliamentary criticisms. Rather, they had learned to deal with setbacks as temporary inconveniences and not to give up their strategic objectives. Their response to emerging constraints was not policy reassessment and reversal but a search, especially by Sharon and Eitan, for means to cope while pursuing the original goals. The policy was gradually, if inconsistently, revised only after the opposition in the cabinet became unsurmountable and the United States applied heavy pressure to adjust it.

Begin and Sharon also shared some of the personality traits associated with risk taking. Both had aggressive, domineering personalities and a high need for power and achievement, and both had proved their courage. Both showed symptoms of narcissism and saw themselves as men of destiny (Benziman, 1985: 264; Sofer, 1988: 106). They were aware of their charismatic appeal and aimed to build up and sustain their heroic image through commitment to high-risk, high-value foreign policies.[152] This accounted not only for the optimism and confidence with which they approached the risks of the Lebanon intervention but also for their persistence in pursuing, and preoccupation with, unrealistic objectives, their inability to recognize and admit mistakes, and their tendency to ignore dissonant information. These personality features contributed to Israel's policy entrapment.

Over the years Begin and Sharon had suffered what they perceived as humiliating setbacks that denied them respect and positions of power commensurate with their skills and contributions to society. Begin was for many years a political pariah. Even after he won electoral victories, he still felt unable to gain his due measure of respect from the political left (Perlmutter, 1987: 334–35). He believed that his and his followers' contributions and sacrifices in the struggle for independence, as well as his unselfish decades-long role as the loyal opposition, had gone unappreciated. He sensed that his proven commitment to democratic values and to a peaceful solution of the Arab-Israeli conflict, one that guarantees Israel's security on acceptable terms, remained suspect to a large portion of the left political elite and their supporters, even after he engineered the peace treaty with Egypt.[153] The resignation of two senior members of his first cabinet, Defense Minister Weizman and Foreign Minister Dayan, only increased his sense that he was still stigmatized as a warmonger and not fully accepted as a national leader with a bipartisan base. These affronts apparently resulted in narcissistic injury to his self-esteem and increased his tendency to make dramatic policies that would heal it.

Sharon, too, felt deeply humiliated because he had not been appointed to the top military position of chief of staff in the 1970s (Benziman, 1985: 123–25). He felt that his proven outstanding skills and achievements on the battlefield made him better qualified for the position than any other appointee. Sharon was determined to prove that he was the best person to deal with complex security problems in peacetime (e.g., evacuating the settlers from the Sinai) and wartime (e.g., eliminating the PLO threat from Lebanon). The similarity between these two leaders in important personality attributes and their shared worldview reinforced and strengthened their functional bond of commitment to a risky policy that served personality-based needs and drives.[154]

FROM INDIVIDUAL RISK-TAKING PREFERENCES TO GOVERNMENT POLICY

My analysis thus far has focused on the risk-taking preferences of three key decisionmakers. But Israel's formal and informal institutional structure made cabinet approval imperative for their preferences to become actual policy. Within the cabinet several ministers, some from Begin's party and some from other coalition parties, were opposed to, or, at best, reluctant to authorize, the initial intervention. Opposition to the policy became more pronounced as Israel became entangled in the Lebanese web. It is important, therefore, to determine how and why the risk-taking preferences of the few came to be shared by the cabinet and then sustained in spite of resurgent disenchantment.

In broad terms, what happened between December 1981 and June 1982 was that an active and persistent minority in the cabinet—Begin, Sharon, and to a lesser degree Foreign Minister Yitzhak Shamir—persuaded a majority to support a risky option. Yet this preference change was not fully accompanied by conversion of the majority to the minority's position; the shift was mostly a matter of compliance.[155] As the war escalated beyond the limits that were initially approved, those who complied but were not converted were put in an ambivalent situation. They repeatedly fluctuated between authorizing additional military escalation and, as suspicion of Sharon grew, withholding authorization for it. By authorizing the intervention the cabinet became at least in part responsible for the consequences. Commitment, responsibility, and accountability resulted in gambling for resurrection: sequential authorizations of incremental military escalation. Yet, not having internalized the conviction that the expanded war was necessary and justified, these cabinet members became more willing to challenge the judgments of a defense minister who had been unable to deliver a fast, decisive victory. Ambivalence caused inconsistent behavior. But exactly how did this come about?

The decisionmaking forum in the Lebanon case could be described as two layered. The core of the decisionmaking group consisted of a faction of three: Begin, Sharon, and Eitan. The initiative for the policy and its details were decided by agreement within this faction. But because of Begin's personal preferences for broad consensus on important security issues, the cabinet as a whole served as a sounding board and forum. The cabinet was crucial in providing consent and approval. It either concurred with a policy initiative, which was then implemented by the defense minister and the chief of staff, or it rejected the policy, a move that was either final or signified a delay while its supporters waited for another opportune occasion to raise the issue. By opposing a policy, the cabinet forced the defense minister and chief of staff to abandon it, make it less risky, or reframe it to make it more acceptable. Even in a modified and milder form, the power of policy implementation gave Sharon and Eitan extensive leverage in shaping and executing policies, especially those framed in vague terms, to suit their wishes more closely than the cabinet's intentions. Often they manipulated defense information, of which they had a near monopoly. Information manipulation resulted in unauthorized raising of the riskiness of the actual policy beyond the acceptability levels set by the cabinet, but the cabinet members were at first unaware of the deviation and were later persuaded or forced to go along as they faced a fait accompli (Shamir, 1994: 168).

Although Begin was the senior member in the Begin-Sharon-Eitan faction, he was at a disadvantage in the informal forum that shaped and framed key national military policy issues. He never had training or experience in commanding troops on the battlefield, nor had he acquired experience in managing national military-strategic issues, as his Labor Party predecessors had. He felt inadequate in these matters compared with Sharon and Eitan, whom he considered heroes and seasoned generals, and he often deferred to them on military matters. His deference made Sharon very influential in military decisions—on occasion, the most influential member in the faction. Most of his views were shared in essence by the chief of staff, which made his position even stronger. Begin's attitude toward Sharon was a mix of suspicion, admiration, and sentimentality. In constituting his second government he allowed admiration for Sharon's military genius and a loyalty embedded in past close relationships between their families in Brisk (Poland) to dominate his choice of defense minister (Weizman, 1981: 121). Begin admired Sharon's military skills but did not fully trust him, which was one reason for not appointing him to the position of defense minister in his first government.

But Begin had much respect for and confidence in Chief of Staff Eitan. This was to some degree based on his admiration for his military experience and heroic image. But it also stemmed from the last few months of Begin's first government, when his first defense minister, Ezer Weizman, resigned and Begin held the position himself, in addition to the prime ministership. The man who practically ran the ministry of defense for him was Eitan. The close relationship then forged between Begin and Eitan consequently meant that Eitan's support for Sharon's policy preferences reinforced Sharon's influence over Begin.

Sharon's appointment as defense minister created obstacles in information flow and information sharing within the defense apparatus and between the defense apparatus and the prime minister's office. He cultivated the National Security Unit of the defense ministry, which acted independently of the IDF's General Staff, especially on political-strategic issues. The influence of the General Staff declined further as Sharon caused the retirement of some senior officers who might have challenged his views.[156] Mordechai Zippori, a retired brigadier general and a moderate deputy defense minister in Begin's first government, was removed from office. He was appointed minister of communications but at Sharon's request was not appointed to the Ministerial Committee on Defense. Then Begin's trusted military adjutant, the veteran Brigadier-General Ephraim Foran, who was well connected in the IDF and had served as an independent source of information, decided to retire. His position was filled by the young Lieutenant Colonel Azriel Nevo, who, being much less experienced and not as well placed in terms of status and connections, could not command Begin's trust or, therefore, be as effective as his predecessor (Benziman, 1985: 265–66; Schiff and Ya'ari, 1984: 33–34). All these developments allowed Sharon to monopolize the flow and interpretation of information. Begin "was being kept in the dark by Sharon, who had managed to become his major contact and source of information about the war and the comings and goings of the IDF. He would report constantly to Begin, much in the manner of a regimental duty officer—details, not substance" (Perlmutter, 1987: 384).

In fact, it could be argued that for all practical purposes the three-man faction evinced most of the pathologies of groupthink as far as the Lebanon policy was concerned. These included overoptimism about the chances of success, selective attention to information, a profound belief that the intervention in Lebanon was morally and practically right, and a view that the enemy (the PLO) was extremely evil. These symptoms eventually gave the faction members a strong shared sense of common purpose. Begin supported Sharon unwaveringly at first, even when it became clear that Sharon was misleading the government, or at least taking liberties with the cabinet (Schiff and Ya'ari, 1984: 382). Indeed, Begin's unfailing support undermined his credibility at home and abroad. The role of this faction declined in importance as the intervention became protracted and costly, and Begin's trust in Sharon waned. As the siege of Beirut proceeded, the locus of decisionmaking moved from the faction to the cabinet, exposing Begin and his government to increasingly blunt domestic and foreign criticism.

The initially inconsistent performance of the cabinet as a watchdog over the congruence between policy directives and policy implementation has to do with the relationship between Begin and his cabinet. Begin was used to absolute control over his party in spite of occasional challenges to his leadership. He did not imagine that controlling the cabinet would be different. He knew that he was held in awe by his ministers. He belonged to the generation of those who founded the Jewish state. After signing the peace agreement with Egypt, he was also internationally recognized and associated with a historic breakthrough in Middle East politics. Although he attempted to run the cabinet in a democratic fashion, he also knew that if he threw his weight behind a decision, the members of his government would not oppose him.[157] As one of his ministers said: "We were carried away by him. This was group dynamics that we could not resist" (Schiff, 1984: 70). Well aware of his power over the cabinet, Begin could not imagine that he would lose control over the Lebanon policy.

As a policy manager, Begin had great confidence in his intuition and experience, little use for systematic policy planning, and little experience in managing large bureaucracies (Ben Meir, 1987: 137–38). He did not understand or anticipate problems of policy implementation or the potential discontinuities between decisions and the way they can be manipulated in the implementation stage. Thus, he underestimated the extent to which complex political-military decisions could get out of control if not tightly supervised. Faced with admitting that the Lebanon policy was out of hand, Begin preferred gambling for resurrection instead.

The dominant role of the prime minister in the Israeli cabinet is based on informal as well as formal resources. By law, the prime minister has preeminence over the other members of his or her government. By resigning, a prime minister can force the resignation of the rest of the government. Either new elections or the establishment of a new government is then required, and that government may be constituted of new individuals. By threatening to resign, the prime minister may force uncooperative ministers or members of parliament to fall into line. In addition, in an amendment to the law that came into force in 1981, the prime minister has full authority to fire any minister at any time. Still, the prime minister's power had limits. Because, under Begin, the government was a coalition government and the two large parties were almost equal in parliamentary representation, the coalition partners had great leverage, and the smaller coalition members could bring down the government. When Begin introduced a plan for attacking western Beirut in July 1982, and the senior representative of the National Religious Party—another member of the state's founding generation and highly respected by Begin—Interior and Religious Affairs Minister Yoseph Burg, hinted that this could lead to a crisis in the coalition, Begin backed off (Schiff and Ya'ari, 1984: 261–62). Their leverage limited the prime minister's ability to apply his formal authority to fire disobedient ministers that were not from his party. On the other hand, formal power and authority were greatly enhanced by the personal stature, authority, and respect that Begin inspired. Although a combination of formal and informal power seemed to be the reason that cabinet members conformed with, or at least did not forcefully challenge, the prime minister's policy preference on Lebanon once it became obvious how committed he was, the relationship between Begin and his cabinet was more complex than appearances would suggest.

Several, rather than one, cross-cutting group-induced influences were operating. Begin, the senior decisionmaker, demonstrated tendencies toward both imposition and collegial consultation. When both Begin and Sharon, particularly the latter, wanted to use the cabinet decisionmaking process in their attempts to share the blame for adverse outcomes, the approach was only partly successful because the Israeli public came to associate the Lebanon war with Sharon and Begin, so in the final account they were held most responsible, no matter what the decisionmaking procedure. The search for legitimacy through shared responsibility required that two conditions be met: that the cabinet vote on decisions and that a convincing majority for high-risk decisions exist. These conditions, as we have seen, imposed constraints on the timing and substance of the policies eventually approved. The debate that accompanied the exploration of policy options often showed that many of the cabinet members were uncomfortable with the policy and were reluctant to share responsibility. The emergence by August 1982 of a more cohesive opposition minority group of ministers who questioned the rationale, wisdom, morality, and execution of the government's policy was a clear indication of that reluctance. As the persuasive power of the dissenters within the cabinet waxed, the persuasive powers of Sharon, and, to a lesser extent, Begin, waned. Some imposition was possible, and Begin used that method on occasion, but its effectiveness decreased over time, either because of Begin's ill health or because of the growing public disaffection with and criticism of his Lebanon policy. The dissidents in the cabinet therefore increasingly took their cues from reference groups and individuals (including family members) outside the cabinet.

Both dissidents and the main supporters of the intervention saw cabinet meetings as opportunities for information exchange, but they had different purposes in mind. Sharon's control over most of the relevant information and its suppliers allowed him to share it selectively and manipulatively with other cabinet members, in order to coax from them approval for escalation. The dissidents in the cabinet, who lacked military experience (except for one) and access to systematic expert advice, attempted with limited success to force Sharon and Eitan to share information. Not being proficient in interpreting military plans, they overlooked important points. They failed to realize, for example, that the flanking maneuver by the IDF against the Syrian army in the Beqa'a during the first week of the intervention was bound to provoke rather than avoid a military confrontation. The Syrian response led to full-scale fighting.

The diverse forces of deceit, compliance, and defiance that operated in the cabinet resulted in mixed outcomes. Neither the launching nor the escalation of the intervention beyond the 40-kilometer limit was prevented. But plans to attack and occupy western Beirut and physically destroy the PLO were replaced by a more modest and much less risky diplomatic option mediated by U.S. envoy Philip Habib. Even after the evacuation of the PLO from Beirut, the moderates could not insist that the IDF withdraw from Lebanon. Begin was able to secure cabinet agreement on a goal—a written agreement on peace and security with the government of Lebanon—impractical though it was in the face of Syrian opposition.

The performance of the cabinet also calls attention to leader-follower interactions. Leaders are not necessarily either inspirational or transactional. Begin seemed different to different members of the cabinet. He was respected by all but was viewed as an inspirational leader mainly by members of his party, in particular, by old loyalists from his Irgun days. This group, known as the Fighting Family, included, for example, Chaim Corfu, minister of transportation. To others, Begin was more of a transactional leader, a very influential one, but not one to determine their preferences. Rather than comply with his wishes as a matter of course, they would decide their preferences on the basis of utilitarian and even opportunistic considerations. Likud Party members, who knew that their political future depended on Begin, had to calculate the boundaries of his displeasure, which depended on the costs that they were willing to incur and on their power position within the party. Similar calculations were necessary for cabinet members from other coalition parties. The most salient example was Interior and Religious Affairs Minister Burg. The important role of his party, the National Religious Party, in the coalition government, his independent power base, and his relationship of mutual respect with Begin, which dated back many years, allowed him to be the most consistent and explicit critic of the Lebanon policy from its introduction at the cabinet meeting of December 1981.

Thus, a number of cabinet members did not feel compelled to support Begin on every decision, even when decisions were clearly important to him. But because of his status and the respect in which he was held, ministers had tacit self-imposed rules regarding acceptance and rejection of his suggestions. When he obviously felt very strongly about a case and expressed his position emotionally (e.g., the annexation of the Golan Heights), he could eventually win the support of a large majority. In such instances, most of the ministers thought that they had to go along out of respect or out of fear for their position. In discussions of other, openly debated issues when the ministers felt they could and did oppose the prime minister, Begin would continue to bring up subjects about which he felt strongly. There would develop an attrition process in which those who initially opposed Begin would eventually shift to his position. As he kept bringing up the same or a modified version of the original proposition, those who had initially opposed him eventually shifted to his position. Such was the process that led to the Lebanon intervention.

Between December 20, 1981, when the cabinet rejected Begin's Big Pines proposal, and June 1982, Begin built up commitment for an extensive military operation. Every time the cabinet considered military action and decided against it, those opposing the operation found it more difficult to stick to their position. Finally, on May 16 the cabinet refused to approve immediate action but under Sharon's pressure resolved that another violation of the cease-fire would result in a large-scale military response (Shiffer, 1984: 87). Thus, after the attempt to assassinate Israel's ambassador in London, most of the ministers felt that they could not continue to oppose Begin's proposal to bomb PLO targets in Beirut. They knew that the PLO was very likely to react by shelling Israel's northern settlements, which would force the cabinet to approve a ground operation that many of the ministers did not want. As one of them explained: "We said no so many times, and now in light of such an international drama [the assassination attempt] and the prime minister's emotional outburst we could not oppose him" (Naor, 1986: 44). Apparently Begin, an experienced politician with a keen comprehension of leader-follower relationships, understood this behavior pattern. He was sure that, given time, he could induce the ministers to approve his proposals, even if they opposed him at first. By influencing others without imposing his views blatantly, he could sustain his image as an open-minded democratic leader despite repeated attempts by the political opposition to discredit his autocratic leadership style.

Having failed on several occasions to gain the decisive majority in the cabinet that Begin believed necessary for a major military offensive, the prime minister and the defense minister introduced as an alternative the modified Operation Peace for Galilee. In this plan the IDF was specifically instructed not to attack the Syrian army. After the operation was launched, Sharon again requested, and received on June 9, authorization to attack the Syrian army. But the request was reframed as a defensive move; in the original plan, in contrast, that attack on the Syrian army was framed as an opening offensive move. The change in plans that was required by the reframed limited and defensive objectives of Operation Peace for Galilee entailed increased prospects of considerable risks. The IDF gave up the chance of achieving strategic surprise; and if the Syrians had decided to attack first, the possible consequences could have been substantial IDF casualties and a political victory for the PLO, even if it lost its infrastructure in south Lebanon. Syria would gain regional prestige and leverage in Lebanon, and a weakened Israel would lose its deterrent posture and its influence in Lebanon. To induce the cabinet to accept the risk of attacking the Syrian army once the intervention was under way, Begin and Sharon wisely shifted the reference point of the discussion in the cabinet to the gains made in the first few days of the operation (e.g., defeat of the PLO in south Lebanon, seizure of arms caches) and to the fact that achieving the main objectives of the operation was within reach, which triggered the endowment effect. With an endowment effect the value of an asset increases when it becomes part of an endowment already held; people then require a much higher price for giving it up than they would have paid to acquire it. In the Israeli case, the substantially risk-averse cabinet shifted toward acceptance of greater risk. The same process occurred repeatedly during the siege of Beirut. The withdrawal of PLO and Syrian military units from Beirut was used to rationalize keeping the IDF in place until a formal security agreement was signed. It was argued that the IDF needed to be present to prevent chaos and factional war. Otherwise, it was implied, the gains from the war would be lost, and the casualties would have been for nothing.[158]

As the limited intervention turned into a protracted war of attrition, cognizance of the cabinet's sensitivity to casualties led Sharon to frame the policy tradeoffs as choices between the potential gains from submitting to international and domestic pressures and the potential costs—mainly increased casualties—of not taking the military measures that he suggested. A majority among the cabinet members found these choices so compelling that they reluctantly approved measures with which they did not feel comfortable and that ended up expanding the war. Sharon used this ploy repeatedly and usually successfully. But it became clear eventually that piecemeal advances, bombing, and tightening the siege of Beirut were not bringing an end to the fighting and were in fact causing additional casualties. The casualties-avoidance argument seemed hollow and manipulative, which it was.

The cabinet ministers, now feeling distrustful of and hostile to the defense minister, refused to entertain further requests for approval of military operations and scrutinized his every move for signs of deviation from cabinet decisions. This confrontational approach was not simply a recognition that the policy was wrong or that they had been misled. It was partly the result of two developments in group interactions. One was that by August 1982 Begin's backing of Sharon was obviously wavering. The unanticipated large and growing number of casualties had a devastating psychological and emotional effect on Begin (Naor, 1993: 319–22). The other was that the faction within the cabinet that was displeased with Sharon's maneuvering, now led by Deputy Prime Minister David Levy, had reached a high degree of internal cohesion (Feldman and Rechnitz-Kijner, 1984: 39–40; Schiff and Ya'ari, 1984: 278–79). The challenges to Sharon's management of the war became more frequent and more resolute as the balance of power within the cabinet changed.

The initially strong impact of the framing of a problem is particularly notable in the Lebanon case for several reasons. The cabinet lacked a professional advisory organ to analyze, assess, and coordinate information, policy options, and operational plans and then supervise implementation of selected policies from a broad national perspective rather than a narrow bureau-organizational one. These functions were dominated by the IDF and the Defense Ministry. The IDF controls strategic planning, policy development, and intelligence assessment (Ben Meir, 1987: 103–13). The chief of military intelligence is responsible for overall assessments at the tactical, operational, and strategic levels and for preparation of the national intelligence assessments. As such, the person in this position has traditionally acted as the de facto intelligence adviser to the prime minister and the cabinet and, at the same time, been accountable to the chief of staff and the defense minister, both of whom could turn down his assessments. Thus the chief of staff and the defense minister could decide which information would reach the cabinet and which would not, and they could use this power to manipulate the decisionmaking process in conformity with their preferences; this was the case with the Lebanon intervention decision.

In the absence of systematic cabinet staff work before decisions were made, ministers found themselves unprepared to challenge the positions of the key players in the national security issue-areas who monopolized the information.[159] Cabinet decisions were often disproportionately influenced by the way the defense establishment framed issues, and cabinet members (besides the defense minister) were not cognizant of the full implications of a particular operation as it was framed. The cabinet also lacked the informational resources to request a reassessment of plans and their underlying assumptions or to counter confidently with an alternative plan (Ben Meir, 1987: 85–103).[160] Only when the policy and political costs became intolerably high was the situation seriously reassessed. The monopolization of the national security issue-area was most pronounced when the minister of defense and the chief of staff were in agreement and had the support of the prime minister. In those circumstances, dissenting views were not likely to have an impact. In fact, Begin and Sharon himself initially used Sharon's and Eitan's military experience and reputations to imply that their representation of the threats and opportunities posed by the situation in Lebanon was the only valid one, and this strengthened their monopolization of the decisionmaking process—at least until the siege of Beirut and the Sabra and Shatilla massacres.

As a decisionmaking group, the Begin government gave evidence of occasional groupthink processes combined with a variety of other group dynamics, including minority effects and leader-follower effects with both transactional and inspirational elements. The leader-follower effects led to compliance and cognitive closure on some occasions; on others, group members showed surprising independence of mind. And although much research on small-group behavior has been focused on the relative importance of informational versus normative influences, the politics of group dynamics has received too little attention. What the behavior of the Israeli cabinet demonstrates is that, when there is adequate stimulus and even when there is no premediation, a coalition may emerge within the group that affects group preferences by changing the distribution of power. In some cases, such a coalition consists of like-minded members. In other cases, an ad hoc sharing of views and interests is the foundation for a temporary alliance among individuals who otherwise have little in common. Such was the dovish group that emerged in the cabinet in August 1982 as a counterweight to what they perceived as manipulative mismanagement of the Lebanon policy. In sequential decision situations, these members learned who held similar views and how to use the formal and informal decisionmaking procedures to advantage. Their preferences were the result of neither informational influences nor normative influences per se but were based on a common cause. If such an alliance becomes permanent, a faction is created.

These findings pose an interesting challenge to experimental group research, which is often focused on single-process assumptions. On the whole, this case analysis provides a complex multivariate situation-contingent explanation of the risk acceptability of the Israeli cabinet that emerged from the interaction between key decisionmakers and the decisionmaking group. To attempt to determine the relative importance of each variable is futile because of the interactive relationships among variables and because of changes in the relative weight of variables as events unfold. Assessing the theoretical framework as a whole through its overall explanatory power makes more practical sense.

Conclusions

Unlike the three successful short-term interventions discussed in Chapter 6, the two interventions in this chapter were protracted and ended in failure. Decisionmakers in both Washington and Jerusalem were explicitly warned of many of the principal risks of intervention—even more so when the question of escalation came up. In neither case can risk taking be explained as an inadvertent result of ignorance; it has to be viewed as a conscious choice. In a comparison of the two interventions, some observations confirm conclusions reached in Chapter 6, some modify earlier conclusions, and others provide new insights.

• All foreign military interventions are driven by multiple interest rationales. In both Vietnam and Lebanon national and personal reputational interests were also important inputs, and in the Vietnam case they were critical in the decision calculus. The importance of national reputation is not surprising given that the interveners were powerful global and regional actors and therefore had a broad foreign policy agenda and aspirations that went well beyond solving the immediate problem. They thought of the potential implications of lack of resolve for other issues, sometimes much more important issues. From this perspective, the immediate policy problems acquired an importance greater than the threatened tangible interests. They became in decisionmakers' minds potential test cases and precedents. But the Vietnam and Lebanon cases clearly show that when a policy is driven by reputational interests it is difficult to sustain its legitimacy with the public when it becomes costly and the prospects of success are dim. As the war costs mounted and success seemed uncertain in Vietnam and Lebanon, it became difficult for the average citizen to understand why so much should be sacrificed for something as intangible and unreal as national reputation.

• Because of the perceived importance of reputation, rituals of moderation and gradualism can be counterproductive, even causing much broader and more violent intervention. The deployment of combat troops in South Vietnam and Lebanon was preceded by an intense use of air power. It is ironic that air warfare, used to avoid the need for direct combat intervention by ground forces, became the imperative for that very strategy. In fact, because bombing was not only a coercive measure but also a highly visible message of commitment, reversing the policy once it failed became extremely difficult. Troops had to be deployed in ground warfare because the alternative was perceived to be a considerable loss of reputation, a perception that fed American and Israeli fears that other vital interests might be challenged. Thus, a solution to the problem became part of the problem itself. Moreover, by exposing the civilian population of the target state to the hardships and suffering that result from bombing, the later ground operations were less likely to be effective. By then, the enemy's population had rallied 'round the flag; its readiness to resist was higher and its confidence greater that it could survive and even prevail. Willingness to fight and confidence in victory, combined with hostility for the country dropping the bombs and causing the suffering, made the enemy's breaking point much more difficult to reach, thus raising the costs of war to the intervener well above the levels initially anticipated.

• In both cases, the inappropriateness of the force structure and the inadequacy of doctrine and strategy played important roles in the sluggish military performance. And in both cases, a piecemeal approach to applying force, as in Vietnam, and to defining objectives, as in Lebanon, reduced risk perception, increased risk acceptability, and contributed to policy failure.

• Both cases demonstrate a similar lack of concern for economic risks. This is somewhat surprising, especially in the Vietnam case, where the intervention was not expected to be very short-term. But leaders in Washington and Jerusalem did not consider economic costs an important reason not to intervene. Economic risks became very salient, however, as both interventions got out of hand.

• In both cases the rally-'round-the-flag effect in the intervener's country lasted longer than made sense based on the casualty levels and other costs but not nearly as long as the national decisionmakers believed and hoped it would last. And in both cases legitimacy for the means used was lost well ahead of legitimacy for the ends.

• The clients—the South Vietnamese, for the United States, and and the Christian Lebanese, for Israel—turned out to be unmanageable in spite of their heavy dependence on their patrons. In fact, the clients became military and political burdens. Not only was their military performance dismal, but their norms of behavior toward the civilian population defied the humanistic norms of the interveners' societies, made the civilian populations antagonistic toward the interv ing power, and undermined the domestic and international legitimacy of the policy. All this indicates that the behavior of a client is a major risk dimension that should be taken into account well before a commitment is made.

• In the short-term interventions, as well as the protracted interventions, external factors were relatively marginal deterrents in the decision to intervene, although they affected the way the intervention was executed. In both protracted cases, the weight of international factors in the decision process became much more important as the intervention stretched out, although it was a slow process.

• A friendly third party (in the Lebanon case, the United States) can effectively align itself with a rival superpower (the Soviet Union) that is unfriendly to the intervener in order to restrain the intervener. At the same time, a third party that has little leverage with the intervener (the Soviet Union in Lebanon) can indirectly affect its behavior by applying pressure on the intervener's more risk-averse ally (the United States). In Lebanon, the United States was more amenable to pressure than its ally, Israel. Once the rivalry between extraregional powers (in the Vietnam case, China and the Soviet Union) was injected into the war, the decisions made by these powers were driven to a large degree by their own rivalry rather than by the behavior of the intervener, but that rivalry affected both the intervener's and the target's risk-taking preferences.

• The timing of risk recognition makes a difference in terms of response. Risks that, if recognized *before* the decision to intervene, might have weighed against intervention are likely to have much less impact if recognized *after* the decision is made and the commitment to the decision is established by action (e.g., the late recognition of the importance of economic costs and international factors).

• Inaccurate time assessments had adverse effects in both Vietnam and Lebanon. Decisionmakers did not foresee the actual length of time the intervention would have to be sustained to gain the desired effects. This misestimate meant that risk evaluations were much too low to prevent entanglement.

• In the Vietnam case, the principal adversary of the United States—North Vietnam—was willing to pay an extremely high price to achieve its policy objectives. The regime was willing and able to impose the costs of this commitment on its population. The United States had limited objectives and conducted a policy to suit its limited goals, yet the U.S. government could not impose the costs of even these limited goals on its population. Still, it found itself escalating its costs because of the unlimited commitment of the adversary. In the Lebanon case, in contrast, the principal actors—Israel, Syria, and the PLO—all had limited goals for which they were willing to pay a limited, even if considerable, price. This major difference between the two cases helps explain why the intensity and duration of the Israeli intervention were much more limited than the U.S. intervention in Vietnam.

• The two chief decisionmakers in the two interventions depended on the same sources of validation in evaluating competing risk assessments: person-based and belief-based validation. They persistently ignored dissonant information that cast doubt on the efficacy of their policies until their confidence in at least one of these validation criteria was badly shaken. Not surprisingly, confidence in formerly trusted advisers (person-based validation) collapsed first, rather than confidence in core beliefs, which are deeply entrenched and not easily given to change.

• Faced with highly complex decision situations, the key decisionmakers in Washington and Jerusalem relied on cognitive simplifiers—especially historical analogies—to bring an illusion of clarity to opaque situations. Yet such analogies were not a decisive influence, because they could not provide detailed policy prescriptions; they mainly provided a general direction and frame for analysis.

• Lyndon Johnson, like Ronald Reagan in Chapter 6, seems to provide further confirmation that extroverts who are unsure of their own value are, more than any other decisionmaker type, likely to turn to military power as a means of satisfying the need for self-affirmation of their leadership capabilities.

• Both cases demonstrate that groupthink is less frequent than is believed to be the case. Although members of both decisionmaking groups occasionally tended to conform, the cluster of symptoms associated with groupthink was missing. Groupthink can be observed, however, in the powerful factions within the broader decisionmaking groups, and these factions were important in bringing about foreign military intervention.

• The comparison between the Vietnam and Lebanon cases shows the importance of the institutional context within which decisions are made. Although the role of group processes is evident in both cases, the differences are substantial. In the Vietnam case, there were attempts at persuasion and manipulation within the group competing for influence over the top decisionmaker, the president. Such attempts were made in the Lebanon case, too. In addition, the top decisionmaker, the prime minister, had to take power politics within the decisionmaking group into account, because in the parliamentary system coalition building and maintenance are important. Coalition members could challenge his position, which could threaten to change the political balance of power and delegitimize his policy; he therefore could be forced to change his policy and risk-taking preferences even against his wishes. We see that the dual-path theory (of individual- and group-dominated decisionmaking) constructed in this book is necessary to capture the main contingent paths to risk taking. This conclusion needs, however, to be qualified as explained by the next observation.

• Although institutions mattered in the Vietnam and Lebanon cases, their impact was in part overridden by individuals. Competent, experienced politicians like Johnson and Begin could manipulate and use institutional arrangements to serve their purposes, particularly in support of policies that were temporarily popular or at least accepted by the public. Once policies were perceived to be too costly or acutely problematic and were challenged domestically, then institutions took on new importance, because those opposed to the policies could use those institutions in working against a leadership committed to the policies. In this context, institutions were used by an active opposition to delegitimize the challenged policies and make their implementation difficult or even impossible. How much institutions mattered changed over time, and with policy stages going from policy initiation, through challenge, to policy termination.

• In the Vietnam and Lebanon cases, the highest civilian authorities (e.g., president, minister of defense) frequently micromanaged the military operation at the implementation phase. The field commanders resented the interference, and it adversely affected the effective military management of the war, even if it was not the only reason for military failures.

• Like Bush, Johnson—another cautious president—had the distinctive cognitive operational style associated with risk aversion. Both were competent tacticians in the political arena who focused on process. Johnson managed to keep his Great Society legislation alive while engaging in a full-scale war in Vietnam. He encouraged tame criticism while emphasizing loyalty and collegiality, and he had a hands-on management style.

CHAPTER 8

Conclusions and Implications

In this book a sociocognitive approach to foreign policy decision-making is taken in order to address critical questions concerning the formation of risk judgments and preferences, with an emphasis on risk as choice rather than circumstance. The theoretical analysis of risk taking is tested in an acute issue-area characterized by risk taking—foreign military intervention—through the empirical comparative analysis of five case studies: U.S. interventions in Vietnam, Grenada, and Panama; Soviet intervention in Czechoslovakia; and Israeli intervention in Lebanon. These cases, representing events that took place over 30 years or so, have common threads as well as elements of diversity, which allow for rich comparisons. Thus, the analysis links the theoretical and the empirical. Theory is used to make sense of evidence, and evidence is used to validate and refine theory. In the following sections I elaborate on the most important findings presented in earlier chapters and add observations concerning the sociocognitive perspective on risk taking in foreign policy decisionmaking, the variable-sets that explain the formation of risk judgments and preferences, and foreign military intervention decisions.

The Sociocognitive Perspective

Risk-taking behavior is an important, even critical component of decisionmaking, especially with high-priority issues. It is not only integral to the essence of policy formulation; it also serves and is driven by potent motives. Hence, risk assessments and dispositions are constructed, not revealed. The riskier the decision, the more reflective it is of the importance of the issue and the decisionmakers' skills, competence, personality attributes (e.g., courage, wisdom), and fitness for higher position—or so the decisionmakers and others believe. That being the case, feedback from a risky decision like foreign military intervention serves several significant functions besides being a response to an acute policy problem requiring a solution. The salience and vividness of intervention decisions cause political decisionmakers to believe that the decisions are defining expressions of the quality of their performance and that the consequences of making or avoiding the decisions will be a measure of how history judges them. Individuals and groups involved in risky tasks are thus seeking feedback for a variety of motives (Robinson and Weldon, 1993). These include self-assessment; self-validation; improvement of individual and group performance in situations involving accountability; improvement of morale and self-confidence; increase in individual and group status by conveying the impression of being highly competent, daring, and capable of meeting serious challenges; and establishment of a public identity for individuals and groups that is distinct from the identity of predecessors (prior governments or leaders).

Risk taking is not necessarily, as implied by the term, a choice between a risky option and a nonrisky option. In many cases the choice among policy options with different risk textures is motivated by a decisionmaker's taste for particular risks, even if the expected values of the competing options are similar. There is nothing to indicate that being risk averse is, as a rule, preferable to being risk acceptant, or that risk acceptance is preferable to risk aversion. To be sure, there is no reason to believe that either caution or daring produces consistently superior policy consequences. Decisionmakers who for reasons of cognitive style or political expediency follow one or the other pattern consistently and thoughtlessly are equally liable to end up harming their country's interests. Risk taking has to be contingent on the situation, and it requires wisely applied, rather than mechanically calculated, assessments and choices. Understanding what drives and biases such judgments and preferences is critical for high-quality decisionmaking.

The results of this study suggest the value of departing from a rational choice approach for a more complex and better nuanced sociocognitive approach to risk.[1] The latter is substantially different from the rational choice approach, which has dominated the study of risk in foreign policy decisionmaking. First, unlike the rational choice approach, the sociocognitive approach emphasizes the importance of process in shaping outcomes. Risk-taking preferences are not cold cognitions. They do not just happen as a calculated rational outcome; rather, they are constructed and evolve.[2] How outcomes evolve is not marginal or trivial because, as shown throughout this book, process shapes outcomes. The formation of risk-taking preferences could thus best be viewed as resulting from a convergence, a process whereby multiple factors—situational-contextual, cultural, social, and individual— come into play and interact.[3] The effects of the multiple factors accumulate over time, and not necessarily in a linear fashion, causing decisionmakers to tilt first to one side and then to the other, to procrastinate, and eventually, out of choice or necessity, to decide on their risk-taking preferences. Once their preferences are formed, those holding them will not be readily responsive to information that brings their validity into question; cognitive closure is the likely result.

Second, unlike the rational choice approach, in which risk dispositions are inferred from behavior, the sociocognitive approach recognizes that behavior and disposition are often not equivalents. Actual risk-taking behavior differs from preferences because of the effects of social, contextual, and other intervening variables. Thus risk-averse decisionmakers (e.g., George Bush, Lyndon Johnson) may find themselves making risky decisions because they believe that they have no choice but to do so. The interaction between judgment and choice produces an interesting phenomenon. Risk-averse policymakers may take more risk, and risk-acceptant or risk-seeking decisionmakers may be observed to take less risk, than one would expect from their known risk dispositions. Biased judgment affects choice. Overoptimistic judgments of risk lead even risk-averse decisionmakers to take more risk than their risk disposition would lead them to take. Overpessimistic judgments of risk lead risk-acceptant decisionmakers to take less risk than their risk disposition would lead them to take. The methodological implication is important: one should not infer risk disposition exclusively from observation of the subject's behavior.

A research program embedded in a sociocognitive approach poses serious empirical and methodological problems relating to data availability, data generation, data processing, and the lack of parsimony. But these problems are not unique to foreign policy analysis and have been attended to in other issue-areas by psychologists, sociologists, and environmental hazards experts. As can be gleaned from the case studies in this book, there is no reason to assume that these problems pose insurmountable obstacles to researchers in the foreign policy field. A sociocognitive approach that allows for a contingency theory of risk taking, compared with the more parsimonious rational-choice-based theories, results in more detailed, richer, and more nuanced explanations and predictions. Rather than just telling us what risk orientations are likely to be, it also informs us about the nature of risk and the process by which risky decisions that may be person-, situation-, or action-specific are taken.[4] Although the cost of using a contingency-oriented theoretical construct is high, I believe that the benefits justify and more than compensate for the larger investment of resources.[5]

A sociocognitive approach attends to decisionmakers' flaws in coping with complexity, uncertainty, and ambiguity. And that attention is reflected in this study of risk in foreign policy decisionmaking. Risk as defined here contains three elements: the possibility of adverse outcomes, the probability that the outcomes will occur, and the validity of probability and value estimates. This definition seems to address effectively the intuitive concerns of decisionmakers facing risky decisions as revealed in the case studies. Yet decisionmakers rarely think systematically about risk or have a clear conception of it.

The lack of precision in conceptualization affects the communication of risk between individual decisionmakers, within groups, and across institutions. Better communication would require the establishment of a standard language of risk to be used by all participants or, alternatively, a research unit whose task it is to translate idiosyncratic expressions of risk by advisers and policy-support organizations into the commonly shared language of risk and then to distribute the assessments in this format to key decisionmakers before they make a policy choice. Either way, there is an urgent need to raise awareness among and train decisionmakers and their advisers in the grammar of risk. In the absence of such training, they will be much less able to figure out the stakes, generate focused and communicable advice, and assess effectively the merits of policy advice received from different sources. Policy analysts and decisionmakers in other policy areas, like business, technology, and the environment, share a more precise language of risk and are well ahead of those in the foreign and security policy areas in this respect.

In the absence of a well-defined concept of risk or the ability to assess risk systematically, due to lack of information or skills, how do decisionmakers approach the task of risk assessment and how do they form risk-taking preferences? According to the normative prescriptions of rational choice theory, decisionmakers identify and add up all the risk dimensions related to a particular policy option and then compare the aggregates with an imaginary threshold of risk acceptability or with the aggregate risk of alternative policy options. But the five case studies in this book suggest that this is not what happens. In practically all the cases, decisionmakers behaved as cognitive misers, using various measures to reduce complexity and increase clarity and certainty—even beyond the degree warranted by the available information.[6] These measures included use of historical analogies, heuristics, and schemas. In all cases, the assessment of the validity of information was dominated by belief- and person-based validation. Epistemic and situation-based validation were less common. The preference for the first two types of validation was shared by risk-averse and risk-acceptant decisionmakers. It, too, is best understood in terms of decisionmakers' tendency to be cognitive misers. These two forms of validation allow relatively quick assessments of information by criteria that are easily available.

The use of shortcuts in processing information and making choices is evident in both low-risk and high-risk situations, indicating that decisionmakers are not necessarily motivated by the higher stakes involved in high-risk decisions to search for and apply improved information-processing strategies; that is, they may apply nonoptimal strategies even when they are aware that the stakes are high and even when they have had adequate time to consider the situation (e.g., Vietnam and Lebanon). It can be argued that in these cases, better trained decisionmakers, who would have been more sensitive to process-quality issues and would have known what to look for, might have been more attentive to the shortcomings and biases involved in applying cognitive shortcuts. In the absence of such training, the decisionmakers did what came naturally and were also more vulnerable to social influences that adversely affected judgment.

Decisionmakers wish to be rational, but in what they consider a practical manner. Decisionmakers who uphold practical rationality may even acknowledge and be aware of the value of normative rationality and its indispensability for accountability. But they also recognize that for complex decisions in highly uncertain situations the demanding criteria of normative rationality cannot be strictly followed. There is therefore a need to reduce complexity to a level that can be handled according to principles resembling rational choice. Complexity can be reduced by applying such noncompensatory principles as "elimination by aspect" or "single-mindedness"; that is, decisionmakers can concentrate on one or very few attributes to the exclusion of all others (Mintz, 1993; Tversky, 1972).

In the same vein, there is also evidence indicating that "as the absolute cost of war increases, the importance of the relative gains diminishes and may ultimately become irrelevant to the decision for war or peace" (Lebow, 1984: 181). This is particularly so where the objectives to be achieved are not perceived as affecting the existence of the state but are nevertheless important, as is the case with foreign military intervention. No matter what the anticipated and actual gains, the anticipated and actual losses are the critical inputs for a decision to intervene, and in cases of protracted intervention, these calculations become the key to the choice between policy persistence and policy termination. These strategies help reduce the dimensions of the problem and the considered policy options to a very small, manageable number. From that point on, decisionmakers will proceed in a manner that is more in tune with normative rational theory. This approach allows decisionmakers to think of risk in a coherent, focused way, rather than trying to figure out the aggregate level of risk represented by a confusing array of risk dimensions, many of them nonquantifiable and extremely difficult to add up, as required by rational choice prescriptions.

Formation of Risk Judgments and Risk Preferences

Having identified and discussed the broad characteristics of the sociocognitive approach to risk, I now turn to a discussion of findings, first those relating to the central theme of the book—formation of risk judgment and risk-taking preference—and then those concerning the risks of foreign military intervention and their management. Because many of the arguments and findings have already been presented in Parts I and II, I amplify here only the most salient ones and concentrate on aspects that have not been discussed.

Let me first recapitulate the main points of the theoretical analysis of risk-taking behavior, from the decisionmaker's awareness and recognition of a stimulus, through its processing, to the crystallization of a risk assessment and a preference (see Figure 2 in Chapter 3). To start with, this study does *not* concern itself with routinized and regularized risks—the types of risk characterized by repetitiveness, which are encountered on a regular basis, and with which the involved decisionmakers are familiar. Such risks typify situations encountered by military officers engaged in military operations for which they have been trained, by experienced scientists working in laboratories with known dangerous biological or chemical substances; and by engineers working with hazardous but well-understood technologies, such as those found in nuclear reactors. Such recurrent substantial risks are posed in situations that are highly structured; and because they are regularly experienced and quite familiar, a network of social institutions, regulations, and proven heuristics have been generated to absorb the danger with an effectiveness that gives individuals confronting these risks a justified sense of confidence in what they are doing. It even allows them to approach risks without having to systematically process and assess every circumstance separately and usually with a repertoire of criteria concerning the contingencies under which they are to accept or avoid risks. The situations on which this book focuses are quite different, not necessarily in the magnitude of the risks involved, but in the attributes of the risks: they are not routine, not conventional, not repetitive, and therefore not regularized. The challenge posed by the policy problems presenting these risks defy simple definitions of the situation as either opportunity or threat. The risks associated with these problems are complex and ill defined and occasionally do not have a readily anticipated end-state. That is, a policy solution may only create a new problem, possibly one even more pernicious than the one it was intended to solve.

Faced with a risky situation that poses a policy challenge and stimulates a quest for an appropriate response, the decisionmaker must first define the situation as a threat or an opportunity, but there is no easy way to do so. The initial perception of the situation emerges through a process of framing. Problem framing, whether externally predetermined (by culture- and organization-generated concepts) or determined by the decisionmaker dealing with the problem, is the first stab at constructing a broad definition of the problem, as well as forming a general idea about risk-taking preferences. Next comes more detailed processing; at this stage, a number of variable-sets shape both risk assessment and risk acceptability. The initial definition of the situation triggers cultural concepts and norms concerning what the outcomes might be; how to view the uncertainty of these outcomes; and how valid the primary assessments of outcome and probability are. Although cultural concepts may provide general and readily accessible assessments of some aspects of the problem, decisionmakers require more specific information and insights. These are provided by the contextual variables, including the vividness and salience of various risk dimensions of the situation; the effects of planning; and the implicit implications of commitment to a particular course of action.

As the decisionmaker becomes better acquainted with the problem, he or she applies personal skills and constructs to gain further understanding and to help decide whether to accept or avoid risk. These skills and constructs include beliefs, values and stereotypes, motivations, personality traits, and preferred heuristics. But the individual is embedded in, and operates in, an organizational setting. Organizations are not only a source of information but are also Janus faced with regard to preferences. Though inherently conservative and risk averse, organizations can also, under specific conditions, become a force for risk taking. Within the usually competitive organizational setting, the main arena for the exchange of views among players and attempts at influencing other players is the small group. There, both individual risk assessments and risk acceptability can be adjusted under the powerful combined impact of group attributes and the effects of social interaction among the members. Where decisionmaking is dominated by an individual (IDDM), individual-level variables are more important in shaping an overall risk assessment and risk acceptability than group-level variables. Where decisionmaking is collective (GDDM), risk assessments and risk acceptability are group dominated.

The converging influences of multiple variable-sets result in an overall risk assessment that includes judgments of adverse outcomes and their costs, the probability that these outcomes will occur, and the perceived veridicality of outcome and probability assessments. Risk-taking preferences are the product of the perceived risks (that is, their nature and the decisionmakers' taste for them) as influenced by individual, group, and cultural factors.[7] Four of the explanatory variable-sets—context, culture, the individual, and the group—affect risk assessment directly. Another variable-set, the organizational setting, influences risk assessment only through the mediation of individual and group variables. Three of the five variable-sets—culture, the individual, and the group—have direct effects on risk-taking preferences. Two other variable-sets—context and organizational setting—affect risk-taking preferences mostly through the mediation of the individual, the group, and the risk assessment that precedes the choice of risk preference.

The central conclusion from the case studies concerns the validity of the theoretical construct (introduced in Chapters 2 and 3) for explaining risk assessment and risk-taking preferences in diverse contingencies. When applied to five different cases—representing short-term and protracted interventions; low-risk, moderate-risk, and high-risk situations; open and closed societies; and different types of governance systems—the theoretical construct was highly effective in explaining consistently the evolution of risk assessment, as well as the formation of risk-taking preferences, policy entrapment, and preference change. The efficacy of the theoretical deductive explanation holds across all cases and allows for insights that go beyond the information given, while providing a highly plausible interpretation of the cases.

Furthermore, the analysis of the cases demonstrates the validity of the dual-path theoretical construct of the formation of risk judgments and preferences introduced in Chapter 3. It was possible and useful to identify each case of decisionmaking as either individual dominated (IDDM) or group dominated (GDDM). At the same time, it is evident that a complex multivariate theory is required to capture fully the reasons that decisionmakers take or avoid specific risks and the ways their behavior persists or changes. A single variable, even if it is pivotal, can explain only part of the variance in risk judgments and risk-taking preferences. Risk judgments and preferences do not usually happen in a flash of inspiration, especially when the stakes are high and decisionmakers are aware of their accountability. Decisionmakers contemplate, hesitate, anticipate, rethink, and change their minds before finally making a decision. This process may be driven by the dominant variable, but the final form the decision takes is affected by a dual process of convergence and synergism, whereby the accumulated effects of the pivotal variable-sets in conjunction with the reinforcing effects of other variables, as explicated in Chapter 3, enables the emergence of a risk preference—whether acceptance, neutrality, or aversion. It is the *convergent synergism* of variables rather than a single factor, important as it may be, that accounts for the variance in outcomes.

Irrespective of the magnitude of risk (low, moderate, or high), the same explanatory variables for risk judgment and risk acceptability operate in all the cases, indicating perhaps that the general attributes of the risky situation (complex, ill defined, and wicked versus regularized and routinized), rather than the magnitude of risk per se, determine which factors affect risk assessment and risk acceptability. The magnitude of risk, as perceived by decisionmakers, matters mostly with regard to the prospects for success. When the risks are perceived by decisionmakers to be low to moderate, and the national interest is perceived to be vital, even if contested domestically, there is a tendency to commit the overwhelming military force required to achieve the policy objectives and to send with this declaration of commitment a clear message of determination. The result is likely to be a successful intervention. In high-risk situations, on the other hand, the perceived costs (military, economic, domestic-political, and international)—especially when the vitality of the national interests involved is contested domestically—are likely to arouse attempts to limit the risks of intervention through gradualism and other measures. An operational approach that combines piecemeal commitment of force with an implicit message to the adversary and third parties of less than complete determination to achieve the objectives of intervention is less likely to result in a successful intervention. As intervention becomes protracted, the likelihood of success, however low, declines further.

Several more observations and modifications to the deductive theory that are based on the comparison across cases suggest themselves. In all five cases, prospect theory provides plausible explanations and predictions—but only for decisionmakers' general readiness to take risks even when they are initially risk averse.[8] It does not explain or predict the specific form and intensity of risks chosen (e.g., foreign military intervention compared with other risky options). Nor does it predict how long decisionmakers will sustain the costs of risky policies. More important, in looking at the explanatory variables across cases, one is struck by the minor part played by mood, stress, and culture in affecting risk judgments and preferences. The literature on stress and culture would lead readers to believe in a greater centrality for these factors than is evident in the cases.

In fact, of the variable-sets that explain risk-taking preferences, the cultural set seems to have the least robust effect. All else being equal, it is hard to imagine that there would have been different risk-taking preferences if just the cultural effects had been different in any of the cases analyzed. Still, in all the cases, cultural factors reinforced risk perceptions and preferences shaped by other factors. One possible explanation for the relatively minor role of cultural factors in foreign policy decisions (compared with, for example, decisions concerning environmental hazards) is that foreign policy hinges on the predominance of decisionmaking by individuals or small groups. Because of the elitism of the foreign policy domain, it is less likely to reflect directly cultural preferences than other, less elitist and more populist domains. The really important effects of culture are latent, long-term, and indirect. These cultural influences shape institutional structures, norms of social behavior, and decisionmaking procedures in groups, and important beliefs held by individuals, which in turn affect risk-taking behavior. But these influences cannot readily be correlated causally with specific events, only with general long-term trends.[9] Stress was also found to play a very limited direct role in shaping risk-taking behavior in the cases described here, although it affected behavior by narrowing the decisionmakers' analytical perspective and limiting the number of relevant options considered. But it is not difficult to imagine that even in the absence of stress or under different cultural premises, the U.S. Vietnam policy between 1964 and 1968 and U.S. policies toward Grenada and Panama, Soviet policy toward Czechoslovakia, and the Israeli Lebanon policy would have been pretty much the same.

Some interesting observations can be made about contextual variables. Individuals who think in an abstract and conceptual manner are less affected by vividness and salience effects and more likely to attend to base-rate information than individuals who are not so trained or inclined. Thus, base-rate information seems to have differential influences on decisionmakers. Robert McNamara and his main advisers, for example, early on accepted the idea that the U.S. war effort in Vietnam was not producing the anticipated and desired results. Trained in systems analysis and other base-rate-oriented methods of assessing the correlation between policy inputs and outputs, McNamara accepted the validity of reports and analyses that cast doubt on the efficacy of strategic bombing and the deployment of many combat troops. Other members of the decisionmaking community were less liable to attend to base-rate information that challenged their initial beliefs about which policies might work.

People are rarely fully aware of, or sensitive to, the process and extent of self-commitment. They believe that their commitments are more reversible and flexible than they are. Individuals often do not realize that at best their flexibility is more in the realm of the instruments for policy implementation than in realm of the policies themselves. Not realizing how difficult it is to disengage, they often incrementally entangle themselves in policies from which they believe they can remove themselves at will. In some ways, policymakers resemble addicts who tell themselves, and others, that they can break the habit any time—but of course they are wrong. Risky policies have a powerful hold over their initiators. They and their office become defined by the policies and especially by their outcomes. Entrapment is, then, a much more serious risk than decisionmakers tend to believe.

The literature on entrapment stresses the self-serving causes or errors in judgment that underlie it. But the cases of protracted intervention call attention to the importance of self-imposed moral obligations as a source of entrapment, even when decisionmakers are keenly aware of the adverse consequences of their policies. One could say that decisionmakers become entrapped by their own normative legitimation arguments and the expectations that these arguments activate at home and abroad. Thus, a change of government does not necessarily lead to termination of a failing policy, because the new group of decisionmakers may be bound by moral obligations created by their predecessors. These obligations commit the intervener as a nation, not just the decisionmakers as individuals. Not fulfilling them is assumed to reflect on the national identity, not just on the government that made the commitment.

Implicit in this attitude is the view that political-strategic motivations are temporary and context bound, whereas moral obligations are permanent and are not bound by time or context. This attitude is reinforced by the concern that decisionmakers in the public service feel for their moral image, notwithstanding how immoral their behavior is, if only for reasons of political expediency. Moral obligations cannot, therefore, be terminated, only cautiously and incrementally phased out. Phasing out, unlike termination, is time extensive. Unforeseen events and developments could prevent a linear progress toward phasing out the commitment in question. This is true not only for interventions for which the main declared motivation is humanitarian (e.g., U.S. intervention in Somalia) but also for interventions driven by political-strategic motivations (e.g., Israel's intervention in Lebanon). When moral issues become part of the legitimation process or emerge as dilemmas during the intervention, the intervener feels accountable for the manner in which they are resolved, which slows the termination process.

In the Lebanon intervention, for example, when the question of withdrawal came up, one of the strongest arguments against a complete, instant Israeli withdrawal was that it would leave a power vacuum and lead to a bloody communal war, making it morally irresponsible for Israel to terminate its presence quickly, even if the costs of phased withdrawal were likely to be high. Similarly, disengagement from Vietnam had to be incremental not only because it was important not to project an image of an American defeat but also because there was a sense of moral responsibility for South Vietnam as a whole and for the hundreds of thousands of South Vietnamese who had cooperated with the United States and were likely to be subject to reprisals.

A "decent interval" was needed for the sake of moral decency, as well as for practical reasons.

Because decisionmakers have so much trouble identifying their hidden biases and recognizing that they are becoming committed or entrapped, antidotes to these omissions are not incorporated in even the most detailed plans. Decisionmakers are even less aware of, and capable of empathizing with, their adversaries' commitments, entrapments, and biases in planning. Because in international politics all risky situations are competitive and involve interaction with adversaries, it is, however, important for the focal actor to understand the adversary's behavior in terms of policy biases that could lead the adversary to entrapment in a costly policy. Such understanding is necessary for anticipating an adversary's moves and for realistically assessing potential responses to the focal actor's policy, which is intended to affect the adversary's behavior in a desired direction.

Are there risk-averse or risk-acceptant personality types in the sense that behavioral preferences can be regularly predicted from these characterizations? The case studies do not allow such an argument to be validated at the national leadership level because of the constraints imposed by the environment in which leaders operate, especially constraints associated with accountability. It may still be true that some leaders are more (or less) cautious than others, but caution or boldness does not necessarily translate into a consistent behavior pattern. Lyndon Johnson, George Bush, and Leonid Brezhnev can be considered risk averse, whereas Ronald Reagan and Menachem Begin can be considered risk acceptant, but none show a strong pattern of common personality characteristics. There are dissimilarities within the two groups and similarities across the groups in attributes and cognitive styles. The conclusion to be drawn is that leaders who are more (or less) risk averse are distinguished not on a single or a few personality attributes but by a package of more general characteristics. What, then, are the defining characteristics of a risk-averse, as compared with a risk-acceptant, profile?

Risk induces a compelling need for control in risk-averse decisionmakers. Their attention is therefore focused on the policy process. They are deeply concerned with their ability to prevent escalation and apprehensive of the consequences of a failure to do so. If their aspirations allow for limited or delayed gratification (e.g., Brezhnev, Bush), the prospective costs of potential uncontrolled escalation loom much larger than the benefits from possible policy successes. To the extent that the status quo is satisfactory, even if not gratifying, a risky policy will thus be framed as a gamble for gains and is likely to be rejected unless the terms of its framing are changed and the risky policy comes to be viewed as a matter of loss avoidance. Risk-acceptant and especially risk-seeking decisionmakers focus on policy outcomes. They are confident, even if not sure, that escalation can be controlled, and they tend to have levels of aspiration that require short-term gratification (e.g., Begin, Reagan). The costs of not achieving important policy objectives to which they are committed loom larger than the remote risks of uncontrolled escalation. To the extent that the status quo is not gratifying, the incongruence between the status quo and their aspired objectives will thus be viewed as a loss. Risky policies to change this state of affairs will be acceptable as measures to avoid loss.

Both risk-averse and risk-acceptant decisionmakers may apply gradualism in defining their objectives and applying a policy. But gradualism will result from different sources. For risk-averse decisionmakers, gradualism promises control and reflects their actual policy preferences. For risk-acceptant decisionmakers, gradualism reflects concession to external or domestic constraints, more often the latter, which do not allow them to target all-out objectives and apply all-out policies. Risk-acceptant decisionmakers will opt for all-out objectives and policies in the absence of such constraints, unlike risk-acceptant decisionmakers.

These profiles are, however, better indicators of dispositions than actual behavior, because of institutional constraints on behavior. The cases suggest that institutions do matter and may even act as powerful constraints on personal preferences. The relative ease with which Reagan imposed on a reluctant military and Congress his preference for intervention in Grenada, compared with the difficulties encountered both by Brezhnev in dealing with the inner circle in the Politburo and by Begin in imposing his preference for intervention in Lebanon on his cabinet, is best explained by the different institutional contexts in which these men operated.

Institutions also matter when it comes to getting out of policy entrapment. A comparison between Begin and another entrapped head of state, Johnson, shows this clearly. Because foreign policy was set by the centralized authority in Washington, dissonant information and negative feedback about Vietnam was not influential in inducing a major policy revision until after the Tet offensive in January 1968. In contrast, although the Israeli prime minister has the most influence over foreign policy, Begin was also constrained by a political system that made him only first among equals. He had to be attuned to the views of coalition partners and even to members of his own party who opposed his position in cabinet meetings or parliamentary debates. Within a few months Begin was forced to be acutely aware of the deficiencies of his intervention policy and to consider phased de-escalation.

Institutional structures and procedures thus affect information processing and consequently risk assessment. And as a result, they also influence risk acceptability and the revision of risk-taking preferences. But as shown in Chapter 7, how much institutions matter could change during the policy cycle; from policy initiation, through challenge, to policy termination. In other words, although institutions matter and can affect risk-taking behavior, they can be manipulated. How much they matter is specific to time and situation. Institutional remedies to control risk-taking behavior are therefore not necessarily effective at all times and under all circumstances.

Finally, a close analysis of cases in which groups were important does not validate the potency of the groupthink phenomenon, which has gained wide popularity in explaining foreign policy fiascoes. There is little evidence for the existence of a groupthink syndrome in the decisionmaking group in Jerusalem during the Lebanon intervention or in Washington during the Johnson administration. To the degree that a groupthink syndrome was evident, it took place within small but influential policy factions that indirectly affected policy outcomes. The evidence does not refute the existence of the groupthink phenomenon but does suggest a more cautious analysis of fiascoes before hastily attributing their causes to groupthink. These cases also confirm the need for a more complex treatment of the role of groups and group processes in decisionmaking and risk taking than is usual—as was suggested by the theory.

Notably and specifically, there is no evidence in the cases to indicate that groups perform on the average better, or are more astute decisionmakers, than individuals. Whether an individual or a group will cope more effectively with risky decisions depends on the attributes of the individual or the group, as elaborated in Chapter 3. Hence the widespread conventional wisdom that decisions by committee are far better suited for dealing with complex situations is not validated by the theoretical logic or empirical analyses in this study. In fact, neither the circumstances in which political decisionmaking groups are formed, nor the context in which they usually operate and the imperatives that drive their members, guarantee that they will be composed of the most qualified people for the task or develop the most appropriate procedures for dealing with their tasks. Yet group decisionmaking often encourages a false sense of confidence that group decisions reflect broader wisdom, a comprehensive canvasing of options and in-depth deliberations of all the relevant aspects of the problem. But this is not necessarily the case.

Foreign Military Intervention Decisions

A policy of foreign military intervention is directed at three audiences simultaneously: the target state, observing states, and the intervener's domestic public. To the target of intervention it is mainly an act of commitment and coercion. To other observing state actors it is a message of vigor and resolve delivered mainly to influence their expectations and behavior in future encounters with the intervener. In terms of temporal reference points, then, intervention indicates present priorities and interests. It defines and signals future concerns and lines that adversaries should not cross, and in some cases it is intended to compensate for past policy miscalculations and failures. To the intervener's domestic public, intervention is a statement of political assertiveness and self-confidence by its political and military leadership, and thus it is basically an attempt at impression management.

In practical terms, how these three audiences perceive the intervener's action is more important than how the intervener intends it to be seen. The target state may respond by compliance or defiance. Third parties may respond with various changes in attitude toward the intervener: they may challenge the intervener and make their own statement of potency and resolve to influence the intervener's behavior, or they may adopt a submissive position to pacify the intervener. The domestic public may view the policy's outcomes as a test of their leaders' skills and competence, one that will determine in their view whether those leaders deserve to be trusted and allowed to continue in power. The problem in crafting a successful intervention policy is in trying to balance and satisfy these three substantially different sets of policy goals and to make them converge within the frame of a single policy. Sooner or later this is bound to turn into a juggling act that becomes more difficult to keep up as the intervention stretches on—hence the salience of the incorporation and explicit treatment of time within the policy objectives.

Time is usually part of the planning of intervention only in the most technical sense. Military operations have a given time frame, but time is not usually treated as a causal agent. And once the time frame begins to come undone as the intervention becomes protracted, time often falls completely out of the utility equation. It is therefore important to include time as an independent variable from the start in calculating the tangible and intangible costs of the decision to intervene, and it is particularly important when interventions become protracted.

The explicit and specific definition of objectives is critical if an intervention is to succeed without introducing new and unacceptable risks. Well-defined objectives allow the differentiation between success and failure, conduce mid-term assessments of progress, allow realistic time assessments for meeting the intervention objectives, and are a necessary condition for developing an appropriate exit strategy (Tellis, 1996). A specific definition of objectives also improves the chances of identifying the appropriate means-ends nexus and thus of coming up with an accurate and realistic assessment of the risks of failure. A fuzzy definition of objectives increases the chances of an optimistic bias that will result in a mismatch of objectives and capabilities.

Specific policy objectives are derived from perceived threats to national interests. What stands out in comparing the five cases is that all the decisions to intervene were driven by multiple interests. The reason is easy to understand. To pursue the policy and political risks of intervention, decisionmakers had to be convinced that the stakes were high enough to justify the risks. Multiple interests served to justify risk taking. Because decisionmakers were sensitive to domestic opposition, they preferred a variety of instrumental justifications that would appeal to a broad spectrum of domestic audiences. In all cases, too, there was clear evidence of the importance of intangible reputational interests on the part of the state and its leader as a cause of intervention. Reputational interests can motivate intervention from the very beginning, or they can be by-products of the intervention policy; either way, they can contribute to policy persistence and entrapment. The power discrepancy between intervener and target makes failure more salient to the prospective intervener than it would be in an interactive relationship between equals. Being the more powerful party, the intervener expects success and views failure as particularly humiliating. The sense of humiliation is reinforced by the intervener's belief that third parties will observe the outcome and draw conclusions that could have adverse implications for the future; these third parties might come to believe that they, too, can challenge the intervener and win. Reputational interests, therefore, are inherent in practically every intervention situation and become more emphasized the more protracted the intervention is.

Reputational interests are, however, abstract concepts not easily grasped by the public, for they reflect adverse long-term consequences of policy failure but do not point to immediate concrete adverse outcomes. As justifications for a costly policy, they become less effective as the intervention becomes protracted and the costs inhibitive and as public opinion becomes more important for sustaining the intervention policy. Only an inspirational leader can maintain the public's support for reputational objectives in the long run by elevating them above self-interest and by motivating followers to make sacrifices to attain them.

In all cases the communicative function of military operations was evident, although the targets were different in each of the five case studies. The weight of the communication function compared with other objectives was also uneven across the five cases. But as a rule, the higher the importance of reputational interests, the more pronounced the communicative value of risk-taking behavior. Of the cases discussed, the Vietnam policy was the one in which the communicative value was most salient, especially that of Operation Rolling Thunder (Gibson, 1986: 319–34). The targets of communication are not just the foreign actors but also domestic audiences in the intervener's society. For example, Bush's Panama decision was supposed to communicate to the American public that their president was not a wimp. Similarly, the Czechoslovakia intervention was a message to Soviet dissidents about the limits of the regime's tolerance. In the Grenada and Lebanon cases, as in the Vietnam case, the targets of communication were mostly foreign actors, although in the former two cases the communicative objective was of relatively low importance. Forming and implementing a policy that can serve the communication objective equally well for the different audiences and, at the same time, preserve the instrumental effectiveness of the policy is, as the case studies demonstrate, difficult and on occasion impossible. Yet decisionmakers are not always aware of these inherent contradictions and their implications for the risks entailed by intervention.

The comparison of the cases shows that the most important operational risk dimensions for foreign military intervention are those related to force structure, appropriate strategy, and domestic legitimacy. The effects of economic risks are usually minimal, because the assumption is made that the intervention will be short. Decisionmakers become painfully aware of the impact of economic risks when intervention becomes protracted and the economic costs threaten to undermine both the sustainability of the military efforts and the public support for and domestic legitimacy of the policy. Of the first three mentioned risk dimensions, decisionmakers were most keenly attentive to force structure and legitimacy.

In connection to force structure, the focus was usually on the availability of troops and hardware, with less concern shown for whether the available troops were trained and equipped for the unique circumstances and specific environment in which they were to operate. Decisionmakers, unless they had extensive military experience, were not fully aware of the importance of designing a strategy that would effectively put combat forces to use and take advantage of the asymmetry of power between intervener and target. Anticipated casualties were an important concern to decisionmakers, especially in societies where individuals' welfare and right to life take precedence over most state interests; but this very real concern was not well addressed in the preintervention stage. For reasons explained in Chapter 2 and confirmed by the case studies, decisionmakers are not comfortable with detailed discussions and projections of the potential for casualties, even with high-risk interventions. Ironically, the political and psychological unease in dealing with the subject, which is a reflection of its importance, may inadvertently reduce the operational salience of casualties in considering foreign military intervention, as is demonstrated especially in the Vietnam and Lebanon cases. Eventually, the number of casualties sneaks up on decisionmakers and becomes an unavoidable concern.

Decisionmakers' attention to legitimacy is first and foremost centered on its cognitive aspects. They believe that if they can convince the domestic public that for reasons of national interest intervention has to be undertaken and that they know how to deal with it effectively and at an acceptable cost, the support of the public is likely to be forthcoming and will trigger a collective sense of patriotism. Normally one would expect that the establishment of legitimacy would precede risky policies. Yet, because intervention policies are often contested, it is not unusual for decisionmakers, especially self-confident ones, to reverse the order of things and follow the dictum "Nothing succeeds like success." Thus they implement the policy even if they have not been able to legitimize it; they hope for quick success and expect that success will legitimize the policy. If implementation does not result in success or if success is very costly, however, the question of accountability is likely to haunt them much more than if they had established legitimacy first.

Somewhat surprising is the relatively low importance attributed by decisionmakers to international laws and norms. The marginal attention given by the top decisionmaker in each case is incommensurate with the inordinate amount of attention focused in the academic literature on international law and norms concerning intervention. Decisionmakers pay much more attention to the law of the land than to the law of nations.[10] In the cases analyzed, the decisionmakers were concerned mostly with the constitutionality of their decisions, not with their normative or legal standing in international law. International law was attended to, if at all, mostly by mid-level bureaucrats in foreign affairs ministries, in the departments responsible for cloaking the intervention in legality. In none of the five cases did I discern a significant concern among top decisionmakers with the implications of international norms and laws when they made their decisions to intervene, if they believed that intervention served the national interest. I doubt very much that intervention would have been prevented if someone among their advisers had made a case against intervention based on international illegality and the violation of the target's sovereignty. This is not to say that decisionmakers were not interested in using legal arguments to refute international criticisms.

The preceding points should be qualified by differentiating between successful and failing interventions. In a successful intervention, the combination of patriotism and success causes normative violations to be quickly forgotten. If, however, the outcomes are perceived as failures or if the intervention is protracted, normative aspects of intervention increase in prominence and can eventually delegitimize the policy. When normative legitimacy becomes an issue over the course of a protracted intervention, the aspects of the policy that first come under attack are the means, not the ends. This point was clearly demonstrated in both Vietnam and Lebanon. The differential effects on legitimacy of policy means, compared with policy ends, are apparently the result of the different views held by the public on means and ends. Ends are cognitively linked to beliefs about important national interests and often reflect cultural values. Ends, like all higher-level beliefs and values, are relatively resistant to doubt. Not so for the means used to achieve national objectives. The choice of means reflects instrumental cause-and-effect beliefs. Such beliefs are relatively vulnerable to the test of practical validation and therefore open to delegitimization.

Irrespective of regime type, all cases indicate the importance of support for the intervention by the top military commanders. Because the risks involved in committing troops are preferably shared by the political and military leadership, the position of the military commanders becomes a critical input in deciding the level of acceptable risk. Support from the military is a necessary, if not sufficient, condition for intervention, especially when the political leadership has doubts about the wisdom of intervention. Support is apparently less important when the political leadership is driven by strongly held beliefs in the efficacy of force, as was the case in the Grenada and Lebanon interventions. In this case, the reluctance of the military leadership to intervene is not likely to prevent intervention.

The case studies do not support simplistic views of the military—military officers as individuals and the military as an institution—always urging the extensive use of military power to facilitate the search for promotions, prestige, and funding. In three cases (Lebanon, Grenada, and Panama) the military tended to express doubt and was reluctant to get involved in foreign military intervention in the absence of a clear and acute strategic threat—in two cases much more so than the civilian decisionmakers. Several reasons can be adduced to explain this counterintuitive finding. First, military decisionmakers understand better than civilians the complexities, uncertainties, and potential risks of large-scale military intervention. In most cases, experience and expertise reduce their confidence in control over events and in the linear connection between control over resources and control over outcomes. They can more easily anticipate potential causes of failure and unexpected costs, and they can more easily imagine the disruption of meticulously planned operations. Aware of the the costs of accountability and the possibility of scapegoating by politicians, military commanders and institutions can see that it is in their interest to avoid the extensive use of military power unless absolutely necessary, especially if they have experienced failure in the past.

A second reason for military caution has to do with the traditional distrust between politicians and generals. The former consider generals narrow-minded, lacking in an understanding of political complexities and nuances. The latter are suspicious of politicians' motives and judgment. They believe that politicians are often inconsistent, self-serving, and untrustworthy. No matter what politicians say, they are likely to attempt to politicize military operations. They will interfere and will not allow professional military personnel to run the operation in the best and most efficient military fashion. They will impose constraints that increase the risks and decrease the chances of successful outcomes. At the first sign of trouble, political or otherwise, they will interfere in the management of the operation and will avoid taking responsibility for failure, leaving the military to take the blame. This line of reasoning, combined with an acute awareness of the possibility of uncontrollable costs, can make the choice of foreign military intervention relatively unattractive to the military compared with more cautious options.

The Vietnam, Czechoslovakia, and Lebanon cases show how easy it is for decisionmakers from an intervening country to miscalculate the effectiveness of their allies in the target state, even when they have long experience in dealing with them. This observation seems to hold for both hostile and friendly interventions. The miscalculations arise from misperceptions of the actual position and support that their local allies enjoy, and is often based on information, provided by the local allies, that exaggerates their domestic support in order to encourage a reluctant intervener to view the operation as less risky than it actually is. Once committed to intervention, interveners are often manipulated by their local allies and have less control over them than they had expected to have.

Third parties act both as incentives for and as constraints on intervention decisions, depending on the decision stage. Before intervention, international factors act mainly as reputation-related incentives. Interveners are concerned that by not taking action they might risk adverse effects on their reputation, and this adds to other incentives to intervene. It is relatively rare, however, that international factors deter intervention. At most, they modify decisions about the scope and style of the intervention. Only with time, as the intervention becomes protracted, does learning result and international factors come to be considered important constraints.

The riskiness of foreign military intervention suggests that the full range of risks should be considered and understood in advance. Intervention is like riding a tiger. If one does not intend to complete the ride, one should not climb on the tiger's back in the first place. In other words, intervention should not be undertaken unless the intervener is prepared to stay the course and commit all the necessary resources in a concentrated, cost-effective manner rather than incrementally. Once the decision to intervene is made, politicians should allow the military to execute the decision in the most efficient military fashion and should interfere in the process as little as possible.

Even though intervention seems a temptingly simple solution to a complex problem, decisions to intervene should be made with the utmost caution and only as the final resort. If intervention becomes unavoidable, it is preferable to limit it in time, taking into account that even the results of a successful intervention tend to be eroded in the long term (Blechman and Kaplan, 1978: 88–91; Foster, 1983; Jentleson, 1992). If intervention is believed to be unavoidable, then, a series of short-term interventions with well-defined goals and targets has an advantage over a protracted intervention. Such a "pulse strategy" involves a sequence of forceful, decisive, short-term interventions and disengagements. As soon as the positive consequences of the prior intervention are eroded, another act of short-term intervention is undertaken if the situation requires it. In the long run, a pulse strategy is both more effective and less costly, even if in the short-run it may be tempting to extend the intervention after an initial success or two and settle the problem once and for all. There are several justifications for this argument:

1. A short-term success reinforces both the domestic and the external legitimacy of intervention as an instrument of foreign policy.

2. A successful short-term intervention limits the costs in casualties and destruction as well as the financial costs. Even if another intervention is required, sequential costs are easier to absorb than continuous ones. A consequence is a reduced likelihood that the fatigue factor will set in and affect the national determination to apply similar measures in the future.

3. A successful short-term intervention, unaffected either by loss of legitimacy or by the fatigue factor, will generate credibility for the intervener's tacit or explicit threat of a possible future intervention. The threat deters an adversary who might be tempted to take advantage of the briefness of the intervention by making it clear that repeated interventions are a possibility. By reducing the opponent's uncertainty about future interventions, the probability of successful deterrence is improved.[11]

Interventions that did not accomplish all their set goals are commonly viewed as failures. This is not necessarily correct. An intervention is an attempt to coerce or deter a second party. Even if not fully successful, it imposes a high cost not only on the intervener but also on the target. Thus it sends a signal to other potential intervention targets of the intervener's resolve and warns that the cost to the target of achieving its policy objectives could be much higher than anticipated if the target triggers an intervention response from a powerful actor, even if the intervener does not fully succeed in preventing the target from achieving its objectives. Precedents of even less than fully successful interventions and a credible threat of repeated interventions affect the cost-benefit calculus of adversaries. This hidden benefit of intervention has largely been ignored. It is associated with the shadow of the future. To make future events an active concern of potential adversaries, the intervener must establish a credible reputation for resolve. It is thus better not to intervene at all than to intervene hesitantly or half-heartedly. Holding back could signal lack of determination, decreasing the deterrence effect. America's recent half-hearted intervention in Somalia and its procrastination in intervening in Haiti and Bosnia, for example, have undermined the reputation acquired in Grenada, Panama, and the Gulf War.

Finally, although the subject falls outside the domain of this study, it is worth noting the parallels between the analysis of foreign military intervention and what is becoming an important and recurrent phenomenon in world politics—domestic military intervention, that is, the state's use of extensive coercive military power against minority groups that wish to exercise the right of self-determination in one form or another, within a well-defined territory. As exemplified by Russia's intervention against the Chechneyans, Iraq's intervention against the Kurds, and Sri Lanka's intervention against the Tamils, the consequences are not limited to the domestic destabilizing effects. The results have spillover effects beyond the borders of the directly involved state and could have serious international ramifications. The use of military power in these contexts poses policy and political risks quite similar to those confronted in foreign military interventions. The superior military capabilities of the state are not readily translated into a quick decisive victory for the government. The risks encountered by the government are comparable to those facing an intervener in a foreign country, including substantial numbers of casualties, rising economic costs, domestic disaffection, some questions of policy legitimacy, and unilateral or multilateral third-party involvement. One important difference between foreign and domestic intervention is that the sovereignty norm supports the claim of legitimate governments to the right to establish their authority over dissenting groups. Uninvited foreign interventions, on the other hand, are difficult to legitimize normatively in the absence of strong arguments for human rights violations; and even then, humanitarian interventions stand on uncertain legal grounds. The theoretical analysis in Chapters 4 and 5 is relevant for domestic military interventions, then, but a systematic application will require an entirely separate study and is beyond the objectives and scope of this book.

Risks have always been the persistent companions of foreign policy decisionmakers, and the formation of risk-taking preferences their burden. The end of the Cold War did not bring about a more regulated and regimented international environment with less complexity and uncertainty. Rather, the removal of the many formal and informal regulations that had been tied to the bipolar structure resulted in even greater difficulties for those charged with anticipating policy outcomes or imposing constraints on would-be violators of international norms. The old bipolar order and its learned certainties have not yet been replaced, and the potential sources of risk in the new international environment have only increased.

The ill-defined nature of foreign policy problems has made decisionmakers' treatment of risk unstructured and, in many cases, implicit rather than explicit. Policy analysts on their part have shown a clear preference for applying rational choice theory, which, though simple and useful in imposing structure on complexity and uncertainty, is inadequate for explaining decisionmakers' actual behavior. More important, such an approach is also inadequate for policy prescription, because it straitjackets leaders into decisionmaking behavior that is, most of the time, impractical in terms of their skills, the nature of the information available to them, and the decision environment in which they operate.

Improving risk assessment and the formation of risk-taking preferences requires improving the introspective capabilities of decisionmakers. The better the decisionmakers understand the factors that could bias their risk assessments and preferences, the more likely they are to be realistically cautious and skeptical about unwarranted assessments and preferences.

It is also important that participants in the decisionmaking process understand the source and nature of the information with which they are provided. They must know whether disagreements in assessing the risks associated with a particular policy are embedded in different informational bases or whether they are judgmental, that is, stemming from different interpretations of the same information. This kind of awareness, though very important, is extremely difficult to induce without extensive and systematic training at all levels of a decisionmaking system.[12]

A theory that realistically addresses the complexities of foreign policy risk-taking on a contingency basis is thus crucial for foreign policy decision analysts. The interdisciplinary theory presented here clearly serves this purpose, and its proven relevance to real-life cases takes us beyond the limited external validity of laboratory research and beyond the problematic internal validity of rational choice theory and suggests a useful way of approaching the issues of risk judgment and the formation of risk-taking preferences. The considerable risky challenges that governments face in the changing world order will provide ample opportunities to apply the findings of the study to the task of coping with these challenges. At the same time, the lessons learned in managing these risks will prove valuable in validating and improving the theory, producing a unity of theory and praxis for mutual gains.

Reference Matter

Notes

Chapter 1

1. A telling example is Walt's (1991) comprehensive review of security studies, which ignores the absence of risk and risk-taking issues from security studies. A more recent review of the literature on foreign policy analysis (Hudson with Vore, 1995) likewise fails to address the subject in spite of its obvious relevance. At the other extreme we find the controversial position of some sociologists who suggest that risk is emerging as the key organizing principle in late modern society (Beck, 1992; Giddens, 1990).

2. Three important methodological reasons make the gambling metaphor inadequate as a representation of social risk. First, the nature of uncertainty in real-world decisions is distinctly different from the explicit gambling devices applied in the experimental operationalization of risky decisions. Unlike the explicit uncertainty of gambling devices, in the real world beliefs about uncertain events are typically ill defined and latent and extend to the underlying data-generation process itself. Second, the effects that indeterminate and unstable real-world contexts have on choice are unlike those in the stable contexts of gambling experiments. Third, the assumption that utilities and probabilities combine independently to determine the expected utility of risky options is unrealistic. It is more likely that payoffs systematically affect probabilities, especially when available information is ambiguous (Einhorn and Hogarth, 1986).

3. An interesting attempt to incorporate risk judgments and preferences into an analysis of the adversaries' decisionmaking calculus in the Gulf War of 1991 can be found in Davis and Arquilla (1991a, 1991b). But in their studies the concept of risk is understood intuitively and loosely rather than being defined rigorously. The authors also assume a direct and unmitigated relation between the definition of the situation and actual behavior and ignore the mediating role of social influences on both the formation of risk-taking preferences and risk-taking behavior.

4. See McKeown (1992: 417), March and Shapira (1987), and Shapira (1995), who also argue for a situation-specific analysis of risk-taking preferences.

5. For a discussion of the problems posed by intervention for the two main tradi-

tional perspectives on international politics, "realism" and "liberalism," see Krasner (1995).

6. Little (1987) clusters the literature on intervention into two schools: behavioralist and traditionalist. If we apply this classification to the four categories, the behavioralist school would encompass all studies included in the first three categories, the traditionalist approach the normatively oriented studies in the fourth category. According to the traditionalist view, "intervention cannot be meaningfully discussed simply as a behavioral phenomenon. It can only be understood in relation to the norm of non-intervention" (Little, 1987: 54).

7. In fairness to the author, it should be noted that he probably never intended his study to be applied to international actors other than the United States.

8. Even in those studies approaching intervention from the linkage politics point of view, which has a focus on transnational and international linkages between domestic groups in the strife-torn society and the intervener's society, "internal belligerents" and "social groups" are spoken of mostly in the abstract as unitary terms. The analysis of incentives and opportunities for intervention is based on the tacit assumption that, given the opportunity and the incentives, intervention will causally result (e.g., Eley, 1972; Mitchell, 1970)—an assumption that is not necessarily true, and one that does not explain the scope and nature of interventions, how they are decided on, initiated, escalated, and terminated.

9. These three risk types are defined and discussed in Chapter 2.

10. Former U.S. Secretary of Defense Caspar Weinberger formulated six tests to apply when military intervention is contemplated: the engagement must be vital to U.S. national interests or to those of U.S. allies; the decisionmakers must commit troops to the intervention wholeheartedly and with the clear intention of winning; the objectives must be clearly defined, and there must be a precise understanding of how military forces can accomplish them; the relation between the objectives and the size, composition, and disposition of forces must be constantly reassessed and adjustments must be constantly made; there must be reasonable assurance of domestic support for the commitment of combat forces abroad; and the use of combat forces must come as a last resort (*New York Times*, Nov. 29, 1984). Although these may seem sensible tests, they are practically useless as guidelines: the criteria are too general to provide a direction for policy-relevant judgments in specific situations. For example, the only interest that falls uncontested within the category of vital interests is survival, but surely Weinberger did not mean to imply that combat forces are to be used only for survival. It would be even more difficult to apply this criterion when considering intervention on behalf of an ally. In many cases, the question of what is vital is likely to be given different interpretations depending on worldview, nor is the question unrelated to subjective judgments of cost. Winning is another vague concept; short of achieving all intended objectives, the judgment of whether and when victory is obtained is highly subjective and often self-serving. Weinberg assumes that adjustment of force size and composition can be easily accomplished, ignoring the cognitive and organizational difficulties of altering major force deployment decisions that involve high personal stakes and puts reputations on the line. These tests also fail to take into account the dynamism of large-scale foreign military interventions, for some objectives emerge in the process of intervention and are not determined in advance. Similarly, public support may fluctuate during the course of intervention in response to costs, duration, and level of success, making it difficult to assess support accurately before the commitment to intervene is made. For further critical discussion and review of the Wein-

berger Doctrine, see Gacek (1994: 265–67), Newland and Johnson (1990), Sabrosky (1990), and Twining (1990).

11. This methodology reflects and combines the advantages of both the deductive and the inductive approaches to theory building and avoids most of the pitfalls encountered when using each separately (see Achen and Snidal, 1989; Brecher, 1993; George and Smoke, 1989; Ragin, 1987, 1994; Sjoberg et al., 1991; Vaughan, 1992; Yin, 1989).

12. Researchers working in the field of technological and environmental risk have similarly recognized the need for an integrative multivariate interdisciplinary approach to risk, a recognition that can be found in the studies of risk amplification (Kasperson, 1992; Kasperson et al., 1988; Renn et al., 1992; Slovic, 1992).

13. For cross-disciplinary critical perspectives on the functional disutility of the prevailing conventional wisdom in the social sciences, which favors the principle of parsimony, and for advocacy of the alternative context-dependent theories, see also Bromiley and Curley (1992); Clarke (1992); Ferguson and Mansbach (1991); Hirschman (1985); Sitkin and Pablo (1992); Vertzberger (1990: 20–21, 342–43); Walker (1991: 39 n. 24). The excessive concern with parsimony can be traced to social scientists' practice of indiscriminately learning by analogy with the hard sciences. Cross-disciplinary fertilization is an enriching and useful practice, but only if it is a source of creative rather than imitative learning. "The social sciences, seeking objectivity, legitimacy and predictability, set out to embrace the traditional methods of the physical and natural sciences. But they did so at a time when physicists, biologists, and mathematicians, concerned about disparities between their theories and the reality they were supposed to characterize, were abandoning old methods in favor of new ones . . . the 'soft' sciences become 'harder' just as the 'hard' sciences were becoming 'softer' " (Gaddis, 1992–93: 54). For a cultural explanation of the emphasis on parsimony in Western scientific communities, see Motokawa (1989). According to this view, Western scientists emphasize parsimony, but Asian scientists do not, and this is because of the differences in worldviews. "This idea of simpler is better is not intuitively obvious. It is clearly the reflection of one-God religion. There must be the ultimate Rule in the universe and that is God's will. Everything has evolved from this Rule and the apparent diversity of rules can be reduced to this final, pure, and crystallized Rule. Eastern science, on the other hand, has no such frugality in rules. This is probably because there is no one god, but there are many Buddhas and thus many rules. That leads Eastern science to stress not the uniformity and the similarity but the differences and specificity" (p. 494).

14. For a general critical view of the external validity of the results from laboratory simulations of decisionmaking, see Ebbesen and Konečni (1980); Shapira (1995). Specifically, Higbee (1972) and Holsti (1989: 36–37) have suggested that decisionmaking in the real world of international conflict differs from laboratory experiments in several important ways: (1) research ethics limit the range of risk that can be tested in the laboratory as well as the duration of exposure to risk; (2) potential gains and losses from real-world decisions are real and at the same time much greater than those produced in the laboratory; (3) the personal involvement of the decisionmaker is much greater in real-world decisions; and (4) the uncertainty of gains and losses in the real world is based on simultaneously interacting multiple determinants rather than on a single one, such as the roll of dice or a choice among outcomes with known probabilities, as is the case in experimental environments. In other words, most experimental problems are cognitively simple, unlike real-life problems. Similarly, in assess-

ing the work of mathematically inclined experimental psychologists, Lopes states: "It also fails to consider (much less explain) the motivational and emotional factors that give risky choice its experiential texture: the hopes and fears that give us in due measure both purpose and pause" (1987: 263). In the same vein, Levy (1992b) and Shafir (1992) caution against the indiscriminate transfer of prospect theory from the laboratory to the analysis of political decisionmaking. Suedfeld and Tetlock are therefore correct in suggesting that "findings in laboratory situations—and even in the most convincing field experiments—can best be thought of as hypothesis generating, not hypothesis testing, in the study of high-level decision making" (1992: 59).

15. The methodology occasionally involves the use of counterfactual argumentations; for discussion of the associated problems, see Fearon (1991); Hawthorn (1993); Roese and Olson (1995); Sherman and McConnell (1995); Tetlock and Belkin (1996).

16. Judgment and choice are different tasks, even though the differences are often blurred; see Billings and Scherer (1988); Einhorn and Hogarth (1981). For a discussion of whether political judgment is rational-scientific, intuitive-noninferential, or both, see Steinberger (1993). Advocates of the rational-scientific approach claim that truth in politics is a truth of reason, as is the case in any science. Advocates of the intuitive-noninferential approach claim that political judgment is about the elucidation of meaning, not about the pursuit of some previously unknown truth. The integrative position, and the one taken in this study, is that the two approaches combined "describe the nature of political judgment in its fullness and depth" (Steinberger, 1993: 283).

Chapter 2

1. The distinction between real and perceived risks raises a number of important questions regarding the epistemological defensibility and practical analytic value of this argument. These questions are discussed in Vertzberger (1990: 35–41), where I espouse the broader realist philosophical view that some truths are distinct from available present evidence and may therefore remain unknown to the observer. These unobservable structures cause observable phenomena or behaviors (see also Little, 1993). Other works in international relations presenting a similar position are, for example, Brecher et al. (1969); Holsti (1965); Sprout and Sprout (1956, 1962). A different view is adopted by phenomenologists. See, for example, Hollis and Smith (1990); Onuf (1989). For a psychologists' view on the ability of experimental testing to distinguish reality from perception, see Taylor and Brown (1988). And for a comprehensive philosophical treatment of this problem as it applies to technological hazards, see Shrader-Frechette (1991).

2. For a review of various social science approaches to the concept of risk, see Bradbury (1989); Renn (1992).

3. See the discussion of the future of risk research in Turner (1994); see also Dror (1994).

4. For a comprehensive discussion of uncertainty and its effects on behavior, see Berger and Gudykunst (1991); Dror (1988); Nutt (1989); Yates and Stone (1992b).

5. This argument is in disagreement with the view that in complex decision situations, subjective probability is more salient than value and has more influence over choice (Dickson, 1978). That finding is based on laboratory results; in reality, judgments of value and probability are not independent of each other. "Probabilities and values are not easy to keep apart. The same influences that edit (or bias) probabilities in certain ways will also edit values" (Douglas and Wildavsky, 1982: 89).

6. Risk, ambiguity, and uncertainty often muddy the ordering of preferences when the causal sequence of events cannot be clearly specified and particularly when the causal path from cause to (adverse) consequence cannot be defined with certainty. In such situations, individuals tend to depend on their social environment (group, organization) as a source of information to validate their beliefs (Kobrin, 1982: 134).

7. The terms *cost* and *risk* are sometimes confused. Riskiness is an attribute of the policy, whereas cost is an attribute of an outcome of a risky policy.

8. In probabilistic terms, perceived risk increases (1) when the probability of undesirable consequences increases and the probability of desirable consequences decreases and (2) when the variance of probabilities is greater, that is, when the probability of extreme outcomes increases. In utility terms, perceived risk increases (1) as the consequences become more undesirable and (2) as the distribution of possible consequences becomes more slanted toward undesirable consequences. In terms of the interaction of probabilities and utilities, an alternative with high negative SEU will be rated as riskier than one with low negative SEU (Milburn and Billings, 1976).

9. For discursive purposes I shall treat the three components separately, although they are inherently intertwined.

10. People have little confidence in intermediate-level subjective probability estimates and tend not to use them in decisionmaking. They also have trouble in understanding and interpreting information about low-probability events, so small probabilities are particularly prone to biasing (Freudenburg, 1988; March and Shapira, 1987; Peterson and Lawson, 1989; Sjöberg, 1979; Stern, 1991: 106). Because of these biases, low-probability outcomes are ignored regardless of how significant their consequences are. The biases also result in more attention being paid to physical certainties, concrete events, and well-specified causal relations at the expense of the less tangible dimensions of a problem. Situations involving low-probability high-consequence events pose a particularly difficult set of problems: (1) the related underlying assumptions cannot be challenged because they have not been tested, and (2) the data set is very small, therefore probability estimates are nonfalsifiable, and (3) low probability estimates are especially vulnerable to the inaccuracies entailed when the calculations fail to take unforeseen events into account (Freudenburg, 1992).

11. The emphasis on confidence in the validity of probability and value estimates is congruent with the more general view presented by Kruglanski (1989) about a cognitive validation stage in the epistemic process of knowledge acquisition. See also Ganzach (1994), who emphasizes the effect of perceived information validity on the extremity of judgment.

12. For a discussion of epistemic-based validation, see Ziman (1978); and for a discussion of person-, belief-, and situation-based validation, see Vertzberger (1990: 76–82).

13. For a discussion of the meaning and effects of policy acceptability, see Farnham (1995); see also Shafir et al. (1993).

14. According to Fagley and Miller (1987), the framing bias may be less robust than has been believed. This view is supported by Heimer (1988: 508), who argues that framing effects are most influential in situations of intermediate importance. Hence, conditions and factors that determine whether the framing bias will occur must be clearly specified. For an opposing view that confirms the robustness of the framing bias and posits that even a small change in the framing of a problem may result in significant preference shifts, see McDaniel and Sistrunk (1991); Schurr (1987); Sitkin and Weingart (1996).

15. Historical analogies can be used for this purpose; see Taliaferro (1994). For a typology of reference points, see Yates and Stone (1992b: 7–8).

16. Framing effects are the main cause of aversion to loss when the status quo serves as a reference point. Thus, manipulating the reference point can produce reversal of preferences (Tversky and Bar-Hillel, 1983). In a culturally regulated issue-area, problem framing reflects cultural biases (Tse et al., 1988). In any case, the effects of framing are likely to be substantial if the issue is unfamiliar to the decisionmaker (Plous, 1993: 53–54).

17. McDaniel and Sistrunk (1991) recognize the robustness of the framing effect but raise doubts about the validity of prospect theory for social dilemmas. Other studies (Osborn and Jackson, 1988; Thaler and Johnson, 1990) came up with results exactly the opposite of those predicted by prospect theory. For an interesting attempt to reconcile the seeming contradictions, see Sitkin and Pablo (1992). Several studies provide convincing arguments for the relevance and applicability of the theory to foreign policy and economic decisionmaking; see Jervis (1992); Levy (1992a, 1992b, 1996); Stein (1992); Weyland (1996). For a note of caution, see Boettcher (1995).

18. One possible explanation is that with several equivalent outcomes it seems that very little extra effort or luck will tilt the balance in favor of a particular alternative. This assumption does not take into account that the same line of reasoning applies to other alternatives as well (Teigen, 1988).

19. For a general discussion of trade-off avoidance by decisionmakers and the resultant irrational consistency, see Jervis (1976: 128–42). Decisionmakers believe that the policy they favor is less risky than alternative policies on several logically independent value dimensions. They focus on one set of risk calculations and then adjust all other risk calculations to fit the conclusions inferred from the focused-on risks (Jervis, 1982–83: 22–23). New information that should have turned their attention to other risk sets is then assimilated into the prior calculations through misperception or misinterpretation. The implications for a particular dilemma—the assessment of the value of human life—are discussed in Jou et al. (1996); Teuber (1990: 241–43). The more general dilemma is expressed by Morton: "The reason why risky choices are so hard, and why different people can reasonably have different attitudes to them, is that we rank some things as incomparably worse than others. Typically there are irreversible disasters. Death is the central example, but there are many others; dishonor, the failure of one's life work, bankruptcy. Once these have gone wrong, nothing will make them right again, or compensate for them. The lack of compensation is what matters most. It underlines a reluctance to reason in terms of average long-term outcomes. . . . So our reluctance to reason probabilistically on some topics and our tendency to value some things incomparably more than others are two sides of the same coin. The problem this poses for decision making is also two-sided, though. On the one hand there are dilemmas about giving a weight to risks of death and other disasters, and on the other hand there is the general problem of how to give less crucial goods any weight at all in comparison" (1991: 109–10).

20. Miller et al. (1990) suggest a number of factors that influence which counterfactuals are most likely to come to mind when framing. First, it is easier to imagine routine actions than exceptional ones. Second, it is easier to generate images of events that follow from actions than images of events that follow from inaction. Third, the greater the number of alternative courses of action considered or available to the decisionmaker before a decision, the more likely he or she is to take responsibility for the consequences. Fourth, decisionmakers who believe in the foreseeability of unfortunate consequences will feel responsible for those consequences. Fifth, the smaller the

change required in the decisionmaker's mental causal model to produce an alternative consequence, the more available this consequence becomes and the greater the affective reaction to the event is. And finally, the more difficult it is to imagine the multiple ways similar outcomes could happen, the more surprising an outcome seems.

21. For an instructive application and discussion of these problems in the context of the seizure of the American intelligence ship *Pueblo* on the high seas in January 1968 by warships of the Democratic Republic of Korea, see Strauch (1971).

22. The study of the communication of risk is still in its early stages. For a review of alternative approaches to the effective communication of risk, see Fischhoff (1994); Fischhoff et al. (1993).

Chapter 3

1. For a discussion of the problematic nature of the state of the art, see Turner (1994), especially his elaboration of what he calls the ironies of risk (pp. 148–50).

2. Both the taking and avoidance of risk can result in errors of two general types. A Type I error is committed by avoiding risks that should have been taken. A Type II error is committed by accepting risks that should have been avoided. The latter type is easier to detect post facto because the consequences of action provide clues to what should not have been done. It is much more difficult to detect a Type I error even post facto. Judgments of what should have been done are based on counterfactuals, which are usually extremely difficult to verify.

3. When decisionmakers have to consider decision options that have distinct, large, and immediate positive utilities, but that also have nondistinct, small, and delayed negative utilities, the options will look very attractive. Alternatively, when these attributes of positive and negative utilities are reversed, the decision options will look particularly unattractive (Björkman, 1984, 1987).

4. This argument is consistent with the finding that decisionmakers may take risks if their assessment of a particular situation seems to justify risk taking, even if their past record shows that their gambles did not work out as well as anticipated. "Decisions to undertake new ventures are based primarily on beliefs about individual events, rather than about overall base rates" (Griffin and Tversky, 1992: 431).

5. For a broad review of the nature and shortcomings of planning, see Mintzberg (1994). For a typology and discussion of the implications for planning of decisionmakers' agreement or disagreement over goals and over the availability or unavailability of proven solutions, see Christensen (1985).

6. This is likely to happen when decisionmakers facing new circumstances do not know how much risk they are prepared to take. They are then likely to be exposed to the effect of "minimal compliance," which "consists of inducing people to take initially small, seemingly inconsequential steps along a path that ultimately will lead them to take much larger and more consequential actions" (Ross and Nisbett, 1991: 50). This inducement can come from another person using framing manipulations to stimulate the first small step (Ross and Nisbett, 1991: 45–58). Such a strategy can be used, for example, by advisers attempting to push a head of government toward a policy that they are not otherwise sure he or she is likely to adopt. But it can also derive from a situational context that is perceived to demand small steps involving low-risk behavior. Increasingly the decisionmakers may be manipulated to accept growing levels of risk. In new administrations people who have no previous experience with decisionmaking responsibilities at the national level become vulnerable to manipulative attempts that compel them to make increasingly riskier decisions to prove their tough-

ness. Gradually they might find themselves trapped into escalating the level of risk taking even though the policy is failing. On the other hand, if the policies seem to work, they may learn and internalize risk-seeking attitudes and are likely in the future to prefer risk taking to risk avoidance.

7. Although Downs (1992) argues forcefully that failed interventions may have had more to do with unpredictable low-probability events than human errors in calculations, he concedes: "It would be foolish to argue that incompetence and self-interest were unimportant" (p. 291).

8. This behavior can be explained by the "playing with the house money" effect (Thaler and Johnson, 1990), according to which people tend to accommodate to past gains quickly and revise their reference point upward. Prior gains thus induce people to be more risk acceptant. Overoptimistic assessments can be traced to the illusion of control over future events. Once committed to a course of action, people exaggerate the extent of their control over unfolding events. Consequently, they are willing to persist when persistence is a mistake (Tyler and Hastie, 1991: 80).

9. If entrapment is a learning deficiency, there is reason to question the view that, at least in the short run, people learn more from failure than from success; see Levy (1994: 304–6).

10. For discussion of causes and behavior patterns that block learning by the government, see Bovens and 't Hart (1996); Etheredge (1985: 95–106); Etheredge and Short (1983). For comprehensive discussions and reviews of various approaches to learning, see Levy (1994); Tetlock (1991). For a survey of issues related to risk taking in a developmental perspective, see Fischhoff (1992); and for an analysis, from a learning perspective, of why people exhibit greater risk aversion for gains than for losses, see March (1996).

11. A society or country may be culturally pluralistic; that is, different groups may hold diverse and competing worldviews and attitudes toward risk (Thompson et al., 1990). When cultural diversity is represented in a decisionmaking group, it could effect the group's aggregate risk perceptions and preferences (Jackson et al., 1995).

12. For discussions of the difficulties in isolating, defining, and analyzing cultural effects on individual, group, organization, and government behavior, as applied to environmental risks and the validity issues that they raise, see Johnson (1991); Kleinhesselink and Rosa (1991); McDaniels and Gregory (1991); Vaughan and Nordenstam (1991).

13. To the extent that different societies have dissimilar needs for uncertainty avoidance (Hofstede, 1984: 139–45), a preference for risk aversion is more likely to be found where the need for uncertainty avoidance is high.

14. The term *cultural syndrome* was suggested by Harry Triandis, who defined it as "a set of elements of subjective culture organized around a theme" (1993: 156).

15. This argument raises a question: Are beliefs an expression of deeper personality attributes, or are they instead merely a reflection of cultural socialization, experience, or deductive thinking?

16. This discussion has an interesting implication: when candidates for high office make commitments during their election campaign that involve policies higher or lower in risk than those of the incumbent, and then do not follow up on their commitments, they did not necessarily lie but may instead have misestimated their own risk disposition.

17. Beliefs and motivations are not unrelated. Beliefs may activate deep motivations, because of the interactive relation between the basic needs for affiliation, achievement,

and power, on the one hand, and operational-code beliefs, on the other. People tend to adopt beliefs compatible with their core needs (Tyler and Hastie, 1991; Walker, 1983), which are acquired in the early stages of life. Stimuli that activate these beliefs may also activate the needs underlying them. The following discussion of the illusion of control illustrates the motivation-belief interaction.

18. For a review and discussion of the sources and consequences of the bias toward optimism in such situations, see Taylor and Brown (1988).

19. A recent study found that "the consequences of information for judgment depend on how subjects construe their relationship to those involved in the risk situation. When this relationship involves common category membership, the information will receive attention and will therefore affect judgment" (Staple et al., 1994: 15).

20. This optimistic view of the effects of stress is based on substantial evidence drawn from multiple case studies (Brecher, 1993).

21. There is some evidence indicating (1) that different types of risk taking (physical versus social) may have different antecedent traits (Levenson, 1990) and (2) that these traits are, on the average, more common in males than females (Arch, 1993).

22. Experts familiar with the use of statistics, and in particular with expected utility theory, tend to be significantly more calculative than intuitive in their decisions, compared with people who are not acquainted with those methods and theories. Lay people are more likely to judge gambles by simplifying the task, by acting as cognitive misers and focusing more on certain risks than on others; they develop decision rules that are more incongruent with normative rules than are those used by trained decisionmakers. In short, statistical knowledge increases the propensity of decisionmakers to use analytic decision rules (Schoemaker, 1980). For a more skeptical view of experts' success in applying their skills to real-life political and strategic problems, see Fischhoff (1991).

23. When uncertainty is ignored, decisionmakers make "something out of nothing" by misinterpreting random data and make "too much from too little" by misinterpreting incomplete and unrepresentative data. See Gilovich (1991) for a good description of these tendencies.

24. The following discussion draws extensively, but with many modifications, on Vertzberger (1990: 144–56).

25. For a review of the underlying mechanisms that explain the hindsight bias, see Hawkins and Hastie (1990).

26. The availability of counterfactuals in cause-effect pairs is weighted on the side of effect; that is, alternatives to the effects are more available than alternatives to the causes. People tend to presuppose cause, so alternatives consist mainly of cases in which the same cause is followed by variable effects (Kahneman and Miller, 1986). Overall, then, people consider a more limited variety of future outcomes than they normatively should because they tend to assume that causes are relatively constant and prefer to ignore the possibility that both causes and effects could change, producing many more outcomes than they consider probable. For a multiple perspectives discussion of the sources of biases in counterfactual thinking in world politics, see Tetlock and Belkin (1996).

27. Even if aware of cases in which other leaders attempted and failed to implement a similar policy at a high cost to themselves and to their countries, decisionmakers may argue that they themselves have more skills and relevant experience: the same will not happen to them. See also Slovic et al. (1981).

28. This will be the case because decisionmakers do not allow enough uncertain-

ties and concerns about possible errors to enter into considerations of the situational context (Dunning et al., 1990; Vallone et al., 1990). Minimizing them is likely to increase preferences for risk taking and foster a "damn-the-torpedo syndrome" (Spangler, 1980).

29. Group-level risk perceptions and preferences are in essence social cognitions of risk held by the group and formed through a sociocognitive process. This interactive process combines individual cognitive operations and social interactions (the latter include communicating, sharing information, and attempting to influence other members' cognitions); see Klimoski and Mohammed (1994); Larson and Christensen (1993); Levine et al. (1993).

30. For discussions and useful literature reviews of majority and minority effects, see Kitayama and Burnstein (1994); Levine and Russo (1987); Maass et al. (1996). Research comparing minority and majority effects found that minorities tend to produce conversion—that is, profound and lasting attitudes, perceptions, and preferences that are generalized to new settings—whereas majorities are more likely to elicit compliance with attitudes, perceptions, and preferences that are confined to the original setting. For a discussion of the cognitive and motivational factors that explain these findings, see Maass et al. (1987).

31. The relative impact of social interaction compared with other influences may well be culture dependent. In collectivist cultures social interaction will be much more important that in individualistic cultures (see Hofstede, 1984: 148–75; Smith and Bond, 1993: 151–56; Triandis, 1993; Triandis, 1995: 107–44; Tse et al., 1988).

32. Group schemes refer to two main dimensions: whether group discussion is open and informal or formal and controlled and what kind of leadership style there is. On the relevance of various group schemes for conformity demands and the reaction of individual members to such pressure by shifting or not shifting their risk-taking preferences, see Forgas (1981); Wehman et al. (1977); Zajonc et al. (1972).

33. The polarization phenomenon is not altered by the presence of an official leader in the group if he or she allows free discussion (Lilienthal and Hutchinson, 1979). The discussion among group members is more important for causing a group-induced preference shift than the nature of the affective bond among group members (Vinokur et al., 1975; Wallach and Kogan, 1965; Yinon and Bizman, 1974).

34. The four explanations are summarized in Dion et al. (1970). The following discussion expands on these explanations by adapting and reinterpreting them to refer not only to the risky-shift phenomenon but also to the more general phenomenon of group polarization.

35. Although I emphasize fear of failure as a motive, those who are mediocre performers and unsure whether they can succeed on their own can also take pride in, and share in, the group's success (Forsyth and Kelley, 1994).

36. Responsibility for outcomes may be unevenly distributed, depending on who (1) play leadership roles, (2) hold central positions within the status hierarchy, (3) have particular task competencies, (4) are assigned responsibility for the performance of a particular task (Leary and Forsyth, 1987).

37. The term *persuasion* is used here to mean not only a cognitive process but also a social motivation. The persuasion explanation, in its present form, is congruent with the finding by Wallach et al. (1968) that there is no evidence that commitment to high-risk taking in and by itself results in greater persuasiveness. The contention in polarization theory is that both the high-risk taker and the highly cautious decisionmaker have powers of persuasion in different contexts.

38. Conformity does not necessarily imply compliance with the majority. We know of many instance of the so-called minority effect. If the majority always prevailed, societies would rarely change, and we know that in history minorities were often able to change existing trends and convert the majority to their views. To be effective, the minority has to have a distinctive behavioral style that attracts the majority's attention and then inspires doubts within the majority about its own position. The minority must project consistency, certainty, confidence, and credibility; it thereby focuses attention on its position both when the majority already has a fixed, well-defined position and when it does not. A minority can project these qualities because its members respond to the perceived external threat to their unpopular position with loyalty (in-group cohesiveness) and commitment (Gerard, 1985; Moscovici, 1976: 68–93). The minority's willingness to face conflict rather than avoiding it is what makes it effective.

39. Entrapment by justification is somewhat similar to the process of self-perception, as demonstrated by Larson's (1985) observations about the Truman administration. Harry Truman's vacillation and hesitation in interpreting the Soviet Union's intentions from 1944 to 1947 were transformed into conviction that it was aggressive when he had to justify aid to Greece in Congress and realized the unavoidable need to aid Europe. Having made the Truman Doctrine speech that portrayed the situation in Greece as part of a global struggle between democracy and totalitarianism, he adopted beliefs about the Soviet Union that justified the perception of threat.

40. Group self-confidence causes a decline in efforts to search for and attend to feedback, especially from sources outside the group (Robinson and Weldon, 1993: 77–78).

41. Groupthink has a number of symptoms: (1) overoptimism prevails, thus encouraging the taking of unwarranted risks; (2) members collectively attempt to rationalize group evaluations so they can ignore information that threatens the evaluation; (3) some group members make themselves responsible for protecting the group from dissonant information; (4) members profoundly believe in their righteousness as a group; (5) opponents are conceived in stereotypic terms and regarded as extremely evil and stupid (the supposed stupidity of opponents also serves to justify overoptimism); (6) group members censor their own doubts about the accuracy of the judgments and evaluations shared by the group; (7) direct pressures are brought to bear on deviating group members; and (8) a shared perception about the existence of group consensus emerges (Janis, 1982: 174–77, 256–59). These symptoms may negatively affect both the accuracy of the definition of the situation adopted by the group and the possibility of its adjustment to reality. They can also cause deindividuation, which leads the group not only to avoid weighing alternatives and risks and seeking outside opinion but also to relinquish exercise of personal judgments and disregard moral and ethical norms (Diener, 1977; Swap, 1984: 74–75; 't Hart, 1990: 66–72).

42. For discussion of the reason for the allure of groupthink and suggestions for broadening the scope of the variables incorporated in order to produce a more general model, see Aldag and Fuller (1993); Fuller and Aldag (1997); 't Hart (1990); 't Hart et al. (1997). Attempts to verify Janis's groupthink hypothesis provided only partial support. Yet none of these studies has been "(1) comprehensive enough to include all of the variables of Janis' theoretical framework, and (2) careful enough to use a comparatively advantageous mode of measurement, depending on the nature of the variables" (Park, 1990: 238). Tetlock et al. (1992) found that among the antecedent conditions for groupthink cited by Janis (1982), structural and procedural faults of the organization were the most important, and that cohesiveness and situa-

tional stressors are neither necessary nor sufficient causes of groupthink (cf. Schafer and Crichlow, 1996). Cohesiveness might be important only when group members are insecure and have a need for approval. It seems to me, however, that these conclusions should be approached with caution and that an explanation of the asymmetry found concerning the importance of such causes as cohesion and stress compared with other variables has to do with a compensation effect. That is, where some of the other causes of groupthink are unusually robust (e.g., structural and procedural faults), groupthink results, even in the absence of the remaining causes, because of the substitution effect (i.e., the robustness of some variables substitutes for the absence of others). Although Janis was correct in identifying the causes of groupthink, he has not paid enough attention to the existence of multiple paths to groupthink, paths that depend on the intensity of the antecedent conditions. This interpretation fits with the view of groupthink as a process theory that considers all antecedents as states, not variables; see Mohamed and Wiebe (1996).

43. Faction membership does not necessarily imply a similarity of views on all matters, only a congruence of some key beliefs, attitudes, and schemata shared by some members with greater intensity than by other group members—a phenomenon common in subcultures (Harris, 1994). In some cases the sharing of views occurs among members representing different organizations and can be traced to a common macro-organizational culture. For example, representatives from the three military services (army, air force, and navy) share an organizational macroculture, even though their microcultures are quite different. This common culture is enough to form the initial basis for factional links, which will cognitively differentiate faction members from other members of the larger group. A similar phenomenon has been observed in industrial organizations; see Abrahamson and Fombrun (1994).

44. Explanations for greater confidence in group-based assessments are both cognitive and motivational. "Indeed, when the group agrees, knowing that each member has different information may enhance their confidence that they have identified the alternative favored by the preponderance of the information. That is, group members may use an information sampling heuristic: the more diverse the sets of information that group members have, the more confident one can be that the consensus choice is supported by the weight of the available evidence" (Stasser, 1992: 62; see also Priem et al., 1995). From the motivational perspective, group decisionmaking involves the investment of more resources (time and effort) than individual decisionmaking. Group members, because they feel compelled to correlate the amount of resources with the quality of the product, are therefore likely to be more confident about their decisions (Mayseless and Kruglanski, 1987; Sniezek, 1992).

45. Things would be very simple if cultural values had only one source, such as the national culture. But decisionmakers are socialized by membership in multiple cultures whose values toward risk taking are incongruent with each other or even with those of the national culture (Vertzberger, 1990: 194–200). For example, foreign ministers, who are almost always members of the group making political-military decisions, are members of both their national society and the national decisionmaking unit and hence are attuned to societal values of risk or caution. But they are also members of a world elite, whose view of astute statesmanship emphasizes caution. It is not unusual for decisionmakers to be especially attuned to the values of the world elite subculture or to strike some compromise between conflicting values.

46. This was the case with the Soviet risk-taking propensity in foreign policy after Stalin (Ross, 1984).

47. It was shown, for example, that the explanation emphasizing the leadership role of more risk-acceptant members of a group cannot by itself satisfactorily account for a risky shift in group decisionmaking (Hoyt and Stoner, 1968). This finding reinforces the view that polarization often does not have one but multiple causes whose nature depends on the conditions in a particular group.

48. Normative and informational influence modes differ on five counts: (1) the mechanism of influence, (2) assumptions about human nature and needs (approval versus knowing and understanding), (3) assumptions about the correctness of a judgment (objective versus subjective), (4) the mode of influence (socioemotional versus cognitive), (5) the focus on judgment (on the judgments of others versus the component parts of judgment, i.e., the facts and arguments) (Kaplan, 1987).

49. Issue-type effects (intellective versus judgmental) were shown to be particularly robust when the decision rule was one of unanimity (Kaplan and Miller, 1987).

50. Degree of cognitive complexity is measured by the complexity of the cognitive rules used to process information. The more cognitively complex decisionmakers are better able to deal with information that requires a new approach or subtle distinctions.

51. The statement about the importance of the distribution of personalities in the group should be modified by reference to the power distribution in the group, which determines whose personality attributes, and therefore which attributes, will be important in determining group preferences and how important the attributes will be. For a useful general discussion of the effects of group diversity on decisionmaking, see Jackson et al. (1995).

52. The list of positive consequences of the unanimity rule, which draws on Gaenslen (1993), should be treated with caution. It is based on the assumption that all group members have equal access to policy-relevant information and that seekers of unanimity hence need to invest a great deal of effort in convincing all members to support a particular policy. But frequently a few group members control the flow of information, and the others depend on them for access. The privileged few can achieve their objectives with much less effort and in a shorter time and consequently without the favorable results described above. For a systematic exploration of the pathologies associated with the unanimity and majority rules, see Hermann (1993).

53. Manipulations are more likely to succeed under procedural majority rule, or where the "procedural script" (Burnstein and Berbaum, 1983) is not clearly defined, as is the case with recently created or inexperienced groups (Stern, 1997; Stern and Sundelius, 1997).

54. For a useful typology of sources of power in a group, see Raven (1990).

55. When facing a failed policy, groups have an advantage over individual decisionmakers in that they provide group-serving attributions. Individuals who make self-serving attributions have to be concerned with the response of other key decisionmakers, who may challenge their attributions or even point out that they are self-serving. In groups that make group-serving attributions, members do not have to be concerned about a challenge from other decisionmakers, especially in cohesive groups. A common preferred group-serving attribution for failure, if it can be suggested credibly, is to blame it on extra-group factors and causes (Leary and Forsyth, 1987: 175–76). Collective cognitive endeavors by the unsuccessful group generate multiple reasons for not attributing failure to the group; they make group-serving attributions more convincing and thus inhibit learning. Group-serving attributions also help to preserve group cohesion by preventing infighting over the responsibility for failure.

56. These arguments are in line with prospect theory. For discussion of the rele-

vance of prospect theory to decisionmaking in group contexts, see Kameda and Davis (1990); McGuire et al. (1987); Whyte (1989); Whyte and Levi (1994).

57. Because of the high degree of risk involved, intervention decisions require a bureaucratic consensus, even if it does not apply to all aspects of the intervention, such as timing, means, and objectives (Brands, 1987–88). Decisionmakers can manipulate the basic consensus for intervention to provide broader legitimacy than the actual scope of consensus would suggest.

58. Institutionalized procedures and practices of choice are important. As the stakes involved in a decision increase in value, rituals of correct decisionmaking come to play a critical part of establishing the quality and legitimacy of the decision to relevant constituencies. These rituals reassure the participants and those affected by the decision that it had been made intelligently and reflectively (March and Shapira, 1992).

59. The distinction between process control and outcome control in an organization is suggested by Ouchi (1977).

60. For a useful review of the literature on obstacles to organizational learning, see Huber (1991).

Chapter 4

1. In spite of the considerable parsimony, intellectual merit, and methodological sophistication of the SEU approach to decisionmaking, its practical relevance and validity have been questioned both on general grounds (e.g., Kahneman and Tversky, 1979) and with specific regard to its applicability to foreign policy decisionmaking (e.g., Patchen, 1988: 104–6; Saris and Gallhofer, 1984). In SEU models there are well-defined problems with probabilities and utilities that refer to distinct and well-defined outcomes. The models do not, therefore, accurately reflect the manner in which real people think in the real world (Björkman, 1984: 45; see also Brehmer, 1987; Lopes, 1981; March and Shapira, 1987; Shapira, 1995). For comprehensive critical reviews of EU and SEU models that show their limited usefulness for real-world decisionmaking, see Follert (1981); Green and Shapiro (1994); Payne (1973); Schoemaker (1980, 1982, 1989). These researchers emphasize that people often fail to maximize expected utility. See also Chapter 8, note 1.

2. In this study I deal with collective military intervention initiated by an international organization (e.g., Connaughton, 1992; Day, 1986; Doyle, 1986; Luard, 1986) only to the extent that it is utilized by the intervening power to manage the political risks of unilateral intervention (e.g., the United States in the Korean War, 1950).

3. This excludes shelling from ships at sea, bombing or firing missiles on another country's territory from planes or missile sites, or providing close combat support to another country's forces in battle—for example, by transporting its military units to the combat zone (e.g., the U.S. Air Force flying units of the South Vietnamese army into battle). But I will discuss such acts when they were a stage in what turned out to be large-scale foreign military intervention.

4. These definitions draw on other, earlier definitions of intervention; they contain elements of the following but are not fully congruent with any: Guelke (1974); Little (1987); Pearson (1974a, 1974b); Pearson and Baumann (1974); Rosenau (1969); Schwarz (1970: 83–86); Thomas (1985: 21); Tillema (1989; 1992); Young (1968). Like many other organizing concepts, foreign military intervention is not exclusive in its applicability. The same use of coercive military force might be analyzed from different perspectives. In this case, what intervention has in common with the generic

phenomenon of war is the use of large-scale violence. But intervention has its own distinguishing attributes (e.g., goals, decisionmakers' expectations, and temporal perceptions), it is directed at the political authority structure, unlike the case in classic wars. "The intervener must thus seek not only to defeat the adversary on the battlefield, but also build political support for his local ally" (Jentleson and Levite, 1992: 6).

5. For a discussion of the size, cost, and quality problems posed by an all-volunteer army, see Fabyanic (1977).

6. For an application of this argument to the use of the U.S. Rapid Deployment Force in the Persian Gulf, see Record (1981: 22–26).

7. For a discussion of the difficulty of defining the notion of proxy intervention, see Dunér (1981, 1985).

8. This argument accords with the findings that affective motives are important inputs in decisions to intercede in secessionist minorities' struggles for independence (Heraclides, 1990: 371–73).

9. For a comparative discussion of presidential and parliamentary institutional frameworks and the implications for the ability to govern, see Weaver and Rockman (1993a, 1993b).

10. See, for example, five out of six cases in Levite et al. (1992); see also Downs (1992).

11. These losses are related either directly to the policy or indirectly to decisions to change national priorities and shift resources from one sector (e.g., the economy) to another sector (e.g., the military).

12. The effectiveness of veto points and their implications for policy choice and implementation can be measured by their number, the veto procedures (e.g., whether a simple majority or more than a simple majority is required), and the extent to which a veto is complete, permanent, and nonappealable (Weaver and Rockman, 1993a: 26; Weaver and Rockman, 1993b: 449).

13. Although decisionmakers will attempt to maintain the domestic coalition in support of the intervention policy, they may not always succeed in doing so. A broad coalition, in particular, is unlikely to hold together when there is political risk and when key members of the coalition sense that they will have insufficient influence on the policy or its implementation. In that case they are likely to prefer an exit option— and causing the demise of the coalition that supported the policy (Lamborn, 1991: 65–66, 347–48).

14. Political capabilities are the reason why the decision to withdraw from Afghanistan was considered in 1983, but the decision to withdraw was not reached until 1987, and actual withdrawal was not completed until 1989. For details, see Mendelson (1993).

15. The juxtaposition of "entrapment" and "abandonment" follows that of Snyder (1984).

16. Symbol manipulation is one of the most potent instruments for obtaining legitimacy. Centralized political systems have greater flexibility and control over symbol manipulation than decentralized systems do. Still, the consequences may be ultimately harmful to the centralized regime, whereas decentralized systems do not incur the adverse consequences that easy manipulation can bring. For detailed discussions of symbol manipulation, see Bar-Siman-Tov (1990, 1994); Merelman (1966).

17. For a review of psychological and sociological approaches to trust and credibility and their implications for risk communication, see Renn and Levine (1991).

18. After 1945 the United States attempted to secure domestic and external legiti-

macy by giving all its interventions (with the exception of the Lebanon intervention in 1958) at least a facade of multilateralism, however thin, unlike before World War II, when unilateralism was a part of its isolationist doctrine, which mandated the avoidance of alliances (Carpenter, 1989). It can be argued that the less involved in policy the various American administrations attempted to keep the legislative branch, the more important became the role of token multilateralism as a symbol of the legitimacy and correctness of a policy.

19. Thus the four major powers—the United States, the Soviet Union, Britain, and France—have tended since 1945 to act according to the transparent legal norms for intervention that evolved between World War I and World War II, with only a few exceptions (Schraeder, 1989b; Tillema and Van Wingen, 1982; Van Wingen and Tillema, 1980).

20. For a discussion of the process by which military effectiveness is affected by social structures, see Rosen (1995).

21. The need for both domestic and international legitimacy is embedded in the noncongruence between state and society and in the nature of states as organizations that operate in both the international and domestic political arenas. The greater the domestic and external legitimacy that the policy of intervention enjoys, the less dependent the intervening state is on coercive or other compensatory strategies. As legitimacy wanes, domestic and international critics have to be co-opted or controlled. Both strategies require that resources be invested in buying support or compensating for the lack of it by coercing critics; see Mastanduno et al. (1989). For the means of dealing with international criticism to reduce its effectiveness, see Stenelo (1984).

22. For a discussion of the effects of ideology on risk taking in Soviet foreign policy, see Adomeit (1973).

23. A more detailed discussion of the effects of time and stress on risk taking and the quality of intervention decisions can be found in Chapter 3.

24. Decisionmakers show greater reluctance to commit themselves to a hostile intervention in another country's domestic dispute than to interfere to affect specific policies, because in the first case the probability of a stalemated and protracted intervention seems greater (Pearson and Baumann, 1974).

25. The ambiguity in measures of effectiveness can, nonetheless, be used constructively by decisionmakers who are interested in terminating an intervention but are heavily invested in it. Ambiguity sometimes allows them to declare victory and pull out without losing too much face or credibility.

26. "An intervention is successful in the short term when its military and immediate political objectives are satisfied. Its success in the longer term must be judged in terms of the durability of the political solution which has resulted, the degree to which the internal forces against whom the intervention was aimed remain active, the nature of the political and strategic advantages accruing to the intervener, and the extent to which intervention has resulted in long-term and costly commitments of the intervener's military forces and other resources" (MacFarlane, 1985: 32).

27. The perceived power discrepancy between intervener and target can be compared with the effect that slack has on organizational behavior. That is, the perception of having resources in excess of current aspirations reduces fear of failure and acts as an incentive for risk taking (MacCrimmon et al., 1986; March, 1988); it also induces the emergence of an inevitable-win bias. A more general point in the same vein is made by Kahneman and Lovallo (1993). They argue that as a rule decision-

makers are risk averse, but at the same time they tend to optimistically underestimate the actual risks of failure.

28. These optimistic expectations sometimes result from the suppression of dissent. Governments in serious disputes take measures to curb dissenters' access to relevant information in order to secure support for a policy that might be contested. The military, when it becomes involved in an armed dispute, suppresses criticism from within its ranks because it has similar concerns and because criticism could threaten organizational morale and stability. The military also claims exclusive expertise to prevent outsiders from evaluating its performance. This approach is facilitated by the recognized need for secrecy. By depriving critics of information for these and other reasons, decisionmakers may eventually end up with overoptimistic assessments of the probabilities and outcomes of the policies that they implement (Pape, 1992: 433–34).

29. Entrapment can be traced back to an earlier phase. Intervention tends to emerge in a process that starts with limited involvement in the domestic affairs of a foreign country. The intervener intends to project an image of commitment to bringing about a particular change in its target state's politics (e.g., regime reform) or international behavior. The failure of milder means to produce the required results could then force the foreign power to intervene in order to preserve its credibility not only with this particular target state but also with other clients, potential clients, and adversaries that observe the intervener's behavior, especially if the intervener has broad regional or global interests that could require its involvement elsewhere in the future. It is then that the shadow of the future causes the intervener to cross from involvement into foreign military intervention. See also Haass (1994: 81); Jentleson et al. (1993).

30. This should not be read as a sweeping statement that cost tolerance is always more important than military capabilities in deciding outcomes. There is no clear-cut evidence that willingness to suffer always gives a decisive edge to the party that has superior cost tolerance over the party that is more powerful militarily. On the contrary, evidence from research shows that the party superior in strength but inferior in tolerance is favored to win (Rosen, 1972). Three points may qualify this conclusion. First, the more cost-tolerant actors win in a fair number of cases. Second, the more powerful party cannot bring its superior military force to bear in many cases because of domestic or international constraints, so the advantages shifts to the weaker but more cost-tolerant party. Finally, it is an open question whether time (which for the weaker party means protracted willingness to suffer) benefits the militarily stronger party or the cost-tolerant weaker one.

Chapter 5

1. For discussions of tacit rules of behavior and superpower spheres of influence, see Keal (1983); Kratochwil (1989). When a major power had expressed commitments through alliances and the establishment of military bases, those commitments are likely to deter hostile intervention by other external powers in the client state (Pearson and Baumann, 1984).

2. For a similar view of the limits of structural realism, see the discussions in Christensen (1993); George (1993: 108–13); and Schroeder (1994).

3. On the importance of nonstructural variables, see Buzan et al. (1993); Jervis (1993); Nye (1988); Ruggie (1983); Snyder (1993); and Snyder and Jervis (1993).

4. Some of the following discussion was stimulated by Hoffmann (1986); Huth et al. (1992); Kaplan (1964); Pelz (1991); Piotrowski (1989); and Rosenau (1969).

5. Huntington (1992) suggests the term *uni-multipolar* to describe the current system (p. 6), but I find this to have an internal contradiction in meaning.

6. There is a counterargument: that the insulation of internal instability from external intervention, a condition that characterized the nineteenth-century balance-of-power system, was probably largely a reflection of the separation of domestic politics from international politics. If, however, a balance-of-power system should reemerge, it is not clear whether domestic politics would again be insulated from international intervention.

7. For discussions of buffer states and their relations with great powers, see Partem (1983); Ross (1986).

8. An intervening state that is a member of a tight bloc will presume a high likelihood of receiving support from the other members of the bloc, thus reducing the perceived risk that members of another bloc will counterintervene. But when the target of intervention is itself a member of a rival tight bloc, the intervening state will presume a higher probability of third-party counterintervention, which will increase the risk for the intervener. Even in a tight bloc, members are less likely to risk intervention if they are risk averse than if they are risk acceptant (Scarborough, 1988). For a systematic quantitative treatment of the use of force by bloc members during the Cold War years (1945–91), see Tillema (1994).

9. Modelski argues that "every internal war creates a demand for foreign intervention" (1964: 20). Yet his definition of intervention includes forms that are not necessarily large-scale military intervention. He himself adds the caveat that "the demand may not always be satisfied" (p. 20)—that is, potential intervening actors may not want to seize the opportunity for intervention. In fact, Forman argues that "because of the immense costs of general war, particularly but not exclusively in this century, declaration of general war tended to be made with great reluctance, and not particular frequency. . . . Either one or both of the potential interveners, reluctant to wage general war, will decide to withdraw or to abstain from acting at all for fear that the opponent in the system is determined to have its own way in the civil war regardless of the risk of general war" (1972: 1122).

10. On how and when a concert-type governance institution will be effective and on its relevance for our times, see Clark (1989); Elrod (1976); Holsti (1992); Jervis (1985a); Kupchan and Kupchan (1991); Miller (1994); Rosecrance (1963, 1992).

11. Whether nuclear balance is a structural or a process attribute is debatable. Waltz (1979) takes the first position, Nye (1988) the latter. It is my view that the variable could be either, depending on the context in which it is used.

12. Both explicit and tacit military guarantees to nonnuclear nations that were not bloc members, as the cost of their adherence to the nonproliferation treaty, increased the likelihood of friendly interventions by the guarantor power (Yarmolinsky, 1968: 234).

13. Although this argument seems to be supported by evidence from specific cases, it does not preclude the use of force by democracies. Aggregate studies of war proneness found that democracies are no less war prone than other polities (Chan, 1984; Weede, 1984) and that democracies used limited military force (short of war) against other democracies (Kegley and Hermann, 1995).

14. For discussions of the increase in Soviet intervention capabilities and the advantages and disadvantages of Soviet capabilities compared with U.S. capabilities, see Menon (1986); Porter (1984).

15. For a general analysis of the influence of moral norms on foreign policy, see,

for example, McElroy (1992). Arguments on the importance of norms in affecting international behavior as incentives independent of self-interest and power are suggested by Goertz and Diehl (1992). According to their typology, the nonintervention norm is "decentralized." In other words, (1) it is characterized by conflict between the norm and self-interest; (2) the sanctioning power is diffused—that is, there is no central sanctioning body; and (3) value and moral aspects are important. For discussions of the rationale for the nonintervention norm in international politics and reviews of its evolution, see Little (1975: 15–32); Moore (1969); Smith (1989); Thomas (1985); Vincent (1974); Walzer (1977). Two explanations are offered for the emergence of the nonintervention norm. One holds that it is a purely moral-ethical principle. The other maintains that it is a practical rule embedded in self-interest, derived from experience, and wisely designed to discourage states from getting involved in unstable and risky situations.

16. Among these are the League of Nations Covenant, the Convention on the Rights and Duties of States (which applies to the Western Hemisphere, promulgated in 1933), the United Nations Charter, the Charter of the Organization of American States, and the Declaration on the Inadmissibility of Intervention in the Domestic Affairs of States and the Protection of Their Independence and Sovereignty (1965). International law has, however, come to permit armed intervention in cases involving, for example, important humanitarian considerations, the exercise of the right of self-defense, or an explicit, willful invitation by the legitimate government of a state. For discussions of these exceptions, see Haas (1993); Joyner (1989); Tanca (1993). But scholars disagree on the degree of permissiveness that should apply to humanitarian intervention—compare Greenwood (1993), Pastor (1993), Roberts (1993), Scheffer (1992), and Walzer (1977), who take a more permissive view, with Haas (1993), who advises caution, and Jackson (1993) and Kratochwil (1995), who argue that recent practice offers only very limited evidence to support the claim that norms have shifted and now give humanitarian considerations precedence over sovereign rights in the justification for military intervention. Mapel (1991), who also takes a much less permissive view, argues that even humanitarian considerations do not justify armed intervention because soldiers' right to life is at least as fundamental a moral right as any that might be invoked to justify intervention. Skepticism with regard to the viability of the doctrine of humanitarian intervention on grounds of its impracticality, even if it is normatively justified, has also been expressed by Hedley Bull and John Vincent. For a discussion of their views, see Wheeler (1993). In contrast to the normative relativist position on intervention is the argument by Morgenthau that "it is futile to search for an abstract principle which would allow us to distinguish in a concrete case between legitimate and illegitimate intervention . . . intervention cannot be limited by abstract principles, let alone effectively outlawed by a United Nations resolution" (1967: 430). For an interesting case discussion of the justification for nonintervention in the case of the international and domestic crisis that was caused by the military overthrow of the legally elected government in Fiji in 1987, see Thakur (1990).

17. Pearson and Baumann (1989) found in their study of military intervention in sub-Saharan Africa that arms imports and military expenditure were correlated with external intervention.

18. Klare takes a worst-case approach to this problem as it applies to U.S. policy. He states: "It is safer to view most arms transfers as a potential proliferation risk rather than as an assured asset for U.S. national security" (1992: 16).

19. One possibility is that the surplus production capability will be converted for

civilian uses. This solution is not practical in the short term and would require large investments of resources that are not readily available. For discussions of conversion-related problems, see Adelman and Augustine (1992); Renner (1992).

20. The classifications of arms producers and suppliers into three tiers draws on Krause's excellent discussion. "First-tier states innovate at the technological frontier and possess an across-the-board production capability for sophisticated weapons. They possess the largest domestic markets and research and development establishments, and although they are the dominant exporters, they do not depend on exports for the survival of their industries. Second-tier states produce weapons at or near the technological frontier and adapt them to their needs but possess much smaller R&D, domestic procurement and production bases. Their industries hence depend heavily on exports or subsidies and are limited by the length of their production runs from producing arms as cheaply as first-tier states. Third-tier producers reproduce existing technologies but are unable to innovate. Their industries are often enclaves in a less industrialized economy, and major political and economic investments are needed to override these disadvantages. Third-tier suppliers' share of global production and exports is limited, and their comparative export advantage lies in specialized niches for unsophisticated weapons" (1992: 212).

21. Neuman argues that the effect of suppliers' diversification has not dramatically reduced superpower leverage in conflicts in Third World countries: "Through a delicate system of tacit rewards for restraints and the threat of punishment for infractions, both superpowers have been able to 'stanch the flow' and modify the level of armed hostilities in various regions of the world" (1987a: 140). I find this argument true only in some cases in which the suppliers were highly dependent on one of the superpowers, and even then, suppliers could bypass a superpower's demands for restraints by channeling arms through third parties. As the number of potential suppliers grows, the major powers' chances of stanching the flow decrease.

22. Suppliers' diversification may have an incentive for longer wars, which in some cases involved change in bloc orientation (Neuman, 1986: 9–15). For detailed information on diversification of supplies of arms to the Third World in the 1980s, see Grimmett (1990); Laurance (1992); Krause (1992).

23. For detailed discussions of these trends toward autonomy and gains in leverage, see Brzoska and Pearson (1994); Grimmett (1990); Husbands (1990); Klare (1987); Krause (1992); Menon (1986); Neuman (1987a, 1987b); Ross (1991). Arms sales are motivated both by political-strategic considerations and by economic need and the greed of suppliers, all of which result in decisions to relax the legal restrictions on the arms trade that most suppliers nominally have. Constraints have been limited to some very specific technologies and customers. Economic necessities take precedence over political prudence owing to domestic pressures, as was the case, for example, with the large weapon sales to oil producers after the hikes in oil prices in the 1970s. On the way this combination of motives makes the regulation of the arms trade extremely difficult, see Hartung (1992); Mussington (1994).

24. The more hierarchical the power structure, the more relevant the globalist perspective on foreign policy. The more fractured the global power structure, the more relevant the regional perspective on foreign policy. For a discussion of the two perspectives, see Doran (1989).

25. For a suggestive discussion of factors affecting the strategic behavior of regional hegemons (or aspiring hegemons), see Myers (1991a, 1991b). The arguments concerning the effects of structural systemic variables in the international system apply to the subsystem level as well. In regions dominated by a single local power there will, as

a rule, be a great deal of actual and threatened hostile and friendly intervention behavior by that power. A bipolar region will have many fewer interventions because of the effect of mutual deterrence, whereas multipolar regions are likely to have more interventions by local powers, mostly of the friendly type, for these will be considered less risky than hostile interventions (Pearson and Baumann, 1984). But at the subsystem level another factor should also be noted. The more equal the distribution of power among participants in local conflicts, the fewer local invitations there will be for extraregional intervention. The more unequal the distribution of power, the higher the likelihood that a weaker party may be forced to invite an extraregional power to intervene militarily to compensate for the power imbalance—thus providing a window of opportunity for interested extraregional parties. Yet such intervention opportunities carry both high military risks and the political risks of antagonizing regional powers. The regional powers may perceive extraregional intervention as interference in their own sphere of influence, thus costing the intervener a worsening of relations with those regional powers. This risk must be considered when assessing the costs of intervention compared with the possible benefits, and it may act as an additional constraint on extraregional intervention despite an increase in opportunities. As I argued earlier, the probability of mobilizing direct military intervention has significantly decreased. Other types of extraregional involvement, such as military and economic aid, have not diminished, however.

26. For statistical evidence that supports these conclusions, see Pearson et al. (1994) and Tillema (1989, 1992); and for some specific regional illustrations, see Klare (1990) and Feste (1992), as well as MacFarlane's (1984) discussion of interventions in Africa. The conclusions are also reinforced by arguments about the different implications of the end of the Cold War for the core states in the international system, whose conflicts are not likely to be resolved unilaterally through the use of force, as compared with the states on the periphery, where force will remain a common instrument of statecraft (Goldgeier and McFaul, 1992).

27. Empirical evidence shows that external aid for a U.S. Third World adversary reduced the chances of American success in certain small wars. This trend was not reversed by the support of U.S. allies for the American war effort (Engelhardt, 1989; Yadav, 1989).

28. For a discussion of the problems associated with a strategy of horizontal escalation, such as choice of targets, goal setting, choice among trade-offs, and risks of counterhorizontal escalation, see Esptein (1987: 30–43).

29. This is the essence of the "risk-theory strategy," first advanced by the German admiral Alfred Von Tirpitz. It is "a strategy employed by a weaker power against a stronger. Its essence is simple: by increasing its capabilities, a state seeks to render its adversary unwilling to risk a confrontation out of fear that, even in victory, his forces will be sufficiently depleted to leave him open to defeat by another of his enemies" (Rock, 1988: 343). To avoid this adverse situation, an intervening power must consider in advance whether it can be averted by mobilizing more domestic resources to increase available military capabilities, by making alliance arrangements, or by seeking conciliation with other adversaries and thus eliminating the chance that they may take advantage of its temporary vulnerability.

30. For a comprehensive comparison of deterrence and reassurance strategies, see Stein (1991).

31. Even risk-averse decisionmakers tend to be optimistic in assessing situations; see Kahneman and Lovallo (1993).

32. Mercer (1993) argues that reputations for having or not having resolve are

self-attributed rather than other-attributed, meaning that the adversary often does not infer future behavior from the other party's past behavior; therefore, concerns with reputational interests tend to be overvalued and dysfunctional. This argument is probably only partly valid. It is probably correct that a *single* behavior does not form a reputation, but a repetitive sequence of behaviors is very likely to form a reputation. Furthermore, even a single behavior can create a reputation, depending on how it is explained by observers. A single behavior that confirms prior-held expectations about an actor, if it is vivid enough, will transform those expectations into a reputation. It follows that an actor wishing to form a reputation through a single behavior should attempt to manipulate the observer's explanation for that behavior.

33. I am grateful to Richard Immerman for discussing this point with me. The studies concerning these two examples are too many to mention, particularly those on the Johnson administration. I shall cite only a few important ones. For analyses of the deliberations of the Eisenhower administration, see Anderson (1991); Arnold (1991); Artaud (1990); Billings-Yun (1988); Burke et al. (1989); Herring and Immerman (1984); Immerman (1987). For discussions of the Johnson administration's decisionmaking process, see Berman (1982, 1989); Burke et al. (1989); Gibbons (1989, 1995); Gibson (1986); Halberstam (1972); Herring (1994); Kahin (1987); Schandler (1977); and Chapter 7 in this book. A comparison based on a content analysis of decisionmaking processes relating to Korea in 1950, Indochina in 1954, and the Cuban missile crisis in 1962 shows a clear tendency on the part of decisionmakers in the Indochina case, as opposed to those in the other cases, to devote attention to analyzing the external environment (Anderson, 1987: 301–3).

34. For a theoretical discussion of the distinction between three types of misperception—cognizance, relevance, and evaluation—see Vertzberger (1990: 44–45).

35. Franck and Weisband correctly claim that in a highly competitive international system like a bipolar system, adherence to the underlying norm of reciprocity could result in unintended negative consequences. "The notion that something one superpower has previously asserted by word and/or deed tends to inhibit it from preventing the other superpower from subsequently asserting the same right is a manifestation of the idea of *reciprocity*" (1971: 129). Therefore, when Lyndon Johnson justified American intervention in the Dominican Republic in 1965 in terms of the doctrine of limited sovereignty, he unintentionally made it possible for the Soviets to use the Brezhnev Doctrine to justify their intervention in Czechoslovakia in 1968.

Chapter 6

1. This brief overview draws mainly on Glass (1985) and Thorndike (1989). For a more detailed account of the background and evolution of events, see Adkin (1989); Burrowes (1988); Lewis (1987); O'Shaughnessy (1984); Pastor (1985, 1990); Payne et al. (1984).

2. As early as March 1983, in the Strategic Defense Initiative speech, President Reagan indicated the threat to U.S. commercial interests posed by the Grenada situation, and emphasized that more than half of all U.S. oil imports pass through the Caribbean (Weinberger, 1990: 105).

3. Secretary George Shultz described the consequences: "If we said no to those [Eastern Caribbean] people, we wouldn't be worth a plugged nickel" (1993: 336).

4. On the problems posed by the War Powers Resolution, see Hall (1991: 167–210); Rubner (1985–86); Schissler (1988a).

5. For an elaboration of the argument that the American students' lives were not endangered by the situation in Grenada, see Adkin (1989: 100–101, 263–64); Kenworthy (1984); Quigley (1992: 204–11). There is no evidence to support the notion that Reagan was purposefully deceitful and that he knew that the students and other U.S. citizens faced no danger. These arguments (Kenworthy, 1984) fail to take innocent misperceptions into account. Reagan and other U.S. officials did misperceive the level of threat to U.S. citizens in Grenada—a misperception that can be attributed at least in part to Reagan's intense dislike for the Grenadian leftist regime. They also probably honestly believed that the threat was real. Much of the fear for the safety of the students was fed by intelligence reports suggesting that those in charge in Grenada, especially General Austin, were ruthless. For supportive evidence for this interpretation, see Beck (1993: 102–3, 106).

6. Uhlig (1985: 96) takes the position that a marines-only operation would have been much more costly in terms of time, casualties, risk to the American students, and destruction in Grenada. Similarly, General John Vessey argues that "forces used in Urgent Fury were chosen based on their capability to fulfill the mission. . . . The Marine Amphibious Unit (MAU) was used because of its proximity to the area. . . . The Rangers and other special forces were chosen because of their unique capability to secure airfields, rescue hostages, and attack selected point targets. Based on the enemy situation . . . the 82nd Airborne was included to ensure an adequate combat power ratio and permit early redeployment of the special forces and the MAU to fulfill pending commitments" (Cypher, 1985b: 101).

7. The Grenadian casualties were heavier: 45 killed, 337 wounded. Of those killed, 24 were civilians, 21 of whom were killed in an accidental bombing of a mental hospital. Cuban casualties were also heavier than the American casualties: 24 killed in action, 59 wounded, and some 600 taken as prisoners (Weinberger, 1990: 124).

8. The Pentagon had a Grenada intervention contingency plan (No. 2360). Unlike the plan that was implemented, it gave overall command to U.S. Forces Caribbean and on-scene command to the commander of the 18th Airborne.

9. During the intervention U.S. forces uncovered a large number of government documents that confirmed many of the earlier suspicions about the regional destabilizing intention of the Grenada regime. The documents also revealed that the extent of Grenada's web of agreements with the Soviet Union, Cuba, and other Soviet bloc countries was much broader than suspected. For a collection of these documents, see Seabury and McDougall (1984). See also U.S. Department of State and Department of Defense (1984).

10. The strategic intelligence picture of the overall size of the PRA forces improved on October 21. Opinions were, however, divided over how much resistance was to be expected. Critical tactical intelligence was still unavailable or inaccurate (Adkin, 1989: 140).

11. One source of concern to the Joint Chiefs, in light of what happened in Beirut, was domestic legitimacy. General John Vessey called the president's attention to the fact that "there is a potential public opinion downside to this because of what happened to the marines" (Meese, 1992: 218). Reagan, however, attempted to gain leverage for his position by arguing that "events in Lebanon and Grenada, though oceans apart, are closely related" because both were examples of the communist threat that could be traced to Moscow and its network of surrogates and terrorists (Payne et al., 1984: 161).

12. The argument that there is no other practical way is always important in estab-

lishing cognitive legitimacy. In the case of Grenada, it seems that Reagan had managed in his briefing on October 24 to bring congressional leaders around to "an acceptance that we really had no alternative" (Weinberger, 1990: 119). This assessment was somewhat optimistic. In fact, views in Congress were divided (Burrowes, 1988: 88–90).

13. For a discussion of the arguments for and against the legality of the intervention and the rebuttal of the former, see Waters (1986).

14. This last argument was probably a fabrication. For details of the steps taken in anticipation of international criticism, see Shultz (1993). Ambassador Jeane Kirkpatrick and the State Department had a serious difference of opinion regarding the legal argumentation for the intervention. Kirkpatrick wanted to invoke an anticipatory self-defense argument. The State Department was emphatically against it out of concern that it would set a precedent that could undermine Article 51 of the U.N. Charter, which limits permissible recourse to self-defense. Allowing the argument for anticipatory self-defense would make it easier for aggressive states to resort to force and justify its use; the argument could even be used to legitimize other Afghanistans (Beck, 1993: 188–96; Gerson, 1991: 222–33).

15. For a content analysis of the coverage of the crisis by three major newspapers—*New York Times*, *Washington Post*, and *Chicago Tribune*—see Nacos (1990: 157–81).

16. Although the exact details were not known at the time, the PRG had far-reaching plans for building up its military capabilities. According to Adkin, "by 1986 the PRAF [People's Revolutionary Armed Forces] would have had 4 regular infantry battalions, over 60 APCs, 108 ZU-23 AA guns, 50 GRAD P launchers, 160 military vehicles, 7 aircrafts, 6 new patrol boats, a workshop repair facility, a large new training base, a school for training Grenadian commanding officers and specialists, together with thousands of tons of equipment, stores, and ammunition. The militia would have grown to no fewer than 14 battalions. . . . The goal was a Grenadian military base in the Caribbean, controlled by Moscow, if necessary through Havana" (1989: 152–53; see also Weinberger, 1990: 132).

17. Members of the OECS included Antigua and Barbuda, Dominica, St. Kitts–Nevis, St. Lucia, Montserrat (still a British dependency), St. Vincent, and Grenada. CARICOM included the OECS countries plus the Bahamas, Barbados, Belize (a British dependency), Guyana, Jamaica, and Trinidad and Tobago. On the different threat perceptions of these two groupings, see Lewis (1990).

18. On October 23 David Montgomery, the deputy British high commissioner in Barbados, flew to Grenada, where he met Governor-General Sir Paul Scoon. According to Adams, that is when Montgomery verbally requested outside assistance to restore law and order. The letter to that effect, drafted by the State Department, was signed by Scoon on October 26 but was backdated to October 24 (Adkin, 1989: 99).

19. These preparations—the invitation from the OECS countries and the letter signed by Scoon—were considered important in anticipating the hostile response of the United Nations to military intervention. At an emergency session on October 28 the Security Council discussed a resolution condemning Operation Urgent Fury. The resolution was supported by eleven members with three abstaining (Britain, Togo, and Zaire) and was vetoed by the United States. On November 2, the General Assembly approved a resolution "deeply deploring" the intervention; it was supported by 108 members with 27 abstaining and 9 in opposition—with a slightly larger margin in favor than for the equivalent resolution passed after the Soviet invasion of Afghanistan. Yet Reagan's response was that "100 nations in the United Nations have not

agreed with us on just about everything that's come before them where we're involved. And, you know, it didn't upset my breakfast at all" (Edel, 1992: 235).

20. Robert McFarlane later described the president's reaction on October 22 to the news that the OECS had formally invited the United States to take the necessary military measures as "very unequivocal. He couldn't wait" (Cannon, 1992: 441).

21. Similarly, the vividness of the failure of the rescue mission in Iran due to lack of backup helicopters caused Defense Secretary Weinberger to focus so much on the risk of not having enough backup troops that he failed to examine the risk that coordination among so many participating organizations might fail owing to the lack of time for preparations.

22. On their rivalry, the diverging views of Weinberger and Shultz, and the sources of their views, see *New York Times Magazine*, Apr. 14, 1985.

23. These allies were not required to have democratic credentials as long as they had a goal in common with the United States, namely, resisting communist authoritarianism (Kegley and Wittkopf, 1982; Pastor, 1987).

24. Hence the emphasis on horizontal escalation. In the case of Soviet-American conflict the United States planned to extend hostilities beyond the initial theater of operations to areas where the Soviet Union was vulnerable (Williams, 1990: 204).

25. For discussions of Reagan's view of the world and its policy expressions, see Edel (1992); Glad (1983); Kegley and Wittkopf (1982); Moore (1990); Shimko (1991); Smith (1988, 1990); Schissler (1988b); Williams (1990). His simplistic black and white view of Soviet-American relations began to change in 1984 (Fisher, 1995).

26. Reagan delegated power but without monitoring the behavior of officeholders. His approach to delegation made his subordinates responsible for coming up with new ideas (Edel, 1992: 150–52). This management style is more typical of risk-acceptant leaders than risk-adverse leaders.

27. The Panamanian air force had only 400 personnel and had no combat aircraft or armed helicopters (Crowell, 1991: 70).

28. For details of air operations, see Crowell (1991: 75–79); McCall (1991); Turner (1991).

29. The risk to American citizens required that the plan take into account the safe evacuation of families of American service men and women and tens of thousands of other U.S. citizens living in Panama City, assuring that they not become hostages in the hands of the PDF (Donnelly et al., 1991: 24–25).

30. Memorandum to Congressman Andrew Jacob, Jr., by the U.S. General Accounting Office, National Security and International Affairs Division, B-240794, Sept. 13, 1990.

31. Broadly speaking, an extremely risk-averse approach to issues will result in reactive policies that lack coherence and direction. This observation is validated by the Bush administration's Panama policy: "The administration lurched into whatever options were presented, no matter how ill considered or contradictory" (Dinges, 1991: 313).

32. For a discussion of the role of affect in frame change, see Farnham (1992).

33. Operation Just Cause was the product of two military men. Thurman developed an operational plan that avoided his predecessor's problem of bureaucratic disagreements among Washington decisionmakers. Powell framed Thurman's plan in a format that made it acceptable to Bush and his advisers, who until the last minute searched for options other than a full-scale foreign military intervention (Donnelly et al., 1991: 397).

34. For detailed discussions of the October coup attempt and the failure of the Bush administration to take advantage of the opportunity offered by the coup to get rid of Noriega, see Buckley (1991: 209–18); Kempe (1990: 369–97).

35. Senior officeholders in democratic states, unlike other risk-averse individuals, can ignore the need to take action when faced with a pressing problem for only a brief period of time. There are three reasons: the public's expectations of officeholders, the media's scrutiny, and officeholders' accountability. The operational style of risk-averse leaders therefore may often be characterized by reluctant hands-on activism.

36. Leaders who are not risk averse may also structure their decisionmaking environment along similar lines, but the motivations will be different.

37. I wish to express my gratitude to Malcolm Byrne of the National Security Archives, Washington, D.C., for providing the documents cited in the following discussion. The documents were compiled by members of the former Czechoslovakia Government Commission to Study the Events of 1967–1970 and translated by Mark Kramer and Ruth Tosek. Mark Kramer also wrote the headnotes that accompany the documents. These documents will be published in the forthcoming reader *The Prague Spring 1968*. The documents are cited here by number.

38. Comprehensive discussions of liberalization and the reforms can be found in Golan (1973) and Skilling (1976).

39. The Czechoslovak crisis is now being reassessed in light of recently released documents from archives in the former Soviet Union and eastern European allies, new memoirs, and other evidence. But, according to one expert, "the evidence that has emerged up to now suggests that, for the most part, the best analyses produced by Western scholars in the pre-glasnost era will stand up very well" (Kramer, 1993: 2). For a review of the new sources, see Kramer (1992b).

40. Some generals considered the Čierna and Bratislava agreements insulting to Soviet national pride.

41. Estimates of the size of the invasion force vary from 250,000 to 750,000 troops and from 400 to 700 combat aircraft (Dupuy, 1970: 129; Skilling, 1976: 713–15).

42. After the completion of the exercises, few troops were withdrawn (according to Prchlík, only 1,500). The troops that were withdrawn were mostly reservists, but their units stayed on and were supplied with new recruits (Dawisha, 1984: 274).

43. The Czechoslovak military had at its disposal 175,000 soldiers organized in fourteen divisions with 2,700 tanks and 2,000 armored personnel carriers. There were also approximately 425,000 trained reservists available for mobilization on short notice. The air force had 690 combat aircraft (Dupuy, 1970: 130).

44. Soviet intelligence, however, faced a problem as a growing number of conservatives were purged from the Czechoslovak security service and party apparat, thus depriving the Soviet leadership of critical information sources, especially about the debate in the highest level of Czechoslovak leadership.

45. In spite of these preparations the logistical system failed to function smoothly, and the invading forces lacked food, water, and fuel (Valenta, 1980: 135).

46. "Senator Mansfield's Memo to the President," Aug. 19, 1968, LBJ Library, National Security Files (NSF), Memos to the President, Walt Rostow, vol. 91, Aug. 11–21, 1968, Box 38.

47. The CIA did not exclude an invasion; see Rostow's "Memorandum for the President," Aug. 22, 1968, LBJ Library, Cabinet Papers, Cabinet Meeting, Aug. 22, 1968.

48. "Notes on Emergency Meeting of the National Security Council," Aug. 20, 1968, LBJ Library, Tom Johnson's Notes on Meetings, Box 3.

49. Czechoslovak citizens fluent in Russian engaged the occupation soldiers in debates about the legitimacy of intervention. They were easily able to appeal to the soldiers' sense of right and wrong and defend the post-January regime in Prague by using Marxist-Leninist arguments (Eidlin, 1980: 128).

50. Economic sanctions were considered and rejected in the Moscow meeting of May 8. It was felt that these measures would only encourage Czechoslovakia to turn to the West (Dawisha, 1984: 100).

51. Shelest had been acting as an informal liaison with the hard-line members of the CPCz (Kramer, 1993: 3).

52. For a discussion of the core decisionmaking group in the Politburo, as well as other participants, see also Dawisha (1984: 315, 357–62).

53. This letter was signed by Alois Indra, Drahmír Kolder, Antonín Kapek, Oldřich Švestka, and Vasil Bil'ak. It was translated into English by Kramer (1992a). The two letters were found in Soviet archives in a folder marked "Never to Be Opened."

54. For a detailed description of the Czechoslovak hard-liners' attempt to effect a political coup, see Brezhnev's speech at the meeting of the Warsaw Five in Moscow on August 18 (NSA, Document no. 84, 1994).

55. For a comprehensive discussion of the role of ideology in the maintenance of Soviet influence in eastern Europe, see Valdez (1993).

56. In November 1968, in a conversation with Bohumil Šimon, a CPCz functionary, Brezhnev is reported to have said: "If I hadn't voted in the Politburo for military intervention, what would have happened? You almost certainly would not be sitting here. And I probably wouldn't be sitting here either" (Mlynář, 1980: 163). See also Musatov (1992: 38).

57. Brezhnev's deep concern on that account is evident from the following remark, made when Kirill Mazurov returned from Prague to Moscow a week after the invasion. According to Mazurov, Brezhnev was in a state of euphoria and said, "Thank God, everything is over," expressing relief that there had been no military clash between the Czechoslovak and Soviet armies (NSA, Document no. 96, 1994).

58. The consolidation of his position apparently did not take place until 1969–70 (Breslauer, 1982: 194).

59. This pessimistic view was not uncontested; see Lenczowski (1982); Wohlforth (1993).

60. By 1968 the U.S. pursuit of a losing policy in Vietnam could be interpreted by observers as an indication of a capacity for recklessness, which raised the level of the perceived risk of a rigorous U.S. response to an invasion of Czechoslovakia. Worth noting in that context is McGeorge Bundy's assessment in 1965 of the effects of Operation Rolling Thunder, the strategic bombing of North Vietnam; he thought that even if that policy would not coerce North Vietnam, it would still enhance the image of U.S. resolve in the international system (see Chapter 7).

61. Washington instructed NATO headquarters to be extra cautious and avoid all activities that might be interpreted as even remotely provocative (Dawisha, 1984: 253; Windsor and Roberts, 1969: 68–69, 87–88).

62. See, for example, the official Soviet explanation for the invasion, published in *Pravda* on August 22 and translated in NSA, Document no. 99, 1994.

63. For a discussion of Suslov's grand strategy and his views on China's threat, see Anderson (1993: 115–20).

64. This position had nothing to do with sympathy for the Czechoslovak leadership or their reforms, as we can see from his rude behavior in meetings with the Czechoslovak leaders (Dubček with Hochman, 1993).

65. It should be noted that both Gromyko and Grechko participated in the critical Politiburo meetings on July 20–21.

66. This position diverges from a bureaucratic-political paradigm (e.g., Valenta, 1975) and agrees with the broader approaches advocated by Dawisha (1980a), Simes (1975), and Williams (1996).

67. In the social construction of risk, framing has two forms, the first of which usually precedes the other. The general frame classifies the problem as belonging to a particular class. The specific frame then ties the current context and events to issues and life experiences that are specific to the decisionmaker and from which follow the requirements for action (Beach et al., 1992).

68. Four important players were directly involved in Hungary in 1956. Suslov was then a member of the Soviet Presidium, Andropov was ambassador to Budapest, Marshal Konev was commander in chief of the Warsaw Pact, and Marshal Grechko was commander of Soviet forces in East Germany.

69. For a discussion of how historical analogies have consistently played a role in U.S. presidents' decisions to intervene in Latin America after 1945, see Hybel (1990).

Chapter 7

1. For a concise discussion of the Johnson administration's attempts to reach a negotiated settlement, see Immerman (1994).

2. On the evolution of the American commitment to Vietnam, see Davidson (1988); Duiker (1994); Gardner (1988).

3. For critical reviews of the domino theory and its premises, see Glad and Taber (1990); MacDonald (1993–94); Slater (1993–94).

4. According to Assistant Secretary of Defense John McNaughton, the reputational interests were by far the most important interests at stake (*Pentagon Papers*, 1971, 3: 695). See also Freedman (1996).

5. See Johnson's address at Johns Hopkins University in April 1965 (U.S. Government, 1966a: 394–99).

6. Sullivan (1985: 25–50) argues that by late 1963 and for the next three and a half years the intangible interests, which had a strong affective symbolic element in them, became more dominant than security interests and apparently drove escalatory decisions.

7. The "military approach" required various degrees of military force. The "political approach" emphasized negotiation with the NLF, accepting whatever results could be achieved and withdrawing from Vietnam; see Strong (1992: 88–96). The most prominent proponents of the political approach in the administration at that time were Vice President Hubert Humphrey, Undersecretary of State George Ball and, to a lesser extent, the president's friend and adviser Clark Clifford; see Ball (1982); Berman (1992); DiLeo (1991); and Gibbons (1995, 19–20).

8. Playing an elaborate numbers game, Military Assistance Command Vietnam (MACV) could tally 172 allied battalions to the enemy's 72, roughly a 2.4:1 force ratio. Taking into account such force multipliers as mobility and firepower, MACV found this force ratio acceptable in the conventional war that it saw evolving in Vietnam (Shapley, 1993: 339).

9. For trenchant, detailed criticism of the manner in which force ratios were calculated and projected, see Krepinevich (1986).

10. The selective draft system with these attributes fostered more opposition than did the all-inclusive draft system applied during the Korean War (Cohen, 1986: 283).

11. The need for trained personnel in Vietnam and the need to train new recruits in the continental United States led by mid-1966 to a drawdown of about 40,000 soldiers posted in Europe. These were soldiers with specific qualifications and skills, so the effects on the preparedness of units for battle in Europe was much larger than the number indicates (Smith, 1991: 366–67). According to General Lymen Lemnitzer, the U.S. commander in Europe, by 1967 the U.S. Army in Europe was "damn near nonexistent" (Perry, 1989: 176).

12. Having only a one-year tour of duty also acted as a disincentive to study the culture, language, and society of Vietnam, adversely affecting the quality of intelligence work (Krepinevich, 1986: 229).

13. On the Vietnam policy of the Kennedy administration, see Cannon (1989); Gelb with Betts (1979); Gibbons (1986); Hammer (1988); Herring (1986); Newman (1992); Smith (1985).

14. On the de-emphasis on counterinsurgency and the failure of the pacification strategy, see Cable (1986); Gibbons (1995: 468–78); Herring (1994); Race (1976); Shafer (1988).

15. The three emphases concerning the bombing objectives represented the competing advocacy of three policy groups (the civilian advisers, the air force, and the army); see Pape (1990: 113–16).

16. Prohibited from attack at that time were most areas of Hanoi and Haiphong, as well as areas along the North Vietnamese–Chinese border. These constraints were incrementally removed, and by late fall of 1967 "there were very few targets of any military or industrial value that had not been bombed and either destroyed or damaged" (Thompson, 1980: 42).

17. It is estimated that only 0.3 percent of North Vietnam's civilian population was killed in strategic bombing, compared with 3.0 percent of Japan's and 1.6 percent of Germany's during the Second World War (Pape, 1990: 126).

18. From 1965 to 1968 North Vietnam received from its two allies approximately $600 million in economic aid and $1 billion in military assistance. This compared very favorably with the estimated loss of $370 million in damages from bombing (Pape, 1990: 129).

19. William Bundy confirms that by mid-1965 "the senior civilians—having exercised tight political constraints up that point—pretty much handed the military conduct of the war in the South over to the command in the field. Not so with the bombing of the North" (LBJ Library, William Bundy manuscript, 1971: ch. 27, p. 28).

20. The failure to define the military mission adversely affected morale, because "the maintenance of effective morale requires that soldiers be given clearly defined objectives, which if accomplished, provide a sense of purpose for the operation in which they engage" (Gacek, 1994: 217).

21. This harsh judgment is confirmed by McNamara himself; see McNamara with VanDeMark (1995: 203, 210–12).

22. For a detailed discussion of the uneasy relationship between the civilians in the Department of Defense and the military, see Perry (1989).

23. A view similar to Hughes's was expressed by the CIA director on July 21, 1965, in a meeting with the president. At the same meeting General Wheeler insisted that the Vietcong would have to "come out and fight" (LBJ Library, Meeting Notes File, Memorandum for the Record, Meeting on Vietnam, July 21, 1965, Box 1).

24. For a discussion of the optimism in MACV headquarters in Saigon, see Wirtz (1991: 111–20).

25. For discussions of the communist strategy, see, for example, Elliot (1993) and

Kolko (1985: 182–87). For a brief review of competing explanations for the American military failure, see Davidson (1993).

26. Fixed targets in North Vietnam were chosen from a list that expanded from 94 in 1965 to 242 in 1966 to 359 in 1967.

27. In addition to the coordination problems, Rolling Thunder missions faced other difficulties that affected their efficiency, such as poor intelligence reports on mission accomplishments, difficult weather conditions between September and April (during the winter monsoon season), and effective North Vietnamese air defenses. Furthermore, the attack aircraft in service (F-105 Thunderchiefs, F-4 Phantoms, and A-4 Skyhawks) were inoperable in poor weather and highly vulnerable to ground fire (Clodfelter, 1989; Momyer, 1978). U.S. losses in fixed-wing aircraft rose from 175 in 1965 to 280 in 1966 and reached 326 in 1967 (Littauer and Uphoff, 1972: 283).

28. According to Wirtz (1991: 100–101), Westmoreland changed his mind when he encountered resistance from both South Vietnam and South Korea, which had the largest contingent of troops (22 combat maneuver battalions in 1967) in the Free World Military Assistance forces in Vietnam. The successful Korean War experience with a unified command led Dean Rusk to support a unified American–South Vietnamese command, but Westmoreland refused and McNamara did not want to push the matter because he already had too many sharp disagreements with the JCS (Rusk, 1990: 452).

29. For a discussion of the Great Society legislation between 1964 and 1968, see Schneider (1988).

30. This does not imply that the incident was provoked on purpose. For detailed discussions of the incident and its ramifications, see Austin (1971); Goulden (1969); McNamara with VanDeMark (1995: 127–43); Windchy (1971).

31. For the full text of the resolution, see Porter (1979: 307).

32. It is ironic that Senator Fulbright, a key proponent of this broad authorization, became one of the bitterest critics of U.S. intervention in Vietnam.

33. For a description of techniques used by the president and his assistants to counter or preempt congressional opposition to his policy, see Gibbons (1989: 127–30).

34. A Gallup Poll released at the end of August 1965 showed that 57 percent approved of the way the administration was handling the situation in Vietnam (compared with 52 percent in mid-July and only 48 percent in mid-June), 25 percent disapproved, and 18 percent had no opinion (Gibbons, 1989: 444).

35. Reflecting years later, McGeorge Bundy admitted that the administration was not prepared for the strength of the antiwar movement. In spite of the seemingly rosy picture, there were some warning signs of what might happen. Although public opinion polls had initially shown strong support for the bombing, support faded away when the practical consequences seemed to be rather disappointing and the moral consequences looked increasingly reprehensible. In March 1965 a Gallup Poll showed that 42 percent of Americans supported an expansion of the war effort, whereas 41 percent favored immediate negotiations. Johnson was facing mounting criticism from liberals, who saw the war as interfering with social and economic reform at home. Voices of dissent were also heard in Congress. Senators Frank Church and George McGovern, who just a few months earlier had supported the Gulf of Tonkin Resolution, denounced the bombing campaign (VanDeMark, 1991: 119–20).

36. See Gibbons (1995); Herring (1993a, 1994). International public opinion was never as important as domestic public opinion. Still, it had an impact and was a factor, for example, in the 1966 decision to limit the escalation of bombing (Small, 1988: 86).

37. For comprehensive discussions of the antiwar movement and the media, see DeBenedetti with Chatfield (1990); Hallin (1986); Joseph (1993); LaFeber (1994); Turner (1985); Vasquez (1976); Wells (1994); Zaroulis and Sullivan (1984).

38. In August 1965 public support of the Vietnam policy stood at 61 percent. As it became clear that the war would be neither short nor inexpensive, support declined to around 50 percent. In the spring of 1967 it fell permanently below 50 percent (Lorell et al., 1985: 20). A Gallup Poll in August 1967 found that only 33 percent of those surveyed approved of Johnson's handling of the war; 54 percent disapproved, and 13 percent had no opinion (Barrett, 1993: 70). One interpretation of the high rate of disapproval that is congruent with the results of other surveys indicating that public opinion did not support unilateral withdrawal is that a majority of the disapproving respondents supported escalation to achieve a quick and decisive victory (Schwarz, 1994).

39. The consensual view among the administration's legal advisers was that the president had full power to use U.S. forces in combat and that the Gulf of Tonkin Resolution gave the president a broad latitude of authority. In fact, they argued, he would be in a better position to exercise this authority without constraint if he did *not* ask Congress for approval. In Congress the question of a new resolution to replace or supplement the Gulf of Tonkin Resolution was raised, but there was very little effective pressure for it at this stage. There was, however, considerable support for further consultation with Congress by the president and his key advisers before they involved the United States in a larger war (Gibbons, 1989: 288–89, 459).

40. Summers (1982) believes that a declaration of war would have committed the public to supporting the war fully. It also would have allowed for the mobilization of reserves.

41. An excellent discussion and elaboration of the points of view of those who supported and those who opposed the war is offered in Levy (1991).

42. The caution was always there, and the risk of a wider war was on Johnson's and his advisers' minds, so certain military options were ruled out. These included using nuclear weapons, invading North Vietnam, destroying the dike system, bombing the civilian population, striking at lines of communication near the China border, mining North Vietnamese ports, and stepping up operations in, or invading, Laos and Cambodia (Gelb with Betts, 1979: 264–65). Some of these restrictions were later removed by the Johnson administration, and most, but not all, of the rest by the Nixon administration.

43. On the debate between hawks and doves in the Soviet Politburo concerning whether to negotiate, what military aid to supply, and what risks the Soviets faced in Vietnam, see Anderson (1993: 131–43, 177–78); Zimmerman (1981).

44. See Gibbons (1989: 412–13); McNamara with VanDeMark (1995: 172–73); U.S. Department of State (1996a: 254–57); U.S. Department of State (1996b: 105). Rusk's assessment that the Sino-Soviet conflict could not be used to U.S. advantage but would become an incentive for both to increase assistance to North Vietnam (Smith, 1991: 69) turned out to be correct. His concern about Chinese intervention was strongly influenced by his Korean War experience and peaked when it was discovered the Chinese Mig fighters were based at Phuc Yen airfield in North Vietnam.

45. The deployment of American combat troops to South Vietnam was less risky in terms of triggering third-party counterintervention than strategic bombing was, because it did not pose the immediate danger of a military confrontation between Americans and Chinese. An invasion of North Vietnam would have been a different matter.

For a detailed discussion of the extent of the PRC's security commitments to North Vietnam, see Jian (1995).

46. The total number of allied troops was 22,404 in 1965, 52,566 in 1966, 59,450 in 1967, and 65,802 in 1968 (Schandler, 1977: 353).

47. For analyses of the role of two of the closest U.S. allies, Japan and Australia, see Havens (1987) and Pemberton (1987).

48. For a discussion of these and other initiatives, see Thies (1980).

49. Some influential senators like Church and Morse advocated having the United Nations call for a cease-fire and help bring North Vietnam to the negotiation table. This approach was supported by the U.S. ambassadors to the United Nations, first Adlai Stevenson and then Arthur Goldberg, and by the assistant secretary of state for international organizations affairs, Harlan Cleveland (Gibbons, 1989: 308–17; see also Geyelin, 1966: 202–9; LBJ Library, William Bundy manuscript, 1972: ch. 31, pp. 15–22; U.S. Department of State, 1996b: 231–32).

50. The relationship with the Saigon government posed a constant problem for American policymakers. The Saigon government was extremely sensitive to any expressed American readiness to negotiate with North Vietnam. It more often than not viewed any such suggestion as a step toward abandonment by the United States. American decisionmakers believed that its concerns about abandonment increased the instability of the regime and reduced its determination to pursue the pacification program vigorously (Simons, 1994).

51. On Ball's persistent attempts to prevent the escalation of both the air war and the ground war and on the uncanny accuracy of his predictions, see Ball (1982: 392–402); DiLeo (1991: 64–92); U.S. Department of State (1996a: 252–61).

52. Ball's attempts to convince the president and his advisers that the French experience in Indochina was the relevant analogy failed. On June 30, 1965, McGeorge Bundy prepared a memo refuting the relevance of the French analogy at the president's request. See U.S. Department of State (1996b: 79–85).

53. See Kearns (1976: 252); Khong (1991, 1992). Among Johnson's advisers Vice President Hubert Humphrey and, to a degree, Clark Clifford were preoccupied in 1965 with the potentially adverse consequences for public support of deploying combat troops. The salience of this risk dimension for Humphrey lay in the Korean analogy. The Korean War cost Truman and the Democrats their public support, as well as the next election. Humphrey, fresh to the Senate in the early 1950s, was a staunch supporter of Truman's Fair Deal program, which the Korean War and the lost elections undermined. Clifford was Truman's strategist during his first term. Both regretted Truman's political fate and thus were keenly aware of the domestic risks of a land war in Asia. Other advisers had also learned from the Korean War experience, which had affected most of them directly, but they were not attentive to the domestic lessons (Neustadt and May, 1986: 161–64; Humphrey, 1976: 322–24).

54. On the sources of misplaced expectations about breaking the will of the enemy, see Shultz (1978); Mueller (1980); Vought (1982).

55. For a recent analysis of limited war in American foreign policy, see Gacek (1994). For a discussion of coercive diplomacy and its variants, see George and Simons (1994a, 1994b).

56. For an account of the effect of the "Kennedy mystique" on the Johnson presidency, see Henggeler (1991). According to Henggeler, "Johnson's effort to link his controversial foreign policies to John Kennedy was in part intended to maintain public support for the war and to neutralize wavering liberals" (p. 175)

57. The JCS were unhappy with his choice and wanted instead "to take positive, prompt, and meaningful military action" (*Pentagon Papers*, 1971, 3: 173).

58. The four geopolitical models of U.S. security are "Fortress America," "Western Hemisphere," "Atlantic," and "Globalist."

59. On Johnson's need for reassurance, see Steinberg (1996). This need was evidently one of the reasons for Johnson's regular consultations with Eisenhower between 1963 and 1967 (Brands, 1985).

60. See LBJ Library, Tom Johnson's Notes on Meetings, Oct. 18, 1967, Box 1.

61. LBJ Library, Tom Johnson's Notes on Meetings, Sept. 26, 1967, Box 1.

62. The idea that only a larger investment of resources was needed is evident, for example, in the confidence expressed by General Wheeler in the efficacy of the ground strategy (Berman, 1982: 108) and by Admiral David McDonald, chief of naval operations, on the efficacy of bombing (Valenti, 1975: 341).

63. The need for self-affirmation made Johnson ignore the doubtful quality of the Wise Men's advice. William Bundy describes his own response to these "quickie consultations with outsiders" as a "slightly queasy feeling" (LBJ Library, William Bundy manuscript, 1971: ch. 27, p. 20).

64. A notable example was McGeorge Bundy. His association with Johnson, unlike his association with Kennedy, was uncomfortable. He no longer had free access to the Oval Office, as he used to. Their "distant partnership" reflected their differences in social background, education, and areas of interest (VanDeMark, 1991: 12–14).

65. Johnson was impatient with those who disagreed with him, but his impatience was distributed unevenly. He was often willing to listen to views different from those expressed by those he respected, trusted, or liked. In any case, "out of a sense of awe, friendship, responsibility, hubris, and even fear, Johnson's advisers did not often cross him" (Schandler, 1977: 326).

66. According to McNamara, "an orderly, rational approach was precluded by the 'crowding out,' which resulted from the fact that Vietnam was but one of a multitude of problems we confronted" (McNamara with VanDeMark, 1995: 277). The validity of this explanation is questionable if we consider the centrality of the Vietnam issue.

67. Not raising important issues also reflected Johnson's lack of self-confidence, which came to be expressed in a growing obsession with secrecy. He was preoccupied with suspicions of disloyalty and conspiracy and therefore preferred small informal decisionmaking forums (Berman, 1982: 4; Berman, 1989; Kearns, 1976: 319). Besides being secretive, he was highly sensitive to criticism. Those who worked for him understood this well, and it inhibited critical review of his decisions.

68. In part this was also the result of the organizational change instituted by McNamara in the defense-planning process. By January 1963 the organizational restructuring of the Department of Defense had brought Planning Programming Budgeting Systems (PPBS) into operation. PPBS allowed for a more centralized defense decisionmaking process, but it also increased the influence of the Department of Defense at the expense of the Department of State in issues that had a substantial military dimension. To influence planning required that the inputs, whether they were policy proposals, priorities, or criteria for comparing alternatives, be expressed in a format acceptable to the system and inserted into the system at specific stages in the development of a program. For the Department of State to influence military policies, its own PPBS staff would have had to participate in the defense-programming process with the Department of Defense. The Department of State, which did not have that capacity, found that it became more difficult than before for it to influence the

planning and implementation of policies with a military dimension (Palmer, 1978: 106–8).

69. Berman (1989: 12) describes the attitudes of the JCS toward the civilians in the Pentagon in 1967–68 as "open revolt." See also Perry (1989: 169, 172).

70. Johnson's basic attitude toward the military was ambivalence. "The president was torn with respect to his military commanders. The president often repeated his mentor Sam Rayburn's observation that political leaders ought to accept military advice—otherwise a lot of money had been wasted training commanders at West Point. On the other hand, he held a deeply rooted suspicion that the military mind knew only how to bomb and escalate" (Berman, 1989: 41). It may well be the case that in spite of his ambivalence, when he had to assess McNamara's pessimistic opposition to the professional advice of the field commander, General Westmoreland, he followed the advice of another general and president whom he greatly respected—President Eisenhower. Eisenhower strongly advised him in 1965 to back his field commander and trust him because demonstrated confidence in the field commander would make the commander much more effective; see LBJ Library, Meeting Notes File, Memorandum of Meeting with the President, Feb. 17, 1965, Box 1. See also Zaffiri (1994).

71. Among the several prominent individuals who expressed their reservations about the deployment of combat troops and escalation were General Maxwell Taylor, Hubert Humphrey, Senator Mike Mansfield, Clark Clifford, and George Ball (Burke at al., 1989: 166–73). Also, some of the most respected experts on Vietnam in the administration, such as Marshall Green, Paul Kattenburg, Edward Rice, Louis Sarris, and Allen Whiting, argued consistently against American intervention (Hilsman, 1996).

72. For a cogent argument supporting a responsible advisory process that emphasizes the quality and integrity of the advisory process over loyalty to superiors and uses the 1965 troop deployment decision as a case study, see Burke (1984). On the adverse effects of loyalty, see also White (1968: 215–17).

73. See Westmoreland's (1976: 120–21) biting assessments of the senior civilian officials in the Pentagon.

74. Of the two main intelligence suppliers, the CIA developed a reputation for relatively dispassionate analysis; see, for example, the CIA director's memo to the president dated August 29, 1967, in which he questions the premise underlying U.S. involvement in the war (McNamara with VanDeMark, 1995: 291–94). The DIA was more inclined to assess the effects of the bombing and the ground strategy positively. Being largely staffed by military personnel at the upper level and having its director reporting to the JCS and, through the JCS, to the secretary of defense, the DIA had a considerable incentive to distort performance assessments. Career hopes were a powerful inducement to provide information that would assist the service that the officers came from or, when they were faced with competing demands from different masters, to compromise the substance of the assessment to suit all (Gallucci, 1975: 66–67). On the biases toward optimism of U.S. government agencies reporting the progress of the war, see also FitzGerald (1972: 363–68).

75. General Bruce Palmer attributes the Joint Chiefs' persistence to the "can do" spirit with which they were imbued (B. Palmer, 1984: 46).

76. For example, the competition between air force and navy for targets and sorties forced an allocation of sortie figures between the two services; figures were allotted for each two-week unit. The competition led to the flying of senseless missions—in bad weather conditions or even with no bomb loads (Gibson, 1986: 357–63).

77. Taylor himself supported an intensive bombing campaign, which he thought would have the desired effect on North Vietnam only if there was an "inexorable continuity of attack repeated day after day without interruption" (1972: 334, 342).

78. Between 1965 and 1968 about 990 aircraft were lost in Rolling Thunder. The economic costs of lost aircraft totaled over $6 billion. This estimate does not include operational costs, logistical base support, and costs of training new pilots to replace casualties. More than 800 airmen were killed or captured (Gibson, 1986: 380; Tilford, 1993: 86).

79. Sharp was one of the strongest advocates of all-out strategic bombing; see Sharp (1978).

80. These reports may have correctly reflected the physical impact of bombing, but if an enemy cares little about the destruction inflicted on it, then the cost the enemy pays is virtually zero. Putting a high cost tag on the enemy's behavior becomes meaningless. In fact, the cost to an attacker that cares about the price paid for causing the damage to the enemy will then be much higher than the cost to the enemy.

81. For an explanation of why direction did not come from the top civilian and military leadership in Washington, see Herring (1994: 36–43).

82. Johnson was urged by some of his advisers to use his authority as commander-in-chief and take charge of the war, but to no avail (Herring, 1993a: 78).

83. The readiness of the JCS to escalate the war far more than the civilians was in part due to their much lesser concern about the domestic constituency.

84. For a description of the debate and decisionmaking process within the PLO regarding the withdrawal from Beirut, see Khalidi (1988).

85. Government decision no. 676 of June 5, 1982; see Tamir (1988: 160).

86. See, for example, the interview with Foreign Minister Yitzhak Shamir on Israeli television on June 20, 1982, in Medzini (1990: 66–68). For a discussion of Syrian perceptions of Israel's short-term and long-term intentions, see Tlas (1988).

87. For a detailed review of Israel's military doctrine, see Levite (1988). For a discussion of Labor and Likud approaches to security and the use of force, see Inbar (1991) and Peleg (1987, 1988).

88. See the exchange between Menachem Begin and Labor Party leader Shimon Peres in *Ma'ariv*, Aug. 20, 1982; see also Rabin (1983: 19–31).

89. The first phase of the war in Lebanon—eliminating the threat of a Palestinian presence within 40 kilometers from the northern border—could be explained to the public as a defensive operation, so it had broad legitimacy. The second phase—the siege of Beirut and the fight with the Syrian army—could not be explained in those terms and was therefore seriously contested.

90. The PLO anticipated a large-scale military operation based on intelligence sources in Israel, as well as France, the Soviet Union, the United States, Egypt, and Lebanon, but there was uncertainty about the scope and timing (Avi-Ran, 1987: 153–73). For criticism of the deficiencies of the Palestinian preparations, see Khalidi (1988: 73–79).

91. For details on the military balance between Israel and Syria in 1982, see Heller et al. (1983).

92. As early as September 1981, the Northern Command had begun to update plans for an invasion of Lebanon and had ordered that units be trained in battle modes appropriate for the terrain there (Schiff and Ya'ari, 1984: 93–94, 98, 109).

93. See, for example, the letter to Sharon from Major General Avraham Rotem on the eve of his retirement (Yaniv, 1987: 130–32).

94. For discussions of combat operations in all three sectors, see Davis (1987: 75–106); Kober (1995, 397–429); Wald (1992).

95. Israeli casualties in the fighting up to September 1982 totaled 368 dead and 2,383 wounded. By the time Israeli forces had completed their withdrawal from Lebanon in June 1985 these numbers had risen to 654 dead and 3,873 wounded (Gabriel, 1984: 235–36; Yaniv, 1987: 245).

96. For a detailed discussion of this situation, see Yaniv (1987).

97. The resentment over these interferences is still evident many years later; see the interview with Major General Amir Drori (*Ma'ariv*, weekend magazine, July 1, 1994).

98. The aversion to risk taking has found expression, for example, in the almost complete avoidance of night operations, for which the IDF had acquired a reputation in the past (Schiff and Ya'ari, 1984: 169).

99. At that time the operation was planned for May 17, but the launch date was postponed at the last minute for reasons other than Saguy's warning.

100. In some instances, the roads were blocked, so transporting these supplies required the use of helicopters, thus diverting the helicopters from essential combat-related operations. At other times, in the eastern sector, for example, the scarcity of fuel slowed progress (Kober, 1995: 417; Schiff and Ya'ari, 1984: 194, 212).

101. For a discussion of the social and economic costs of Israel's military mobilization system in times of war, see Kimmerling with Backer (1985).

102. Estimates of the cost of the Lebanon policy between June 1982 and the final withdrawal in June 1985 vary widely. It is particularly difficult to assess the indirect costs stemming from loss of production. Barkai (1986), to cite one estimate, claims that the direct costs were $1.5 billion and the indirect costs $5 billion. Another more conservative estimate is that the direct costs of the war were $700 million, the costs of the occupation $650 million, and the loss of GNP $800 million (Yaniv, 1987: 314 n. 81).

103. A minority government with the support of fewer than 61 members of the Knesset is possible but is considered highly unstable.

104. The parliamentary majority increased during the war when a small right-wing hawkish party, Hatehia, joined the coalition on July 27, 1982.

105. For example, the three-day delay between the beginning of the intervention and the military operation against the Syrian forces allowed the Syrians to bring in reinforcements, reducing the force ratio advantage from 4:1 in Israel's favor to only 2:1 (Kober, 1995: 410).

106. According to Big Pines, airborne forces were to seize key road junctions before the Syrians could take defensive positions. According to the scaled-down version of the plan, armored forces would have to advance on narrow mountain roads, clearing their way through stubborn enemy resistance to seize these areas (Shiffer, 1984: 87).

107. For discussions of Labor Party views of the operation, see also Inbar (1991: 125–30); Schiff and Ya'ari (1984: 111–12); Yaniv (1987: 120–27).

108. For 1981 election results and the distribution of seats in the Knesset, see Scammon (1986: 266–67).

109. On the implications of this attribute, see Barzilai (1992); Horowitz (1987).

110. Sharon was well aware that these claims delegitimized his policy and caused him heavy political damage. He therefore did not miss an opportunity to emphasize

that everything he did was with the consent and authorization of the cabinet (*Ma'ariv*, Sept. 17, 1982; Sharon, 1989: 463–64; *Yediot Aharonot*, June 10 and 17, 1983).

111. For a discussion of the various extraparliamentary opposition groups and their positions, see Yishai (1983–84).

112. See, for example, Begin's position in a meeting with U.S. Senator John Glenn in February 1982 (Shiffer, 1984: 80–81).

113. According to Minister of Energy Yitzhak Berman's account, nine voted for, eight voted against, and three others were not present. Half of those who opposed the attack were Likud ministers (*Ma'ariv*, June 5, 1992).

114. The issue of Lebanese civilian casualties concerned the Israeli government from the start as a potential source of delegitimation, so leaders emphasized the measures taken to limit civilian casualties. For example, at a press conference on June 11, 1982, Sharon said: "Tremendous effort was put by our people to secure the lives of civilian population, and I think that the steps that were taken by us would not have been taken, and were not taken, by any other nation on earth. No one would have done what we did here, and we paid for it. But we found, being a small nation, being Jews, we found that that was what we were expecting from ourselves to do, and therefore we did it" (Medzini, 1990: 23, 192). For a critical analysis of exaggerated accounts and distorted stories in the U.S. media, see Muravchik (1983).

115. This found expression in the sharp decline in the personal popularity of Begin and Sharon and in the public's assessment of the quality of government performance. By June 1983 some 50 percent of Likud voters and 89 percent of Labor voters opposed the war (Barzilai, 1992: 210).

116. Sharon admitted in a June 1983 interview that not anticipating the difficulties of establishing and maintaining domestic legitimacy was one of his two major errors in judgment (*Yediot Aharonot*, June 17, 1983).

117. For a discussion of Israel's evolving strategy for dealing with the constraints posed by the patron-client relationship with the United States, see Bar-Siman-Tov (1987).

118. In an interview in 1987, Lieutenant General Rafael Eitan made an interesting retrospective argument to justify the delay in attacking the Syrian army. According to him, if Israel had first attacked and defeated the Syrian army in Lebanon, the Soviets would have been forced to get involved much earlier by pressuring the United States to stop the fighting. Had that happened, Israel might have been prevented from destroying PLO forces. Therefore it made sense to attack the PLO first (*Ma'ariv*, June 2, 1987). There is, however, no other evidence that this was indeed the line of reasoning that guided the decisionmakers. In fact, Major General Avigdor Gal, Northern Corps commander during the war, argued that one of the major strategic mistakes in the war was not attacking the Syrian army first or at least simultaneously with the attack on the PLO. Such a strategy could have resulted, according to him, in much more substantial achievements (*Ma'ariv*, June 2, 1987).

119. A typical statement of confidence in Israel's strategic importance to the United States can be found in Begin's remarks on that matter to the press on June 23, 1982, after returning from the meeting with Reagan: "This [U.S.-Israel relationship] is not a one-way street. We do a lot for you and also in recent battles we did a lot for the United States" (Medzini, 1990: 80).

120. The implications of this argument are that the intervention would have taken place even if the United States had voiced its objection in a more determined fashion.

The visits of Sharon and the chief of military intelligence to Washington prior to the intervention should therefore be interpreted mainly as ritualistic. The visits were to inform Israel's strategic partner of Israeli intentions, not to secure American endorsement

121. Aligned with Haig was Jeane Kirkpatrick, U.S. ambassador to the United Nations. Opposed were Vice President Bush, Chief of White House Staff James Baker, Secretary of Defense Weinberger, and National Security Adviser William Clarke. For discussions of the alignments in the administration of Reagan's first term that affected U.S. Middle East policy, see Spiegel (1985: 401–7); Tanter (1990: 120–24, 210–13).

122. See, for example, his phone call to the Israeli Foreign Office on February 12, 1981 (Shiffer, 1984: 78), and his informal conversation with the minister of energy two months later (*Ma'ar.*, June 5, 1992). The Americans were also aware of the broad military outlines of the operation (Haig, 1984: 332–35).

123. Alexander Haig said repeatedly that the United States would understand a large-scale Israeli military operation only if it were in proportional response to an "internationally recognized provocation" (1984: 326–27, 332; see Tucker, 1982).

124. The use of legalistic arguments also served Begin's prized principle of following the rule of international law in dealing with international affairs (Sofer, 1988: 142).

125. This was the language used by the president in his letter to Begin on August 3 (Shultz, 1993: 61).

126. For a discussion of the Reagan Peace Plan, see Quandt (1993: 344–48); Reagan (1990: 430–35); Shultz (1993: 85–100).

127. During the PLO evacuation from Beirut, when Israel delayed the evacuation, the U.S. Navy received orders on one occasion to open fire on IDF forces if necessary so that a ship carrying PLO members could sail out of the harbor (Schiff and Ya'ari, 1984: 280–81). Another incident in February 1983 involved an encounter between an Israeli tank and a U.S. Marines unit (Yaniv, 1987: 166–67).

128. For example, in late June, sensing the change in the administration's mood and the growing criticism of Israel, the Saudis advised the PLO to harden its position on being disarmed and withdrawing from Lebanon (Haig, 1984: 343–44).

129. Syria is not treated as a third party but as a primary target actor, as is clear in the preceding analysis.

130. As was, for example, the case shortly after the invasion on June 8, when the United States prevented the condemnation of Israel in the Security Council (Haig, 1984: 339).

131. According to Secretary Shultz, by August 1982 Begin became deeply concerned that "the U.S.-Israel relationship was coming apart at the seams" (Shultz, 1993: 70–71).

132. According to Ross (1990), the number of Soviet personnel required was 5,000.

133. For example, the risk of a Soviet response to Israel's intervention in Lebanon is not mentioned at all in the memoirs of Defense Minister Sharon (1989).

134. See, for example, his remarks at a CBS interview on June 21, 1982 (Medzini, 1990: 74–75).

135. U.S. economic and military aid to Egypt in 1981–82 came to $2 billion and was supposed to increase 20 percent in 1983–84. U.S. arms sales to Egypt between 1979 and 1982 amounted to $3 billion (Greenwald, 1988: 106).

136. The texts of all Security Council resolutions and draft resolutions mentioned in this section can be found in Medzini (1990).

137. Begin told the Knesset Committee on Foreign and Defense Affairs that "Israel has never entered an Arab capital and will not do so in the case of Beirut" (Feldman and Rechnitz-Kijner, 1984: 16). He was embarrassed to find out that when he issued this statement, IDF forces were already in Beirut.

138. They were wrong. Eitan was correct in thinking that the only difference between Big Pines and the approved limited operation was in the way it would begin.

139. On one occasion he described Sharon as "a skilled craftsman"; on other occasions he referred to him as "a superb defense minister" and as "a brilliant general, vicious man" (Sofer, 1988: 205, 206). When Sharon suggested a deep flanking move against the Syrian army that would bring the Syrians into the war—against the explicit initial intention of the government—Begin did not realize the far-reaching consequences of the maneuver. He even exclaimed to his cabinet that it was "a tactic worthy of Hannibal" (Schiff and Ya'ari, 1984: 146).

140. This is a particularly salient point because of Begin's high sensitivity to casualties (Sofer, 1988: 70).

141. Disillusionment with the optimistic assessment of casualties was first manifested in the Beaufort battle, which took place during the first week of the intervention. The ruined Beaufort fortress had important strategic and symbolic value to both Israelis and Palestinians. Begin was told that there were no casualties. The next day he arrived with Sharon and journalists to inspect the place and was shocked to hear that six Israeli soldiers had been killed in the fighting (Schiff and Ya'ari, 1984: 162).

142. Eitan was uneasy about engaging the Syrian army in battle. He wanted to concentrate on the PLO but never confronted Sharon directly on that issue (Schiff and Ya'ari, 1984: 123–24).

143. For trenchant criticism of the unclear definition of the objectives of the operation by a member of the Begin cabinet, Energy Minister Yitzhak Berman, see *Ma'ariv*, weekend magazine, June 5, 1992. See also the interview with the then commander of the Northern Command, Major General Drori (*Ma'ariv*, weekend magazine, July 1, 1994).

144. For claims that the threat to the northern settlements was removed, see, for example, Sharon's interviews on August 14 and August 29, 1982, and Begin's Knesset speech of September 8 in what was then thought to be the concluding debate on the operation (Medzini, 1990: 157–58, 161–62, 194–96). Begin concluded his speech by saying, "We won. Happy are we that we have attained this victory" (p. 196). A few days later, IDF forces had to enter western Beirut after the Sabra and Shatilla massacres. A month later, the Israeli government declared the conditions for pulling out from Lebanon, which amounted to extending the original objectives (*New York Times*, Oct. 11, 1982). Israel was entrapped in Lebanon for almost another two years.

145. Sharon suspected that the main goal of Philip Habib, President Reagan's special envoy to the Middle East, was to get an agreement that would preserve the PLO's reputation. Habib used the term *evacuation* for the withdrawal of the PLO from Beirut, for example, but Sharon insisted on using the term *expulsion* (Schiff and Ya'ari, 1984: 272–73).

146. The May 17 agreement was endorsed by seventeen cabinet ministers, and two others voted against it. The agreement called for the mutual withdrawal of Syria and Israel from Lebanon and to the establishment of security arrangements between Lebanon and Israel. For the full text of the agreement, see Medzini (1990: 401–14).

147. For detailed discussions of the historical and ideational background of Revisionism and Neo-Revisionism, see Peleg (1987); Seliktar (1988); Sofer (1988).

148. On the effects of Begin's early life experiences, see Haber (1978); Perlmutter (1987); Silver (1984).

149. On one occasion, for example, when responding to a European view of the Arab-Israeli conflict, Begin said: "I am going to remind the Europeans of the Holocaust and the rivers of Jewish blood. I'll talk even to the Dutch. They helped protect the Jews. But I will tell them we don't want our daughters to leave behind books like *The Diary of Anne Frank*" (Peleg, 1987: 67).

150. In a letter to Reagan written during the war in Lebanon, Begin used an analogy to describe his feelings: "I feel like a prime minister empowered to instruct a valiant army facing Berlin, where, among innocent civilians, Hitler [meaning Arafat] and his henchmen hide in a bunker deep under the surface" (Peleg, 1987: 67).

151. As was also exemplified by one of his heroes, Giuseppe Garibaldi (Sofer, 1988: 107–8). According to Sofer, "Begin always exhibited the self-awareness of a historical hero convinced of his ability to shape reality and affect its course" (p. 98).

152. It is difficult to ascertain to what extent their beliefs and values were a product of their personality attributes or independent of them and the result of learning and socialization.

153. This was one of Begin's main motives for asking Dayan to join his first cabinet, and it also explains why it was important to him to convey the impression of a leader who did not impose his views on cabinet members and who listened to their views, although he also demonstrated that he could impose his preferences on the cabinet when he wanted to.

154. It can be convincingly argued that the same preference for risk seeking accounts for Begin's quick reaction to Sadat's peace initiative and for his proactive response and initiative in inviting Sadat to Jerusalem, in spite of the concerned views of the defense establishment.

155. Some of those who shifted toward the minority position felt that they were emotionally and psychologically coerced into doing so.

156. For criticism of the meekness of the General Staff under Sharon, see the interview with Major General Saguy in *Ma'ariv*, weekend magazine, Feb. 28, 1992.

157. This was proved with the unanimous passage of the Golan Annexation Law in December 1981. Some ministers privately considered the law, which made the occupied territory of the Golan Heights part of Israel, unnecessarily provocative and risky because of a potential Syrian military response, as well as because of the likely response of the international community. There was special concern for the response of the United States, Israel's main ally. The Syrians were deterred from taking military action, but the United States took punitive action against Israel by suspending a strategic Memorandum of Understanding that was signed shortly before the event and by initiating sanctions involving military and economic assistance.

158. This effect will be particularly robust where the costs of the asset are high or particularly painful, and there will be a motivated bias toward overvaluation of the asset (Kahneman et al., 1990; Kahneman and Tversky, 1979).

159. The government can establish issue-specific cabinet committees with broad authority and access to information. Begin did not set up a defense committee for six months after the 1981 election. Once established, it did not meet even once between June 1982 and the end of that year (Ben Meir, 1987: 121).

160. One important reason that cabinet meetings were not occasions for information sharing—contrary to the normative expectations expressed in literature on small-

group behavior—was the secretive tradition in Israeli decisionmaking, especially in discussions of national security decisions (Ben Meir, 1987: 118).

Chapter 8

1. For detailed, trenchant criticism of the pathologies of rational choice theory, see Green and Shapiro (1994). There are also numerous other studies emphasizing the limitations of rational choice theory; a few recent ones are Euben (1995); Cook and Levi (1990); Hogarth and Kunreuther (1995); Kelley (1995); Lane (1995); Monroe (1995); Zuckert (1995). See also Chapter 4, note 1.

2. For a review of decisionmaking literature in which the importance of reality construction is emphasized, see Payne et al. (1992).

3. For a somewhat similar but more general statement about how essential it is to study the influences of organizations, individuals, and groups on one another, as well as why such studies are relatively rare, see Mowday and Sutton (1993).

4. For example, Steinberg (1996) points out that, although both Eisenhower and Johnson belonged to the same generational cohort and used the Korean War as their principal analogy, the two men drew very different lessons from that analogy and opted for very different policies.

5. There is, in some of the literature on risk taking, a tendency to reduce the causes and attributes of risky decisions to personality dispositions. The complex process that generated the decisions is thus ignored. It is important not to succumb to explanations that attribute policies to a single causal set, such as the role of individual leaders and their personality attributes (Trimpop, 1994; Yukl, 1989: 265–67).

6. For discussions of individuals as cognitive misers and the implications of that view for foreign policy decisionmaking, see Fiske and Taylor (1984); Vertzberger (1990).

7. Although for analytic purposes I have separated the effects on risk judgments from the effects on risk-taking preferences, it is in reality difficult to separate the effects of the explanatory variable-sets for risk assessment from their impact on risk acceptability. Risk assessment and risk acceptability are intertwined, and determining them probably takes place simultaneously. Nor do the variable-sets always operate in a strictly sequential order. In that sense this explanation of risk-taking behavior has the flavor of "garbage can decisionmaking processes" (Cohen et al., 1972; March, 1994: 198–206).

8. As prospect theory predicts, loss avoidance is a prime motivation for risk taking. In fact, one way to identify risk seekers among policymakers is by their disposition to take high risks not only to avoid losses but also to make gains.

9. It can be argued on methodological grounds that if I had chosen more culturally extreme cases, my conclusions with regard to the secondary role of cultural variables might have been different. But I avoided extreme cases on purpose, because they would have been less representative of the population of cases as a whole and would therefore have reduced the external validity of the conclusions flowing from the comparative analysis.

10. A similar conclusion was also reached by an observer of U.S. international behavior concerning matters of high politics. The same observer, Kreisberg (1986), is cautiously optimistic, however, that greater attention will be paid to international law in the future.

11. For a similar argument on the effect of deterrence in general, see Altfeld (1985).

The convincing power of such threats may be increased by a public commitment to repeat the intervention, which will add credibility to the intervener's policy. Because if
the intervener's leaders do not carry out the threats, they will bear substantial political costs due to their domestic constituency's response; see Fearon (1994).

12. Underlying this argument is the assumption that the quality of the decisionmaking process determines the quality of outcomes. For validation of this assumption, see Birren and Fisher (1990); Dean and Sharfman (1996); Vertzberger (1990).

Bibliography

Abelson, R. P. 1959. "Modes of Resolution of Belief Dilemmas." *Journal of Conflict Resolution*, 3: 343–52.

Abelson, R. P., and A. Levi. 1985. "Decision Making and Decision Theory." In G. Lindzey and E. Aronson, eds., *Handbook of Social Psychology: Theory and Methodology*, vol. 1 (3rd ed.), pp. 231–309. New York: Random House.

Abrahamson, E., and C. J. Fombrun. 1994. "Macrocultures: Determinants and Consequences." *Academy of Management Review*, 19: 728–55.

Achen, C. H., and D. Snidal. 1989. "Rational Deterrence Theory and Comparative Case Studies." *World Politics*, 41: 143–69.

Adelman, L. K., and N. R. Augustine. 1992. "Defense Conversion: Bulldozing the Management." *Foreign Affairs*, 71(2): 26–47.

Adkin, M. 1989. *Urgent Fury: The Battle for Grenada*. Lexington, Mass.: D. C. Heath.

Adler, S. 1980. "Risk-Making Management." *Business Horizons*, 23 (Apr.): 11–14.

Adomeit, H. 1986. "Soviet Crisis Prevention and Management: Why and When Do the Soviet Leaders Take Risks?" *Orbis*, 30: 42–64.

———. 1982. *Soviet Crisis Behavior: A Theoretical and Empirical Risk-Taking and Analysis*. London: Allen and Unwin.

———. 1973. *Soviet Risk-Taking and Crisis Behavior: From Confrontation to Coexistence?* Adelphi Paper no. 101. London: International Institute for Strategic Studies.

Akerlof, G. A. 1991 "Procrastination and Obedience." *American Economic Review*, 81: 1–19.

Aldag, R. J., and S. R. Fuller. 1993. "Beyond Fiasco: A Reappraisal of the Groupthink Phenomenon and a New Model of Group Decision Processes." *Psychological Bulletin*, 113: 533–52.

Alicke, M. D., and E. Largo. 1995. "The Role of the Self in the False Consensus Effect." *Journal of Experimental Social Psychology*, 31: 28–47.

Allison, G. T., and M. Halperin. 1972. "Bureaucratic Politics: A Paradigm and Some Policy Implications." In R. Tanter and R. Ullman, eds., *On Theory and Policy in International Relations*, pp. 40–79. Princeton, N.J.: Princeton University Press.

Allison, G. T., E. May, and A. Yarmolinsky. 1970. "Limits to Intervention." *Foreign Affairs*, 48: 245–61.

Altfeld, M. F. 1985. "Uncertainty as a Deterrence Strategy: A Critical Assessment." *Comparative Strategy*, 5: 1–26.

Anderson, C. A. 1983. "Abstract and Concrete Data in the Perseverance of Social Theories: When Weak Data Lead to Unshakable Beliefs." *Journal of Experimental Social Psychology*, 19: 93–108.

Anderson, C. A., M. R. Lepper, and L. Ross. 1980. "Perseverance of Social Theories: The Role of Explanation in the Persistence of Discredited Information." *Journal of Personality and Social Psychology*, 39: 1037–49.

Anderson, D. L. 1991. *Trapped by Success: The Eisenhower Administration and Vietnam, 1953–1961*. New York: Columbia University Press.

Anderson, P. A. 1987. "What Do Foreign Policy Decision Makers Do When They Make a Foreign Policy Decision? The Implications for the Comparative Study of Foreign Policy." In C. F. Hermann, C. W. Kegley, and J. N. Rosenau, eds., *New Directions in the Study of Foreign Policy*, pp. 285–308. Boston: Allen and Unwin.

———. 1985. "Deciding How to Decide in Foreign Affairs: Decision-Making Strategies as Solutions to Presidential Problems." In G. C. Edwards, S. A. Shull, and N. C. Thomas, eds., *The Presidency and Public Policy Making*, pp. 151–72. Pittsburgh, Pa.: University of Pittsburgh Press.

Anderson, R. D. 1993. *Public Politics in an Authoritarian State: Making Foreign Policy During the Brezhnev Years*. Ithaca, N.Y.: Cornell University Press.

Andriole, S. J. 1985. "A Decision Theoretic Analysis of the Reagan Administration's Decision to Invade Grenada." In P. M. Dunn and B. W. Watson, eds., *American Intervention in Grenada: The Implications of Operation "Urgent Fury,"* pp. 73–88. Boulder, Colo.: Westview Press.

Anyanwu, R. A. 1976. "Great Power Intervention in Small Power Conflicts (1945–1975): An Investigation of International Conflict Outcomes." Ph.D. diss., University of Pennsylvania.

Arch, E. C. 1993. "Risk Taking: A Motivational Basis for Sex Differences." *Psychological Reports*, 73: 3–11.

Arkes, H. R., and C. Blumer. 1985. "The Psychology of Sunk Costs." *Organizational Behavior and Human Decision Processes*, 35: 124–40.

Arkes, H. R., L. T. Herren, and A. M. Isen. 1988. "The Role of Potential Loss in the Influence of Affect on Risk-Taking Behavior." *Organizational Behavior and Human Decision Processes*, 42: 181–93.

Arnold, J. R. 1991. *The First Domino: Eisenhower, the Military, and America's Intervention in Vietnam*. New York: William Morrow.

Artaud, D. 1990. "Conclusion." In L. S. Kaplan, D. Artaud, and M. R. Rubin, eds., *Dien Bien Phu and the Crisis of Franco-American Relations, 1954–1955*, pp. 269–74. Wilmington, Del.: SR Books.

Asch, S. E. 1958. "Effects of Group Pressure upon the Modification and Distortion of Judgments." In E. E. McCoby, T. M. Newcomb, and E. L. Hartley, eds., *Readings in Social Psychology*, pp. 174–83. New York: Holt, Rinehart and Winston.

Aspaturian, V. V. 1970. "The Soviet Union and Eastern Europe: The Aftermath of the Czechoslovak Invasion." In I. W. Zartman, ed., *Czechoslovakia Intervention and Impact*, pp. 15–46. New York: New York University Press.

Auerbach, Y. 1989. "Legitimation for Turning-Point Decisions in Foreign Policy: Israel Vis-á-Vis Germany 1952 and Egypt 1977." *Review of International Studies,* 15: 329–40.

Austin, A. 1971. *The President's War: The Story of the Tonkin Gulf Resolution and How the Nation Was Trapped in Vietnam.* Philadelphia: Lippincott.

Avi-Ran, R., ed. 1987. *The War of Lebanon—Arab Documents,* vol. 1: *The Road to War.* Tel Aviv: Ma'arachot (in Hebrew).

———. 1986. *Syrian Involvement in Lebanon (1975–1985).* Tel Aviv: Ma'arachot (in Hebrew).

Baird, I. S., and H. Thomas. 1985. "Toward a Contingency Model of Strategic Risk Taking." *Academy of Management Review,* 10: 230–43.

Baker, J. A., with T. M. De Frank. 1995. *The Politics of Diplomacy: Revolution, War and Peace, 1989–1992.* New York: G. P. Putnam's Sons.

Ball, G. W. 1982. *The Past Has Another Pattern: Memoirs.* New York: W. W. Norton.

Ball, M. A. 1992. *Vietnam-on-the-Potomac.* New York: Praeger.

Bar, M. 1990. *Red Lines in Israel's Deterrence Strategy.* Tel Aviv: Ma'arachot (in Hebrew).

Baral, J. K. 1978. *The Pentagon and the Making of U.S. Foreign Policy: A Case Study of Vietnam, 1960–1968.* New Delhi: Radiant.

Barber, J. D. 1977. *The Presidential Character: Predicting Performance in the White House* (2nd ed.). Englewood Cliffs, N.J.: Prentice Hall.

Bar-Hillel, M. 1980. "The Base-Rate Fallacy in Probability Judgments." *Acta Psychologica,* 44: 211–33.

Baritz, L. 1985. *Backfire: A History of How American Culture Led the U.S. into Vietnam and Made Us Fight the Way We Did.* New York: William Morrow.

Barkai, H. 1986. "Reflections on the Economic Costs of the Lebanon War." *Jerusalem Quarterly,* no. 37: 95–106.

Barnet, R. J. 1968. *Intervention and Revolution: The United States in the Third World.* New York: World Publishing Company.

Baron, J. 1994. "Reference Points and Omission Bias." *Organizational Behavior and Human Decision Processes,* 39: 475–98.

Baron, J., and J. C. Hershey. 1988. "Outcome Bias in Decision Evaluation." *Journal of Personality and Social Psychology,* 54: 569–79.

Barrett, D. M. 1993. *Uncertain Warriors: Lyndon Johnson and His Vietnam Advisers.* Lawrence: University Press of Kansas.

———. 1988. "The Mythology Surrounding Lyndon Johnson, His Advisers, and the 1965 Decision to Escalate the Vietnam War." *Political Science Quarterly,* 101: 637–63.

Bar-Siman-Tov, Y. 1994. *Israel and the Peace Process, 1977–1982: In Search of Legitimacy for Peace.* Albany: State University of New York Press.

———. 1990. "Peace as a Significant Change in Foreign Policy: The Need for Legitimacy." *Jerusalem Journal of International Relations,* 12: 13–30.

———. 1987. *Israel, The Superpowers, and the War in the Middle East.* New York: Praeger.

———. 1984. "The Strategy of War by Proxy." *Cooperation and Conflict,* 19: 263–73.

Bar-Tal, D. 1990. *Group Beliefs: A Conception for Analyzing Group Structure, Processes, and Behavior.* New York: Springer-Verlag.

Barzilai, G. 1992. *A Democracy in Wartime: Conflict and Consensus in Israel.* Tel Aviv: Sifriat Poalim (in Hebrew).

Bass, B. M. 1985. *Leadership and Performance Beyond Expectations*. New York: Free Press.

Bavly, D., and E. Salpeter. 1984. *Fire in Beirut: Israel's War in Lebanon with the PLO*. New York: Stein and Day.

Bazerman, M. H., T. Giuliano, and A. Appelman. 1984. "Escalation of Commitment in Individual and Group Decision-Making." *Organizational Behavior and Human Performance*, 33: 141–52.

Beach, L. R., T. R. Mitchell, T. S. Paluchowski, and E. H. Van Zee. 1992. "Image Theory: Decision Framing and Decision Deliberation." In F. Heller, ed., *Decision-Making and Leadership*, pp. 172–88. Cambridge, Eng.: Cambridge University Press.

Beck, R. J. 1993. *The Grenada Invasion: Politics, Law, and Foreign Policy Decisionmaking*. Boulder, Colo.: Westview Press.

Beck, U. 1992. *Risk Society: Towards a New Modernity*. London: Sage.

Belbutowski, P. M. 1996. "Strategic Implications of Cultures in Conflict." *Parameters*, 26(1): 32–42.

Beloff, M. 1970. "Reflections on Intervention." In M. Beloff, ed., *The Intellectual in Politics*, pp. 225–34. London: Wiedenfeld and Nicholson.

Ben-Meir, Y. 1987. *National Security Decisionmaking: The Israeli Case*. Tel Aviv: Hakibbutz Hameuchad (in Hebrew).

Benziman, U. 1985. *Sharon: An Israeli Caesar*. New York: Adama Books.

Ben Zur, H., and S. J. Breznitz. 1981. "The Effect of Time Pressure on Risky Choice Behavior." *Acta Psychologica*, 47: 89–104.

Ben-Zvi, A. 1993. *The United States and Israel: The Limits of the Special Relationship*. New York: Columbia University Press.

Berger, C., and W. B. Gudykunst. 1991. "Uncertainty and Communication." In B. Dervin and M. J. Voigt, eds., *Progress in Communication Science*, vol. 10, pp. 21–66. Norwood, N.J.: ABLEX.

Berman, L. 1992. "From Intervention to Disengagement: The United States in Vietnam." In A. E. Levite, B. W. Jentleson, and L. Berman, eds., *Foreign Military Intervention: The Dynamics of Protracted Conflict*, pp. 23–64. New York: Columbia University Press.

———. 1989. *Lyndon Johnson's War: The Road to Stalemate in Vietnam*. New York: W. W. Norton.

———. 1982 *Planning a Tragedy: The Americanization of the Vietnam War*. New York: W. W. Norton.

Berman, L., and B. W. Jentleson. 1991. "Bush and the Post–Cold War World: New Challenges for American Leadership." In C. Campbell and B. A. Rockman, eds., *The Bush Presidency: First Appraisals*, pp. 93–128. Chatham, N.J.: Chatham House Publishers.

Betts, R. K. 1988. "Nuclear Peace and Conventional War." *Journal of Strategic Studies*, 11: 79–95.

———. 1977. *Soldiers, Statesmen, and Cold War Crises*. Cambridge, Mass.: Harvard University Press.

Beyth-Marom, R. 1982. "How Probable Is Probable? A Numerical Translation of Verbal Probability Expressions." *Journal of Forecasting*, 1: 257–69.

Bilder, R. B. 1981. *Managing the Risks of International Agreement*. Madison: University of Wisconsin Press.

Billings, R. S., and L. L. Scherer. 1988. "The Effects of Response Mode and Importance on Decision-Making Strategies: Judgment Versus Choice." *Organizational Behavior and Human Decision Processes*, 41: 1–19.

Billings-Yun, M. 1988. *Decisions Against War: Eisenhower and Dien Bien Phu, 1954.* New York: Columbia University Press.

Birren, J. E., and L. M. Fisher. 1990. "The Elements of Wisdom: Overview and Integration." In R. J. Sternberg, ed., *Wisdom: Its Nature, Origins, and Development*, pp. 317–32. Cambridge, Eng.: Cambridge University Press.

Björkman, M. 1987. "Time and Risk in the Cognitive Space." In L. Sjöberg, ed., *Risk and Society: Studies of Risk Generation and Reaction to Risk*, pp. 13–35. London: Allen and Unwin.

———. 1984. "Decision Making, Risk Taking and Psychological Time: Review of Empirical Findings and Psychological Theory." *Scandinavian Journal of Psychology*, 25: 31–49.

Blaylock, B. K. 1985. "Risk Perception: Evidence of an Interactive Process." *Journal of Business Research*, 13: 207–21.

Blechman, B. M., and S. S. Kaplan. 1978. *Force Without War: U.S. Armed Forces as a Political Instrument.* Washington, D.C.: Brookings Institution.

Bobrow, D. B. 1977. "Communications, Command, and Control: The Nerves of Intervention." In E. P. Stern, ed., *The Limits of Military Intervention*, pp. 101–20. Beverly Hills, Calif.: Sage.

Boettcher, W. A. 1995. "Context, Methods, Numbers, and Words: Prospect Theory in International Relations." *Journal of Conflict Resolution*, 39: 561–83.

Booth, K. 1979. *Strategy and Ethnocentrism.* London: Croom Helm.

Borgida, E., and N. Brekke. 1981. "The Base Rate Fallacy in Attribution and Prediction." In J. H. Harvey, W. J. Ickes, and R. F. Kidd, eds., *New Directions in Attribution Research*, vol. 3, pp. 63–95. Hillsdale, N.J.: Lawrence Erlbaum.

Bovens, M., and P. 't Hart. 1996. *Understanding Policy Fiascoes.* New Brunswick, N.J.: Transaction Publishers.

Bradbury, J. A. 1989. "The Policy Implications of Differing Concepts of Risk." *Science, Technology and Human Values*, 14: 381–99.

Brands, H. W., Jr. 1987–88. "Decisions on American Armed Intervention: Lebanon, Dominican Republic, and Grenada." *Political Science Quarterly*, 102: 607–24.

———. 1985. "Johnson and Eisenhower: The President, the Former President, and the War in Vietnam." *Presidential Quarterly*, 15: 589–601.

Brecher, M. 1993. *Crises in World Politics: Theory and Reality.* Oxford, Eng.: Pergamon Press.

Brecher, M., with B. Geist. 1980. *Decisions in Crises: Israel, 1967 and 1973.* Berkeley: University of California Press.

Brecher, M., B. Steinberg, and J. Stein. 1969. "A Framework for Research on Foreign Policy Behavior." *Journal of Conflict Resolution*, 13: 75–101.

Brecher, M., J. Wilkenfeld, and S. Moser. 1988. *Crises in the Twentieth Century: Handbook of International Crises*, vol. 1. Oxford, Eng.: Pergamon Press.

Brehmer, B. 1987. "The Psychology of Risk." In W. T. Singleton and J. Hovden, eds., *Risk and Decisions*, pp. 25–39. Chichester, Eng.: John Wiley and Sons.

Breslauer, G. W. 1982. *Khrushchev and Brezhnev as Leaders: Building Authority in Soviet Politics.* London: Allen and Unwin.

Briggs, C. S. 1990. *Operation Just Cause—Panama, December 1989: A Soldier's Eyewitness Account.* Harrisburg, Pa.: Stackpole Books.

Brockner, J. 1992. "The Escalation of Commitment to a Failing Course of Action: Toward Theoretical Progress." *Academy of Management Review*, 17: 39–61.

Brockner, J., and J. Z. Rubin. 1985. *Entrapment in Escalating Conflicts: A Social Psychological Analysis*. New York: Springer-Verlag.

Bromiley, P., and S. P. Curley. 1992. "Individual Differences in Risk Taking." In J. F. Yates, ed., *Risk-Taking Behavior*, pp. 87–132. Chichester, Eng.: John Wiley and Sons.

Brown, R. W. 1965. *Social Psychology*. New York: Free Press.

Brun, W., and K. H. Teigen. 1988. "Verbal Probabilities: Ambiguous, Context-Dependent, or Both?" *Organizational Behavior and Human Decision Processes*, 41: 390–404.

Brunsson, N. 1985. *The Irrational Organization*. Chichester, Eng.: John Wiley and Sons.

Brzoska, M., and F. S. Pearson, 1994. "Developments in the Global Supply of Arms: Opportunity and Motivations." *Annals of the American Academy of Political and Social Science*, 535: 58–72.

Buckley, K. 1991. *Panama: The Whole Story*. New York: Simon and Schuster.

Budescu, D. V., and T. S. Wallsten. 1985. "Consistency in Interpretation of Probabilistic Phrases." *Organizational Behavior and Human Decision Processes*, 36: 391–405.

Buehler, R., D. Grifin, and M. Ross. 1994. "Exploring the 'Planning Fallacy': Why People Underestimate Their Task Completion Times." *Journal of Personality and Social Psychology*, 67: 366–81.

Bueno de Mesquita, B. 1985. "Toward a Scientific Understanding of International Conflict: A Personal View." *International Studies Quarterly*, 29: 121–36.

———. 1981. *The War Trap*. New Haven, Conn.: Yale University Press.

Bukszar, E., and T. Connolly. 1988. "Hindsight Bias and Strategic Choice: Some Problems in Learning from Experience." *Academy of Management Journal*, 31: 628–41.

Bull, H., ed. 1986. *Intervention in World Politics*. Oxford, Eng.: Clarendon Press.

Burgin, E. 1992. "Congress, the War Powers Resolution, and the Invasion of Panama." *Polity*, 25: 217–42.

Burke, J. P. 1984. "Responsibilities of Presidents and Advisers: A Theory and Case Study of Vietnam Decision-Making." *Journal of Politics*, 46: 818–44.

Burke, J. P., and F. I. Greenstein with L. Berman and R. Immerman. 1989. *How Presidents Test Reality: Decisions on Vietnam, 1954 and 1965*. New York: Russell Sage Foundation.

Burlatsky, F. 1989. "Leonid Brezhnev, Reflections on Leadership." *Soviet Life*, Feb.: 37–41.

Burns, J. M. 1978. *Leadership*. New York: Harper and Raw.

Burnstein, E., and M. L. Berbaum. 1983. "Stages in Group Decision Making: The Decomposition of Historical Narratives." *Political Psychology*, 4: 531–61.

Burnstein, E., and S. Katz. 1971. "Individual Commitment to Risky and Conservative Choices as a Determinant of Shifts in Group Decisions." *Journal of Personality*, 39: 564–80.

Burrowes, R. A. 1988. *Revolution and Rescue in Grenada*. New York: Greenwood Press.

Buzan, B., C. Jones, and R. Little. 1993. *The Logic of Anarchy: Neorealism to Structural Realism*. New York: Columbia University Press.

Cable, L. E. 1993. "The Operation Was a Success, but the Patient Died: The Air War in Vietnam, 1964–1969." In D. E. Showalter and J. G. Albert, eds., *An American Dilemma: Vietnam, 1964–1973*, pp. 109–58. Chicago: Imprint.

―――. 1991. *Unholy Grail: The U.S. and the Wars in Vietnam, 1965–68*. London: Routledge.

―――. 1986. *Conflict of Myths: The Development of American Counterinsurgency Doctrine and the Vietnam War*. New York: New York University Press.

Cameron, B., and J. L. Myers. 1966. "Some Personality Correlates of Risk Taking." *Journal of General Psychology*, 74: 51–60.

Campbell, C. 1991. "The White House and Cabinet Under the 'Let's Deal' Presidency." In C. Campbell and B. A. Rockman, eds., *The Bush Presidency: First Appraisals*, pp. 185–222. Chatham, N.J.: Chatham House Publishers.

Campbell, J. C. 1989. "The Soviet Union, the United States, and the Twin Crises of Hungary and Suez." In Wm. R. Louis and R. Owen, eds., *Suez 1956: The Crisis and Its Consequences*, pp. 233–53. Oxford, Eng.: Clarendon Press.

Cannon, L. 1992. *President Reagan: The Role of a Lifetime*. New York: Simon and Schuster.

Cannon, M. W. 1989. "Raising the Stakes: The Taylor-Rostow Mission." *Journal of Strategic Studies*, 12: 125–65.

Carpenter, T. G. 1989. "Direct Military Intervention." In P. J. Schraeder, ed., *Intervention in the 1980s: U.S. Foreign Policy in the Third World*, pp. 131–44. Boulder, Colo.: Lynne Rienner.

Cartwright, D. 1971. "Risk Taking by Individuals and Groups: An Assessment of Research Employing Choice Dilemmas." *Journal of Personality and Social Psychology*, 20: 361–78.

Carus, W. S. 1985. "Military Lessons of the 1982 Israel-Syria Conflict." In R. E. Harkavy and S. G. Neuman, eds., *The Lessons of Recent Wars in the Third World*, pp. 261–80. Lexington, Mass.: Lexington Books.

Cervone, D., and P. K. Peake. 1986. "Anchoring, Efficacy and Action: The Influence of Judgmental Heuristics on Self-Efficacy Judgments and Behavior." *Journal of Personality and Social Psychology*, 50: 492–501.

Chaiken, S., A. Liberman, and A. H. Eagly. 1989. "Heuristic and Systematic Information Processing Within and Beyond the Persuasion Context." In J. S. Uleman and J. A. Bargh, eds., *Unintended Thought*, pp. 212–52. New York: Guilford Press.

Chan, S. 1984. "Mirror, Mirror on the Wall . . . Are the Free Countries More Pacific?" *Journal of Conflict Resolution*, 28: 617–48.

Chanda, N. 1986. *Brother Enemy: The War After the War*. San Diego, Calif.: Harcourt Brace Jovanovich.

Chapman, L. J., and J. P. Chapman. 1969. "Illusory Correlation as an Obstacle to the Use of Valid Psychodiagnostic Signs." *Journal of Abnormal Psychology*, 74: 271–80.

Chen, M. 1992. *The Strategic Triangle and Regional Conflicts: Lessons from the Indochina Wars*. Boulder, Colo.: Lynn Rienner.

Christensen, K. S. 1985. "Coping with Uncertainty in Planning." *Journal of the American Planning Association*, 51: 63–73.

Christensen, T. J. 1993. "Conclusion: System Stability and the Security of the Most Vulnerable Significant Actor." In J. Snyder and R. Jervis, eds., *Coping with Complexity in the International System*, pp. 329–56. Boulder, Colo.: Westview Press.

Clark, D. A. 1990. "Verbal Uncertainty Expressions: A Critical Review of Two Decades of Research." *Current Psychology: Research and Reviews*, 9: 203–35.

Clark, I. 1989. *The Hierarchy of States: Reform and Resistance in the International Order*. Cambridge, Eng.: Cambridge University Press.

Clarke, L. 1992. "Context Dependency and Risk Decision Making." In J. F. Short and L. Clarke, eds., *Organizations, Uncertainties, and Risk*, pp. 27–38. Boulder, Colo.: Westview Press.

Clifford, C., with R. Holbrooke. 1991. *Counsel to the President*. New York: Random House.

Clodfelter, M. 1989. *The Limits of Air Power: The American Bombing of North Vietnam*. New York: Free Press.

Cohen, E. A. 1992. "Dynamics of Military Intervention." In A. E. Levite, B. W. Jentleson, and L. Berman, eds., *Foreign Military Intervention: The Dynamics of Protracted Conflict*, pp. 261–84. New York: Columbia University Press.

———. 1986. "Constraints on America's Conduct of Small Wars." In S. E. Miller, ed., *Conventional Forces and American Defense Policy*, pp. 277–308. Princeton, N.J.: Princeton University Press.

———. 1985. *Citizens and Soldiers: The Dilemmas of Military Service*. Ithaca, N.Y.: Cornell University Press.

Cohen, M. D., J. G. March, and J. P. Olsen. 1972. "A Garbage Can Model of Organizational Choice." *Administrative Science Quarterly*, 17: 1–25.

Cohen, R. 1990. *Culture and Conflict in Egyptian-Israeli Relations: A Dialogue of the Deaf*. Bloomington: Indiana University Press.

Cohen, W. I. 1980. *Dean Rusk*. Totowa, N.J.: Cooper Square.

Collins, B. E., and M. F. Hoyt. 1972. "Personal Responsibility-for-Consequences: An Integration and Extension of the 'Forced Compliance' Literature." *Journal of Experimental Social Psychology*, 8: 558–93.

Connaughton, R. 1992. *Military Intervention in the 1990s: A New Logic of War*. London: Routledge.

Cook, K. S., and M. Levi, eds. 1990. *The Limits of Rationality*. Chicago: University of Chicago Press.

Cooper, C. 1970. *The Lost Crusade: America in Vietnam*. New York: Dodd, Mead.

Cooper, J. 1971. "Personal Responsibility and Dissonance: The Role of Foreseen Consequences." *Journal of Personality and Social Psychology*, 18: 354–63.

Cooper, R. N. 1986. "Managing Risks to the International Economic System." In R. N. Cooper, ed., *Economic Policy in an Interdependent World*. Cambridge, Mass.: MIT Press.

Corbin, R. M. 1980. "Decisions That Might Not Get Made." In T. S. Wallsten, ed., *Cognitive Processes in Choice and Decision Behavior*, pp. 47–67. Hillsdale, N.J.: Lawrence Erlbaum.

Cottam, M. L. 1994. *Images and Intervention: U.S. Policies in Latin America*. Pittsburgh, Pa.: University of Pittsburgh Press.

Cross, J. G., and M. J. Guyer. 1980. *Social Traps*. Ann Arbor: University of Michigan Press.

Crowell, L. 1991. "The Anatomy of Just Cause: The Forces Involved, the Adequacy of Intelligence, and Its Success as a Joint Operation." In B. W. Watson and P. G. Tsouras, eds., *Operation Just Cause: The U.S. Intervention in Panama*, pp. 67–104. Boulder, Colo.: Westview Press.

Crozier, B., ed. 1987. *The Grenada Documents*. London: Sherward Press.

Curley, S. P., J. F. Yates, and R. A. Abrams. 1986. "Psychological Sources of Ambiguity Avoidance." *Organizational Behavior and Human Decision Processes*, 38: 230–56.

Cypher, D. 1985a. "Grenada: Indications, Warning, and the U.S. Response." In P. M. Dunn and B. W. Watson, eds., *American Intervention in Grenada: The Implications of Operation "Urgent Fury,"* pp. 45–54. Boulder, Colo.: Westview Press.

———. 1985b. "Urgent Fury: The U.S. Army in Grenada." In P. M. Dunn and B. W. Watson, eds., *American Intervention in Grenada: The Implications of Operation "Urgent Fury,"* pp. 99–108. Boulder, Colo.: Westview Press.

Dagget, S. 1989. "Government and the Military Establishment." In P. J. Schraeder, ed., *Intervention in the 1980s: U.S. Foreign Policy in the Third World*, pp. 161–74. Boulder, Colo.: Lynne Rienner.

Dake, K. 1992. "Myths of Nature: Culture and the Social Construction of Risk." *Journal of Social Issues*, 48(4): 21–37.

D'andrade, R. 1990. "Some Propositions About the Relations Between Culture and Human Cognition." In J. W. Stigler, R. A. Shweder, and G. Herdt, eds., *Cultural Psychology: Essays on Comparative Human Development*, pp. 65–129. Cambridge, Eng.: Cambridge University Press.

Daniels, R. V. 1982. "The Two Faces of Brezhnev." *New Leader*, Nov. 29: 6–8.

David, C.-P. 1996. "Who Was the Real George Bush? Foreign Policy Decision-Making Under the Bush Administration." *Diplomacy and Statecraft*, 7: 197–220.

David, S. R. 1989. "Why the Third World Matters." *International Security*, 14(1): 50–85.

———. 1987. "The Use of Proxy Forces by Major Powers in the Third World." In S. G. Neuman and R. E. Harkavy, eds., *The Lessons of Recent Wars in the Third World*, vol. 2, pp. 199–226. Lexington, Mass.: D. C. Heath.

Davidson, P. B. 1993. "The American Military's Assessment of Vietnam, 1964–1992." In D. E. Showalter and J. G. Albert, eds., *An American Dilemma: Vietnam, 1964–1973*, pp. 53–61. Chicago: Imprint.

———. 1988. *Vietnam at War: The Victory, 1946–1975*. Navato, Calif.: Presidio Press.

Davis, J. H. 1992. "Some Compelling Intuitions About Group Consensus Decisions, Theoretical and Empirical Research, and Interpersonal Aggregation Phenomena: Selected Examples, 1950–1990." *Organizational Behavior and Human Decision Processes*, 52: 3–38.

Davis, M. A., and P. Bobko. 1986. "Contextual Effects on Escalation Processes in Public Sector Decision-Making." *Organizational Behavior and Human Decision Processes*, 37: 121–38.

Davis, M. T. 1987. *40 Km into Lebanon: Israel's 1982 Invasion*. Washington, D.C.: National Defense University Press.

Davis, P. K., and J. Arquilla. 1991a. *Deterring or Coercing Opponents in Crisis: Lessons from the War with Saddam Hussein*. Santa Monica, Calif.: Rand Corporation, R-4111-JS.

———. 1991b. *Thinking About Opponent Behavior in Crisis and Conflict: A Generic Model for Analysis and Group Discussion*. Santa Monica, Calif.: Rand Corporation, N-3322-JS.

Davis, W. L., and E. J. Phares. 1967. "Internal-External Control as a Determinant of Information-Seeking in a Social Influence Situation." *Journal of Personality*, 35: 547–61.

Dawisha, A. I. 1980. *Syria and the Lebanon Crisis*. London: Macmillan.

Dawisha, K. 1984. *The Kremlin and the Prague Spring*. Berkeley: University of California Press.

———. 1980a. "The Limits of the Bureaucratic Politics Model: Observations on the Soviet Case." *Studies in Comparative Communism*, 13: 300–46.

———. 1980b. "Soviet Security and the Role of the Military: The 1968 Czechoslovak Crisis." *British Journal of Political Science*, 10: 341–63.

Day, A. R. 1986. "Conclusions: A Mix of Means." In A. R. Day and M. W. Doyle, eds., *Escalation and Intervention: Multilateral Security and Its Alternatives*, pp. 152–71. Boulder, Colo.: Westview Press.

Dean, J. W., and M. P. Sharfman. 1996. "Does Decision Process Matter?" A Study of Decision-Making Effectiveness." *Academy of Management Journal*, 39: 368–96.

DeBenedetti, C., with C. Chatfield. 1990. *An American Ordeal: The Antiwar Movement of the Vietnam Era*. Syracuse, N.Y.: Syracuse University Press.

Deibel, T. L. 1992. "Strategies Before Containment: Patterns for the Future." *International Security*, 16(4): 79–108.

———. 1991. "Bush's Foreign Policy: Mastery and Inaction." *Foreign Policy*, no. 84: 3–23.

Demchak, C. C. 1991. *Military Organizations, Complex Machines: Modernization in the U.S. Armed Services*. Ithaca, N.Y.: Cornell University Press.

Dickson, G. C. A. 1981. "A Comparison of Attitudes Towards Risk Among Business Managers." *Journal of Occupational Psychology*, 54: 157–64.

Dickson, J. W. 1978. "Perception of Risk as Related to Choice in a Two-Dimensional Risk Situation." *Psychological Reports*, 43: 1059–62.

Diener, E. 1977. "Deindividuation: Causes and Consequences." *Social Behavior and Personality*, 15: 143–55.

DiLeo, D. L. 1991. *George Ball, Vietnam, and the Rethinking of Containment*. Chapel Hill, N.C.: University of North Carolina Press.

Dinges, J. 1991. *Our Man in Panama*. New York: Random House.

Dion, K. L., R. S. Baron, and N. Miller. 1970. "Why Do Groups Make Riskier Decisions Than Individuals?" In L. Berkowitz, ed., *Advances in Experimental Social Psychology*, vol. 5, pp. 305–77. New York: Academic Press.

Donnelly, T., M. Roth, and C. Baker. 1991. *Operation Just Cause: The Storming of Panama*. New York: Lexington Books.

Doran, C. F. 1989. "The Globalist-Regionalist Debate." In P. J. Schraeder, ed., *Intervention in the 1980s: U.S. Foreign Policy in the Third World*, pp. 45–59. Boulder, Colo.: Lynne Rienner.

Dornberg, J. 1974. *Brezhnev: The Masks of Power*. New York: Basic Books.

Doty, W. G. 1986. *Mythography: The Study of Myths and Rituals*. Tuscaloosa: University of Alabama Press.

Douglas, M. 1990. "Risk as a Forensic Resource." *Daedalus*, 119(4): 1–16.

———. 1985. *Risk Acceptability According to the Social Sciences*. Occasional Reports on Current Topics, no. 11. New York: Russell Sage Foundation.

Douglas, M., and A. Wildavsky. 1982. *Risk and Culture*. Berkeley: University of California Press.

Downey, H. K., and J. W. Slocum. 1975. "Uncertainty: Measures, Research and Sources of Variation." *Academy of Management Journal*, 18: 562–78.

Downs, G. W. 1992. "The Lessons of Disengagement." In A. E. Levite, B. W. Jentleson, and L. Berman, eds., *Military Intervention: The Dynamics of Protracted Conflict*, pp. 285–300. New York: Columbia University Press.

Doyle, M. W. 1986. "Introduction." In A. R. Day and M. W. Doyle, eds., *Escalation and Intervention: Multilateral Security and Its Alternatives*, pp. 1–10. Boulder, Colo.: Westview Press.

Dror, Y. 1994. "Statecraft as Prudent Risk Taking: The Case of the Middle East Peace Process." *Journal of Contingencies and Crisis Management*, 2: 126–35.

———. 1988. "Uncertainty: Coping with It and with Political Feasibility." In H. J. Miser and E. S. Quade, eds., *Handbook of Systems Analysis: Craft, Issues, and Procedural Choices*, pp. 247–81. New York: John Wiley and Sons.

———. 1986. *Policymaking Under Adversity*. New Brunswick, N.J.: Transaction Books.

———. 1971. *Crazy States*. Lexington, Mass.: Heath Lexington Books.

Drummond, H. 1995. "De-escalation in Decision Making: A Case of a Disastrous Partnership." *Journal of Management Studies*, 32: 265–81.

Dubček, A., with J. Hochman. 1993. *Hope Dies Last: The Autobiography of Alexander Dubček*. New York: Kodansha.

Duffy, M., and D. Goodgame. 1992. *Marching in Place: The Status Quo Presidency of George Bush*. New York: Simon and Schuster.

Duiker, W. J. 1994. *U.S. Containment Policy and the Conflict in Indochina*. Stanford, Calif.: Stanford University Press.

Dunér, B. 1987. *The Bear, the Cubs and the Eagle: Soviet Bloc Interventionism in the Third World and the U.S. Response*. Aldershot, Eng.: Gower.

———. 1985. *Military Intervention in Civil Wars: The 1970s*. Aldershot, Eng.: Gower.

———. 1983a. "The Intervener: Lone Wolf or . . . ? Cooperation Between Interveners in Civil Wars." *Cooperation and Conflict*, 18: 197–213.

———. 1983b. "The Many-Pronged Spear: External Military Intervention in Civil Wars in the 1970s." *Journal of Peace Research*, 20: 59–72.

———. 1981. "Proxy Intervention in Civil Wars." *Journal of Peace Research*, 18: 353–61.

Dunn, P. M., and B. W. Watson, eds. 1985. *American Intervention in Grenada: The Implications of Operation "Urgent Fury."* Boulder, Colo.: Westview Press.

Dunning, D., D. W. Griffin, J. D. Milojkovic, and L. Ross. 1990. "The Over-confidence Effect in Social Prediction." *Journal of Personality and Social Psychology*, 58: 568–81.

Dupuy, T. N. 1970. *The Almanac of World Military Power*. Dunn Loring, Va.: T. N. Dupuy.

Ebbesen, E. B., and V. J. Konečni. 1980. "On the External Validity of Decision-Making Research: What Do We Know About Decisions in the Real World?" In T. S. Wallsten, ed., *Cognitive Processes in Choice and Decision Behavior*, pp. 21–45. Hillsdale, N.J.: Lawrence Erlbaum.

Ebinger, C.1976. "External Intervention in Internal War: The Politics and Diplomacy of the Angolan War." *Orbis*, 20: 669–99.

Eckstein, H. 1975. "Case Study and Theory in Political Science." In F. I. Greenstein and N. W. Polsby, eds., *Handbook of Political Science*, vol. 7, pp. 79–138. Reading, Mass.: Addison-Wesley.

Edel, W. 1992. *The Reagan Presidency: An Actor's Finest Performance*. New York: Hippocrene Books.

Eidlin, F. H. 1984. "Misperception, Ambivalence, and Indecision in Soviet Policy-Making: The Case of the 1968 Invasion of Czechoslovakia." *Conflict*, 5: 89–117.

———. 1981. " 'Capitalization,' 'Resistance' and the Framework of Normalization: The August 1968 Invasion of Czechoslovakia and the Czechoslovak Response." *Journal of Peace Research*, 18: 319–32.

———. 1980. *The Logic of "Normalization."* Boulder, Colo.: East European Monographs.

Einhorn, H. J., and R. M. Hogarth. 1986. "Decision-Making Under Ambiguity." *Journal of Business*, 59: S225–50.

———. 1981. "Behavioral Decision Theory Processes of Judgment and Choice." In M. R. Rosenzweig and L. W. Porter, eds., *Annual Review of Psychology*, vol. 32, pp. 52–88. Palo Alto, Calif.: Annual Reviews.

———. 1978. "Confidence in Judgment: Persistence of the Illusion of Validity." *Psychological Review*, 85: 395–416.

Eisenhower, D. D. 1963. *Mandate for Change, 1953–1956*. New York: Doubleday.

Eitan, R., with D. Goldstein. 1985. *A Soldier's Story*. Tel Aviv: Ma'ariv Library (in Hebrew).

Eley, J. W. 1972. "Toward a Theory of Intervention." *International Studies Quarterly*, 16: 245–56.

Elliot, A. J., and P. G. Devine. 1994. "On the Motivational Nature of Cognitive Dissonance: Dissonance as Psychological Discomfort." *Journal of Personality and Social Psychology*, 67: 382–94.

Elliot, D. W. P. 1993. "Hanoi's Strategy in the Second Indochina War." In J. S. Werner and L. D. Huynh, eds., *The Vietnam War: Vietnamese and American Perspectives*, pp. 66–94. Armonk, N.Y.: M. E. Sharpe.

Ellsberg, D. 1971. "The Quagmire Myth and the Stalemate Machine." *Public Policy*, 19: 217–74.

———. 1961. "Risk, Ambiguity and the Savage Axioms." *Quarterly Journal of Economics*, 75: 643–69.

Elrod, R. B. 1976. "The Concert of Europe: A Fresh Look at an International System." *World Politics*, 28: 159–74.

Elster, J. 1989. *Solomonic Judgments: Studies in the Limitations of Rationality*. Cambridge, Eng.: Cambridge University Press.

Engelhardt, M. J. 1989. "America Can Win, Sometimes: U.S. Success and Failure in Small Wars." *Conflict Quarterly*, 9(1): 20–35.

Epstein, J. M. 1987. *Strategy and Force Planning: The Case of the Persian Gulf*. Washington, D.C.: Brookings Institution.

Etheredge, L. S. 1985. *Can Governments Learn? American Foreign Policy and Central American Revolutions*. New York: Pergamon Press.

Etheredge, L. S., and J. Short. 1983. "Thinking About Government Learning." *Journal of Management Studies*, 20: 41–58.

Euben, R. 1995. "When Worldviews Collide: Conflicting Assumptions About Human Behavior Held by Rational Actor Theory and Islamic Fundamentalism." *Political Psychology*, 16: 157–78.

Evans, J. St. B. T. 1989. *Bias in Human Reasoning: Causes and Consequences*. Hove, Eng.: Lawrence Erlbaum.

Evron, Y. 1987. *War and Intervention in Lebanon*. Baltimore, Md.: Johns Hopkins University Press.

Fabyanic, T. A. 1981. "Conceptual Planning and the Rapid Deployment Joint Task Force." *Armed Forces and Society*, 7: 343–65.

———. 1977. "Manpower, Military Intervention, and the All-Volunteer Force." In

E. P. Stern, ed., *The Limits of Military Intervention*, pp. 281–302. Beverly Hills, Calif.: Sage.

Fagley, N. S., and P. M. Miller. 1987. "The Effects of Decision Framing on Choice of Risky vs. Certain Options." *Organizational Behavior and Human Decision Processes*, 39: 264–77.

Farnham, B. 1995. "The Impact of the Political Context on Foreign Policy Decision-Making." Paper prepared for the 36th International Studies Association Annual Convention, Chicago, Feb. 21–25.

———. 1992. "Roosevelt and the Munich Crisis: Insights from Prospect Theory." *Political Psychology*, 13: 205–35.

———. 1990. "Political Cognition and Decision-Making." *Political Psychology*, 11: 83–111.

Fearon, J. D. 1994. "Domestic Political Audiences and the Escalation of International Disputes." *American Political Science Review*, 88: 577–92.

———. 1991. "Counterfactuals and Hypothesis Testing in Political Science." *World Politics*, 43: 169–95.

Feldman, S. 1992. "Israel's Involvement in Lebanon: 1975–1985." In A. E. Levite, B. W. Jentleson, and L. Berman, eds., *Foreign Military Intervention: The Dynamics of Protracted Conflict*, pp. 129–61. New York: Columbia University Press.

Feldman, S., and H. Rechnitz-Kijner. 1984. *Deception, Consensus and War: Israel in Lebanon*. Paper no. 27. Tel Aviv: Jaffee Center for Strategic Studies, Tel Aviv University.

Ferguson, Y., and W. Mansbach. 1991. "Between Celebration and Despair: Constructive Suggestions for Future International Theory." *International Studies Quarterly*, 35: 363–86.

Feste, K. A. 1992. *Expanding the Frontiers: Superpower Intervention in the Cold War*. New York: Praeger.

Festinger, L. 1957. *Conflict, Decision and Dissonance*. Stanford, Calif.: Stanford University Press.

Finer, H. 1964. *Dulles over Suez: The Theory and Practice of His Diplomacy*. Chicago: Quadrangle Books.

Fischhoff, B. 1994. "What Forecasts (Seem to) Mean." *International Journal of Forecasting*, 10: 387–403.

———. 1992. "Risk Taking: A Developmental Perspective." In J. F. Yates, ed., *Risk-Taking Behavior*, pp. 133–62. Chichester, Eng.: John Wiley and Sons.

———. 1991. "Nuclear Decisions: Cognitive Limits to the Thinkable." In P. Tetlock, J. L. Husbands, R. Jervis, P. C. Stern, and C. Tilly, eds., *Behavior, Society and Nuclear War*, vol. 2, pp. 110–92. New York: Oxford University Press.

———. 1983. "Strategic Policy Preferences: A Behavioral Decision Theory Perspective." *Journal of Social Issues*, 39(1): 133–60.

———. 1982. "For Those Condemned to Study the Past: Heuristics and Biases in Hindsight." In D. Kahneman, P. Slovic, and A. Tversky, eds., *Judgment Under Uncertainty: Heuristics and Biases*, pp. 335–51. Cambridge, Eng.: Cambridge University Press.

———. 1975. "Hindsight ≠ Foresight: The Effect of Outcome Knowledge on Judgment Under Uncertainty." *Journal of Experimental Psychology: Human Perception and Performance*, 1: 288–99.

Fischhoff, B., and R. Beyth. 1975. " 'I Knew It Would Happen'—Remembered Probabilities of Once-Future Things." *Organizational Behavior and Human Performance*, 13: 1–16.

Fischhoff, B., A. Bostcom, and M. J. Quadrel. 1993. "Risk Perception and Communication." In J. S. Omenn, J. E. Fielding, and L. B. Lave, eds., *Annual Review of Public Health*, vol. 14, pp. 183–203. Palo Alto, Calif.: Annual Reviews.

Fischhoff, B., S. Lichtenstein, P. Slovic, S. L. Derby, and R. L. Keeney. 1981. *Acceptable Risk*. Cambridge, Eng.: Cambridge University Press.

Fischhoff, B., P. Slovic, and S. Lichtenstein. 1978. "Fault Trees: Sensitivity of Estimated Failure Probabilities to Problem Representation." *Journal of Experimental Psychology: Human Perception and Performance*, 4: 330–44.

———. 1977. "Knowing with Certainty: The Appropriateness of Extreme Confidence." *Journal of Experimental Psychology: Human Perception and Performance*, 3: 552–64.

Fisher, B. A. 1995. "The Reagan Reversal: America's Soviet Policy, 1981–1985." Manuscript, University of Toronto.

Fiske, S. T., and S. E. Taylor. 1984. *Social Cognition*. Reading, Mass.: Addison-Wesley.

FitzGerald, F. 1972. *Fire in the Lake: The Vietnamese and the Americans in Vietnam*. Boston: Little, Brown.

Flanagan, E. M. 1993. *Battle for Panama: Inside Operation Just Cause*. Washington, D.C.: Brassey's.

Follert, V. 1981. "Risk Analysis: Its Application to Argumentation and Decision-Making." *Journal of the American Forensic Association*, 18: 99–108.

Forgas, J. P. 1995. "Mood and Judgment: The Affect Infusion Model (AIM)." *Psychological Bulletin*, 117: 39–66.

———. 1989. "Mood Effects on Decision Making Strategies." *Australian Journal of Psychology*, 41: 197–214.

———. 1981. "Responsibility Attribution by Groups and Individuals: The Effects of the Interaction Episode." *European Journal of Social Psychology*, 11: 87–99.

Forman, E. M. 1972. "Civil War as a Source of International Violence." *Journal of Politics*, 34: 1111–34.

Forsyth, D. R., and K. N. Kelley. 1994. "Attribution in Groups: Estimation of Personal Contributions to Collective Endeavors." *Small Group Research*, 25: 367–83.

Foster, G. D. 1983. "On Selective Intervention." *Strategic Review*, 6(4): 48–63.

Franck, T. W., and E. Weisband. 1971. *World Politics: Verbal Strategy Among the Superpowers*. New York: Oxford University Press.

Freedman, L. 1996. "Vietnam and the Disillusioned Strategist." *International Affairs*, 72: 133–51.

Freudenburg, W. R. 1992. "Heuristics, Biases, and the Not-so-General Publics: Expertise and Error in the Assessment of Risks." In S. Krimsky and D. Golding, eds., *Social Theories of Risk*, pp. 229–49. Westport, Conn.: Praeger.

———. 1988. "Perceived Risk, Real Risk: Social Science and the Art of Probabilistic Risk Assessment." *Science*, 242: 44–49.

Fuller, S. R., and R. J. Aldag. 1997. "Challenging the Mindguards: Moving Small Group Analysis Beyond Groupthink." In P. 't Hart, E. K. Stern, and B. Sundelius, eds., *Beyond Groupthink: Political Group Dynamics and Foreign Policy-Making*. Ann Arbor: Michigan University Press.

Gabriel, R. A. 1984. *Operation Peace for Galilee*. New York: Hill and Wang.

Gacek, C. M. 1994. *The Logic of Force: Limited Wars in American Foreign Policy*. New York: Columbia University Press.

Gaddis, J. L. 1992–93. "International Relations Theory and the End of the Cold War." *International Security*, 17(3): 5–58.

Gaenslen, F. 1993. "Decision Makers as Social Beings: Consensual Decision Making in Russia, China, and Japan." Paper prepared for the 89th Annual Meeting of the American Political Science Association, Washington, D.C., Sept. 2–5.

———. 1986. "Culture and Decision Making in China, Japan, Russia and the United States." *World Politics*, 39: 78–103.

Gaiduk, I. V. 1995–96. "The Vietnam War and Soviet-American Relations, 1964–73: New Russian Evidence." *Cold War International History Project Bulletin*, nos. 6–7 (winter): 232, 250–58.

Galloway, J. 1970. *The Gulf of Tonkin Resolution*. Rutherford, N.J.: Fairleigh Dickinson University Press.

Gallucci, R. L. 1975. *Neither Peace nor Honor*. Baltimore, Md.: Johns Hopkins University Press.

Ganzach, Y. 1994. "Inconsistency and Uncertainty in Multi-attribute Judgment of Human Performance." *Journal of Behavioral Decision Making*, 7: 193–211.

Gärdenfors, P., and N. E. Sahlin. 1982. "Unreliable Probabilities, Risk-Taking, and Decision-Making." *Synthese*, 53: 361–86.

Gardner, L. C. 1995. *Pay Any Price: Lyndon Johnson and the Wars for Vietnam*. Chicago: Ivan R. Dee.

———. 1988. *Approaching Vietnam: From World War II Through Dienbienphu, 1941–1954*. New York: W. W. North.

Gelb, L. H., with R. K. Betts. 1979. *The Irony of Vietnam: The System Worked*. Washington, D.C.: Brookings Institution.

Gelman, H. 1984. *The Brezhnev Politburo and the Decline of Detente*. Ithaca, N.Y.: Cornell University Press.

George, A. L. 1993. *Bridging the Gap: Theory and Practice in Foreign Policy*. Washington, D.C.: United States Institute of Peace Press.

———. 1986. "The Impact of Crisis-Induced Stress on Decision Making." In F. Solomon and R. Q. Marston, eds., *The Medical Implications of Nuclear War*, pp. 529–52. Washington, D.C.: National Academy Press.

———. 1980a. "Domestic Constraints on Regime Change in U.S. Foreign Policy: The Need for Policy Legitimacy." In O. R. Holsti, R. M. Siverson, and A. L. George, eds., *Change in the International System*, pp. 233–62. Boulder, Colo.: Westview Press.

———. 1980b. *Presidential Decisionmaking in Foreign Policy: The Effective Use of Information and Advice*. Boulder, Colo.: Westview Press.

———. 1979a. "Case Studies and Theory Development: The Method of Structured, Focused Comparison." In P. G. Lauren, ed., *Diplomacy: New Approaches in History, Theory and Policy*, pp. 43–68. New York: Free Press.

———. 1979b. "The Causal Nexus Between Cognitive Beliefs and Decision-Making Behavior: The 'Operational Code' Belief System." In L. S. Falkowski, ed., *Psychological Models in International Politics*, pp. 95–124. Boulder, Colo.: Westview Press.

———. 1969. "The 'Operational Code': A Neglected Approach to the Study of Political Leaders and Decision-Making." *International Studies Quarterly*, 13: 190–222.

George, A. L., and T. J. McKeown. 1985. "Case Studies and Theories of Organizational Decision Making." In R. F. Coulam and R. A. Smith, eds., *Advances in Information Processing in Organizations*, vol. 2, pp. 21–58. Greenwich, Conn.: JAI Press.

George, A. L., and W. E. Simons, 1994a. "Findings and Conclusions." In A. L. George and W. E. Simons, eds., *The Limits of Coercive Diplomacy*, pp. 267–94. Boulder, Colo.: Westview Press.

———, eds. 1994b. *The Limits of Coercive Diplomacy*. Boulder, Colo.: Westview Press.

George, A. L., and R. Smoke. 1989. "Deterrence and Foreign Policy." *World Politics*, 41: 170–82.

Gerard, H. B. 1985. "When and How the Minority Prevails." In S. Moscovici, G. Mugny, and E. von Avermaet, eds., *Perspective on Minority Influence*, pp. 171–86. Cambridge, Eng.: Cambridge University Press.

Gerson, A. 1991. *The Kirkpatrick Mission: Diplomacy Without Apology. America at the United Nations, 1981–1985*. New York: Free Press.

Geyelin, P. 1966. *Lyndon B. Johnson and the World*. New York: Praeger.

Gibbons, W. C. 1995. *The U.S. Government and the Vietnam War: Executive and Legislative Roles and Relationships*, part IV: *July 1965–January 1968*. Princeton, N.J.: Princeton University Press.

———. 1989. *The U.S. Government and the Vietnam War: Executive and Legislative Roles and Relationships*, part III: *January–July 1965*. Princeton, N.J.: Princeton University Press.

———. 1986. *The U.S. Government and the Vietnam War: Executive and Legislative Roles and Relationships*, part II: *1961–1964*. Princeton, N.J.: Princeton University Press.

Gibson, J. W. 1986. *The Perfect War: Technowar in Vietnam*. Boston: Atlantic Monthly Press.

Giddens, A. 1990. *The Consequences of Modernity*. Cambridge, Eng.: Polity Press.

Gigerenzer, G., Z. Swijtink, T. Porter, L. Daston, J. Beatty, and L. Krüger. 1989. *The Empire of Chance: How Probability Changed Science and Everyday Life*. Cambridge, Eng.: Cambridge University Press.

Gilmore, W. C. 1984. *The Grenada Intervention: Analysis and Documentation*. New York: Facts on File Publications.

Gilovich, T. 1991. *How We Know What Isn't So: The Fallibility of Human Reason in Everyday Life*. New York: Free Press.

Glad, B. 1988. "The United States' Ronald Reagan." In B. Kellerman and J. Z. Rubin, eds., *Leadership and Negotiation in the Middle East*, pp. 200–29. New York: Praeger.

———. 1983. "Black-and-White Thinking: Ronald Reagan's Approach to Foreign Policy." *Political Psychology*, 4: 33–76.

Glad, B., and C. S. Taber. 1990. "Images, Learning, and the Decision to Use Force: The Domino Theory of the United States." In B. Glad, ed., *Psychological Dimensions of War*, pp. 56–81. Newbury Park, Calif.: Sage.

Glass, C. 1985. "The Setting." In P. M. Dunn and B. W. Watson, eds., *American Intervention in Grenada: The Implications of Operation "Urgent Fury,"* pp. 1–13. Boulder, Colo.: Westview Press.

Goertz, G., and P. F. Diehl. 1992. "Toward a Theory of International Norms: Some Conceptual and Measurement Issues." *Journal of Conflict Resolution*, 36: 634–64.

Golan, G. 1982–83. "The Soviet Union and the Israel Action in Lebanon." *International Affairs*, 59: 7–16.

———. 1973. *Reform Rule in Czechoslovakia: The Dubček Era, 1968–1969*. London: Cambridge University Press.

Goldgeier, J. M. 1990. "Soviet Leaders and International Crises: The Influence of Domestic Political Experience on Foreign Policy Strategies." Ph.D. diss., University of California, Berkeley.

Goldgeier, J. M., and M. McFaul. 1992. "A Tale of Two Worlds: Core and Periphery in the Post–Cold War Era." *International Organization*, 46: 467–91.

Goldman, E. F. 1969. *The Tragedy of Lyndon Johnson*. New York: Alfred A. Knopf.

Goldstein, W. M. 1990. "Judgments of Relative Importance in Decision Making: Global vs. Local Interpretations of Subjective Weight." *Organizational Behavior and Human Decision Processes*, 47: 313–36.

Golec, J., and M. Tamarkin. 1995. "Do Bettors Prefer Long Shots Because They Are Risk-Lovers or Are They Just Overconfident?" *Journal of Risk and Uncertainty*, 11: 51–64.

Gologor, E. 1977. "Group Polarization in a Non-Risk-Taking Culture." *Journal of Cross-Cultural Psychology*, 8: 331–46.

Gomori, G. 1972. "Hungarian and Polish Attitudes on Czechoslovakia, 1968." In E. J. Czerwinki and J. Piekalkiewicz, eds., *The Soviet Invasion of Czechoslovakia: Its Effects on Eastern Europe*, pp. 107–19. New York: Praeger.

Goulden, J. C. 1969. *Truth Is the First Casualty: The Gulf of Tonkin Affair—Illusion and Reality*. Chicago: Rand McNally.

Graebner, N. A. 1993. "The Scholar's View of Vietnam, 1964–1992." In D. E. Showalter and J. G. Albert, eds., *An American Dilemma: Vietnam, 1964–1973*, pp. 13–52. Chicago: Imprint.

Graff, H. F. 1970. *The Tuesday Cabinet: Deliberation and Decision on Peace and War Under Lyndon B. Johnson*. Englewood Cliffs, N.J.: Prentice Hall.

Gray, C. S. 1986. *Nuclear Strategy and National Style*. Lanham, Md.: Hamilton Press.

Green, D. P., and I. Shapiro. 1994. *Pathologies of Rational Choice Theory: A Critique of Applications in Political Science*. New Haven, Conn.: Yale University Press.

Greenwald, K. D. 1988. "Egypt's Hosni Mubarak." In B. Kellerman and J. Z. Rubin, eds., *Leadership and Negotiation in the Middle East*, pp. 96–117. New York: Praeger.

Greenwood, C. 1993. "Is There a Right of Humanitarian Intervention?" *World Today*, 49(2): 34–40.

Griffin, D., and A. Tversky. 1992. "The Weighing of Evidence and the Determinants of Confidence." *Cognitive Psychology*, 24: 411–35.

Griffin, D. W., and L. Ross. 1991. "Subjective Construal, Social Inference, and Human Misunderstanding." In M. P. Zanna, ed., *Advances in Experimental Social Psychology*, vol. 24, pp. 319–59. San Diego, Calif.: Academic Press.

Grigorenko, P. G. 1982. *Memoires*. New York: W. W. Norton.

Grimmett, R. F. 1990. *Trends in Conventional Arms Transfers to the Third World by Major Supplier, 1982–1989*. Washington, D.C.: Congressional Research Service.

Gromyko, A. 1989. *Memories*. London: Century Hutchinson.

Guelke, A. 1974. "Force, Intervention and Internal Conflict." In F. S. Northedge, ed., *The Use of Force in International Relations*, pp. 99–123. London: Faber and Faber.

Ha'aretz (in Hebrew).

Haas, E. B. 1993. "Beware the Slippery Slope: Notes Toward the Definition of Justifiable Intervention." In L. W. Reed and C. Kaysen, eds., *Emerging Norms of Justified Intervention*, pp. 63–87. Cambridge, Mass.: American Academy of Arts and Sciences.

———. 1990. *When Knowledge Is Power: Three Models of Change in International Organization*. Berkeley: University of California Press.

Haass, R. N. 1994. *Intervention: The Use of American Military Force in the Post–Cold War World*. Washington, D.C.: Carnegie Endowment.

Haber, E. 1978. *Menahem Begin: The Legend and the Man*. New York: Delacorte Press.

Haftendorn, H. 1988. "Toward Reconstruction of American Strength: A New Era in the Claim to Global Leadership?" In H. Haftendorn and J. Schissler, eds., *The Reagan Administration: A Reconstruction of American Strength?* pp. 3–29. Berlin: Walter de Gruyter.

Haig, A. M. 1984. *Caveat: Realism, Reagan, and Foreign Policy*. New York: Macmillan.

Halberstam, D. 1972. *The Best and the Brightest*. New York: Random House.

Hale, A. R. 1987. "Subjective Risk." In W. T. Singleton and J. Hovden, eds., *Risk and Decisions*, pp. 67–85. Chichester, Eng.: John Wiley and Sons.

Hall, D. L. 1991. *The Reagan Wars*. Boulder, Colo.: Westview Press.

Hallin, D. C. 1986. *The "Uncensored War": The Media and Vietnam*. New York: Oxford University Press.

Halperin, M. H., with P. Clapp and H. Kanter. 1974. *Bureaucratic Politics and Foreign Policy*. Washington, D.C.: Brookings Institution.

Hamburg, R. 1977. "Soviet Perspectives on Military Intervention." In E. P. Stern, ed., *The Limits of Military Intervention*, pp. 45–82. Beverly Hills, Calif.: Sage.

Hamilton, D. L. 1979. "A Cognitive-Attributional Analysis of Stereotyping." In L. Berkowitz, ed., *Advances in Experimental Social Psychology*, vol. 12, pp. 53–84. New York: Academic Press.

Hammer, E. J. 1988. *A Death in November: America in Vietnam, 1963*. Oxford, Eng.: Oxford University Press.

Hammond, P. Y. 1992. *LBJ and the Presidential Management of Foreign Relations*. Austin: University of Texas Press.

Harris, S. G. 1994. "Organizational Culture and Individual Sensemaking: A Schema-Bases Perspective." *Organizational Science*, 5: 309–21.

Hartnett, J. J., and R. M. Barber. 1974. "Fear of Failure in Group Risk-Taking." *British Journal of Social Clinical Psychology*, 13: 125–29.

Hartung, W. D. 1992. "Curbing the Arms Trade: From Rhetoric to Restraint." *World Policy Journal*, 9: 219–48.

Hatcher, P. L. 1990. *The Suicide of an Elite: American Internationalists and Vietnam*. Stanford, Calif.: Stanford University Press.

Hathaway, M. R. 1991. "The Role of Drugs in the U.S.-Panamanian Relationship." In B. W. Watson and P. G. Tsouras, eds., *Operation Just Cause: The U.S. Intervention in Panama*, pp. 29–46. Boulder, Colo.: Westview Press.

Havens, T. R. H. 1987. *Fire Across the Sea: The Vietnam War and Japan, 1965–1975*. Princeton, N.J.: Princeton University Press.

Hawkins, S. A., and R. Hastie. 1990. "Hindsight: Biased Judgments of Past Events After the Outcomes Are Known." *Psychological Bulletin*, 107: 311–27.

Hawthorn, G. 1993. *Plausible Worlds: Possibility and Understanding in History and the Social Sciences*. Cambridge, Eng.: Cambridge University Press.

Hayes-Roth, B. 1980. *Estimation of Time Requirements During Planning: Interaction Between Motivation and Cognition*. Santa Monica, Calif.: Rand Corporation, N-1581-ONR.

Heath, C., and A. Tversky. 1991. "Preference and Belief: Ambiguity and Competence in Choice Under Uncertainty." *Journal of Risk and Uncertainty*, 4: 5–28.

Heimer, C. A. 1988. "Social Structure, Psychology, and the Estimation of Risk." In W. R. Scott and J. Blake, eds., *Annual Review of Sociology*, vol. 14, pp. 491–519. Palo Alto, Calif.: Annual Reviews.

Heller, M., D. Tamari, and Z. Eytan, eds. 1984. *The Middle East Military Balance 1984.* Boulder, Colo.: Westview Press.

———, eds. 1983. *The Middle East Military Balance 1983.* Tel Aviv: Jaffee Center for Strategic Studies, Tel Aviv University.

Henderson, J. C., and P. C. Nutt. 1980. "The Influence of Decision Style on Decision Making Behavior." *Management Science*, 26: 371–86.

Henggeler, P. R. 1991. *In His Steps: Lyndon Johnson and the Kennedy Mystique.* Chicago.: Ivan R. Dee.

Hensley, T. R., and G. W. Griffin. 1986. "Victims of Groupthink: The Kent State University Board of Trustees and the 1977 Gymnasium Controversy." *Journal of Conflict Resolution*, 30: 497–531.

Heraclides, A. 1990. "Secessionist Minorities and External Involvement." *International Organization*, 44: 341–78.

Hermann, C. F. 1993. "Avoiding Pathologies in Foreign Policy Decision Groups." In D. Caldwell and T. J. McKeown, eds., *Diplomacy, Force, and Leadership: Essays in Honor of Alexander L. George*, pp. 179–207. Boulder, Colo.: Westview Press.

———. 1990. "Changing Course: When Governments Choose to Redirect Foreign Policy." *International Studies Quarterly*, 34: 3–22.

Hermann, M. G., C. F. Hermann, and J. D. Hagan. 1987. "How Decision Units Shape Foreign Policy Behavior." In C. F. Hermann, C. W. Kegley, Jr., and J. N. Rosenau, eds., *New Directions in the Study of Foreign Policy*, pp. 309–36. Boston: Unwin Hyman.

Herring, G. C. 1994. *LBJ and Vietnam: A Different Kind of War.* Austin: University of Texas Press.

———. 1993a. " 'Cold Blood': LBJ's Conduct of Limited War in Vietnam." In D. E. Showalter and J. G. Albert, eds., *An American Dilemma: Vietnam, 1964–1973*, pp. 62–85. Chicago: Imprint.

———. 1993b. "The Reluctant Warrior: Lyndon Johnson as Commander in Chief." In D. L. Anderson, ed., *Shadow on the White House: Presidents and the Vietnam War, 1945–1975*, pp. 87–112. Lawrence: University Press of Kansas.

———. 1993c. "The Strange 'Dissent' of Robert McNamara." In J. S. Werner and L. D. Huynh, eds., *The Vietnam War: Vietnamese and American Perspectives*, pp. 140–51. Armonk, N.Y.: M. E. Sharpe.

———. 1990. "Peoples Quite Apart: Americans, South Vietnamese and the War in Vietnam." *Diplomatic History*, 14: 1–23.

———. 1987. "The War in Vietnam." In R. A. Divine, ed., *The Johnson Years*, vol. 1, pp. 27–62. Lawrence: University Press of Kansas.

———. 1986. *America's Longest War: The United States and Vietnam, 1950–1975* (2nd ed.). Philadelphia: Temple University Press.

Herring, G. C., and R. H. Immerman. 1984. "Eisenhower, Dulles and Dien Bien Phu: 'The Day We Didn't Go to War' Revisited." *Journal of American History*, 71: 343–63.

Higbee, K. L. 1972. "Group Risk Taking in Military Decisions." *Journal of Social Psychology*, 88: 55–64.

Highhouse, S., and P. Yüce. 1996. "Perspectives, Perceptions, and Risk-Taking Behavior." *Organizational Behavior and Human Decision Processes*, 65: 159–67.

Hill, D. M., and P. Williams. 1990. "The Reagan Presidency: Style and Substance." In D. M. Hill, R. A. Moore, and P. Williams, eds., *The Reagan Presidency: An Incomplete Revolution?* pp. 3–25. New York: St. Martin's Press.

Hilsman, R. 1996. "McNamara's War—Against the Truth: A Review Essay." *Political Science Quarterly*, 111: 151–63.

Hirschman, A. O. 1985. "Against Parsimony: Three Easy Ways of Complicating Some Categories of Economic Discourse." *Economics and Philosophy*, 1: 7–21.

Hirshberg, M. 1993. "The Self-Perpetuating National Self-Image: Cognitive Biases in Perceptions of International Intervention." *Political Psychology*, 14: 77–98.

Hodnett, G., and P. J. Potichynyj. 1970. *The Ukraine and the Czechoslovak Crisis.* Canberra: Australian National University.

Hoffmann, S. 1986. "The Problem of Intervention." In H. Bull, ed., *Intervention in World Politics*, pp. 7–28. Oxford, Eng.: Clarendon Press.

———. 1968. *Gulliver's Troubles, or the Setting of American Foreign Policy.* New York: McGraw-Hill.

———. 1962. "Restraints and Choices in American Foreign Policy." *Daedalus*, 91: 668–704.

Hofstede, G. 1984. *Culture's Consequences: International Differences in Work Related Values* (abridged ed.). Beverly Hills, Calif.: Sage.

Hogarth, R. M., and H. Kunreuther. 1995. "Decision Making Under Ignorance: Arguing with Yourself." *Journal of Risk and Uncertainty*, 10: 15–36.

Hollander, E. P. 1986. "On the Central Role of Leadership Processes." *International Review of Applied Psychology*, 35: 39–52.

Hollis, M., and S. Smith. 1990. *Explaining and Understanding International Relations.* Oxford, Eng.: Clarendon Press.

Holsti, K. J. 1992. "Governance Without Government: Polyarchy in Nineteenth-Century European International Politics." In J. N. Rosenau and E.-O. Czempiel, eds., *Governance Without Government: Order and Change in World Politics*, pp. 30–57. Cambridge, Eng.: Cambridge University Press.

Holsti, O. R. 1989. "Crisis Decision Making." In P. E. Tetlock, J. L. Husbands, R. Jervis, P. C. Stern, and C. Tilly, eds., *Behavior, Society and Nuclear War*, vol. 1, pp. 8–84. New York: Oxford University Press.

———. 1977. *The 'Operational Code' as an Approach to the Analysis of Beliefs Systems.* Final Report to the National Science Foundation. Grant no. SOC-75-15368. Durham, N.C.: Duke University.

———. 1965. "The 1914 Case." *American Political Science Review*, 59: 365–78.

Holsti, O. R., and A. L. George. 1975. "The Effects of Stress on the Performance of Foreign Policy-Makers." In C. P. Cotter, ed., *Political Science Annual*, vol. 6: *Individual Decision Making*, pp. 255–319. Indianapolis, Ind.: Bobbs-Merril.

Hong, L. K. 1978. "Risky Shift and Cautious Shift: Some Direct Evidence on the Culture-Value Theory." *Social Psychology*, 41: 342–46.

Hoopes, T. 1969. *The Limits of Intervention.* New York: David McKay.

Hopple, G., and C. Gilley. 1985. "Policy Without Intelligence." In P. M. Dunn and B. W. Watson, eds., *American Intervention in Grenada: The Implications of Operation "Urgent Fury,"* pp. 55–71. Boulder, Colo.: Westview Press.

Horowitz, D. 1987. "Strategic Limitations of 'A Nation in Arms.'" *Armed Forces and Society*, 13: 277–94.

———. 1985. "Continuity and Change in Israel's Security Conception." In A. Yariv, ed., *War by Choice*, pp. 57–115. Tel Aviv: Hakibbutz Hameuchad (in Hebrew).

Horvath, P., and M. Zuckerman. 1993. "Sensation Seeking, Risk Appraisal, and Risky Behavior." *Personality and Individual Differences*, 14: 41–52.

Hosmer, S. T. 1987. *Constraints on U.S. Strategy in Third World Conflicts*. New York: Crane Russak.

Hosmer, S. T., and T. W. Wolfe. 1983. *Soviet Policy and Practice Toward Third World Conflicts*. Lexington, Mass.: Lexington Books.

Howell, W. C. 1972. "Compounding Uncertainty from Internal Sources." *Journal of Experimental Psychology*, 95: 6–13.

———. 1971. "Uncertainty from Internal and External Sources: A Clear Case of Overconfidence." *Journal of Experimental Psychology*, 89: 240–43.

Hoyt, G. C., and J. A. F. Stoner. 1968. "Leadership and Group Decisions Involving Risk." *Journal of Experimental Social Psychology*, 4: 275–84.

Hoyt, P. D., and J. A. Garrison. 1997. "Political Manipulation Within the Small Group: Foreign Policy Advisers in the Carter Administration." In P. 't Hart, E. K. Stern, and B. Sundelius, eds., *Beyond Groupthink: Political Group Dynamics and Foreign Policy-Making*, pp. 249–76. Ann Arbor: Michigan University Press.

Huber, G. P. 1991. "Organizational Learning: The Contributing Processes and the Literatures." *Organization Science*, 2: 88–115.

Hudson, V. M., with C. S. Vore. 1995. "Foreign Policy Analysis: Yesterday, Today and Tomorrow." *Mershon International Studies Review*, 39: 209–38.

Humphrey, D. C. 1984. "Tuesday Lunch at the Johnson White House: A Preliminary Assessment." *Diplomatic History*, 8: 81–101.

Humphrey, H. H. 1976. *The Education of a Public Man: My Life and Politics*. Garden City, N.Y.: Doubleday.

Hunt, M. H. 1987. *Ideology and U.S. Foreign Policy*. New Haven, Conn.: Yale University Press.

Huntington, S. P. 1992. "America's Changing Strategic Interests." *Survival*, 33(1): 3–17.

Husbands, J. L. 1990. "A Buyer's Market for Arms." *Bulletin of the Atomic Scientists*, 46(4): 14–19.

Huth, P., D. S. Bennett, and C. Gelpi. 1992. "System Uncertainty, Risk Propensity, and International Conflict Among the Great Powers." *Journal of Conflict Resolution*, 36: 478–517.

Hybel, A. R. 1990. *How Leaders Reason: U.S. Intervention in the Caribbean Basin and Latin America*. Cambridge, Mass.: Basil Blackwell.

Illingworth, G. F. 1985. "Grenada in Retrospect." In P. M. Dunn and B. W. Watson, eds., *American Intervention in Grenada: The Implications of Operation "Urgent Fury,"* pp. 129–47. Boulder, Colo.: Westview Press.

Immerman, R. H. 1994. " 'A Time in the Tide of Men's Affairs': Lyndon Johnson and Vietnam." In W. I. Cohen and N. B. Tucker, eds., *Lyndon Johnson Confronts the World: American Foreign Policy, 1963–1968*, pp. 57–97. Cambridge, Eng.: Cambridge University Press.

———. 1987. "Between the Unattainable and the Unacceptable: Eisenhower and Dien Bien Phu." In R. A. Melanson and D. Mayers, eds., *Reevaluating Eisenhower: American Foreign Policy in the 1950s*, pp. 120–54. Urbana: University of Illinois Press.

Inbar, E. 1991. *War and Peace in Israeli Politics: Labor Party Positions on National Security*. Boulder, Colo.: Lynne Rienner.

———. 1989. "The 'No Choice War' Debate in Israel." *Journal of Strategic Studies*, 12: 22–37.

Isen, A. M., and N. Geva. 1987. "The Influence of Positive Affect on Acceptable Level of Risk: The Person with a Large Canoe Has a Large Worry." *Organizational Behavior and Human Decision Processes*, 39: 145–54.

Isen, A. M., B. Means, R. Patrick, and G. Nowicki. 1982. "Some Factors Influencing Decision-Making Strategy and Risk-Taking." In M. S. Clark and S. T. Fiske, eds., *Affect and Cognition*, pp. 243–61. Hillsdale, N.J.: Lawrence Erlbaum.

Isen, A. M., and R. Patrick. 1983. "The Effect of Positive Feelings on Risk Taking: When the Chips Are Down." *Organizational Behavior and Human Performance*, 31: 194–202.

Jackson, R. H. 1993. "Armed Humanitarianism." *International Journal*, 48: 579–606.

Jackson, S. E., and J. E. Dutton. 1988. "Discerning Threats and Opportunities." *Administrative Science Quarterly*, 33: 370–87.

Jackson, S. E., K. E. May, and K. Whitney. 1995. "Understanding the Dynamic of Diversity in Decision-Making Teams." In R. A. Guzzo, E. Salas, and associates, eds., *Team Effectiveness and Decision-Making in Organizations*, pp. 204–61. San Francisco: Jossey-Bass.

Janis, I. L. 1989. *Crucial Decisions: Leadership in Policymaking and Crisis Management*. New York: Free Press.

———. 1982. *Groupthink: Psychological Studies of Policy Decisions and Fiascoes* (2nd ed.). Boston: Houghton Mifflin.

Janis, I. L., and L. Mann. 1977. *Decision-Making: A Psychological Analysis of Conflict, Choice, and Commitment*. New York: Free Press.

Jellison, J. M., and J. Riskind. 1971. "Attribution of Risk to Others as a Function of Their Ability." *Journal of Personality and Social Psychology*, 20: 413–15.

Jentleson, B. W. 1992. "The Pretty Prudent Public: Post-Post-Vietnam American Opinion on the Use of Military Force." *International Studies Quarterly*, 36: 49–74.

———. 1987. "American Commitments in the Third World: Theory vs. Practice." *International Organization*, 41: 667–704.

Jentleson, B. W., and A. E. Levite. 1992. "The Analysis of Protracted Foreign Military Intervention." In A. E. Levite, B. W. Jentleson, and L. Berman, eds., *Foreign Military Intervention: The Dynamics of Protracted Conflict*, pp. 1–22. New York: Columbia University Press.

Jentleson, B. W, A. E. Levite, and L. Berman. 1993. "Protracted Foreign Military Intervention: A Structured, Focused Comparative Analysis." In D. Caldwell and T. J. McKeown, eds., *Diplomacy, Force, and Leadership: Essays in Honor of Alexander L. George*, pp. 229–54. Boulder, Colo.: Westview Press.

———. 1992. "Foreign Military Intervention in Perspective." In A. E. Levite, B. W. Jentleson, and L. Berman, eds., *Foreign Military Intervention: The Dynamics of Protracted Conflict*, pp. 301–25. New York: Columbia University Press.

Jervis, R. 1993. "System and Interaction Effects." In J. Snyder and R. Jervis, eds., *Coping with Complexity in the International System*, pp. 25–46. Boulder, Colo.: Westview Press.

———. 1992. "Political Implications of Loss Aversion." *Political Psychology*, 13: 187–204.

———. 1991. "Domino Beliefs and Strategic Behavior." In R. Jervis and J. Snyder, eds., *Dominos and Bandwagons: Strategic Beliefs and Great Power Competition in the Eurasian Rimland*, pp. 20–50. New York: Oxford University Press.

———. 1985a. "From Balance to Concert: A Study of International Security Cooperation." *World Politics*, 38: 58–79.

———. 1985b. "Pluralistic Rigor: A Comment on Bueno de Mesquita." *International Studies Quarterly*, 29: 145–49.

———. 1982–83. "Deterrence and Perception." *International Security*, 7(3): 3–30.

———. 1976. *Perception and Misperception in International Politics*. Princeton, N.J.: Princeton University Press.

Jian, C. 1995. "China's Involvement in the Vietnam War, 1964–69." *China Quarterly*, no. 142: 356–87.

Johnson, B. B. 1991. "Risk and Culture Research: Some Cautions." *Journal of Cross-Cultural Psychology*, 22: 141–49.

Johnson, E. J., and A. Tversky. 1983. "Affect, Generalization, and the Perception of Risk." *Journal of Personality and Social Psychology*, 45: 20–31.

Johnson, L. B. 1971. *The Vantage Point: Perspectives of the Presidency 1963–1969.* New York: Holt, Rinehart and Winston.

Johnston, A. I. 1995. "Thinking About Strategic Culture." *International Security*, 19(4): 32–64.

Jones, C. D. 1975. "Autonomy and Intervention: The CPSU and the Struggle for the Czechoslovak Communist Party, 1968." *Orbis*, 19: 591–625.

Jones, E. E., and C. A. Johnson. 1973. "Delay of Consequences and the Riskiness of Decisions." *Journal of Personality*, 41: 613–37.

Joseph, P. 1993. "Direct and Indirect Effects of the Movement Against the Vietnam War." In J. S. Werner and L. D. Huynh, eds., *The Vietnam War: Vietnamese and American Perspectives*, pp. 165–84. Armonk, N.Y.: M. E. Sharpe.

Jou, J., J. Shanteau, and R. J. Harris. 1996. "An Information Processing View of Framing Effects: The Role of Causal Schemes in Decision-Making." *Memory and Cognition*, 24: 1–15.

Joyner, C. C. 1989. "International Law." In P. J. Schraeder, ed., *Intervention in the 1980s: U.S. Foreign Policy in the Third World*, pp. 191–204. Boulder, Colo.: Lynne Rienner.

Jungermann, H. 1977. "Cognitive Processes and Societal Risk Taking: Comments." In H. Jungermann and G. De Zeeuw, eds., *Decision Making and Change in Human Affairs*, pp. 37–43. Dordrecht, Neth.: D. Reidel.

Kahin, G. M. 1987. *Intervention: How America Became Involved in Vietnam*. Garden City, N.J.: Doubleday, Anchor Press.

Kahneman, D., J. L. Knetsch, and R. H. Thaler. 1990. "Experimental Tests of the Endowment Effect and the Coase Theorem." *Journal of Political Economy*, 98: 1325–48.

Kahneman, D., and D. Lovallo. 1993. "Timid Choices and Bold Forecasts: A Cognitive Perspective on Risk Taking." *Management Science*, 39: 17–31.

Kahneman, D., and D. T. Miller. 1986. "Norm Theory: Comparing Reality to Its Alternatives." *Psychological Review*, 93: 136–53.

Kahneman, D., P. Slovic, and A. Tversky, eds. 1982. *Judgment Under Uncertainty: Heuristics and Biases*. Cambridge, Eng.: Cambridge University Press.

Kahneman, D., and A. Tversky. 1984. "Choices, Values and Frames." *American Psychologist*, 39: 341–50.

———. 1982a. "The Psychology of Preferences." *Scientific American*, 246: 136–42.

———. 1982b. "The Simulation Heuristic." In D. Kahneman, P. Slovic, and A. Tversky, eds., *Judgment Under Uncertainty: Heuristics and Biases*, pp. 201–8. Cambridge, Eng.: Cambridge University Press.

———. 1979. "Prospect Theory: An Analysis of Decision Under Risk." *Econometrica*, 47: 263–91.

Kameda, T., and J. H. Davis. 1990. "The Function of the Reference Point in Individual and Group Risk Decision Making." *Organizational Behavior and Human Decision Processes*, 46: 55–76.

Kameda, T., and S. Sugimori. 1993. "Psychological Entrapment in Group Decision Making: An Assigned Decision Rule and a Groupthink Phenomenon." *Journal of Personality and Social Psychology*, 65: 282–92.

Kanet, R. E., and E. A. Kolodziej, eds. 1991. *The Cold War as Cooperation*. Baltimore, Md.: Johns Hopkins University Press.

Kanouse, D. E., and L. R. Hanson, Jr. 1971. "Negativity in Evaluations." In E. E. Jones, D. E. Kanouse, H. H. Kelley, R. E. Nisbett, S. Valins, and B. Weiner, eds., *Attribution: Perceiving the Causes of Behavior*, pp. 47–62. Morristown, N.J.: General Learning Press.

Kanwisher, N. 1989. "Cognitive Heuristics and American Security Policy." *Journal of Conflict Resolution*, 33: 652–75.

Kaplan, M. 1987. "The Influencing Process in Group Decision Making." In C. Hendrick, ed., *Group Processes*, pp. 189–212. Newbury Park, Calif.: Sage.

Kaplan, M. A. 1964. "Intervention in Internal War: Some Systemic Sources." In J. N. Rosenau, ed., *International Aspects of Civil Strife*, pp. 92–121. Princeton, N.J.: Princeton University Press.

Kaplan, M. F., and C. E. Miller. 1987. "Group Decision Making and Normative Versus Informational Influence: Effects of Type of Issue and Assigned Decision Rule." *Journal of Personality and Social Psychology*, 53: 306–13.

Karnow, S. 1983. *Vietnam: A History*. New York: Viking Press.

Karp, A. 1994. "The Rise of Black and Gray Markets." *Annals of the American Academy of Political and Social Science*, 535: 175–89.

Kasper, R. G. 1980. "Perceptions of Risk and Their Effects on Decision Making." In R. C. Schwing and W. A. Albers, Jr., eds., *Societal Risk Assessment: How Safe Is Safe Enough?* pp. 71–80. New York: Plenum Press.

Kasperson, R. E. 1992. "The Social Amplification of Risk: Progress in Developing an Integrative Framework." In S. Krimsky and D. Golding, eds., *Social Theories of Risk*, pp. 153–78. Westport, Conn.: Praeger.

Kasperson, R. E., O. Renn, P. Slovic, H. S. Brown, J. Emel, R. Goble, J. X. Kasperson, and S. Ratick. 1988. "The Social Amplification of Risk: A Conceptual Framework." *Risk Analysis*, 8: 177–87.

Kattenburg, P. M. 1980. *The Vietnam Trauma in American Foreign Policy*. New Brunswick, N.J.: Transaction Books.

Kaw, M. 1989. "Predicting Soviet Military Intervention." *Journal of Conflict Resolution*, 33: 402–29.

Keal, P. 1983. *Unspoken Rules and Superpower Dominance*. New York: St. Martin's Press.

Kearns, D. 1976. *Lyndon Johnson and the American Dream*. New York: Harper and Row.

Kegley, C. W., and M. G. Hermann. 1995. "Military Intervention and the Democratic Peace." *International Interactions*, 21: 1–21.

Kegley, C. W., and E. R. Wittkopf. 1982. "The Reagan Administration's World View." *Orbis*, 26: 223–44.

Keinan, G., N. Friedland, and Y. Ben-Porath. 1987. "Decision-Making Under Stress: Scanning of Alternatives Under Physical Threat." *Acta Psychologica*, 64: 219–28.

Kelley, S. 1995. "The Promise and Limitations of Rational Choice Theory." *Critical Review*, 9: 95–106.

Kempe, F. 1990. *Divorcing the Dictator: America's Bungled Affair with Noriega.* New York: G. P. Putnam's Sons.

Kennedy, M. M. 1979. "Generalizing from Single Case Studies." *Evaluation Quarterly*, 3: 661–78.

Kenworthy, E. 1984. "Grenada as Theater." *World Policy Journal*, 1: 635–52.

Khalidi, R. 1988. *Under Siege: PLO Decisionmaking During the 1982 War.* Tel Aviv: Ma'arachot (in Hebrew).

Khong, Y. F. 1992. *Analogies at War: Korea, Munich, Dien Bien Phu, and the Vietnam Decisions of 1965.* Princeton, N.J.: Princeton University Press.

———. 1991. "The Lessons of Korea and the Vietnam Decisions of 1965." In G. W. Breslauer and P. E. Tetlock, eds., *Learning in U.S. and Soviet Foreign Policy*, pp. 302–49. Boulder, Colo.: Westview Press.

Kiesler, C. A. 1971. *The Psychology of Commitment.* New York: Academic Press.

Kimmerling, B., with I. Backer. 1985. *The Interrupted System: Israeli Civilians in War and Routine Times.* New Brunswick, N.J.: Transaction Books.

King, G., R. O. Keohane, and S. Verba. 1994. *Designing Social Inquiry: Scientific Inferences in Qualitative Research.* Princeton, N.J.: Princeton University Press.

Kissinger, H. A. 1969. "Domestic Structure and Foreign Policy." In J. N. Rosenau, ed., *International Politics and Foreign Policy* (rev. ed.), pp. 261–75. New York: Free Press.

Kitayama, S., and E. Burnstein. 1994. "Social Influence, Persuasion and Group Decision Making." In S. Shavit and T. C. Brock, eds., *Persuasion: Psychological Insights and Perspectives*, pp. 175–93. Boston: Allyn and Bacon.

Klare, M. T. 1992. *Conventional Arms Transfers: Exporting Security or Arming Adversaries?* Strategic Concepts in National Strategy Series. Carlisle, Pa.: Strategic Studies Institute, U.S. Army War College.

———. 1990. "Wars in the 1990s: Growing Firepower in the Third World." *Bulletin of the Atomic Scientists*, 46(4): 9–13.

———. 1989. "Subterranean Alliances: America's Global Proxy Network." *Journal of International Affairs*, 43: 97–118.

———. 1987. "The Arms Trade: Changing Patterns in the 1980s." *Third World Quarterly*, 9: 1257–81.

Klausner, S. Z. 1968. "The Intermingling of Pain and Pleasure: The Stress-Seeking Personality in Its Social Context." In S. Z. Klausner, ed., *Why Man Takes Chances*, pp. 137–68. New York: Doubleday.

Klein, W. M., and Z. Kunda. 1994. "Exaggerated Self-Assessment and the Preference for Controllable Risks." *Organizational Behavior and Human Decision Processes*, 59: 410–27.

Klein, Y. 1991. "A Theory of Strategic Culture." *Comparative Strategy*, 10: 3–23.

Kleinhesselink, R. A., and E. A. Rosa. 1991. "Cognitive Representation of Risk Perceptions: A Comparison of Japan and the United States." *Journal of Cross-Cultural Psychology*, 22: 11–28.

Klimoski, R., and S. Mohammed. 1994. "Team Mental Model: Construct or Metaphor?" *Journal of Management*, 20: 403–37.

Klinghoffer, A. J. 1980. *The Angolan War: A Study in Soviet Policy in the Third World*. Boulder, Colo.: Westview Press.

Kober, A. 1995. *Military Decision in the Arab-Israeli Wars, 1948–1982*. Tel Aviv: Ma'arachot (in Hebrew).

Kobrin, S. J. 1982. *Managing Political Risk Assessment: Strategic Response to Environmental Change*. Berkeley: University of California Press.

———. 1979. "Political Risk: A Review and Reconsideration." *Journal of International Business Studies*, 10: 67–80.

Kogan, N., and M. A. Wallach. 1967. "Risk Taking as a Function of the Situation, the Person, and the Group." In G. Mandler, P. Mussen, N. Kogan, and M. A. Wallach, eds., *New Directions in Psychology*, vol. 3, pp. 111–278. New York: Holt, Rinehart and Winston.

———. 1964. *Risk Taking: A Study in Cognition and Personality*. New York: Holt, Rinehart and Winston.

Kolko, G. 1985. *Anatomy of a War: Vietnam, the United States, and the Modern Historical Experience*. New York: Pantheon Books.

———. 1972. "The American Goals in Vietnam." In N. Chomsky and H. Zinn, eds., *The Pentagon Papers: Critical Essays*, vol. 5, pp. 1–15. Boston: Beacon Press.

Komer, R. W. 1986. *Bureaucracy at War: U.S. Performance in the Vietnam Conflict*. Boulder, Colo.: Westview Press.

Koopman, C., R. McDermott, R. Jervis, J. Snyder, and J. Dioso. 1995. "Stability and Change in American Elite Beliefs About International Relations." *Peace and Conflict Journal of Peace Psychology*, 1: 365–82.

Koriat, A., S. Lichtenstein, and B. Fischhoff. 1980. "Reasons for Confidence." *Journal of Experimental Psychology: Human Learning and Memory*, 6: 107–18.

Kowert, P. A., and M. G. Hermann. 1995. "When Prospects Look Dim: Rival Hypotheses to Prospect Theory for Risky Decision Making." Paper prepared for the 36th Annual Meeting of the International Studies Association, Chicago, Feb. 21–25.

Kramer, M. 1993. "The Prague Spring and the Soviet Invasion of Czechoslovakia." *Cold War International History Project Bulletin*, no. 3 (Fall): 2–13, 54–55.

———. 1992a. "A Letter to Brezhnev: The Czech Hardliners' 'Request' for Soviet Intervention, August 1968." *Cold War International History Project Bulletin*, no. 2 (Fall): 35.

———. 1992b. "New Sources on the 1968 Soviet Invasion of Czechoslovakia." *Cold War International History Bulletin*, no. 2 (Fall): 1, 4–13.

Krasner, S. D. 1995. "Sovereignty and Intervention." In G. M. Lyons and M. Mastanduno, eds., *Beyond Westphalia? State Sovereignty and International Intervention*, pp. 228–49. Baltimore, Md.: Johns Hopkins University Press.

———. 1985. "Toward Understanding in International Relations." *International Studies Quarterly*, 29: 137–44.

Kratochwil, F. V. 1995. "Sovereignty as *Dominium*: Is There a Right of Humanitarian Intervention?" In G. M. Lyons and M. Mastanduno, eds., *Beyond Westphalia? State Sovereignty and International Intervention*, pp. 21–42. Baltimore, Md.: Johns Hopkins University Press.

———. 1989. *Rules, Norms, and Decisions*. Cambridge, Eng.: Cambridge University Press.

Krause, K. 1992. *Arms and the State: Patterns of Military Production and Trade.* Cambridge, Eng.: Cambridge University Press.

Kreisberg, P. H. 1986. "Does the U.S. Government Think That International Law Is Important?" *Yale Journal of International Law*, 11: 479–91.

Krepinevich, A. F. 1993. "Vietnam: Evaluating the Ground War, 1965–1968." In D. E. Showalter and J. G. Albert, eds., *An American Dilemma: Vietnam, 1964–1973*, pp. 87–107. Chicago: Imprint.

———. 1986. *The Army and Vietnam.* Baltimore, Md.: Johns Hopkins University Press.

Krimsky, S., and D. Golding, eds. 1992. *Social Theories of Risk.* Westport, Conn.: Praeger.

Kroon, M. B. R., P. 't Hart, and D. Van Kreveld. 1991. "Managing Group Decision Making Processes: Individual Versus Collective Accountability and Groupthink." *International Journal of Conflict Management*, 2: 91–115.

Kroon, M. B. R., D. Van Kreveld, and J. M. Rabbie. 1992. "Group Versus Individual Decision Making: Effects of Accountability and Gender on Groupthink." *Small Group Research*, 23: 427–58.

Krueger, N., and P. R. Dickson. 1994. "How Believing in Ourselves Increases Risk Taking: Perceived Self-Efficacy." *Decision Sciences*, 25: 385–400.

Kruglanski, A. W. 1989. *Lay Epistemics and Human Knowledge: Cognitive and Motivational Bases.* New York: Plenum Press.

Kuipers, B., A. J. Moskowitz, and J. P. Kassirer. 1988. "Critical Decisions Under Uncertainty: Representation and Structure." *Cognitive Science*, 12: 177–210.

Kunda, Z. 1990. "The Case for Motivated Reasoning." *Psychological Bulletin*, 108: 480–98.

Kupchan, C. A. 1992. "Getting In: The Initial Stage of Military Intervention." In A. E. Levite, B. W. Jentleson, and L. Berman, eds., *Foreign Military Intervention: The Dynamics of Protracted Conflict*, pp. 241–60. New York: Columbia University Press.

Kupchan, C. A., and C. A. Kupchan. 1991. "Concerts, Collective Security, and the Future of Europe." *International Security*, 16(1): 114–61.

LaFeber, W. 1994. "Johnson, Vietnam and Tocqueville." In W. I. Cohen and N. B. Tucker, eds., *Lyndon Johnson Confronts the World: American Foreign Policy, 1963–1968*, pp. 31–55. Cambridge, Eng.: Cambridge University Press.

Lamare, J. W. 1991. "International Intervention and Public Support: America During the Reagan Years." In J. W. Lamare, ed., *International Crisis and Domestic Politics*, pp. 7–28. New York: Praeger.

Lamborn, A. C. 1991. *The Price of Power: Risk and Foreign Policy in Britain, France and Germany.* Boston: Unwin Hyman.

———. 1985. "Risk and Foreign Policy Choice." *International Studies Quarterly*, 29: 385–410.

Lamm, H., and N. Kogan. 1970. "Risk Taking in the Context of Intergroup Negotiation." *Journal of Experimental Social Psychology*, 6: 351–63.

Lane, R. E. 1995. "What Rational Choice Explains." *Critical Review*, 9: 107–26.

Langer, E. J. 1975. "The Illusion of Control." *Journal of Personality and Social Psychology*, 32: 311–28.

Lanir, Z. 1985. "The Political Goals and the Military Objectives in Israel's Wars." In A. Yariv, ed., *War by Choice*, pp. 117–56. Tel Aviv: Hakibbutz Hameuchad (in Hebrew).

Lanir, Z., B. Fischhoff, and S. Johnson. 1988. "Military Risk-Taking: C³I and the Cognitive Functions of Boldness in War." *Journal of Strategic Studies*, 11: 96–114.

Lant, T. K. 1992. "Aspiration Level Adaptation: An Empirical Exploration." *Management Science*, 38: 623–44.

Larrabee, F. S. 1984. "Soviet Crisis Management in Eastern Europe." In D. Holloway and J. M. Sharp, eds., *The Warsaw Pact: Alliance in Transition?* pp. 111–38. Ithaca, N.Y.: Cornell University Press.

Larson, D. W. 1985. *Origins of Containment: A Psychological Exploration*. Princeton, N.J.: Princeton University Press.

Larson, J. R., and C. Christensen. 1993. "Groups as Problem-Solving Units: Toward a New Meaning of Social Cognition." *British Journal of Social Psychology*, 32: 5–30.

Laurance, E. J. 1992. *The International Arms Trade*. New York: Lexington Books.

Leary, M. R., and D. R. Forsyth. 1987. "The Attribution of Responsibility for Collective Endeavors." In C. Hendrick, ed., *Group Processes*, pp. 167–88. Newbury Park, Calif.: Sage.

Lebow, R. N. 1984. "Windows of Opportunity: Do States Jump Through Them?" *International Security*, 9(1): 147–86.

Leffler, M. P. 1992. *A Preponderance of Power: National Security, the Truman Administration, and the Cold War*. Stanford, Calif.: Stanford University Press.

Legro, J. W. 1994. "Military Culture and Inadvertent Escalation in World War II." *International Security*, 18(4): 108–42.

Lenczowski, J. 1982. *Soviet Perceptions of U.S. Foreign Policy*. Ithaca, N.Y.: Cornell University Press.

Levenson, M. R. 1990. "Risk Taking and Personality." *Journal of Personality and Social Psychology*, 58: 1073–80.

Levi, A., and J. B. Pryor. 1987. "Use of the Availability Heuristic in Probability Estimates of Future Events: The Effects of Imagining Outcomes Versus Imagining Reasons." *Organizational Behavior and Human Decision Processes*, 40: 219–34.

Levin, I. P., R. D. Johnson, and M. L. Davis. 1987. "How Information Frame Influences Risky Decisions: Between-Subject and Within-Subject Comparisons." *Journal of Economic Psychology*, 8: 43–54.

Levine, J. M., and R. L. Moreland. 1990. "Progress in Small Group Research." In M. R. Rosenzweig and L. W. Porter, eds., *Annual Review of Psychology*, vol. 41, pp. 585–634. Palo Alto, Calif.: Annual Reviews.

Levine, J. M., L. B. Resnick, and E. T. Higgins. 1993. "Social Foundations of Cognition." In L. W. Porter and M. R. Rosenzweig, eds., *Annual Review of Psychology*, vol. 44, pp. 585–612. Palo Alto, Calif.: Annual Reviews.

Levine, J. M., and E. M. Russo. 1987. "Majority and Minority Influence." In C. Hendrick, ed., *Group Processes*, pp. 13–54. Newbury Park, Calif.: Sage.

Levite, A. 1988. *Offense and Defense in Israel: Military Doctrine*. Tel Aviv: Hakibbutz Hameuchad (in Hebrew).

Levite, A. E., B. W. Jentleson, and L. Berman, eds. 1992. *Foreign Military Intervention: The Dynamics of Protracted Conflict*. New York: Columbia University Press.

Levitt, B., and J. G. March. 1988. "Organizational Learning." In W. R. Scott and J. Blake, eds., *Annual Review of Sociology*, vol. 14, pp. 319–40. Palo Alto, Calif.: Annual Reviews.

Levy, D. W. 1991. *The Debate over Vietnam*. Baltimore, Md.: Johns Hopkins University Press.

Levy, J. S. 1996. "Loss Aversion, Framing, and Bargaining: The Implications of Prospect Theory for International Conflict." *International Political Science Review*, 17: 179–95.

———. 1994. "Learning and Foreign Policy: Sweeping a Conceptual Minefield." *International Organization*, 48: 279–312.

———. 1992a. "An Introduction to Prospect Theory." *Political Psychology*, 13: 171–86.

———. 1992b. "Prospect Theory and International Relations: Theoretical Applications and Analytical Problems." *Political Psychology*, 13: 283–310.

Lewis, G. K. 1987. *The Jewel Despoiled*. Baltimore, Md.: Johns Hopkins University Press.

Lewis, V. 1990. "Small States, Eastern Caribbean Security and the Grenada Intervention." In J. Heine, ed., *A Revolution Aborted: The Lessons of Grenada*, pp. 257–63. Pittsburgh, Pa.: University of Pittsburgh Press.

Lewy, G. H. 1978. *America in Vietnam*. New York: Oxford University Press.

Liberson, S. 1992. "Small N's and Big Conclusions: An Examination of the Reasoning in Comparative Studies Based on a Small Number of Cases." In C. C. Ragin and H. S. Becker, eds., *What Is a Case? Exploring the Foundations of Social Inquiry*, pp. 105–18. Cambridge, Eng.: Cambridge University Press.

Lichtenstein, S., B. Fischhoff, and L. D. Phillips. 1982. "Calibration of Probabilities: The State of the Art to 1980." In D. Kahneman, P. Slovic, and A. Tversky, eds., *Judgement Under Uncertainty: Heuristics and Biases*, pp. 306–34. Cambridge, Eng.: Cambridge University Press.

Lilienthal, R. A., and S. L. Hutchinson. 1979. "Group Polarization (Risky Shift) in Led and Leaderless Group Discussions." *Psychological Reports*, 45: 168.

Lipshitz, R. and O. Strauss. 1997. "Coping with Uncertainty: A Naturalistic Decision-Making Analysis." *Organizational Behavior and Human Decision Processes*, 69: 149–63.

Littauer, R., and N. Uphoff, eds. 1972. *The Air War in Indochina* (rev. ed.). Boston: Beacon Press.

Little, D. 1993. "Evidence and Objectivity in the Social Sciences." *Social Research*, 60: 363–96.

Little, R. 1987. "Revisiting Intervention: A Survey of Recent Developments." *Review of International Studies*, 13: 49–60.

———. 1975. *Intervention: External Involvement in Civil Wars*. Totowa, N.J.: Rowman and Littlefield.

Lockhart, C. 1978. "Flexibility and Commitment in International Conflicts." *International Studies Quarterly*, 22: 545–68.

Loewenheim, F. L. 1994. "Dean Rusk and the Diplomacy of Principle." In G. A. Craig and F. L. Loewenheim, eds., *The Diplomats, 1939–1979*, pp. 499–536. Princeton, N.J.: Princeton University Press.

Lopes, L. L. 1987. "Between Hope and Fear: The Psychology of Risk." In L. Berkowitz, ed., *Advances in Experimental Social Psychology*, vol. 20, pp. 255–95. San Diego, Calif.: Academic Press.

———. 1981. "Decision-Making in the Short Run." *Journal of Experimental Psychology: Human Learning and Memory*, 7: 377–85.

Lorell, M., C. Kelley, Jr., with D. Hensler. 1985. *Casualties, Public Opinion, and Presidential Policy During the Vietnam War.* Santa Monica, Calif.: Rand Corporation, R-60-AF.

Lowenthal, R. 1968. "The Sparrow in the Cage." *Problems of Communism*, 17(6): 2–28.

Lowrance, W. W. 1980. "The Nature of Risk." In R. C. Schwing and W. A. Albers, Jr., eds., *Societal Risk Assessment: How Safe Is Safe Enough?* pp. 5–14. New York: Plenum Press.

Luard, E. 1989. *The Blunted Sword: The Erosion of Military Power in Modern World Politics.* London: I. B. Tauris.

———. 1986. "Collective Intervention." In H. Bull, ed., *Intervention in World Politics*, pp. 157–79. Oxford, Eng.: Clarendon Press.

Lunch, W. L., and P. W. Sperlich. 1979. "American Public Opinion and the War in Vietnam." *Western Political Quarterly*, 22: 21–44.

Lyndon, J. E., and M. P. Zanna. 1990. "Commitment in the Face of Adversity: A Value-Affirmation Approach." *Journal of Personality and Social Psychology*, 58: 1040–47.

Lyndon B. Johnson Library. The Papers of Lyndon B. Johnson. Austin, Tex.

Lyons, G. M., and M. Mastanduno. 1993. "International Intervention, State Sovereignty and the Future of International Society." *International Social Science Journal*, no. 138: 517–32.

Ma'ariv (in Hebrew).

Maass, A., C. Volpato, and A. Mucchi-Faina. 1996. "Social Influence and the Verifiability of the Issue Under Discussion: Attitudinal Versus Objective Items." *British Journal of Social Psychology*, 35: 15–26.

Maass, A., S. G. West, and R. B. Cialdini. 1987. "Minority Influence and Conversion." In C. Hendrick, ed., *Group Processes*, pp. 55–79. Newbury Park, Calif.: Sage.

MacCrimmon, K. R., and D. A. Wehrung with W. T. Stanbury. 1986. *Taking Risks: The Management of Uncertainty.* New York: Free Press.

MacDonald, D. J. 1993–94. "Falling Dominoes and System Dynamics: A Risk Aversion Perspective." *Security Studies*, 3: 225–58.

MacFarlane, S. N. 1985. *Intervention and Regional Security.* Adelphi Paper, no. 196. London: International Institute for Strategic Studies.

———. 1984. "Africa's Decaying Security System and the Rise of Intervention." *International Security*, 8(4): 127–51.

Machlis, G. E., and E. A. Rosa. 1990. "Desired Risk: Broadening the Social Amplification of Risk Framework." *Risk Analysis*, 10: 161–68.

Mack, A. 1975. "Why Big Nations Lose Small Wars: The Politics of Asymmetric Conflict." *World Politics*, 27: 175–200.

MacLean, J. 1988. "Belief Systems and Ideology in International Relations: A Critical Approach." In R. Little and S. Smith, eds., *Belief Systems and International Relations*, pp. 57–82. Oxford, Eng.: Basil Blackwell.

Mandel, R. 1987. *Irrationality in International Confrontation.* New York: Greenwood Press.

———. 1984. "The Desirability of Irrationality in Foreign Policy Making: A Preliminary Theoretical Analysis." *Political Psychology*, 5: 643–60.

Mann, L. 1992. "Stress, Affect and Risk Taking." In J. F. Yates, ed., *Risk-Taking Behavior*, pp. 201–29. Chichester, Eng.: John Wiley and Sons.

Maoz, Z. 1990a. "Framing the National Interest: The Manipulation of Foreign Policy Decisions in Group Settings." *World Politics*, 43: 77–110.

———. 1990b. *Paradoxes of War: On the Art of National Self-Entrapment*. Boston: Unwin Hyman.

———. 1989. "Power, Capabilities and Paradoxical Conflict Outcomes." *World Politics*, 41: 239–66.

Mapel, D. R. 1991. "Military Intervention and Rights." *Millennium*, 20: 41–56.

March, J. G. 1996. "Learning to Be Risk Averse." *Psychological Review*, 103: 309–19.

———. 1994. *A Primer on Decision Making: How Decisions Happen*. New York: Free Press.

———. 1988. "Variable Risk Preferences and Adaptive Aspirations." *Journal of Economic Behavior and Organization*, 9: 5–24.

March, J. G., and J. P. Olsen. 1989. *Rediscovering Institutions: The Organizational Basis of Politics*. New York: Free Press.

March, J. G., and Z. Shapira. 1992. "Behavioral Decision Theory and Organizational Decision Theory." In M. Zey, ed., *Decision Making: Alternatives to Rational Choice Models*, pp. 273–303. Newbury Park, Calif.: Sage.

———. 1987. "Managerial Perspectives on Risk and Risk-Taking." *Management Science*, 33: 1404–18.

March, J. G., L. S. Sproull, and N. Tamuz. 1991. "Learning from Samples of One or Fewer." *Organization Science*, 2: 1–13.

Markus, H. R., and S. Kitayama. 1991. "Culture and Self: Implications for Cognition, Emotion, and Motivation." *Psychological Review*, 98: 224–53.

Marquis, D. G., and H. J. Reitz. 1969. "Effects of Uncertainty on Risk Taking in Individual and Group Decisions." *Behavioral Science*, 14: 281–88.

Mason, R. O., and I. I. Mitroff. 1981. *Challenging Strategic Planning Assumptions*. New York: John Wiley and Sons.

Mastanduno, M., D. A. Lake, and G. J. Ikenberry. 1989. "Toward a Realist Theory of State Action." *International Studies Quarterly*, 33: 457–74.

May, E. R. 1973. *"Lessons" of the Past: The Use and Misuse of History in American Foreign Policy*. London: Oxford University Press.

Mayseless, O., and A. W. Kruglanski. 1987. "What Makes You So Sure? Effects of Epistemic Motivation on Judgmental Confidence." *Organizational Behavior and Human Decision Processes*, 39: 162–83.

Mazarr, M. J. 1993. "The Military Dilemmas of Humanitarian Intervention." *Security Dialogue*, 24: 151–62.

Mazursky, D., and C. Ofir. 1990. " 'I Could Never Have Expected It to Happen': The Reversal of the Hindsight Bias." *Organizational Behavior and Human Decision Processes*, 46: 20–33.

McCall, N. L. 1991. "Assessing the Role of Air Power." In B. W. Watson and P. G. Tsouras, eds., *Operation Just Cause: The U.S. Intervention in Panama*, pp. 115–21. Boulder, Colo.: Westview Press.

McCauley, C. 1989. "The Nature of Social Influence in Groupthink: Compliance and Internalization." *Journal of Personality and Social Psychology*, 57: 250–60.

McDaniel, W. C., and F. Sistrunk. 1991. "Management Dilemmas and Decisions: Impact of Framing and Anticipated Response." *Journal of Conflict Resolution*, 35: 21–42.

McDaniels, T. L., and R. S. Gregory. 1991. "A Framework for Structuring Cross-Cultural Research in Risk and Decision Making." *Journal of Cross-Cultural Psychology*, 22: 103–28.

McElroy, R. W. 1992. *Morality and American Foreign Policy*. Princeton, N.J.: Princeton University Press.

McGuire, T. W., S. Kiesler, and J. Siegel. 1987. "Group and Computer-Mediated Discussion Effects in Risk Decision Making." *Journal of Personality and Social Psychology*, 52: 917–30.

McKeown, T. J. 1992. "Decision Processes and Co-operation in Foreign Policy." *International Journal*, 47: 402–19.

McNamara, R. S., with B. VanDeMark. 1995. *In Retrospect: The Tragedy and Lessons of Vietnam*. New York: Times Books.

McNaugher, T. L. 1987. "The Limits of Access: Projecting U.S. Forces to the Persian Gulf." In W. J. Olson, ed., *U.S. Strategic Interests in the Gulf Region*, pp. 173–87. Boulder, Colo.: Westview Press.

Medzini, M., ed. 1990. *Israel's Foreign Relations: Selected Documents, 1982–1984*, vol. 8. Jerusalem: Ministry for Foreign Affairs.

Meese, E. 1992. *With Reagan: The Inside Story*. Washington, D.C.: Regnery Gateway.

Meindle, J. R. 1990. "On Leadership: An Alternative to the Conventional Wisdom." In B. M. Staw and L. L. Cummings, eds., *Research in Organizational Behavior*, vol. 12, pp. 159–203. Greenwich, Conn.: JAI Press.

Mendelson, S. E. 1993. "Internal Battles and External Wars: Politics, Learning, and the Soviet Withdrawal from Afghanistan." *World Politics*, 45: 327–60.

Menges, C. C. 1988. *Inside the National Security Council*. New York: Simon and Schuster.

Menon, R. 1986. *Soviet Power and the Third World*. New Haven, Conn.: Yale University Press.

Mercer, J. 1993. "Independence or Interdependence: Testing Resolve Reputation." In J. Snyder and R. Jervis, eds., *Coping with Complexity in the International System*, pp. 163–89. Boulder, Colo.: Westview Press.

Merelman, R. M. 1966. "Learning and Legitimacy." *American Political Science Review*, 60: 548–61.

Metcalf, J., III. 1986. "Decision Making and the Grenada Rescue Operation." In J. G. March and R. Weissinger-Baylon, eds., *Ambiguity and Command: Organizational Perspectives on Military Decision Making*, pp. 277–97. Marshfield, Mass.: Pitman.

Metselaar, M. V., and B. Verbeek. 1997. "Beyond Decision-Making in Formal and Informal Groups: The Dutch Cabinet and the West New Guinea Conflict." In P. 't Hart, E. K. Stern, and B. Sundelius, eds., *Beyond Groupthink: Political Group Dynamics and Foreign Policy-Making*, pp. 95–122. Ann Arbor: Michigan University Press.

Midlarsky, M. I. 1991. "International Structure and the Learning of Cooperation: The Postwar Experience." In C. W. Kegley, ed., *The Long Postwar Peace: Contending Explanations and Projections*, pp. 105–22. New York: Harper Collins.

Milburn, T. W., and R. S. Billings. 1976. "Decision-Making Perspectives from Psychology: Dealing with Risk and Uncertainty." *American Behavioral Scientist*, 20: 111–26.

Milburn, T. W., and D. J. Christie. 1990. "Effort Justification as a Motive for Continuing War: The Vietnam Case." In B. Glad, ed., *Psychological Dimensions of War*, pp. 236–51. Newbury Park, Calif.: Sage.

Miller, B. 1994. "Exploring the Emergence of Great Power Concerts." *Review of International Studies*, 20: 327–48.

Miller, D. T., W. Turnbull, and C. McFarland. 1990. "Counterfactual Thinking and Social Perception: Thinking About What Might Have Been." In M. P. Zanna, ed., *Advances in Experimental Social Psychology*, vol. 23, pp. 305–31. San Diego, Calif.: Academic Press.

Minix, D. A. 1982. *Small Groups and Foreign Policy Decision-Making*. Washington, D.C.: University Press of America.

Mintz, A. 1993. "The Decision to Attack Iraq: A Noncompensatory Theory of Decision Making." *Journal of Conflict Resolution*, 37: 595–618.

Mintzberg, H. 1994. *The Rise and Fall of Strategic Planning: Reconceiving Roles for Planning, Plans, Planners*. New York: Free Press.

Mitchell, C. R. 1970. "Civil Strife and the Involvement of External Parties." *International Studies Quarterly*, 14: 166–94.

Mlotek, R., and S. Rosen. 1974. "The Calculus of Cost-Tolerance: Public Opinion and Foreign War." In J. D. Ben-Dak, ed., *The Future of Collective Violence: Societal and International Perspectives*, pp. 99–123. Lund, Sweden: Studentlitteratur.

Mlynář, Z. 1980. *Nightfrost in Prague: The End of Humane Socialism*. New York: Karz.

Modelski, G. 1964. "The International Relations of Internal War." In J. N. Rosenau, ed., *International Aspects of Civil Strife*, pp. 14–44. Princeton, N.J.: Princeton University Press.

Mohamed, A. A., and F. A. Wiebe. 1996. "Toward a Process Theory of Groupthink." *Small Group Research*, 27: 416–30.

Momyer, W. W. 1978. *Air Power in Three Wars*. Washington, D.C.: U.S. Government Printing Office.

Monroe, K. R. 1995. "Psychology and Rational Actor Theory." *Political Psychology*, 16: 1–21.

Moore, J. N. 1969. "The Control of Foreign Intervention in Internal Conflict." *Virginia Journal of International Law*, 9: 205–342.

Moore, R. A. 1990. "The Reagan Presidency and Foreign Policy." In D. M. Hill, R. A. Moore, and P. Williams, eds., *The Reagan Presidency: An Incomplete Revolution?* pp. 179–98. New York: St. Martin's Press.

Moreland, R. L., and J. M. Levine. 1992. "The Composition of Small Groups." In E. Lawler, B. Murkovsky, C. Ridgeway, and H. Walker, eds., *Advances in Group Processes*, vol. 9, pp. 237–80. Greenwich, Conn.: JAI Press.

Morgan, T. C., and S. H. Campbell. 1991. "Domestic Structure, Decisional Constraints, and War: So Why Kant Democracies Fight?" *Journal of Conflict Resolution*, 35: 187–211.

Morgenthau, H. J. 1967. "To Intervene or Not to Intervene." *Foreign Affairs*, 45: 425–36.

Morton, A. 1991. *Disasters and Dilemmas: Strategies for Real-Life Decision Making*. Oxford, Eng.: Basil Blackwell.

Moscovici, S. 1976. *Social Influence and Social Change*. London: Academic Press.

Motokawa, T. 1989. "Sushi Science and Hamburger Science." *Perspectives in Biology and Medicine*, 32: 489–504.

Mowday, R. T., and R. I. Sutton. 1993. "Organizational Behavior: Linking Individuals and Groups to Organizational Contexts." In L. W. Porter and M. R. Rosenzweig, eds., *Annual Review of Psychology*, vol. 44, pp. 195–229. Palo Alto, Calif.: Annual Reviews.

Mrozek, D. J. 1989. *Air Power and the Ground War in Vietnam*. Washington, D.C.: Pergamon-Brassey's.

Mueller, J. E. 1980. "The Search for the 'Breaking Point' in Vietnam: The Statistics of a Deadly Quarrel." *International Studies Quarterly*, 24: 497–519.

———. 1973. *War Presidents and Public Opinion*. New York: John Wiley.

Mullins, K., and A. Wildavsky. 1992. "The Procedural Presidency of George Bush." *Political Science Quarterly*, 107: 31–62.

Muravchick, J. 1983. "Misreporting Lebanon." *Policy Review*, no. 23: 11–66.

Murphy, P. J. 1981. *Brezhnev: Soviet Politician*. Jefferson, N.C.: McFarlane.

Musatov, V. 1992. "The Inside Story of the Invasion." *New Times*, Apr.: 37–39.

Mussington, D. 1994. *Understanding Contemporary International Arms Transfer*. Adelphi Paper, no. 291. London: International Institute for Strategic Studies.

Myers, D. G. 1982. "Polarizing Effects of Social Interaction." In H. Brandstätter, J. H. Davis, and G. Stocker-Kreichgauer, eds., *Group Decision Making*, pp. 125–61. London: Academic Press.

Myers, D. G., and S. J. Arenson. 1972. "Enhancement of Dominant Risk Tendencies in Group Discussion." *Psychological Reports*, 30: 615–23.

Myers, D. G., and H. Lamm. 1976. "The Group Polarization Phenomenon." *Psychological Bulletin*, 83: 602–27.

Myers, D. J. 1991a. "Patterns of Aspiring Hegemon Threat Perception and Strategic Response: Conclusions and Directions for Research." In D. J. Myers, ed., *Regional Hegemons: Threat Perception and Strategic Response*, pp. 305–60. Boulder, Colo.: Westview Press.

———. 1991b. "Threat Perception and Strategic Response of the Regional Hegemons: A Conceptual Overview." In D. J. Myers, ed., *Regional Hegemons: Threat Perception and Strategic Response*, pp. 1–29. Boulder, Colo.: Westview Press.

Nacos, B. L. 1990. *The Press, Presidents, and Crises*. New York: Columbia University Press.

Naor, A. 1993. *Begin in Power: A Personal Testimony*. Tel Aviv: Yediot Aharonot (in Hebrew).

———. 1986. *Cabinet at War: The Functioning of the Israeli Cabinet During the Lebanon War (1982)*. Tel Aviv: Lahav (in Hebrew).

National Research Council. 1989. *Improving Risk Communication*. Washington, D.C.: National Academic Press.

National Security Archives. 1994. *The Prague Spring, 1968: Translated Documents*. Washington, D.C. Mimeo.

Neff, D. 1981. *Warriors at Suez: Eisenhower Takes America into the Middle East*. New York: Linden Press.

Neuman, S. G. 1987a. "The Role of Military Assistance in Recent Wars." In S. G. Neuman and R. E. Harkavy, eds., *The Lessons of Recent Wars in the Third World*, vol. 2, pp. 115–55. Lexington, Mass.: D.C. Heath.

———. 1987b. "Third World Military Industries: Capabilities and Constraints in Recent Wars." In S. G. Neuman and R. E. Harkavy, eds., *The Lessons of Recent Wars in the Third World*, vol. 2, pp. 157–97. Lexington, Mass.: Lexington Books.

———. 1986. *Military Assistance in Recent Wars: The Dominance of the Superpowers*. Washington Papers, no. 122. New York: Praeger.

Neustadt, R. E., and E. R. May. 1986. *Thinking in Time: The Uses of History for Decision-Makers.* New York: Free Press.

Newland, S. J., and D. V. Johnson II. 1990. "The Military and Operational Significance of the Weinberger Doctrine." *Small Wars and Insurgencies,* 1: 171–90.

Newman, J. M. 1992. *JFK and Vietnam: Deception Intrigue and the Struggle for Power.* New York: Warner Books.

New York Times.

New York Times Magazine.

Nincic, D. J., and M. Nincic. 1995. "Commitment to Military Intervention: The Democratic Government as Economic Investor." *Journal of Peace Research,* 32: 413–26.

Nisbett, R. E., and L. Ross. 1980. *Human Inference: Strategies and Shortcomings of Social Judgment.* Englewood Cliffs, N.J.: Prentice Hall.

Nutt, P. C. 1989. *Making Tough Decisions: Tactics for Improving Managerial Decision-Making.* San Francisco, Calif.: Jossey-Bass.

Nye, J. S. 1988. "Neorealism and Neoliberalism." *World Politics,* 40: 235–51.

O'Brien, W. V. 1979. *U.S. Military Intervention: Law and Morality.* Washington Papers, no. 68. Beverly Hills, Calif.: Sage.

O'Neill, T. P., with W. Novak. 1987. *Man of the House: The Life and Political Memoirs of Speaker Tip O'Neill.* New York: Random House.

Onuf, N. G. 1989. *World of Our Making: Rules and Rule in Social Theory and International Relations.* Columbia: University of South Carolina Press.

Oren, N. 1982. "Prudence in Victory." In N. Oren, ed., *Termination of Wars,* pp. 147–63. Jerusalem: Magnes Press.

O'Shaughnessy, H. 1984. *Grenada: An Eyewitness Account of the U.S. Invasion and the Caribbean History That Provoked It.* New York: Dodd, Mead.

O'Sullivan, P. 1987. "The Geography of Wars in the Third World." In S. G. Neuman and R. E. Harkavy, eds., *The Lessons of Recent Wars in the Third World,* vol. 2, pp. 33–52. Lexington, Mass.: D. C. Heath.

Osborn, G. K., and W. J. Taylor, Jr. 1981. "The Employment of Force: Political-Military Considerations." In S. C. Sarkesian and W. L. Scully, eds., *U.S. Policy and Low-Intensity Conflict: Potential for Struggles in the 1980s,* pp. 17–47. New Brunswick, N.J.: Transaction Books.

Osborn, R. N., and D. H. Jackson. 1988. "Leaders, Riverboat Gamblers, or Purposeful Unintended Consequences in the Management of Complex, Dangerous Technologies." *Academy of Management Journal,* 31: 924–47.

Ostrom, C. W., and B. L. Job. 1986. "The President and the Political Use of Force." *American Political Science Review,* 80: 541–66.

Otway, H. 1992. "Public Wisdom, Expert Fallibility: Toward a Contextual Theory of Risk." In S. Krimsky and D. Golding, eds., *Social Theories of Risk,* pp. 215–28. Westport, Conn.: Praeger.

Ouchi, W. G. 1977. "The Relationship Between Organizational Structure and Organizational Control." *Administrative Science Quarterly,* 22: 95–113.

Page, M. 1987. "Risk in Defence." In W. T. Singleton and J. Hovden, eds., *Risk and Decisions,* pp. 191–205. Chichester, Eng.: John Wiley and Sons.

Pallak, M. S., S. R. Sogin, and A. Van Zante. 1974. "Bad Decisions: Effects of Volition, Locus of Causality and Negative Consequences on Attitude Change." *Journal of Personality and Social Psychology,* 30: 217–27.

Palmer, B. 1984. *The 25-Year War: America's Military Role in Vietnam*. Lexington: University Press of Kentucky.

Palmer, D. R. 1984. *Summons of the Trumpets*. New York: Ballantine Books.

Palmer, G. 1978. *The McNamara Strategy and the Vietnam War*. Westport, Conn.: Greenwood Press.

Palmlund, I. 1992. "Social Drama and Risk Evaluation." In S. Krimsky and D. Golding, eds., *Social Theories of Risk*, pp. 197–212. Westport, Conn.: Praeger.

Pape, R. A. 1992. "Coercion and Military Strategy: Why Denial Works and Punishment Doesn't." *Journal of Strategic Studies*, 15: 423–75.

Pape, R. A., Jr. 1990. "Coercive Air Power in the Vietnam War." *International Security*, 15(2): 103–46.

Park, W.-W. 1990. "A Review of Research on Groupthink." *Journal of Behavioral Decision Making*, 3: 229–45.

Partem, M. C. 1983. "The Buffer System in International Relations." *Journal of Conflict Resolution*, 27: 3–26.

Pastor, R. A. 1993. "Forward to the Beginning: Widening the Scope for Global Collective Action." In L. W. Reed and C. Kaysen, eds., *Emerging Norms of Justified Intervention*, pp. 133–47. Cambridge, Mass.: American Academy of Arts and Sciences.

———. 1990. "The United States and the Grenada Revolution: Who Pushed First and Why?" In J. Heine, ed., *A Revolution Aborted: The Lessons of Grenada*, pp. 181–214. Pittsburgh, Pa.: University of Pittsburgh Press.

———. 1987. "The Reagan Administration and Latin America: Eagle Insurgent." In K. A. Oye, R. J. Lieber, and D. Rothchild, eds., *Eagle Resurgent? The Reagan Era in American Foreign Policy*, pp. 359–92. Boston: Little, Brown.

———. 1985. "U.S. Policy Toward the Caribbean: Continuity and Change." In P. M. Dunn and B. W. Watson, eds., *American Intervention in Grenada: The Implications of Operation "Urgent Fury,"* pp. 15–28. Boulder, Colo.: Westview Press.

Patchen, M. 1988. *Resolving Disputes Between Nations: Coercion or Conciliation?* Durham, N.C.: Duke University Press.

Paul, D. W. 1971. "Soviet Foreign Policy and the Invasion of Czechoslovakia: A Theory and a Case Study." *International Studies Quarterly*, 15: 159–202.

Payne, A. 1990. "The Foreign Policy of the Peoples' Revolutionary Government." In J. Heine, ed., *A Revolution Aborted: The Lessons of Grenada*, pp. 123–52. Pittsburgh, Pa.: University of Pittsburgh Press.

———. 1984. *The International Crisis in the Caribbean*. Beckenhan, Eng.: Croom Helm.

Payne, A., P. Sutton, and T. Thorndike. 1984. *Grenada: Revolution and Invasion*. New York: St. Martin's Press.

Payne, J. W. 1985. "Psychology of Risky Decisions." In G. Wright, ed., *Behavioral Decision Making*, pp. 3–23. New York: Plenum Press.

———. 1975. "Relation of Perceived Risk to Preferences Among Gambles." *Journal of Experimental Psychology: Human Perception and Performance*, 104: 86–94.

———. 1973. "Alternative Approaches to Decision Making Under Risk: Moments vs. Risk Dimension." *Psychological Bulletin*, 80: 439–53.

Payne, J. W., J. R. Bettman, and E. J. Johnson. 1992. "Behavioral Decision Research: A Constructive Processing Perspective." In M. R. Rosenzweig and L. W. Porter, eds., *Annual Review of Psychology*, vol. 43, pp. 87–131. Palo Alto, Calif.: Annual Reviews.

Payne, J. W., and M. L. Braunstein. 1978. "Risky Choice: An Examination of Information Acquisition Behavior." *Memory and Cognition*, 6: 554–61.

Pearson, F. S. 1974a. "Foreign Military Interventions and Domestic Disputes." *International Studies Quarterly*, 18: 259–89.

———. 1974b. "Geographic Proximity and Foreign Military Intervention." *Journal of Conflict Resolution*, 18: 432–60.

Pearson, F. S., and R. Baumann. 1989. "International Military Intervention in Sub-Saharan African Subsystems." *Journal of Political and Military Sociology*, 17: 115–50.

———. 1984. "Toward a Regional Model of International Military Intervention: The Middle Eastern Experience." *Arms Control*, 4: 187–222.

———. 1974. "Foreign Military Intervention by Large and Small Powers." *International Interactions*, 1: 273–78.

Pearson, F. S., R. A. Baumann, and J. J. Pickering. 1994. "Military Intervention and Realpolitik." In F. W. Wayman and P. F. Diehl, eds., *Reconstructing Realpolitik*, pp. 205–25. Ann Arbor: University of Michigan Press.

Peleg, I. 1988. "The Foreign Policy of Herut and the Likud." In B. Reich and G. R. Kieval, eds., *Israeli National Security Policy: Political Actors and Perspectives*, pp. 55–78. Westport, Conn.: Greenwood Press.

———. 1987. *Begin's Foreign Policy, 1977–1983: Israel's Move to the Right*. Westport, Conn.: Greenwood Press.

Pelz, S. 1991. "Changing International Systems, the World Balance of Power, and the United States, 1776–1976." *Diplomatic History*, 15: 47–81.

Pemberton, G. J. 1987. *All the Way: Australia's Road to Vietnam*. Sydney: Allen and Unwin.

Pentagon Papers: The Defense Department History of United States Decisionmaking on Vietnam. 1971. Senator Gravel ed., vols. 3 and 4. Boston: Beacon Press.

Perlmutter, A. 1987. *The Life and Times of Menachem Begin*. Garden City, N.Y.: Doubleday.

Perry, M. 1989. *Four Stars*. Boston: Houghton Mifflin.

Peterson, S. A., and R. Lawson. 1989. "Risky Business: Prospect Theory and Politics." *Political Psychology*, 10: 325–39.

Pfaltzgraff, R. L., Jr., and J. K. Davis, eds. 1990. *National Security Decisions: The Participants Speak*. Lexington, Mass.: Lexington Books.

Pfeffer, J. 1981. "Management as Symbolic Action: The Creation and Maintenance of Organizational Paradigms." In L. L. Cummings and B. M. Staw, eds., *Research in Organizational Behavior*, vol. 3, pp. 1–52. Greenwich, Conn.: JAI Press.

Phillips, L. D., and C. N. Wright. 1977. "Cultural Differences in Viewing Uncertainty and Assessing Probabilities." In H. Jungermann and G. De Zeeuw, eds., *Decision Making and Change in Human Affairs*, pp. 507–19. Dordrecht, Neth.: D. Reidel.

Pickett, J. R. 1977. "Airlift and Military Intervention." In E. P. Stern, ed., *The Limits of Military Intervention*, pp. 137–50. Beverly Hills, Calif.: Sage.

Piotrowski, H. 1989. "The Structure of the International System." In P. J. Schraeder, ed., *Intervention in the 1980's: U.S. Foreign Policy in the Third World*, pp. 175–89. Boulder, Colo.: Lynne Rienner.

Plax, T. G., and L. B. Rosenfeld. 1976. "Correlates of Risky Decision-Making." *Journal of Personality Assessment*, 40: 413–18.

Plous, S. 1993. *The Psychology of Judgment and Decision Making*. Philadelphia: Temple University Press.

————. 1989. "Thinking the Unthinkable: The Effects of Anchoring on Likelihood Estimates of Nuclear War." *Journal of Applied Social Psychology*, 19: 67–91.

Podhoretz, N. 1982. *Why We Were in Vietnam*. New York: Simon and Schuster.

Porter, B. D. 1984. *The USSR in Third World Conflicts: Soviet Arms and Diplomacy in Local Wars, 1945–1980*. Cambridge, Eng.: Cambridge University Press.

Porter, G., ed. 1979. *Vietnam: The Definitive Documentation of Human Decisions*, vol. 2. Stanfordville, N.Y.: Earl M. Coleman.

Post, J. M. 1993. "Current Concepts of the Narcissistic Personality: Implication for Political Psychology." *Political Psychology*, 14: 99–121.

————. 1991. "The Impact of Crisis-Induced Stress on Policy Makers." In A. L. George, ed., *Avoiding War Problems of Crisis Management*, pp. 471–94. Boulder, Colo.: Westview Press.

Post, J. M., and R. S. Robins. 1993. *When Illness Strikes the Leader: The Dilemma of the Captive King*. New Haven, Conn.: Yale University Press.

Powell, C. L., with J. E. Persico. *My American Journey*. New York: Random House.

Preston, T. 1997. "'Following the Leader': The Impact of U.S. Presidential Style Upon Advisory Group Dynamics, Structure, and Decision." In P. 't Hart, E. K. Stern, and B. Sundelius, eds., *Beyond Groupthink: Political Group Dynamics and Foreign Policy-Making*, pp. 191–248. Ann Arbor: Michigan University Press.

Priem, R. L., D. A. Harrison, and N. K. Muir. 1995. "Structural Conflict and Consensus: Outcomes in Group Decision Making." *Journal of Management*, 21: 691–710.

Putnam, R. D. 1988. "Diplomacy and Domestic Politics: The Logic of Two-Level Games." *International Organization*, 42: 427–60.

Quandt, W. B. 1993. *Peace Process: American Diplomacy and the Arab Israeli Conflict Since 1967*. Washington, D.C.: Brookings Institution.

————. 1984. "Reagan's Lebanon Policy: Trial and Error." *Middle East Journal*, 38: 237–54.

Quayle, D. 1994. *Standing Firm: A Vice Presidential Memoir*. New York: Harper Collins.

Quester, G. H. 1985. "Grenada and the News Media." In P. M. Dunn and B. W. Watson, eds., *American Intervention in Grenada: The Implications of Operation "Urgent Fury,"* pp. 109–27. Boulder, Colo.: Westview Press.

Quigley, J. 1992. *The Ruses for War: American Interventionism Since World War II*. Buffalo, N.Y.: Prometheus Books.

Quinlivan, J. T. 1995–96. "Force Requirements in Stability Operations." *Parameters*, 25(4): 59–69.

Ra'anan, G. D. 1979. *The Evolution of the Soviet Use of Surrogates in Military Relations with the Third World, with Particular Emphasis on Cuban Participation in Africa*. Santa Monica, Calif.: Rand Corporation, P-6420.

Rabin, Y. 1983. *The War in Lebanon*. Tel Aviv: Am Oved (in Hebrew).

Rabinovich, I. 1985. *The War for Lebanon, 1970–1985* (rev. ed.). Ithaca, N.Y.: Cornell University Press.

Rabinovich, I., and H. Zamir. 1982. *War and Crisis in Lebanon, 1975–1981*. Tel Aviv: Hakibbutz Hameuchad (in Hebrew).

Race, J. 1976. "Vietnam Intervention: Systematic Distortion in Policy Making." *Armed Forces and Society*, 2: 377–96.

Ragin, C. C. 1994. *Constructing Social Research*. Thousand Oaks, Calif.: Pine Forge Press.

————. 1987. *The Comparative Method: Moving Beyond Qualitative and Quantitative Strategies*. Berkeley: University of California Press.

Ranyard, R. H. 1976. "Elimination by Aspects as a Decision Rule for Risky Choice." *Acta Psychologica*, 40: 299–310.

Raven, B. H. 1990. "Political Applications of the Psychology of Interpersonal Influence and Social Power." *Political Psychology*, 11: 493–520.

Rayner, S. 1992. "Cultural Theory and Risk Analysis." In S. Krimsky and D. Golding, eds., *Social Theories of Risk*, pp. 83–115. Westport, Conn.: Praeger.

Reagan, R. 1990. *An American Life*. New York: Simon and Schuster.

Reason, J. 1990. *Human Error*. Cambridge, Eng.: Cambridge University Press.

Record, J. 1981. *The Rapid Deployment Force and U.S. Military Intervention in the Persian Gulf*. Cambridge, Mass.: Institute for Foreign Policy Analysis.

Reed, L. W., and C. Kaysen, eds. 1993. *Emerging Norms of Justified Intervention*. Cambridge, Mass.: American Academy of Arts and Sciences.

Reedy, G. 1982. *Lyndon B. Johnson: A Memoir*. New York: Andrews and McMeel.

Reitman, W. R. 1964. "Heuristic Decision Procedures, Open Constraints, and the Structure of Ill-Defined Problems." In M. W. Shelly and G. L. Bryan, eds., *Human Judgment and Optimality*, pp. 282–315. New York: John Wiley.

Remington, R. A. 1971. *The Warsaw Pact: Case Studies in Communist Conflict Resolution*. Cambridge, Mass.: MIT Press.

Renn, O. 1992. "Concepts of Risk: A Classification." In S. Krimsky and D. Golding, eds., *Social Theories of Risk*, pp. 53–79. Westport, Conn.: Praeger.

Renn, O., W. J. Burns, J. X. Kasperson, R. E. Kasperson, and P. Slovic. 1992. "The Social Amplification of Risk: Theoretical Foundations and Empirical Applications." *Journal of Social Issues*, 48(4): 137–60.

Renn, O., and D. Levine. 1991. "Credibility and Trust in Risk Communication." In R. E. Kasperson and P. J. M. Stallen, eds., *Communicating Risks to the Public: International Perspectives*, pp. 175–218. Dordrecht, Neth.: Kluwer.

Renner, M. 1992. *Economic Adjustment After the Cold War: Strategies for Conversion*. Dartmouth, Eng.: Aldershot.

Rice, C. 1984. *The Soviet Union and the Czechoslovak Army, 1948–1983: Uncertain Allegiance*. Princeton, N.J.: Princeton University Press.

Riddell, T. A. 1975. "A Political Economy of the American War in Indo-China: Its Costs and Consequences." Ph.D. diss., American University, Washington, D.C.

Ridley, D. R., P. D. Young, and D. E. Johnson. 1981. "Salience as a Dimension of Individual and Group Risk Taking." *Journal of Psychology*, 109: 283–91.

Rim, Y. 1966. "Machiavellianism and Decisions Involving Risk." *British Journal of Social and Clinical Psychology*, 5: 30–36.

————. 1964. "Personality and Group Decisions Involving Risk." *Psychological Record*, 14: 37–45.

Risse-Kappen, T. 1991. "Public Opinion, Domestic Structure, and Foreign Policy in Liberal Democracies." *World Politics*, 43: 479–512.

Roberts, A. 1993. "Humanitarian War: Military Intervention and Human Rights." *International Affairs*, 69: 429–49.

Roberts, C. 1965. *LBJ's Inner Circle*. New York: Delacourte Press.

Robinson, S., and E. Weldon. 1993. "Feedback Seeking in Groups: A Theoretical Perspective." *British Journal of Social Psychology*, 32: 71–86.

Rock, S. R. 1988. "Risk Theory Reconsidered: American Success and German Failure in the Coercion of Britain, 1890–1914." *Journal of Strategic Studies*, 11: 342–64.

Rockman, B. A. 1991. "The Leadership Style of George Bush." In C. Campbell and B. A. Rockman, eds., *The Bush Presidency: First Appraisals*, pp. 1–36. Chatham, N.J.: Chatham House Publishers.

Roeder, P. G. 1984. "Soviet Policies and Kremlin Politics." *International Studies Quarterly*, 28: 171–93.

Roese, N. J., and J. M. Olson. 1995. "Counterfactual Thinking: A Critical Review." In N. J. Roese and J. M. Olson, eds., *What Might Have Been: The Social Psychology of Counterfactual Thinking*, pp. 1–55. Mahwah, N.J.: Lawrence Erlbaum.

Ropelewski, R. R. 1990. "Planning, Precision and Surprise Led to Panama Successes." *Armed Forces Journal International*, Feb.: 26–32.

Rose, R. 1991. *The Postmodern President: George Bush Meets the World*. Chatham, N.J.: Chatham House Publishers.

Rosecrance, R. N. 1992. "A New Concert of Powers." *Foreign Affairs*, 71(2): 64–82.

———. 1963. *Action and Reaction in World Politics*. Boston: Little, Brown.

Rosen, S. 1972. "War Powers and the Willingness to Suffer." In B. M. Russett, ed., *Peace, War, and Numbers*, pp. 167–83. Beverly Hills, Calif.: Sage.

Rosen, S. P. 1995. "Military Effectiveness: Why Society Matters." *International Security*, 19(4): 5–31.

———. 1982. "Vietnam and the American Theory of Limited War." *International Security*, 7(2): 83–113.

Rosenau, J. N. 1981. "Foreign Intervention as Adaptive Behavior." In J. N. Rosenau, ed., *The Study of Political Adaptation*, pp. 125–47. London: Frances Pinter.

———. 1969. "Intervention as a Scientific Concept." *Journal of Conflict Resolution*, 13: 149–71.

———. 1968. "The Concept of Intervention." *Journal of International Affairs*, 22: 165–76.

Ross, A. L. 1991. "The Arming of the Third World: Patterns and Trends." *SAIS Review*, 11(2): 69–94.

Ross, D. 1984. "Risk Aversion in Soviet Decisionmaking." In J. Valenta and W. C. Potter, eds., *Soviet Decisionmaking for National Security*, pp. 237–51. London: Allen and Unwin.

Ross, D. B. 1990. "Soviet Behavior Toward the Lebanon War, 1982–84." In G. W. Breslauer, ed., *Soviet Strategy in the Middle East*, pp. 99–121. Boston: Unwin Hyman.

Ross, L., D. Greene, and D. House. 1977. "The 'False Consensus' Effect: An Egocentric Bias in Social Perception and Attribution Processes." *Journal of Experimental Social Psychology*, 13: 279–301.

Ross, L., and R. E. Nisbett. 1991. *The Person and the Situation: Perspectives of Social Psychology*. New York: McGraw-Hill.

Ross, T. E. 1986. "Buffer States: A Geographer's Perspective." In J. Chay and T. E. Ross, eds., *Buffer States in World Politics*, pp. 11–28. Boulder, Colo.: Westview Press.

Rostow, W. W. 1972. *The Diffusion of Power: An Essay in Recent History*. New York: Macmillan.

Rotter, A. J. 1987. *The Path to Vietnam*. Ithaca, N.Y.: Cornell University Press.

Rotter, J. B. 1966. "Generalized Expectancies for Internal Versus External Control of Reinforcement." *Psychological Monographs: General and Applied*, 80(609): 1–28.

Rowe, W. D. 1977. *An Anatomy of Risk*. New York: John Wiley and Sons.

Rubinstein, A. Z. 1988. *Moscow's Third World Strategy*. Princeton, N.J.: Princeton University Press.

Rubner, R. 1985–86. "The Reagan Administration, the 1973 War Powers Resolution, and the Invasion of Grenada." *Political Science Quarterly*, 100: 627–48.

Ruggie, J. G. 1983. "Continuity and Transformation in the World Polity: Toward a Neorealist Synthesis." *World Politics*, 35: 261–85.

Rusk, D. 1990. *As I Saw It*. New York: W. W. Norton.

Russett, B. 1990. *Controlling the Sword: The Democratic Governance of National Security*. Cambridge, Mass.: Harvard University Press.

Rystad, G. 1982. *Prisoners of the Past? The Munich Syndrome and Makers of American Foreign Policy in the Cold War Era*. Lund, Sweden: CWK Gleerup.

Sabrosky, A. N. 1990. "Applying Military Force: The Future Significance of the Weinberger Doctrine." *Small Wars and Insurgencies*, 1: 191–201.

Salancik, G. R. 1977. "Commitment Is Too Easy!" *Organizational Dynamics*, 6(1): 62–80.

Sandford, G., and R. Vigilante. 1984. *Grenada: The Untold Story*. Lanham, Md.: Madison Books.

Sandler, S. 1993. *The State of Israel, the Land of Israel: The Statist and Ethnonational Dimensions of Foreign Policy*. Westport, Conn.: Greenwood Press.

Saris, W. E., and I. N. Gallhofer. 1984. "Formulation of Real Life Decisions: A Study of Foreign Policy Decisions." *Acta Psychologica*, 56: 247–65.

Scammon, R. M. 1986. "Appendix: Knesset Election Results, 1949–1981." In H. R. Penniman and D. J. Elazar, eds., *Israel at the Polls, 1981: A Study of the Knesset Election*, pp. 266–67. Bloomington: Indiana University Press.

Scarborough, G. I. 1988. "Polarity, Power and Risk in International Disputes." *Journal of Conflict Resolution*, 32: 511–33.

Schafer, M., and S. Crichlow. 1996. "Antecedents of Groupthink: A Quantitative Approach." *Journal of Conflict Resolution*, 40: 415–35.

Schafer, M., M. D. Young, and S. G. Walker. 1996. "U.S. Presidents as Conflict Managers: The Operational Codes of George Bush and Bill Clinton." Paper prepared for the nineteenth annual scientific meeting of the International Society of Political Psychology, Vancouver, British Columbia, June 30–July 3.

Schandler, H. Y. 1977. *Lyndon Johnson and Vietnam: The Unmaking of a President*. Princeton, N.J.: Princeton University Press.

Scheffer, D. J. 1992. "Toward a Modern Doctrine of Humanitarian Intervention." *University of Toledo Law Review*, 23: 253–93.

Schelling, T. C. 1966. *Arms and Influence*. New Haven, Conn.: Yale University Press.

Schiff, Z. 1984. "Lebanon: Motivations and Interests in Israel's Policy." *Yearbook of World Affairs*, 38: 220–27.

———. 1983. "The Green Light." *Foreign Policy*, no. 50: 73–85.

Schiff, Z., and E. Ya'ari. 1984. *Deceitful War*. Tel Aviv: Schocken (in Hebrew).

Schissler, J. 1988a. "The Impact of War Powers Resolution on Crisis Decision Making." In H. Haftendorn and J. Schissler, eds., *The Reagan Administration: A Reconstruction of American Strength?* pp. 215–29. Berlin: Walter de Gruyter.

———. 1988b. "Political Culture and the Reagan Administration." In H. Haftendorn and J. Schissler, eds., *The Reagan Administration: A Reconstruction of American Strength?* pp. 31–49. Berlin: Walter de Gruyter.

Schmid, A. P., with E. Berends. 1985. *Soviet Military Interventions Since 1945*. New Brunswick, N.J.: Transaction Books.

Schneider, J. C. 1988. "Guns Versus Butter: Vietnam's Effect on Congressional Support for the Great Society." In H. Jones, ed., *The Foreign and Domestic Dimensions of Modern Warfare: Vietnam, Central America, and Nuclear Strategy*, pp. 94–108. Tuscaloosa: University of Alabama Press.

Schoemaker, P. J. H. 1989. "Preferences for Information on Probabilities Versus Prizes: The Role of Risk-Taking Attitudes." *Journal of Risk and Uncertainty*, 2: 37–60.

———. 1982. "The Expected Utility Model: Its Variants, Purposes, Evidence and Limitations." *Journal of Economic Literature*, 20: 529–63.

———. 1980. *Experiments on Decisions Under Risk: The Expected Utility Hypothesis*. Boston: Martinus Nijhoff.

Schoenbaum, T. J. 1988. *Waging Peace and War: Dean Rusk in the Truman, Kennedy and Johnson Years*. New York: Simon and Schuster.

Schraeder, P. J., ed. 1989a. *Intervention in the 1980's: U.S. Foreign Policy in the Third World*. Boulder, Colo.: Lynne Rienner.

———. 1989b. "U.S. Intervention in Perspective." In P. J. Schraeder, ed., *Intervention in the 1980's: U.S. Foreign Policy in the Third World*, pp. 283–97. Boulder, Colo.: Lynne Rienner.

Schroeder, P. 1994. "Historical Reality vs. Neo-realist Theory." *International Security*, 19(1): 108–48.

Schulman, P. R. 1989. "The 'Logic' of Organizational Irrationality." *Administration and Society*, 21: 31–53.

Schulze, K. E. 1996. "Perceptions and Misperceptions: Influences on Israeli Intelligence Estimates During the 1982 Lebanon War." *Journal of Conflict Studies*, 16: 134–52.

Schurr, P. H. 1987. "Effects of Gain and Loss Decision Frames on Risky Purchase Negotiations." *Journal of Applied Psychology*, 72: 351–58.

Schwarz, B. C. 1994. *Casualties, Public Opinion, and U.S. Military Intervention: Implications for U.S. Regional Deterrence Strategies*. Santa Monica, Calif.: Rand Corporation, MR-431-A/AF.

Schwarz, U. 1970. *Confrontation and Intervention in the Modern World*. Dobbs Ferry, N.Y.: Oceana.

Schwarzkopf, N. H., with P. Petre. 1992. *It Doesn't Take a Hero: The Autobiography*. New York: Bantam Books.

Schwenk, C. R. 1986. "Information, Cognitive Biases, and Commitment to a Course of Action." *Academy of Management Review*, 11: 298–310.

———. 1984. "Cognitive Simplification Processes in Strategic Decision-Making." *Strategic Management Journal*, 5: 111–28.

Scodel, A., P. Ratoosh, and J. S. Minas. 1959. "Some Personality Correlates of Decision-Making Under Conditions of Risk." *Behavioral Science*, 4: 19–28.

Scott, A. M. 1970. "Military Intervention by the Great Powers: The Rules of the Game." In W. Zartman, ed., *Czechoslovakia: Intervention and Impact*, pp. 85–104. New York: New York University Press.

Scranton, M. E. 1991. *The Noriega Years: U.S. Panamian Relations, 1981–1990*. Boulder, Colo.: Lynne Rienner.

Seabury, P., and W. A. McDougall, eds. 1984. *The Grenada Papers*. San Francisco: Institute for Contemporary Studies Press.

Seitz, M. E. 1991. "Command, Control, Communications, and Intelligence (C³I)

Factors." In B. W. Watson and P. G. Tsouras, eds., *Operation Just Cause: The U.S. Intervention in Panama*, pp. 105–13. Boulder, Colo.: Westview Press.

Seliktar, O. 1988. "Israel's Menachem Begin." In B. Kellerman and J. Z. Rubin, eds., *Leadership and Negotiation in the Middle East*, pp. 30–48. New York: Praeger.

Sella, A. 1983. "The USSR and the War in Lebanon: Mid 1982." *RUSI*, 128(2): 35–41.

Shackle, G. L. S. 1961. *Decision, Order and Time in Human Affairs*. Cambridge, Eng.: Cambridge University Press.

Shafer, D. M. 1988. *Deadly Paradigms: The Failure of U.S. Counterinsurgency Policy*. Princeton, N.J.: Princeton University Press.

Shafir, E. 1992. "Prospect Theory and Political Analysis: A Psychological Perspective." *Political Psychology*, 13: 311–22.

Shafir, E., I. Simonson, and A. Tversky. 1993. "Reason-Based Choice." *Cognition*, 49: 11–36.

Shamir, Y. 1994. *Summing-Up*. Tel Aviv: Edanim, Yediot Aharonot (in Hebrew).

Shapira, Z. 1995. *Risk Taking: A Managerial Perspective*. New York: Russell Sage Foundation.

Shapley, D. 1993. *Promise and Power: The Life and Times of Robert McNamara*. Boston: Little, Brown.

Sharon, A., with D. Chanoff. 1989. *Warrior: The Autobiography of Ariel Sharon*. Tel Aviv: Steimatzky.

Sharp, U. S. G. 1978. *Strategy for Defeat: Vietnam in Retrospect*. San Rafael, Calif.: Presidio Press.

Sherman, S. J. 1970. "Attitudinal Effects of Unforeseen Consequences." *Journal of Personality and Social Psychology*, 16: 510–20.

Sherman, S. J., and E. Corty. 1984. "Cognitive Heuristics." In R. S. Wyer and T. K. Srull, eds., *Handbook of Social Cognition*, vol. 1, pp. 189–286. Hillsdale, N.J.: Lawrence Erlbaum.

Sherman, S. J., and A. R. McConnell. 1995. "Dysfunctional Implications of Counterfactual Thinking: When Alternatives to Reality Fail Us." In N. J. Roese and J. M. Olson, eds., *What Might Have Been: The Social Psychology of Counterfactual Thinking*, pp. 199–231. Mahwah, N.J.: Lawrence Erlbaum.

Shiffer, S. 1984. *Snow Ball: The Story Behind the Lebanon War*. Tel Aviv: Edanim (in Hebrew).

Shimko, K. L. 1991. *Images and Arms Control: Perceptions in the Reagan Administration*. Ann Arbor: University of Michigan Press.

Shrader-Frechette, K. S. 1991. *Risk and Rationality: Philosophical Foundations for Populist Reforms*. Berkeley: University of California Press.

Shultz, G. P. 1993. *Turmoil and Triumph: My Years as Secretary of State*. New York: Charles Scribner's Sons.

Shultz, R. 1978. "Breaking the Will of the Enemy During the Vietnam War: The Operationalization of the Cost-Benefit Model of Counterinsurgency Warfare." *Journal of Peace Research*, 15: 109–29.

Silver, E. 1984. *Begin: The Haunted Prophet*. New York: Random House.

Simes, D. K. 1975. "The Soviet Invasion of Czechoslovakia and the Limits of Kremlinology." *Studies in Comparative Communism*, 8: 174–80.

Simon, S. W. 1975. "The Role of Outsiders in the Cambodian Conflict." *Orbis*, 19: 209–30.

Simons, W. E. 1994. "U.S. Coercive Pressure on North Vietnam, Early 1965." In A. L. George and W. E. Simons, eds., *The Limits of Coercive Diplomacy* (2nd ed.), pp. 133–73. Boulder, Colo.: Westview Press.

Simonson, I., and P. Nye. 1992. "The Effect of Accountability on Susceptibility to Decision Errors." *Organizational Behavior and Human Decision Processes*, 51: 416–46.

Simonson, I., and B. M. Staw. 1992. "Deescalation Strategies: A Comparison of Techniques for Reducing Commitments to Losing Courses of Action." *Journal of Applied Psychology*, 77: 419–26.

Simpson, C. 1995. *Presidential Directives: National Security Policy During the Reagan-Bush Years*. Boulder, Colo.: Westview Press.

Singh, J. V. 1986. "Performance, Slack and Risk Taking in Organizational Decision Making." *Academy of Management Journal*, 29: 562–85.

Sitkin, S. B., and A. L. Pablo. 1992. "Reconceptualizing the Determinants of Risk Behavior." *Academy of Management Review*, 17: 9–38.

Sitkin, S. B., and L. R. Weingart. 1996. "Determinants of Risky Decision-Making Behavior: A Test of the Mediating Role of Risk Perception and Propensity." *Academy of Management Journal*, 38: 1573–92.

Sjoberg, G., N. Williams, T. R. Vaughan, and A. Sjoberg. 1991. "The Case Study Approach in Social Research: Basic Methodological Issues." In J. R. Feagin, A. M. Orum, and G. Sjoberg, eds., *A Case for the Case Study*, pp. 27–79. Chapel Hill: University of North Carolina.

Sjöberg, L. 1987. "Risk, Power and Rationality: Conclusions of a Research Project on Risk Generation and Risk Assessment in a Societal Perspective." In L. Sjöberg, ed., *Risk and Society: Studies of Risk Generation and Reaction to Risk*, pp. 239–43. London: Allen and Unwin.

———. 1980. "The Risks of Risk Analysis." *Acta Psychologica*, 45: 301–21.

———. 1979. "Strength of Belief and Risk." *Policy Sciences*, 11: 39–57.

Skilling, H. G. 1976. *Czechoslovakia's Interrupted Revolution*. Princeton, N.J.: Princeton University Press.

Slater, J. 1993–94. "The Domino Theory and International Politics: The Case of Vietnam." *Security Studies*, 3: 186–224.

Slovic, P. 1992. "Perception of Risk: Reflections on the Psychometric Paradigm." In S. Krimsky and D. Golding, eds., *Social Theories of Risk*, pp. 117–52. Westport, Conn.: Praeger.

———. 1975. "Choice Between Equally Valued Alternatives." *Journal of Experimental Psychology: Human Perception and Performance*, 1: 280–87.

———. 1964. "Assessment of Risk-Taking Behavior." *Psychological Bulletin*, 61: 220–33.

Slovic, P., B. Fischhoff, and S. Lichtenstein. 1984. "Behavioral Decision Theory Perspectives on Risk and Safety." *Acta Psychologica*, 56: 183–203.

———. 1982a. "Rating the Risks: The Structure of Expert and Lay Perceptions." In C. Hohenemser and J. X. Kasperson, eds., *Risk in the Technological Society*, pp. 141–66. Boulder, Colo.: Westview Press.

———. 1982b. "Response Mode, Framing, and Information-Processing Effects in Risk Assessment." In R. M. Hogarth, ed., *New Directions for Methodology of Social and Behavioral Science: Question Framing and Response Consistency*, pp. 21–36. San Francisco: Jossey-Bass.

———. 1981. "Perceived Risk: Psychological Factors and Social Implications." *Proceedings of the Royal Society* (London), A376: 17–34.

———. 1980. "Facts and Fears: Understanding Perceived Risk." In R. C. Schwing and W. A. Albers, Jr., eds., *Societal Risk Assessment: How Safe Is Safe Enough?* pp. 181–214. New York: Plenum Press.

———. 1977. "Behavioral Decision Theory." In M. R. Rosenzweig and L. W. Porter, eds., *Annual Review of Psychology*, vol. 28, pp. 1–39. Palo Alto, Calif.: Annual Reviews.

———. 1976. "Cognitive Processes and Societal Risk Taking." In J. S. Carrol and J. W. Payne, eds., *Cognition and Social Behavior*, pp. 165–84. Hillsdale, N.J.: Lawrence Erlbaum.

Slovic, P., and S. Lichtenstein. 1971. "Comparison of Bayesian and Regression Approaches to the Study of Information Processing in Judgment." *Organizational Behavior and Human Performance*, 6: 649–744.

———. 1968. "Relative Importance of Probabilities and Payoffs in Risk Taking." *Journal of Experimental Psychology Monographs*, 78 (no. 3, part II): 1–18.

Small, M. 1988. *Johnson, Nixon and the Doves*. Brunswick, N.J.: Rutgers University Press.

———. 1987. "Influencing the Decision-Makers: The Vietnam Experience." *Journal of Peace Research*, 24: 185–98.

Smart, C., and I. Vertinsky. 1977. "Designs for Crisis Decision Units." *Administrative Science Quarterly*, 22: 640–57.

Smith, M. 1990. "The Reagan Presidency and Foreign Policy." In J. Hogan, ed., *The Reagan Years: The Record in Presidential Leadership*, pp. 259–85. Manchester, Eng.: Manchester University Press.

———. 1988. "The Reagan Administration's Foreign Policy, 1981–1985: Learning to Live with Uncertainty." *Political Studies*, 36: 52–73.

Smith, M. J. 1989. "Ethics and Intervention." *Ethics and International Affairs*, 3: 1–26.

Smith, P. B., and M. H. Bond. 1993. *Social Psychology Across Cultures: Analysis and Perspectives*. New York: Harvester Wheatsheaf.

Smith, R. B. 1991. *An International History of the Vietnam War: The Making of a Limited War, 1965–66*, vol. 3. New York: St. Martin's Press.

———. 1985. *An International History of the Vietnam War: The Kennedy Strategy*, vol. 2. New York: St. Martin's Press.

Smoke, R. 1977a. "Analytic Dimensions of Intervention Decisions." In E. P. Stern, ed., *The Limits of Military Intervention*, pp. 25–44. Beverly Hills, Calif.: Sage.

———. 1977b. *War: Controlling Escalation*. Cambridge, Mass.: Harvard University Press.

Sniezek, J. A. 1992. "Group Under Uncertainty: An Examination of Confidence in Group Decision Making." *Organizational Behavior and Human Decision Processes*, 52: 124–55.

Snyder, G. H. 1984. "The Security Dilemma in Alliance Politics." *World Politics*, 36: 461–95.

Snyder, J. 1993. "Introduction: New Thinking About the New International System." In J. Snyder and R. Jervis, eds., *Coping with Complexity in the International System*, pp. 1–23. Boulder, Colo.: Westview Press.

———. 1991a. "Conclusion." In R. Jervis and J. Snyder, eds., *Dominos and Band-*

wagons: Strategic Beliefs and Great Power Competition in the Eurasia Rimland, pp. 276–90. New York: Oxford University Press.

———. 1991b. *Myths of Empire: Domestic Politics and International Ambition*. Ithaca, N.Y.: Cornell University Press.

Snyder, J., and R. Jervis, eds. 1993. *Coping with Complexity in the International System*. Boulder, Colo.: Westview Press.

Snyder, M. 1981. "On the Self-Perpetuating Nature of Social Stereotypes." In D. L. Hamilton, ed., *Cognitive Processes in Stereotyping and Intergroup Behavior*, pp. 183–212. Hillsdale, N.J.: Lawrence Erlbaum.

Snyder, M., and W. B. Swann, Jr. 1978. "Hypothesis-Testing Processes in Social Interaction." *Journal of Personality and Social Psychology*, 36: 1202–12.

Sofer, S. 1988. *Begin: An Anatomy of Leadership*. Oxford, Eng.: Basil Blackwell.

Spangler, M. B. 1980. "Syndromes of Risk and Environmental Protection: The Conflict of Individual and Societal Values." *Environmental Professional*, 2: 274–91.

Spear, J., and P. Williams. 1988. "Belief Systems and Foreign Policy: The Cases of Carter and Reagan." In R. Little and S. Smith, eds., *Belief Systems and International Relations*, pp. 190–208. Oxford, Eng.: Basil Blackwell.

Spechler, D. R. 1988. "The Soviet Union's Leonid Brezhnev." In B. Kellerman and J. Z. Rubin, eds., *Leadership and Negotiation in the Middle East*, pp. 166–99. New York: Praeger.

Spector, R. H. 1993. " 'How Do You Know If You're Winning?' Perception and Reality in America's Military Performance in Vietnam, 1965–1970." In J. S. Werner and L. D. Huynh, eds., *The Vietnam War: Vietnamese and American Perspectives*, pp. 152–64. Armonk, N.Y.: M. E. Sharpe.

Spiegel, S. L. 1985. *The Other Arab-Israeli Conflict: Making America's Middle East Policy, from Truman to Reagan*. Chicago: University of Chicago Press.

Spranca, M., E. Minsk, and J. Baron. 1991. "Omission and Commission in Judgment and Choice." *Journal of Experimental Social Psychology*, 27: 76–105.

Sprout, H., and M. Sprout. 1962. *Foundations of International Politics*. Princeton, N.J.: Van Nostrand.

———. 1956. *Man-Milieu Relationship Hypotheses in the Context of International Politics*. Princeton, N.J.: Center of International Studies, Princeton University.

Staple, D. A., S. D. Reicher, and R. Spears. 1994. "Social Identity, Availability and the Perception of Risk." *Social Cognition*, 12: 1–17.

Stasser, G. 1992. "Pooling of Unshared Information During Group Discussion." In S. Worchel, W. Wood, and J. A. Simpson, eds., *Group Process and Productivity*, pp. 48–67. London: Sage.

Staw, B. M. 1981. "The Escalation of Commitment to a Course of Action." *Journal of Management Review*, 6: 577–87.

———. 1976. "Knee-Deep in the Big Muddy: A Study of Escalating Commitment to a Chosen Course of Action." *Organizational Behavior and Human Performance*, 16: 27–44.

Staw, B. M., and J. Ross. 1989. "Understanding Behavior in Escalation Situations." *Science*, 246: 216–20.

———. 1987. "Behavior in Escalation Situations: Antecedents, Prototypes and Solutions." In L. L. Cummings and B. M. Staw, eds., *Research in Organizational Behavior*, vol. 9, pp. 39–78. Greenwich, Conn.: JAI Press.

Stein, G. J. 1992. "International Co-operation and Loss Avoidance: Framing the Problem." *International Journal*, 47: 202–34.

———. 1991. "Deterrence and Reassurance." In P. E. Tetlock, J. L. Husbands, R. Jervis, P. C. Stern, and C. Tilly, eds., *Behavior, Society, and Nuclear War*, vol. 2, pp. 8–72. New York: Oxford University Press.

Steinberg, B. S. 1996. *Shame and Humiliation: Presidential Decision Making on Vietnam.* Pittsburgh, Pa.: Pittsburgh University Press.

———. 1991. "Shame and Humiliation in the Cuban Missile Crisis: A Psychoanalytic Perspective." *Political Psychology*, 12: 653–90.

Steinberger, P. J. 1993. *The Concept of Political Judgment.* Chicago: Chicago University Press.

Steinbruner, J. D. 1974. *The Cybernetic Theory of Decision.* Princeton, N.J.: Princeton University Press.

Steiner, I. D. 1982. "Heuristic Models of Groupthink." In H. Brandstätter, J. H. Davis, and G. Stocker-Kreichgauer, eds., *Group Decision Making*, pp. 503–24. London: Academic Press.

Steiner, P. P. 1989. "In Collusion with the Nation: A Case Study of Group Dynamics at a Strategic Nuclear Policymaking Meeting." *Political Psychology*, 10: 647–73.

Stenelo, L. G. 1984. *The International Critic.* Boulder, Colo.: Westview Press.

Stern, E. K. 1997. "Probing the Plausibility of Newgroup Syndrome: Kennedy and the Bay of Pigs." In P. 't Hart, E. K. Stern, and B. Sundelius, eds., *Beyond Groupthink: Political Group Dynamics and Foreign Policy-Making*, pp. 153–89. Ann Arbor: Michigan University Press.

Stern, E. K., and B. Sundelius. 1997. "Understanding Small Group Decisions in Foreign Policy: Process Diagnosis and Research Procedure." In P. 't Hart, E. K. Stern, and B. Sundelius, eds., *Beyond Groupthink: Political Group Dynamics and Foreign Policy-Making*, pp. 123–50. Ann Arbor: Michigan University Press.

Stern, P. C. 1991. "Learning Through Conflict: A Realistic Strategy for Risk Communication." *Policy Sciences*, 24: 99–119.

Stewart, P. D., M. G. Hermann, and C. F. Hermann. 1989. "Modeling the 1973 Soviet Decision to Support Egypt." *American Political Science Review*, 83: 35–59.

Stone, D. N. 1994. "Overconfidence in Initial Self-Efficacy Judgments: Effects on Decision Processes and Performance." *Organizational Behavior and Human Decision Processes*, 59: 452–74.

Strauch, R. 1980. *Risk Assessment as a Subjective Process.* Santa Monica, Calif.: Rand Corporation, P-6460.

———. 1971. *The Operational Assessment of Risk: A Case Study of the Pueblo Mission.* Santa Monica, Calif.: Rand Corporation, R-691-PR.

Streufert, S., and S. C. Streufert. 1970. "Effects of Increasing Failure and Success on Military and Economic Risk Taking." *Journal of Applied Psychology*, 54: 393–400.

———. 1968. "Information Load, Time Spent, and Risk Taking in Complex Decision Making." *Psychonomic Science*, 13: 327–30.

Stroebe, W., and C. Fraser. 1971 "The Relationship Between Riskiness and Confidence in Choice Dilemma Decisions." *European Journal of Social Psychology*, 1: 519–26.

Strong, R. A. 1992. *Decisions and Dilemmas: Case Studies in Presidential Foreign Policy Making.* Englewood Cliffs, N.J.: Prentice Hall.

Suedfeld, P., and P. E. Tetlock. 1992. "Psychological Advice About Political Decision Making: Heuristics, Biases, and Cognitive Defects." In P. Suedfeld and P. E. Tetlock, eds., *Psychology and Social Policy*, pp. 51–70. New York: Hemisphere.

Sullivan, M. P. 1985. *The Vietnam War: A Study in the Making of American Policy*. Lexington: University Press of Kentucky.

Summers, H. G., Jr. 1982. *On Strategy: A Critical Analysis of the Vietnam War*. Novato, Calif.: Presidio Press.

Svenson, O. 1991. "The Time Dimension in Perception and Communication of Risk." In R. E. Kasperson and P. J. M. Stallen, eds., *Communicating Risks to the Public: International Perspectives*, pp. 263–85. Dordrecht, Neth.: Kluwer.

Swansbrough, R. H. 1994. "A Kohutian Analysis of President Bush's Personality and Style in the Persian Gulf Crisis." *Political Psychology*, 15: 227–76.

Swap, W. C. 1984. "Destructive Effects of Groups on Individuals." In W. C. Swap and Associates, eds., *Group Decision Making*, pp. 69–95. Beverly Hills, Calif.: Sage.

't Hart, P. 1991. "Groupthink, Risk-Taking and Recklessness: Quality of Process and Outcome in Policy Decision Making." *Politics and the Individual*, 1: 67–90.

———. 1990. *Groupthink in Government: A Study of Small Groups and Policy Failures*, Amsterdam: Swets and Zeitlinger.

't Hart, P., E. K. Stern, and B. Sundelius. 1997. "Foreign Policymaking at the Top: Political Group Dynamics." In P. 't Hart, E. K. Stern, and B. Sundelius, eds., *Beyond Groupthink: Political Group Dynamics and Foreign Policy-Making*, pp. 3–34. Ann Arbor: Michigan University Press.

Taliaferro, J. W. 1994. "Analogical Reasoning and Prospect Theory: Hypotheses on Framing." Paper prepared for the 35th Annual International Studies Association Convention, Washington, D.C., Mar. 29–Apr. 2.

Tamir, A. 1988. *Peace Loving Soldier*. Tel Aviv: Edanim, Yediot Aharonot (in Hebrew).

Tanca, A. 1993. *Foreign Aimed Intervention in Internal Conflict*. Dordrecht, Neth.: Martinus Nijhoff.

Tanter, R. 1990. *Who's at the Helm? Lessons of Lebanon*. Boulder, Colo.: Westview Press.

Tarr, D. W. 1981. "The Employment of Force: Political Constraints and Limitations." In S. C. Sarkesian and W. L. Scully, eds., *U.S. Policy and Low-Intensity Conflict: Potential for Struggles in the 1980s*, pp. 49–67. New Brunswick, N.J.: Transaction Books.

Tatu, M. 1981. "Intervention in Eastern Europe." In S. S. Kaplan, ed., *Diplomacy of Power: Soviet Armed Forces as a Political Instrument*, pp. 205–64. Washington, D.C.: Brookings Institution.

Taylor, M. D. 1972. *Swords and Plowshares*. New York: W. W. Norton.

Taylor, S. C. 1993. "Lyndon Johnson and the Vietnamese." In D. L. Anderson, ed., *Shadow on the White House: Presidents and the Vietnam War*, pp. 113–29. Lawrence: University Press of Kansas.

Taylor, S. E. 1982. "The Availability Bias in Social Perception and Interaction." In D. Kahneman, P. Slovic, and A. Tversky, eds., *Judgment Under Uncertainty: Heuristics and Biases*, pp. 190–200. Cambridge, Eng.: Cambridge University Press.

Taylor, S. E., and J. D. Brown. 1988. "Illusion and Well-Being: A Social Psychological Perspective on Mental Health." *Psychological Bulletin*, 103: 193–210.

Tefft, S. K. 1990. "Cognitive Perspectives on Risk Assessment and War Traps: An Alternative to Functional Theory." *Journal of Political and Military Sociology*, 18: 57–77.

Teigen, K. H. 1988. "When Are Low-Probability Events Judged to Be 'Probable'? Effects of Outcome-Set Characteristics on Verbal Probability Estimates." *Acta Psychologica*, 67: 157–74.

Tellis, A. J. 1996. "Terminating Intervention: Understanding Exit Strategy and U.S. Involvement in Intrastate Conflicts." *Studies in Conflict and Terrorism*, 19: 117–51.

Tetlock, P. E. 1992. "The Impact of Accountability on Judgment and Choice: Toward a Social Contingency Model." In M. P. Zanna, ed., *Advances in Experimental Social Psychology*, vol. 25, pp. 331–76. San Diego, Calif.: Academic Press.

———. 1991. "Learning in U.S. and Soviet Foreign Policy: In Search of an Elusive Concept." In G. W. Breslauer and P. E. Tetlock, eds., *Learning in U.S. and Soviet Foreign Policy*, pp. 20–61. Boulder, Colo.: Westview Press.

———. 1985. "Accountability: The Neglected Social Context of Judgment and Choice." In L. L. Cummings and B. M. Staw, eds., *Research in Organizational Behavior*, vol. 7, pp. 297–332. Greenwich, Conn.: JAI Press.

Tetlock, P. E., and A. Belkin, eds. 1996. *Counterfactual Thought Experiments in World Politics: Logical, Methodological and Psychological Perspectives*. Princeton, N.J.: Princeton University Press.

Tetlock, P. E., and C. McGuire, Jr. 1984. "Cognitive Perspectives on Foreign Policy." In S. Long, ed., *Political Behavior Annual*, vol. 1, pp. 255–73. Boulder, Colo.: Westview Press.

Tetlock, P. E., R. S. Peterson, C. McGuire, S.-J. Chang, and P. Feld. 1992. "Assessing Political Group Dynamics: A Test of the Groupthink Model." *Journal of Personality and Social Psychology*, 63: 403–25.

Teuber, A. 1990. "Justifying Risk." *Daedalus*, 119(4): 235–54.

Thakur, R. 1990. "Non-intervention in International Relations: A Case Study." *Political Science*, 42: 27–61.

Thaler, R. H. 1983. "Illusions and Mirages in Public Policy." *Public Interest*, no. 73: 60–74.

Thaler, R. H., and E. J. Johnson. 1990. "Gambling with the House Money and Trying to Break Even: The Effects of Prior Outcomes on Risky Choice." *Management Science*, 36: 643–60.

Thies, W. J. 1980. *When Governments Collide: Coercion and Diplomacy in the Vietnam Conflict, 1964–1968*. Berkeley: University of California Press.

Thomas, C. 1985. *New States, Sovereignty and Intervention*. New York: St. Martin's Press.

Thompson, J. C. 1980. *Rolling Thunder: Understanding Policy and Program Failure*. Chapel Hill: University of North Carolina Press.

Thompson, J. E., and A. L. Carsrud. 1976. "The Effects of Experimentally Induced Illusions of Invulnerability and Vulnerability on Decisional Risk Taking in Triads." *Journal of Social Psychology*, 100: 263–67.

Thompson, M., R. Ellis, and A. Wildavsky. 1990. *Cultural Theory*. Boulder, Colo.: Westview Press.

Thomson, J. C., Jr. 1968. "How Could Vietnam Happen? An Autopsy." *Atlantic*, 221(4): 47–53.

Thorndike, T. 1989. "Grenada." In P. J. Schraeder, ed., *Intervention in the 1980s: U.S. Foreign Policy in the Third World*, pp. 249–64. Boulder, Colo.: Lynne Rienner.

Tilford, E., Jr. 1993. *Cross Winds: The Air Force's Setup in Vietnam*. Austin: Texas A & M University Press.

Tillema, H. K. 1994. "Cold War Alliance and Overt Military Intervention." *International Interactions*, 3: 249–78.

———. 1992. "Foreign Military Intervention and the Cost of War: International Armed Conflicts, 1945–1988." Paper prepared for the 33rd Annual International Studies Association Convention, Atlanta, Ga., Mar. 31–Apr. 4.

———. 1990. "The Meaning and Restraints of Superpower Intervention." In K. A. Feste, ed., *American and Soviet Intervention: Effects on World Stability*, pp. 23–28. New York: Crane Russak.

———. 1989. "Foreign Overt Military Intervention in the Nuclear Age." *Journal of Peace Research*, 26: 179–95.

———. 1973. *Appeal to Force: American Military Intervention in the Era of Containment.* New York: Thomas Y. Crowell.

Tillema, H. K., and J. R. Van Wingen. 1982. "Law and Power in Military Intervention." *International Studies Quarterly*, 26: 220–50.

Tindale, R. S. 1993. "Decision Errors Made by Individuals and Groups." In N. J. Castellan, Jr., ed., *Individual and Group Decision Making: Current Issues*, pp. 109–24. Hillsdale, N.J.: Lawrence Erlbaum.

Tlas, M., ed., 1988. *The Israeli Invasion of Lebanon.* Tel Aviv: Ma'arachot (in Hebrew).

Triandis, H. C. 1995. *Individualism and Collectivism.* Boulder, Colo.: Westview Press.

———. 1993. "Collectivism and Individualism as Cultural Syndromes." *Cross-Cultural Research*, 27: 155–80.

Trimpop, R. M. 1994. *The Psychology of Risk Taking Behavior.* Amsterdam: North-Holland.

Triska, J. F. 1966. *Patterns and Level of Risk in Soviet Foreign Policy-Making, 1945–1963.* China Lake, Calif.: U.S. Naval Ordnance Test Station.

Trout, B. T. 1975. "Rhetoric Revisited: Political Legitimation and the Cold War." *International Studies Quarterly*, 19: 251–84.

Truman, H. S. 1956. *Memoirs: Years of Trial and Hope*, vol. 2. Garden City, N.Y.: Doubleday.

Tse, D. K., K.-H. Lee, I. Vertinsky, and D. A. Wehrung. 1988. "Does Culture Matter? A Cross-Cultural Study of Executives' Choices, Decisiveness, and Risk Adjustment in International Marketing." *Journal of Marketing*, 52(4): 81–95.

Tuchman, B. W. 1984. *The March of Folly: From Troy to Vietnam.* New York: Alfred A. Knopf.

Tucker, R. W. 1982. "Lebanon: The Case for the War." *Commentary*, 74(4): 19–30.

Tullar, W. L., and D. F. Johnson. 1973. "Group Decision-Making and the Risky Shift: A Trans-national Perspective." *International Journal of Psychology*, 8: 117–23.

Tullock, G. 1974. *The Social Dilemma: The Economics of War and Revolution.* Blacksburg, Va.: University Publications.

Turner, B. A. 1994. "The Future of Risk Research." *Journal of Contingencies and Crisis Management*, 2: 147–56.

Turner, J. W. 1991. "The Adequacy of Logistic Support." In B. W. Watson and P. G. Tsouras, eds., *Operation Just Cause: The U.S. Intervention in Panama*, pp. 123–26. Boulder, Colo.: Westview Press.

Turner, K. J. 1985. *Lyndon Johnson's Dual War: Vietnam and the Press.* Chicago: University of Chicago Press.

Tversky, A. 1972. "Elimination by Aspects: A Theory of Choice." *Psychological Review*, 79: 281–99.

Tversky, A., and M. Bar-Hillel. 1983. "Risk: The Long and the Short." *Journal of Experimental Psychology: Learning, Memory, and Cognition*, 1983: 713–17.

Tversky, A., and D. Kahneman. 1992. "Advances in Prospect Theory: Cumulative Representation of Uncertainty." *Journal of Risk and Uncertainty*, 5: 297–323.

———. 1988. "Rational Choice and the Framing of Decisions." In D. E. Bell, H. Raiffa, and A. Tversky, eds., *Decision Making: Descriptive, Normative, and Prescriptive Interactions*, pp. 167–92. Cambridge, Eng.: Cambridge University Press.

———. 1983. "Extensional Versus Intuitive Reasoning: The Conjunction Fallacy in Probability Judgment." *Psychological Review*, 90: 293–315.

———. 1981. "The Framing of Decisions and the Psychology of Choice." *Science*, 211: 453–58.

———. 1974. "Judgment Under Uncertainty: Heuristics and Biases." *Science*, 185: 1124–31.

———. 1973. "Availability: A Heuristic for Judging Frequency and Probability." *Cognitive Psychology*, 5: 207–32.

———. 1971. "Belief in the Law of Small Numbers." *Psychological Bulletin*, 76: 105–10.

Twining, D. T. 1990. "The Weinberger Doctrine and the Use of Force in the Contemporary Era." *Small Wars and Insurgencies*, 1: 97–117.

Tyler, T., and R. Hastie. 1991. "The Social Consequences of Cognitive Illusions." In M. H. Bazerman, R. J. Lewicki, and B. H. Sheppard, eds., *Research on Negotiation in Organization*, vol. 3, pp. 69–98. Greenwich, Conn.: JAI Press.

Uhlig, F. Jr. 1985. "Amphibious Aspects of the Grenada Episode." In P. M. Dunn and B. W. Watson, eds., *American Intervention in Grenada: The Implications of Operation "Urgent Fury,"* pp. 89–97. Boulder, Colo.: Westview Press.

U.S. Department of State. 1996a. *Foreign Relations of the United States, 1964–1968*, vol. 2: *Vietnam, January–June 1965*. Washington, D.C.: U.S. Government Printing Office.

———. 1996b. *Foreign Relations of the United States, 1964–1968*, vol. 3: *Vietnam, June–December 1965*. Washington, D.C.: U.S. Government Printing Office.

———. 1992. *Foreign Relations of the United States, 1964–1968*, vol. 1: *Vietnam 1964*. Washington, D.C.: U.S. Government Printing Office.

U.S. Departments of State and Defense. 1984. *Grenada Documents: An Overview and Selection*. Washington, D.C.

U.S. Government. 1966a. *Public Papers of the Presidents of the United States: Lyndon B. Johnson, January 1 to May 31, 1965*. Washington, D.C.: U.S. Government Printing Office.

———. 1966b. *Public Papers of the Presidents of the United States: Lyndon B. Johnson, June 1 to December 31, 1965*. Washington, D.C.: U.S. Government Printing Office.

———. 1962. *Public Papers of the Presidents of the United States: John F. Kennedy, January 20 to December 31, 1961*. Washington, D.C.: U.S. Government Printing Office.

U.S. Senate. 1984. *Hearings Before a Subcommittee of the Committee on Appropriations, United States Senate*, 98th Congress, 2nd sess., Mar. 21. Supplemental Appropriations Supporting U.S. Military Actions In and Around Grenada. Washington, D.C.: U.S. Government Printing Office.

Valdez, J. C. 1993. *Internationalism and the Ideology of Soviet Influence in Eastern Europe*. Cambridge, Eng.: Cambridge University Press.

Valenta, J. 1991. *Soviet Intervention in Czechoslovakia, 1968: Anatomy of a Decision* (rev. ed.). Baltimore, Md.: Johns Hopkins University Press.

———. 1980. "From Prague to Kabul: The Soviet Style of Invasion." *International Security*, 5(1): 114–41.

———. 1978. "The Soviet-Cuban Intervention in Angola, 1975." *Studies in Comparative Communism*, 11: 3–33.

———. 1975. "Soviet Decisionmaking and the Czechoslovak Crisis of 1968." *Studies in Comparative Communism*, 7: 147–73.

Valenta, J., and H. J. Ellison, eds. 1986. *Grenada and Soviet\Cuban Policy: Internal Crisis and U.S.\OECS Intervention*. Boulder, Colo.: Westview Press.

Valenti, J. 1975. *A Very Human President*. New York: W. W. Norton.

Vallone, R. P., D. W. Griffin, S. Lin, and L. Ross. 1990. "Overconfident Prediction of Future Actions and Outcomes by Self and Others." *Journal of Personality and Social Psychology*, 58: 582–92.

VanDeMark, B. 1991. *Into the Quagmire: Lyndon Johnson and the Escalation of the Vietnam War*. New York: Oxford University Press.

Van Evera, S. 1990. "Why Europe Matters, Why the Third World Doesn't: American Grand Strategy After the Cold War." *Journal of Strategic Studies*, 13(2): 1–51.

Van Wingen, J., and H. K. Tillema. 1980. "British Military Intervention After World War II: Militance in a Second-Rank Power." *Journal of Peace Research*, 17: 291–303.

Vasquez, J. A. 1976. "A Learning Theory of the American Anti–Vietnam War Movement." *Journal of Peace Research*, 13: 299–314.

Vaughan, D. 1996. *The Challenger Launch Decision: Risky Technology, Culture and Deviance at NASA*. Chicago: University of Chicago Press.

———. 1992. "Theory Elaboration: The Heuristics of Case Analysis." In C. C. Ragin and H. S. Becker, eds., *What Is a Case? Exploring the Foundations of Social Inquiry*, pp. 173–202. Cambridge, Eng.: Cambridge University Press.

Vaughan, E., and B. Nordenstam. 1991. "The Perception of Environmental Risks Among Ethnically Diverse Groups." *Journal of Cross-Cultural Psychology*, 22: 29–60.

Vertzberger, Y. Y. I. 1992. "National Capabilities and Foreign Military Intervention: A Policy Relevant Theoretical Analysis." *International Interactions*, 17: 349–73.

———. 1990. *The World in Their Minds: Information Processing, Cognition, and Perception in Foreign Policy Decisionmaking*. Stanford, Calif.: Stanford University Press.

———. 1984. "Bureaucratic-Organizational Politics and Information Processing in a Developing State." *International Studies Quarterly*, 28: 69–95.

Vickers, G. R. 1993. "U.S. Military Strategy and the Vietnam War." In J. S. Werner and L. D. Huynh, eds., *The Vietnam War: Vietnamese and American Perspectives*, pp. 113–29. Armonk, N.Y.: M. E. Sharpe.

Vincent, R. J. 1974. *Nonintervention and International Order*. Princeton, N.J.: Princeton University Press.

Vinokur, A. 1971. "Review and Theoretical Analysis of the Effects of Group Processes upon Individual and Group Decisions Involving Risk." *Psychological Bulletin*, 76: 231–50.

Vinokur, A., Y. Trope, and E. Burnstein. 1975. "A Decision-Making Analysis of Persuasive Argumentation and the Choice-Shift Effect." *Journal of Experimental Social Psychology*, 11: 127–48.

Vlahos, M. 1987. "Force from the Sea: A Modest Proposal." In W. J. Olson, ed., *U.S. Strategic Interests in the Gulf Region*, pp. 189–202. Boulder, Colo.: Westview Press.

Vlek, C., and P. J. Stallen. 1981. "Judging Risks and Benefits in the Small and in the Large." *Organizational Behavior and Human Performance*, 28: 235–71.

———. 1980. "Rational and Personal Aspects of Risk." *Acta Psychologica*, 45: 273–300.

Voss, J. F., and T. A. Post. 1988. "On the Solving of Ill-Structured Problems." In M. T. H. Chi, R. Glaser, and M. J. Farr, eds., *The Nature of Expertise*, pp. 261–85. Hillsdale, N.J.: Lawrence Erlbaum.

Vought, D. 1982. "American Culture and American Arms: The Case of Vietnam." In R. A. Hunt and R. H. Shultz, Jr., eds., *Lessons from an Unconventional War: Reassessing U.S. Strategies for Future Conflicts*, pp. 158–90. New York: Pergamon Press.

Wald, E. 1992. *The Wald Report: The Decline of Israeli National Security Since 1967*. Boulder, Colo.: Westview Press.

Waldstein, F. A. 1990. "Cabinet Government: The Reagan Management Model." In J. Hogan, ed., *The Reagan Years: The Record in Presidential Leadership*, pp. 54–75. Manchester, Eng.: Manchester University Press.

Walker, S. G. 1990. "The Evolution of Operational Code Analysis." *Political Psychology*, 11: 403–18.

———. 1983. "The Motivational Foundations of Political Belief Systems: A Reanalysis of the Operational Code Construct." *International Studies Quarterly*, 27: 179–201.

Walker, W. O. 1991. "Decision-Making Theory and Narcotic Foreign Policy: Implications for Historical Analysis." *Diplomatic History*, 15: 31–46.

Wallach, M. A., and N. Kogan. 1965. "The Role of Information, Discussion, and Consensus in Group Risk Taking." *Journal of Experimental Social Psychology*, 1: 1–19.

Wallach, M. A., N. Kogan, and R. B. Burt. 1968. "Are Risk Takers More Persuasive Than Conservatives in Group Discussion?" *Journal of Experimental Social Psychology*, 4: 76–88.

Wallach, M. A., and C. W. Wing. 1968. "Is Risk a Value?" *Journal of Personality and Social Psychology*, 9: 101–6.

Wallsten, T. S. 1990. "The Costs and Benefits of Vague Information." In R. M. Hogarth, ed., *Insights in Decision Making: A Tribute to Hillel J. Einhorn*, pp. 28–43. Chicago: University of Chicago Press.

Walt, S. M. 1991. "The Renaissance of Security Studies." *International Studies Quarterly*, 35: 211–40.

Waltz, K. N. 1986. "Reflections on 'Theory of International Politics': A Response to My Critics." In R. O. Keohane, ed., *Neorealism and Its Critics*. New York: Columbia University Press.

———. 1979. *Theory of International Politics*. Reading, Mass.: Addison-Wesley.

Walzer, M. 1977. *Just and Unjust Wars*. New York: Basic Books.

Washington Post.

Waters, M. 1986. "The Invasion of Grenada, 1983, and the Collapse of Legal Norms." *Journal of Peace Research*, 23: 229–46.

Watson, B. W. 1991. "Assessing Press Access to Information." In B. W. Watson and P. G. Tsouras, eds., *Operation Just Cause: The U.S. Intervention in Panama*, pp. 133–51. Boulder, Colo.: Westview Press.

Watson, B. W., and P. G. Tsouras, eds. 1991. *Operation Just Cause: The U.S. Intervention in Panama.* Boulder, Colo.: Westview Press.

Weaver, R. K., and B. A. Rockman. 1993a. "Assessing the Effects of Institutions." In R. K. Weaver and B. A. Rockman, eds., *Do Institutions Matter? Government Capabilities in the United States and Abroad*, pp. 1–41. Washington, D.C.: Brookings Institution.

———. 1993b. "When and How Do Institutions Matter?" In R. K. Weaver and B. A. Rockman, eds., *Do Institutions Matter? Government Capabilities in the United States and Abroad*, pp. 445–61. Washington, D.C.: Brookings Institution.

Weede, E. 1984. "Democracy and War Involvement." *Journal of Conflict Resolution*, 28: 649–64.

Weede, E., and Z. Mannheim. 1978. "U.S. Support for Foreign Governments or Domestic Disorder and Imperial Intervention, 1958–1965." *Comparative Political Studies*, 10: 497–528.

Wehman, P., M. A. Goldstein, and J. R. Williams. 1977. "Effects of Different Leadership Styles on Individual Risk-Taking in Groups." *Human Relations*, 30: 249–59.

Wehrung, D. A., K.-H. Lee, D. K. Tse, and I. B. Vertinsky. 1989. "Adjusting Risky Situations: A Theoretical Framework and Empirical Test." *Journal of Risk and Uncertainty*, 2: 189–212.

Weinberger, C. W. 1990. *Fighting for Peace: Seven Critical Years in the Pentagon.* New York: Warner Books.

Weinstein, N. D. 1980. "Unrealistic Optimism About Future Life Events." *Journal of Personality and Social Psychology*, 39: 806–20.

Weizman, E. 1981. *The Battle for Peace.* New York: Bantam Books.

Wells, T. 1994. *The War Within: America's Battle over Vietnam.* Berkeley: University of California Press.

Wendt, A. E. 1987. "The Agent-Structure Problem in International Relations Theory." *International Organization*, 41: 335–70.

Westmoreland, W. C. 1976. *A Soldier's Report.* Garden City, N.Y.: Doubleday.

Weyland, K. 1996. "Risk Taking in Latin American Economic Restructuring: Lessons from Prospect Theory." *International Studies Quarterly*, 40: 185–208.

Wheeler, N. J. 1993. "Pluralist or Solidarist Conceptions of International Society: Bull and Vincent on Humanitarian Intervention." *Millennium*, 21: 463–87.

Whetten, L. L. 1969. "Military Aspects of the Soviet Occupation of Czechoslovakia." *World Today*, 25: 60–68.

Whipple, C. 1992. "Inconsistent Values in Risk Management." In S. Krimsky and D. Golding, eds., *Social Theories of Risk*, pp. 343–54. Westport, Conn.: Praeger.

White, R. K. 1968. *Nobody Wanted War: Misperception in Vietnam and Other Wars.* Garden City, N.Y.: Doubleday.

Whithey, S. F. 1962. "Reactions to Uncertain Threat." In G. W. Baker and D. W. Chapman, eds., *Man and Society in Disaster*, pp. 93–123. New York: Basic Books.

Whiting, A. S. 1960. *China Crosses the Yalu: The Decision to Enter the Korean War.* Stanford, Calif.: Stanford University Press.

Whyte, G. 1991. "Diffusion of Responsibility: Effects on the Escalation Tendency." *Journal of Applied Psychology*, 76: 408–15.

———. 1989. "Groupthink Reconsidered." *Academy of Management Review*, 14: 40–56.

———. 1986. "Escalating Commitment to a Course of Action: A Reinterpretation." *Academy of Management Review*, 11: 311–21.

Whyte, G., and A. S. Levi. 1994. "The Origins and Function of the Reference Point in Risky Group Decision Making: The Case of the Cuban Missile Crisis." *Journal of Behavioral Decision Making*, 7: 243–60.

Wicklund, R. A., and J. W. Brehm. 1976. *Perspectives on Cognitive Dissonance.* Hillsdale, N.J.: Lawrence Erlbaum.

Wildavsky, A., and K. Dake. 1990. "Theories of Risk Perception: Who Fears What and Why?" *Daedalus*, 119(4): 41–60.

Wilder, D. A., and V. L. Allen. 1978. "Group Membership and Preference for Information About Others." *Personality and Social Psychology Bulletin*, 4: 106–10.

Williams, K. 1996. "New Sources on Soviet Decision Making During the 1968 Czechoslovak Crisis." *Europe-Asia Studies*, 48: 457–70.

Williams, P. 1990. "The Reagan Administration and Defense Policy." In D. M. Hill, R. A. Moore, and P. Williams, eds., *The Reagan Presidency: An Incomplete Revolution?* pp. 199–230. New York: St. Martin's Press.

Willner, A. R. 1984. *The Spellbinders: Charismatic Political Leadership.* New Haven, Conn.: Yale University Press.

Windchy, E. C. 1971. *Tonkin Gulf.* Garden City, N.Y.: Doubleday.

Windsor, P. 1978. "Yugoslavia, 1951, and Czechoslovakia, 1968." In B. M. Blechman and S. S. Kaplan, eds., *Force Without War: U.S. Armed Forces as a Political Instrument*, pp. 440–514. Washington, D.C.: Brookings Institution.

Windsor, P., and A. Roberts. 1969. *Czechoslovakia, 1968: Reform Repression and Resistance.* New York: Columbia University Press.

Winter, D. G., M. G. Hermann, W. Weintraub, and S. G. Walker. 1991. "The Personality of Bush and Gorbachev Measured at a Distance: Procedures, Portraits, and Policy." *Political Psychology*, 12: 215–45.

Wirtz, J. J. 1991. *The Tet Offensive: Intelligence Failure in War.* Ithaca, N.Y.: Cornell University Press.

Witte, E. H., and A. J. Arez. 1974. "The Cognitive Structure of Choice Dilemma Decisions. *European Journal of Social Psychology*, 4: 313–28.

Wohlforth, W. C. 1993. *The Elusive Balance: Power and Perceptions During the Cold War.* Ithaca, N.Y.: Cornell University Press.

Wohlstetter, A. 1968. "Illusions of Distance." *Foreign Affairs*, 46: 242–55.

Wolfe, T. W. 1969. *Soviet Foreign and Defense Policy Under the Brezhnev-Kosygin Regime.* Santa Monica, Calif.: Rand Corporation, P-4227.

Wood, W., and B. Stagner. 1994. "Why Are Some People Easier to Influence Than Others?" In S. Shavit and T. C. Brock, eds., *Persuasion: Psychological Insights and Perspectives*, pp. 149–74. Boston: Allyn and Bacon.

Woodward, B. 1991. *The Commanders.* New York: Simon and Schuster.

Woolley, P. J. 1991. "Geography and the Limits of U.S. Military Intervention." *Conflict Quarterly*, 11(4): 35–50.

Wright, G. N., and L. D. Phillips. 1980. "Cultural Variation in Probabilistic Thinking: Alternative Ways of Dealing with Uncertainty." *International Journal of Psychology*, 15: 239–57.

Yadav, S. S. 1989 "Failed Great Power War and the Soviet Retreat from Afghanistan." *Comparative Strategy*, 8: 353–68.

Yaniv, A. 1987. *Dilemmas of Security: Politics, Strategy and the Israeli Experience in Lebanon.* New York: Oxford University Press.

Yaniv, A., and R. J. Lieber. 1983. "Personal Whim or Strategic Imperative: The Israeli Invasion of Lebanon." *International Security*, 8(2): 117–42.

Yarmolinsky, A. 1968. "American Foreign Policy and the Decision to Intervene." *Journal of International Affairs*, 22: 231–36.

Yates, F. J., and E. R. Stone. 1992a. "Risk Appraisal." In J. F. Yates, ed., *Risk-Taking Behavior*, pp. 51–85. Chichester, Eng.: John Wiley and Sons.

———. 1992b. "The Risk Construct." In J. F. Yates, ed., *Risk-Taking Behavior*, pp. 1–25. Chichester, Eng.: John Wiley and Sons.

Yediot Aharonot (in Hebrew).

Yin, R. K. 1989. *Case Studies Research: Design and Methods* (rev. ed.). Newbury Park, Calif.: Sage.

Yinon, Y., and A. Bizman. 1974. "The Nature of Affective Bonds and the Degree of Personal Responsibility as Determinants of Risk Taking for 'Self and Others.'" *Bulletin of the Psychonomic Society*, 4(2A): 80–82.

Yishai, Y. 1983–84. "Dissent in Israel: Opinions on the Lebanon War." *Middle East Review*, 16(2): 38–44.

Young, M. B. 1991. *The Vietnam Wars, 1945–1990*. New York: Harper Perennial.

Young, O. R. 1974. "Systemic Bases of Intervention." In J. N. Moore, ed., *Law and Civil War in the Modern World*, pp. 111–26. Baltimore, Md.: Johns Hopkins University Press.

———. 1968. "Intervention and International Systems." *Journal of International Affairs*, 22: 177–87.

Yukl, G. A. 1989. *Leadership in Organizations* (2nd ed.). Englewood Cliffs, N.J.: Prentice Hall.

Zacher, M. W. 1992. "The Decaying Pillars of the Westphalian Temple: Implications for International Order and Governance." In J. N. Rosenau and E.-O. Czempiel, eds., *Governance Without Government: Order and Change in World Politics*, pp. 58–101. Cambridge, Eng.: Cambridge University Press.

Zaffiri, S. 1994. *Westmoreland: A Biography of General William C. Westmoreland*. New York: William Morrow.

Zajac, E. J., and M. H. Bazerman. 1991. "Blind Spots in Industry and Competitor Analysis: Implications of Interfirm (Mis)perceptions for Strategic Decisions." *Academy of Management Review*, 16: 37–56.

Zajonc, R. B., R. J. Wolosin, and M. A. Wolosin. 1972. "Group Risk-Taking Under Various Group Decision Schemes." *Journal of Experimental Social Psychology*, 8: 16–30.

Zakheim, D. S. 1986. "The Grenada Operation and Superpower Relations: A Perspective from the Pentagon." In J. Valenta and H. J. Ellison, eds., *Grenada and Soviet/Cuban Policy: Internal Crisis and U.S./OECS Intervention*, pp. 175–85. Boulder, Colo.: Westview Press.

Zaleska, M., and N. Kogan. 1971. "Level of Risk Selected by Individuals and Groups When Deciding for Self and for Others." *Sociometry*, 34: 198–213.

Zaroulis, N., and G. Sullivan. 1984. *Who Spoke Up? American Protest Against the War in Vietnam, 1963–1975*. Garden City, N.Y.: Doubleday.

Zartman, I. W. 1968. "Intervention Among Developing States." *Journal of International Affairs*, 22: 188–97.

Zeelenberg, M., J. Beattie, J. Van der Plight, and N. K. de Vries. 1996. "Consequences of Regret Aversion: Effects of Expected Feedback on Risky Decision-Making." *Organizational Behavior and Human Decision Processes*, 65: 148–58.

Ziman, J. 1978. *Reliable Knowledge: An Exploration of the Grounds for Belief in Science*. Cambridge, Eng.: Cambridge University Press.

Zimmerman, W. 1981. "The Korean and Vietnam Wars." In S. S. Kaplan, ed., *Diplomacy of Power: Soviet Armed Forces as a Political Instrument*, pp. 314–56. Washington, D.C.: Brookings Institution.

Zuckert, C. H. 1995. "On the Rationality of Rational Choice." *Political Psychology*, 16: 179–98.

Index

In this index an "f" after a number indicates a separate reference on the next page, and an "ff" indicates separate references on the next two pages. A continuous discussion over two or more pages is indicated by a span of page numbers, e.g., "57–59." *Passim* is used for a cluster of references in close but not consecutive sequence.

Library of Congress Cataloging-in-Publication Data

Vertzberger, Yaacov.
  Risk taking and decisionmaking : foreign military in-
tervention decisions / Yaacov Y. I. Vertzberger.
    p. cm.
  Includes bibliographical references and index.
  ISBN 0–8047–2747–3 (cloth : alk. paper). — ISBN
  0–8047–3168–3 (pbk. : alk. paper)
    1. Intervention (International law).   2. Interna-
  tional relations—Decision making.   3. Risk-taking
  (Psychology).   I. Title.
  JX4481.V47 1998
  355.6'83—dc21   97–5623
                                                     CIP

 ⊗ This book is printed on acid-free, recycled paper.

Original printing 1998
Last figure below indicates year of this printing:
07 06 05 04 03 02 01 00 99 98